ANNUAL REVIEW OF
FLUID MECHANICS

EDITORIAL COMMITTEE (1982)

J. D. GODDARD
J. L. LUMLEY
C. C. MEI
A. R. SEEBASS III
M. VAN DYKE
G. VERONIS
J. V. WEHAUSEN
T. Y. WU

Responsible for the organization of Volume 14
(Editorial Committee, 1980)

J. F. KENNEDY
J. L. LUMLEY
W. R. SCHOWALTER
A. R. SEEBASS III
M. VAN DYKE
G. VERONIS
J. V. WEHAUSEN
T. Y. WU

Production Editors   E. P. BROWER
                      K. R. DODSON
Indexing Coordinator M. A. GLASS

ANNUAL REVIEW OF FLUID MECHANICS

MILTON VAN DYKE, *Co-Editor*
Stanford University

J. V. WEHAUSEN, *Co-Editor*
University of California, Berkeley

JOHN L. LUMLEY, *Associate Editor*
Cornell University

VOLUME 14

1982

ANNUAL REVIEWS INC.   4139 EL CAMINO WAY   PALO ALTO, CALIFORNIA 94306 USA

ANNUAL REVIEWS INC.
Palo Alto, California, USA

COPYRIGHT © 1982 BY ANNUAL REVIEWS INC., PALO ALTO, CALIFORNIA, USA. ALL RIGHTS RESERVED. The appearance of the code at the bottom of the first page of an article in this serial indicates the copyright owner's consent that copies of the article may be made for personal or internal use, or for the personal or internal use of specific clients. This consent is given on the condition, however, that the copier pay the stated per-copy fee of $2.00 per article through the Copyright Clearance Center, Inc. (21 Congress Street, Salem, MA 01970) for copying beyond that permitted by Sections 107 or 108 of the US Copyright Law. The per-copy fee of $2.00 per article also applies to the copying, under the stated conditions, of articles published in any Annual Review serial before January 1, 1978. Individual readers, and nonprofit libraries acting for them, are permitted to make a single copy of an article without charge for use in research or teaching. This consent does not extend to other kinds of copying, such as copying for general distribution, for advertising or promotional purposes, for creating new collective works, or for resale. For such uses, written permission is required. Write to Permissions Dept., Annual Reviews Inc., 4139 El Camino Way, Palo Alto, CA 94306 USA.

International Standard Serial Number: 0066-4189
International Standard Book Number: 0-8243-0714-3
Library of Congress Catalog Card Number: 74-80866

Annual Reviews Inc. and the Editors of its publications assume no responsibility for the statements expressed by the contributors to this Review.

PRINTED AND BOUND IN THE UNITED STATES OF AMERICA

Annual Review of Fluid Mechanics
Volume 14, 1982

CONTENTS

VILHELM BJERKNES AND HIS STUDENTS, *Arnt Eliassen*	1
SEDIMENT RIPPLES AND DUNES, *Frank Engelund and Jørgen Fredsøe*	13
STRONGLY NONLINEAR WAVES, *L. W. Schwartz and J. D. Fenton*	39
TOPOLOGY OF THREE-DIMENSIONAL SEPARATED FLOWS, *Murray Tobak and David J. Peake*	61
DYNAMICS OF GLACIERS AND LARGE ICE MASSES, *Kolumban Hutter*	87
THE MATHEMATICAL THEORY OF FRONTOGENESIS, *B. J. Hoskins*	131
DYNAMICS OF LAKES, RESERVOIRS, AND COOLING PONDS, *Jörg Imberger and Paul F. Hamblin*	153
TURBULENT JETS AND PLUMES, *E. J. List*	189
GRAVITY CURRENTS IN THE LABORATORY, ATMOSPHERE, AND OCEAN, *John E. Simpson*	213
THE FLUID DYNAMICS OF HEART VALVES: EXPERIMENTAL, THEORETICAL, AND COMPUTATIONAL METHODS, *Charles S. Peskin*	235
THE COMPUTATION OF TRANSONIC POTENTIAL FLOWS, *David A. Caughey*	261
UNSTEADY AIRFOILS, *W. J. McCroskey*	285
LOW-GRAVITY FLUID FLOWS, *Simon Ostrach*	313
THE STRANGE ATTRACTOR THEORY OF TURBULENCE, *Oscar E. Lanford III*	347
DYNAMICS OF OIL GANGLIA DURING IMMISCIBLE DISPLACEMENT IN WATER-WET POROUS MEDIA, *A. C. Payatakes*	365
NUMERICAL METHODS IN FREE-SURFACE FLOWS, *Ronald W. Yeung*	395
INDEXES	
Author Index	443
Cumulative Index of Contributing Authors, Volumes 10–14	451
Cumulative Index of Chapter Titles, Volumes 10–14	453

ANNUAL REVIEWS INC. is a nonprofit corporation established to promote the advancement of the sciences. Beginning in 1932 with the *Annual Review of Biochemistry*, the Company has pursued as its principal function the publication of high quality, reasonably priced *Annual Review* volumes. The volumes are organized by Editors and Editorial Committees who invite qualified authors to contribute critical articles reviewing significant developments within each major discipline. Annual Reviews Inc. is administered by a Board of Directors, whose members serve without compensation.

1982 Board of Directors, Annual Reviews Inc.

Dr. J. Murray Luck, Founder and Director Emeritus of Annual Reviews Inc.
Professor Emeritus of Chemistry, Stanford University
Dr. Joshua Lederberg, President of Annual Reviews Inc.
President, The Rockefeller University
Dr. James E. Howell, Vice President of Annual Reviews Inc.
Professor of Economics, Stanford University
Dr. William O. Baker, *Retired Chairman of the Board, Bell Laboratories*
Dr. Robert W. Berliner, *Dean, Yale University School of Medicine*
Dr. Sidney D. Drell, *Deputy Director, Stanford Linear Accelerator Center*
Dr. Eugene Garfield, *President, Institute for Scientific Information*
Dr. Conyers Herring, *Professor of Applied Physics, Stanford University*
Dr. D. E. Koshland, Jr., *Professor of Biochemistry, University of California, Berkeley*
Dr. William D. McElroy, *Chancellor, University of California, San Diego*
Dr. William F. Miller, *President, SRI International*
Dr. Esmond E. Snell, *Professor of Microbiology and Chemistry, University of Texas, Austin*
Dr. Harriet A. Zuckerman, *Professor of Sociology, Columbia University*

Management of Annual Reviews Inc.

John S. McNeil, Publisher and Secretary-Treasurer
Alister Brass, M. D., Editor-in-Chief
Sharon E. Hawkes, Production Manager
Mickey G. Hamilton, Promotion Manager
Donald S. Svedeman, Business Manager

Publications

PLANNED FOR 1983: Annual Review of Immunology

ANNUAL REVIEWS OF
Anthropology
Astronomy and Astrophysics
Biochemistry
Biophysics and Bioengineering
Earth and Planetary Sciences
Ecology and Systematics
Energy
Entomology
Fluid Mechanics
Genetics
Materials Science
Medicine
Microbiology
Neuroscience
Nuclear and Particle Science
Nutrition
Pharmacology and Toxicology
Physical Chemistry
Physiology
Phytopathology
Plant Physiology
Psychology
Public Health
Sociology

SPECIAL PUBLICATIONS
Annual Reviews Reprints:
  Cell Membranes, 1975–1977
  Cell Membranes, 1978–1980
  Immunology, 1977–1979
Excitement and Fascination of Science, Vols. 1 and 2
History of Entomology
Intelligence and Affectivity, by Jean Piaget
Telescopes for the 1980s

For the convenience of readers, a detachable order form/envelope is bound into the back of this volume.

VILHELM BJERKNES AND HIS STUDENTS

Arnt Eliassen
Institute of Geophysics, University of Oslo, Norway

Vilhelm Bjerknes' influence upon the advance of meteorology was only in part the result of his own scientific contributions in the field. These contributions were significant, and marked by his attitude as a theoretical physicist. However, of equal importance was his remarkable ability to attract and stimulate bright students and create an enthusiastic milieu around him. In this way he brought about a favorable recruitment of able scientists into the fields of meteorology and oceanography, without which these sciences would have been much poorer indeed.

Vilhelm Bjerknes' father, Carl Anton Bjerknes, was professor of mathematics at Norway's only university in Oslo, or Christiania as the city was called at the time. He was exploring the forces between bodies moving in a fluid, and found a striking analogy between these forces and electrostatic and magnetic forces. In particular, he found that two spheres submerged in water, pulsating at the same frequency, would be acted upon by attractive or repellant pressure forces which would satisfy Coulomb's law, except that they would have the opposite direction: attraction would result when the pulsations were in phase, repulsion when the phases were opposite. Moreover, the streamlines in the fluid would have the same shape as the lines of force between electric charges. Likewise, the forces between rotating cylinders in a fluid would be similar to the forces between electric currents.

As a young student, Vilhelm Bjerknes closely followed his father's research and assisted him in hydrodynamic experiments which successfully verified the theory. This was at a time when electromagnetic action at a distance and the existence of an "aether" were much debated, and C. A. Bjerknes thought that he might be on the track of a kind of hydrodynamic theory of electromagnetism.

Vilhelm Bjerknes studied science at the University of Christiania. After graduating he spent two years with Heinrich Hertz at Bonn. During these years he made valuable (but now probably forgotten) contributions to the theory of electromagnetic waves and resonance. In 1893 he came to the University of Stockholm where he two years later was appointed professor of mechanics and mathematical physics.

In Stockholm Vilhelm Bjerknes took up his father's ideas and spent several years trying to carry the theory of the hydrodynamic-electromagnetic analogy to a completion. It was in connection with these investigations that in 1897 he discovered the circulation theorem which carries his name, and which expresses how the circulation of a material closed curve in a fluid will change as a result of baroclinicity (V. Bjerknes 1898). This discovery caused Bjerknes to enter the field of geophysics. He immediately saw that his theorem was just what was needed to explain the formation of circulatory motions in the atmosphere and the ocean, and he set out to exploit the idea.

The classical hydrodynamics at the time was based on Helmholtz' and Kelvin's theorems on vortex conservation (in the absence of viscosity). These conservation theorems were based on a fluid model in which baroclinicity was not possible; even the concept seemed to have been left out of consideration. Either the fluid was assumed homogeneous and incompressible, or, if compressible, it was assumed that the density would depend upon pressure only, and in the same manner for all fluid particles.

Bjerknes realized that such fluid models, which he later termed *autobarotropic,* are too restrictive to represent air or seawater in motion. To explain the maintenance of circulatory motions in the atmosphere or the ocean, a *general fluid model* is required, in which the pressure of a particle does not determine its state, or density. The density of a particle changes not only because of pressure changes, but also as a result of heat sources. Thus a link between the motion and the heat sources was established; the relevant theory is a combination of hydrodynamics and thermodynamics, which Bjerknes called *physical fluid dynamics.*

As an intermediate fluid model, Bjerknes introduced the *piezotropic fluid,* where the density of every particle is a unique function of its pressure, but where this relationship in general is different for different particles. This model permits baroclinicity and is suited for the study of oscillations and waves. Bjerknes' classification of fluid models is very useful, and, in the author's opinion, it is unfortunate that his terminology has not been generally adopted in fluid dynamics.

With the circulation theorem and the concept of a general fluid for which the density would change as a result of heating (and possibly also by change in composition), Bjerknes realized that he could for the first time formulate a complete set of hydro- and thermodynamic equations that govern the processes in the atmosphere. Consequently he proposed (Bjerknes 1904) attacking the problem of weather prediction as an initial-value problem of mathematical physics, where the initial state was to be determined from observations, and the future change from integration of the governing equations. In 1905 he got the opportunity to lecture about this bold program in Washington, D.C. This resulted in a yearly grant from the Carnegie Institution, which he retained for about 35 years, until the Second World War. The money could hardly have found a better use. During the years, it enabled Bjerknes to employ a considerable number of research assistants, all of whom later became well-known geophysicists.

In the years 1893—1896, Fridtjof Nansen and his crew crossed the Arctic Ocean, partly on board the FRAM, and partly on skis. He returned with a wealth of observations and experiences, and also with many unexplained problems. One question which Nansen discussed with Vilhelm Bjerknes was the so-called "dead-water" phenomenon. This is a mysterious ship resistance that occurs suddenly, as if the ship were held back by an evil ghost. The phenomenon had been observed on the FRAM near the Siberian coast. Bjerknes suggested that it was a wave resistance due to internal waves on the interface between relatively fresh surface water and more saline water underneath. He gave the problem to his young Swedish student Vagn Walfrid Ekman, who made a thorough laboratory study of the phenomenon, and also a theoretical investigation, based on the work on surface waves by Stokes and Kelvin. Ekman's results, which were published as part of the scientific results of the FRAM expedition, fully confirmed that the dead-water phenomenon was indeed caused by internal gravity waves.

Today Ekman's name is connected with his analysis of the viscous boundary layer in the presence of rotation. This work, too, emerged from a discussion, in the year 1900, between Nansen and Bjerknes. During FRAM's drifting in the arctic ice, the direction of the ice drift showed a systematic deviation to the right relative to the wind direction. Nansen attributed this deviation to the earth's rotation and correctly concluded that, for the same reason, the deviation angle of the current direction must increase downwards in the sea. Bjerknes proposed that Nansen also give this problem to Ekman. He was called, and came out the same evening with the celebrated Ekman spiral equations.

Ekman was early aware (in a paper from 1906) that his theory was applicable also to the atmospheric boundary layer. He noticed, however, that the angle between the surface wind and the isobars was considerably less than the 45° demanded by his theory. This discrepancy he correctly explained as a result of a downwards decrease of the eddy viscosity in the lowest air layers. He suggested that this could be accounted for by introducing a slip at the earth's surface in the direction of the spiral tangent; this would reduce the angle between the surface wind and the isobars. G. I. Taylor proposed the same theory in 1915, without knowing Ekman's work.

In 1910 Ekman was appointed professor of mechanics and mathematical physics at the University of Lund (Sweden). In his later work on ocean currents, he aimed towards a theory that would combine the effects of surface wind stress, bottom friction, and horizontal pressure gradients.

Another of Bjerknes' students in Stockholm was Johan Wilhelm Sandström. In contrast to Ekman, who had grown up in an academic milieu, Sandström came from a small farm in Northern Sweden. For many years he assisted Bjerknes in calculating tables and graphs for the practical application of the circulation theorem in meteorology and oceanography. Sandström was employed as Bjerknes' first Carnegie assistant. Their cooperation resulted in the first volume (*Statics*) of the work *Dynamic Meteorology and Hydrography*, which appeared in 1910. It contained, among other things, a detailed description of how to construct topographic maps

Vagn Walfrid Ekman

for a set of isobaric surfaces in the atmosphere by starting at the surface and working upwards by successively adding maps of relative topography. The method secures a vertical consistency that is not always present in the methods of analysis used today.

Sandström later worked in the Swedish Meteorological-Hydrological Service. Among his scientific contributions was a theorem which states that the maintainance of a thermal circulation against frictional dissipation requires the heat source to be located at a lower level than the cold source. He went on oceanographic expeditions in the North Atlantic and studied the effect of the warm Gulf Stream on the climate of Western Europe.

In 1907 Vilhelm Bjerknes was called to a chair at the University of Christiania. The next year he employed two young Norwegian science students as his Carnegie assistants: Theodor Hesselberg and Olaf Devik. The latter was succeeded in 1911 by another science student, Harald Ulrik Sverdrup. These three men all became prominent geophysicists in three different fields: Hesselberg in meteorology, Sverdrup in oceanography, and

Harald Ulrik Sverdrup

Devik in hydrology. With the assistance of Hesselberg and Devik, Bjerknes in 1911 published the second volume (*Kinematics*) of *Dynamic Meteorology and Hydrography*. A third volume (*Dynamics*) was planned, but never finished.

Hesselberg and Sverdrup worked for some years in close cooperation, resulting in a series of joint papers, in particular on turbulent friction and variation of wind with height. Much of this work was done at the new Geophysical Institute in Leipzig which was established in 1913 with Bjerknes as Director.

One of Hesselberg's papers from this time is remarkable in that it anticipates the modern theory of the geostrophic momentum approximation. In 1915 Hesselberg went to Oslo to take over as Director of the Norwegian Meteorological Institute, a position he held for forty years. In these later years he studied recent climate variations. Together with his associate B. J. Birkeland, he was one of the first to detect and map the great warming trend of the first half of our century. Hesselberg also did much to further international cooperation which is so important in meteorology, and he was President of The International Meteorological Organization from 1935 to 1947.

Theodor Hesselberg

H. U. Sverdrup's main work during the time he spent in Leipzig was his very thorough meteorological study of the structure of the North Atlantic trade-wind region. In another remarkable paper he estimated the average dissipation of mechanical energy into heat in the atmosphere to be 4 watts m^{-2}, which is not far from modern estimates.

Sverdrup's later career is remarkable indeed. He spent seven years with Roald Amundsen's polar expedition on the MAUD, in charge of the scientific investigations, ten years as professor in Bergen, then twelve years (1936–1948) as Director of Scripps Institution of Oceanography in California, and finally his last nine years in Oslo as Director of the Polar Institute and professor at the University. His scientific contributions cover a broad spectrum of the geophysical sciences, with physical oceanography as the main theme. It is not possible here to describe his many-sided activities; readers are referred to a biographic article by Revelle & Munk (1948).

Let us now return to Vilhelm Bjerknes in Leipzig. When Hesselberg and Sverdrup left, Bjerknes employed two young Norwegians as his next Carnegie assistants; they were his son Jack Bjerknes and Halvor Solberg.

Halvor Solberg

This was during the First World War, and the situation at the Geophysical Institute became difficult. Most of the German students and associates were called to the front. In 1917, Bjerknes left Leipzig and went with his two Carnegie assistants to Bergen where he had been invited to found a Geophysical Institute.

Thirteen years had elapsed since Bjerknes had put forward his ambitious program to forecast the future states of the atmosphere by integrating the governing differential equations. Undoubtedly, he had thought much about how to solve the mathematical problem, but at this stage he had not proceeded very far. The complicated nonlinear equations would not easily disclose the information which they contained. In this situation, Bjerknes began to study the simpler linearized equations. Cyclones, he reasoned, must in their formative stage necessarily be weak and therefore tractable by linear equations. He writes (Bjerknes et al. 1933, p.785, the author's translation): "However, linear equations always seem to have as solutions stable or unstable *wave motions. Also the atmospheric perturbations, the cyclones, must therefore begin as waves.* So far, I had not the slightest idea of how the transition from the as yet unobserved wave to the well-known vortex was to be conceived."

This mysterious transition, however, was soon after found on the weather map. In Bergen, Jack Bjerknes put forward his model of the frontal cyclone. Solberg studied old weather charts over the North Atlantic and could demonstrate the existence of the polar front and the formation of frontal-wave disturbances which grew into frontal cyclones of the type found by Jack Bjerknes. The picture was completed by the Swede Tor Bergeron, who in 1919 joined V. Bjerknes' Bergen team. He found that as the cyclone grew older, its warm sector would collapse or occlude so that finally the cold air would cover the whole cyclone at low levels, while the warm air had been lifted to higher levels. The result was a four-dimensional cyclone model, with a typical change of structure during its life cycle. As Hesselberg once put it: "The cyclone is born as Solberg's initial wave on the polar front, develops into Jack Bjerknes' ideal cyclone, and finally suffers Bergeron's occlusion death."

In an attempt to recruit new meteorologists, Vilhelm Bjerknes went to Sweden and offered jobs in Bergen for interested students for the summer 1919. One of those who responded was Carl-Gustaf Rossby, who at the time was studying mathematics, mechanics, and astronomy in Stockholm. According to Bergeron (1959), in his masterly biographic article about Rossby, it is quite unlikely that Rossby would have become a meteorologist without Bjerknes' intervention. On the other hand, although Rossby got his first meteorological training in Bergen, he did not stay there long, and it may be misleading to call him a student of Vilhelm Bjerknes. Rossby never quite accepted the Bergen ideas with the polar front as the principal weather-producing agent. When the upper-air weather maps became available in the 1930s, Rossby directed his attention towards motion systems of larger scales, in particular the long waves of the upper westerlies. His barotropic model and famous wave formula formed the basis for new ideas which have been of the greatest importance for meteorology and oceanography.

During the 1920s and 1930s the three creators of the polar-front meteorology extended and deepened their concepts and ideas. Jack Bjerknes studied the three-dimensional structure of the frontal cyclones and the air flow in the upper troposphere. From 1940, Jack Bjerknes was professor of meteorology at the University of California, Los Angeles. Under his direction in cooperation with Jörgen Holmboe, the Meteorology Department at UCLA became one of the leading research centers in the field. Holmboe, also, had started his meteorological career as one of Vilhelm Bjerknes' Carnegie assistants; at UCLA, he worked with Jack Bjerknes on cyclone

Tor Bergeron, Carl-Gustaf Rossby, Svein Rosseland (left to right), Vilhelm Bjerknes (standing)

dynamics. For further information about Jack Bjerknes' remarkable research activity, the reader is referred to articles by Charney (1975), Mintz (1975), and Namias (1975).

Tor Bergeron, likewise, became one of the great meteorologists of his generation. After the war he was professor of meteorology at the University of Uppsala, Sweden. One of his many contributions, in particular, was his explanation of how fronts are formed in the atmosphere. Moreover, he laid the basis for air-mass analysis and developed this into a very useful prognostic tool. His work in cloud physics has been of fundamental importance for the development of that field (biographic notes: Eliassen 1978, Blanchard 1978).

Following the empirical discoveries of the Bergen school, Solberg attacked the problem of the growth of frontal waves mathematically, on the basis of linearized equations. He began with a systematic investigation of wave motions. In addition to the known acoustic and internal gravity waves, his equations indicated the existence of a third wave type, provided the fluid was rotating. A physical interpretation of these inertia waves was given in 1929 by Vilhelm Bjerknes and Solberg in a joint paper. Here they demonstrate that a barotropic vortex possesses an internal stability if its specific angular momentum increases with increasing distance from the axis. Strangely enough, they were not aware that Rayleigh had given the same criterion for rotational stability already in 1916. However, Solberg was probably the first to give a mathematical treatment of internal inertia waves.

Solberg proceeded to study waves on a sloping "front" or interface between two zonal air currents of different density. Among several different wave types, he found one which he considered to correspond to a growing cyclone wave. It is difficult today to judge his work. He was not able to satisfy the boundary condition for a level ground which intersects the sloping interface, and introduced, instead, two sloping rigid boundaries parallel with the undisturbed interface. As we now know from the work of Charney & Stern and Pedlosky, such a change of boundary conditions may change the quasi-geostrophic stability properties of the system. However, the growth of frontal waves has later been convincingly demonstrated theoretically by E. Eliasen and Orlanski. Of Solberg's later work, his 1936 analysis of the "symmetric" stability of a baroclinic circular vortex is particularly important.

In 1926, Vilhelm Bjerknes came to the University of Oslo as professor of mechanics and theoretical physics. Together with his Bergen team he published a textbook on physical fluid dynamics which contained a comprehensive account of the results and methods of the Bergen school, including much of Solberg's wave theory, and, in addition, two chapters on the old hydrodynamic-electromagnetic analogies (V. Bjerknes et al. 1933). In this work he had valuable help from his Carnegie assistant Carl Ludvig Godske, who later took over the chair in Bergen that became vacant when Jack Bjerknes went to UCLA. Godske took up agro-meteorology and micro-climatology as his special field of research.

Vilhelm Bjerknes' last Carnegie assistant (from 1935) and collaborator for many years was Einar Høiland. Together they held weekly seminars on meteorology, hydrodynamics, thermodynamics, and statistical physics which attracted inquisitive students. Both Ragnar Fjørtoft and the author of this article were captured for meteorology in this way. After the war, Høiland became professor of hydro- and aerodynamics at Oslo. Like his teacher Vilhelm Bjerknes, he too had a great talent for attracting and inspiring students.

Vilhelm Bjerknes was born in 1862. Biographical articles about him by Bergeron, Devik, and Godske (1962) are found in the *V. Bjerknes Centennial Volume* of *Geophysica Norvegica*. For several years after the war, when Bjerknes was in his eighties, he still came every day to his office in the Astrophysics Building, which also housed the geophysicists at the University of Oslo. He was an ardent listener to seminars and thus gave his encouragement and moral support to the research activity around him. He remained, as he had always been, a source of inspiration to younger generations.

Literature Cited

Bergeron, T. 1959. The young Carl-Gustaf Rossby. In *The Atmosphere and the Sea in Motion* (The Rossby Memorial Volume), ed. B. Bolin, pp. 51–55. New York: The Rockefeller Institute Press

Bergeron, T., Devik, O., Godske, C. L. 1962. *Geof. Publ.* 24: 7–25 (*V. Bjerknes Centennial Volume*)

Bjerknes, V. 1898. Über die Bildung von Circulationsbewegungen und Wirbeln in reibungslosen Flüssigkeiten. *Videnskabsselskapets Skrifter. I Math. Naturv. Klasse, No. 5*

Bjerknes, V. 1904. Das Problem der Wettervorhersage, betrachtet vom Standpunkte der Mechanik und der Physik. *Meteor. Zeitschr.* 21:1–7

Bjerknes, V., Bjerknes, J., Solberg, H., Bergeron, T. 1933. *Physikalische Hydrodynamik*, pp. 777–90. Berlin: Springer. 797 pp.

Blanchard, D. C. 1978. Tor Bergeron and his autobiografic note. *Bull. Am. Meteorol. Soc.* 59: 389–92

Charney, J. G. 1975. Jacob Bjerknes—an appreciation. In *Selected papers of Jacob Aall Bonnevie Bjerknes*, ed. M. G. Wurtele, pp. 11-13. North Hollywood, Calif: Western Periodicals Company. 606 pp.

Eliassen, A. 1978. Tor Bergeron. *Bull. Am. Meteorol. Soc.* 59: 387–89

Ekman, V. W. 1906. Beiträge zur Theorie der Meeresströmungen. *Ann. Hydrogr. Marit. Meteor.* 1906: 1–50

Mintz, Y. 1975. Jacob Bjerknes and our understanding of the atmospheric general circulation. In *Selected papers of J. Bjerknes,* ed. M. G. Wurtele, pp. 14–15. North Hollywood, Calif: Western Periodicals Co. 606 pp.

Namias, J. 1975. The contributions of J. Bjerknes to air-sea interaction. In *Selected Papers of J. Bjerknes*, ed. M. G. Wurtele, pp. 16–18. North Hollywood, Calif: Western Periodicals Co. 606 pp.

Revelle, R., Munk, W. 1948. Harald Ulrik Sverdrup—an appreciation. *J. Marine Res.* 7:127–31.

SEDIMENT RIPPLES AND DUNES

Frank Engelund and Jørgen Fredsøe

Institute of Hydrodynamics and Hydraulic Engineering, Technical University of Denmark, Lyngby, Denmark

INTRODUCTION

It is a general observation that some of the most appealing and fruitful research fields are found at the borderlines of traditional research areas. One example of this is the study of sediment transport and the behavior of alluvial channels, where important contributions have been made by geologists, oceanographers, and civil engineers.

One of the basic problems is that when a flow is confined by boundaries composed of noncohesive sediment, the interaction between flow and boundary molds the geometry of the channel and, hence, determines the hydraulic roughness. Further, the rate of sediment transport, which is another quantity of fundamental importance, depends to a large extent on the hydraulic resistance developed by the bed configuration. The dual problem of predicting flow resistance and sediment transport as well as attempting to explain the general behavior of alluvial channels is the subject of this review.

Many of the aspects are far from being clarified in a satisfactory way, and it has not been possible to avoid controversial issues. In such cases the authors have tried to state the situation objectively, but without trying to hide their own point of view.

To restrict the length of the paper, pneumatic transport of sediment is not mentioned. Neither do we treat important special subjects such as gravel-bed streams and rivers with very graded sediment or very high content of wash load.

BED CONFIGURATIONS

Usually a plane stream bed will be unstable. It tends to break up and take one of the configurations shown in Figure 1, which is a picture based on the first extensive and systematic investigations made in a sufficiently wide flume (Fort Collins, Colorado; see Simons & Richardson 1961 and Guy et al. 1966). The most important bed forms are ripples, dunes, plane bed, and antidunes.

Ripples

When the tractive force is increased to the point where sediment transport starts, the bed will be unstable. In the case of fine sediment, ripples are formed, while coarse sediments will usually form dunes.

"Ripples" is the name given to small triangular sand waves, usually shorter than about 0.6 meter and not higher than about 60 mm, formed in the bed.

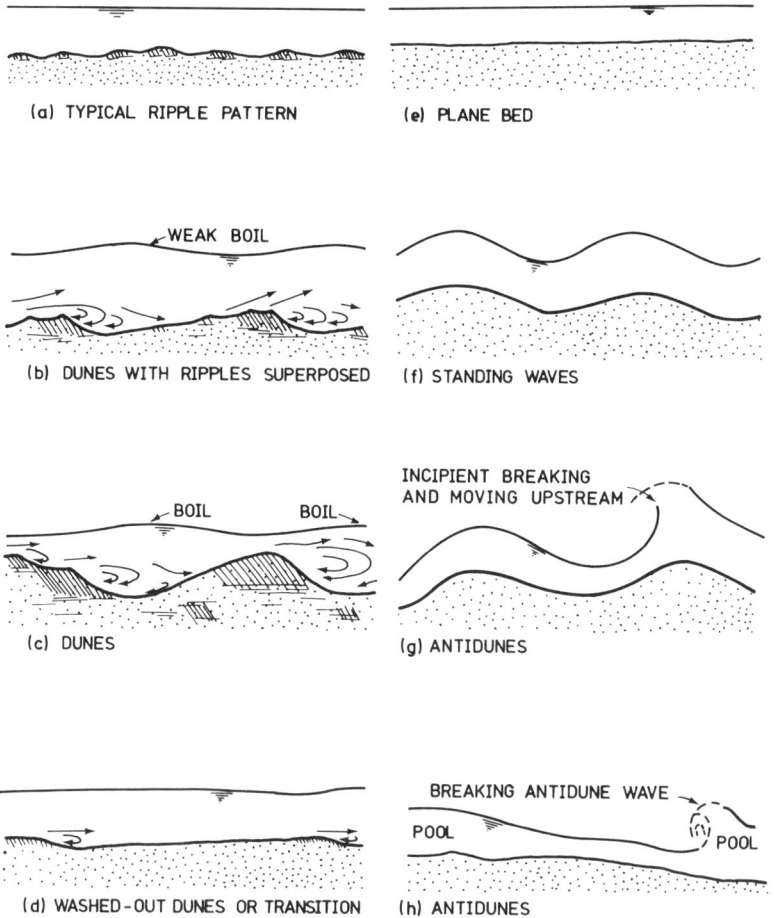

Figure 1 Typical bed forms in order of increased stream power.

At relatively small flow velocities a viscous sublayer of thickness

$$\delta = \frac{11.6\,v}{U_*} \tag{1}$$

is formed. v is the kinematic viscosity of the fluid and U_* is the friction velocity, $\sqrt{\tau_0/\rho}$. This "hydraulically smooth" situation occurs when δ is smaller than the sediment size d. It is assumed (Simons & Richardson 1961) that ripples are formed if a viscous layer is present when the critical tractive force is just surpassed, while dunes are formed if the bed is hydraulically rough.

The ripple length depends on the sediment size (and other parameters), but is essentially independent of the water depth. Three-dimensional ripple patterns have been discussed by Raudkivi (1976).

Dunes

Dunes are the large, more or less irregular sand waves usually formed in natural streams. This is by far the most important bed form in practical river engineering.

The longitudinal profile of a dune is roughly triangular, with a mild and slightly curved upstream surface and a downstream slope approximately equal to the angle of repose (see Figure 2).

Flow separation occurs at the crest, reattachment in the trough, so that bottom rollers are formed at the lee side of each dune. Above this a zone of violent free turbulence is formed in which a large production (and dissipation) of turbulent energy takes place. Near the reattachment sediment particles are moved by turbulence, even when the local shear stress is below its critical value (Raudkivi 1963).

On the upstream side of the dune the shear stress moves sediment particles uphill until they pass the crest and eventually become buried in the bed for a period. As sediment is moved from the upstream side and deposited on the lee side of the dune, the result is a slow, continuous downstream migration of the dune pattern.

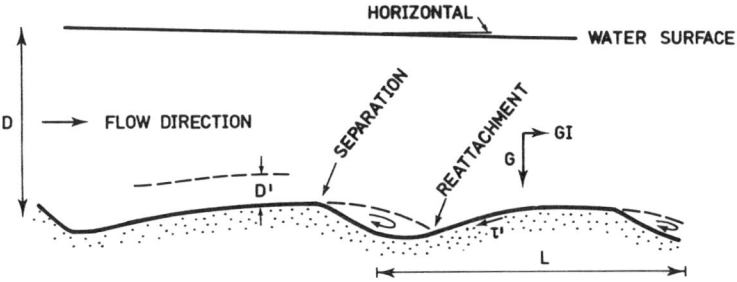

Figure 2 Longitudinal profile of dunes (exaggerated vertical scale).

Transition and Plane Bed

For increased stream power the dunes tend to wash out, i.e. to become much longer and flatter and finally disappear. This happens at Froude numbers below 1, i.e. in subcritical flow, and marks the transition to the so-called upper flow regime. This change from dunes to flat bed means a rather drastic reduction of both hydraulic resistance and water depth, which may be a major practical problem for river navigation.

Antidunes

Further increase in stream power leads from transition and plane bed to the so-called antidunes and related configurations. In this case the longitudinal bed profile is nearly sinusoidal, and so is the water surface, but usually with a much larger amplitude (Figure 1f and g).

At higher Froude numbers the amplitude of the surface profiles often tends to grow until breaking occurs. After breaking, the amplitude may be small for a time and then the process of growth and breaking is repeated.

The name "antidune" indicates the fact that the bed and surface profiles are moving upstream, particularly just before breaking. Figure 1h illustrates an extreme form of antidunes occurring at high Froude numbers.

An overview of bedforms in sand-bed streams is obtained if we imagine an experiment in a flume in which the discharge (or stream power) is gradually increased, as shown in Figure 3. The ordinate is the total bed shear stress τ_0 and the abscissa is the mean velocity V. In the case of a fixed (immobile) bed, the relation between τ_0 and V would be as the dotted

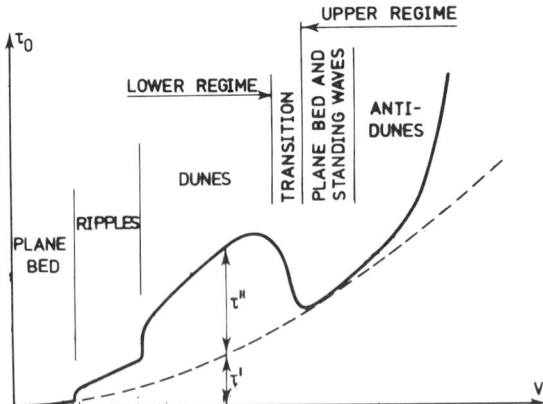

Figure 3 Relation between total bed shear stress τ_0 and flow velocity V for different bed forms.

curve in Figure 3, i.e. close to a second-order parabola, corresponding to the expression

$$\tau_0 = \frac{1}{2} f\rho V^2, \qquad (2)$$

which defines the friction factor f.

The occurrence of ripples and dunes obviously implies a considerable increase of the hydraulic resistance. On the other hand, plane bed, standing waves, and weak antidunes bring the resistance back to skin friction only. The total shear stress τ_0 is the sum of the pure skin friction τ' and a contribution τ'', which originates from the normal stresses acting on the bed (cf. form drag). The problem of estimating the effective shear stress τ' has not yet been solved in a completely satisfactory way. The method currently used was originally suggested by Einstein (1950), in a way slightly different from that presented here.

Close to the crest the flow is converging, which means that the flow attains the character of a boundary layer with thickness D', (see Figure 2). In these circumstances the mean velocity V can approximately be given as

$$\frac{V}{U'_*} = 6 + 2.5 \ln \frac{D'}{k}, \qquad (3)$$

where k is the roughness, and

$$U'_* = \sqrt{gD'I}, \qquad (4)$$

where I is the energy gradient (slope).

At first sight it would be natural to assume the roughness k to be equal to the mean sediment diameter d. However, the irregular surface prevailing during bed-load movements gives a somewhat larger value, so that k must be taken to be about $2.5\,d$. When V and d are known, D' and U'_* can be calculated from Equations (3) and (4).

For later use, the dimensionless version of U_*, often called the Shields parameter, is defined as

$$\theta = \frac{U_*^2}{(s-1)gd} = \frac{\tau_0}{\rho g(s-1)d}, \qquad (5)$$

where g is the acceleration of gravity and s is the relative density of the sediment.

The two main questions associated with flow in alluvial rivers are the following:

1. What is the hydraulic resistance of the channel?
2. What is the rate of sediment transport?

These two questions are more closely related than most scientists realize. The sediment transport creates ripples or dunes, which in turn are responsible for a major part of the hydraulic resistance. Hence, a change of the sediment transport rate will usually change the resistance, and vice versa. Nevertheless, in most cases research is concerned with one of the two main aspects, leaving the other one unconsidered.

THE HYDRAULIC RESISTANCE

In order to predict the stage-discharge relation for an alluvial river, it is necessary to know the hydraulic resistance. The diagram in Figure 3 represents the outcome of observations for one specific sediment size. Take another sediment and the result is likely to be similar, but different in detail. Then the question arises: Is it possible to introduce such changes of scale that a universal relationship is obtained? If so, the dimensional variables τ_0 and V from Figure 3 are likely to be replaced by dimensionless versions. Such a universal relationship would be of great practical importance, because we would then be able to go the other way round and predict the hydraulic resistance, and from this the stage-discharge relation.

Several methods for calculating the hydraulic resistance of alluvial channels have been suggested, and the topic has been discussed at various international hydraulics meetings. For a presentation of the methods that have been discussed throughout the years, reference is made to Vanoni (1975) and Raudkivi (1976). Here, we refer only to methods that are currently under discussion.

The first rational approach to the problem was probably that of Einstein & Barbarossa (1952). They postulated a functional relationship of the type

$$\frac{V}{U''_*} = F(\theta') \tag{6}$$

in which V is the mean flow velocity, while U''_* is defined by the above-mentioned division of the shear stress,

$$U''_* = \sqrt{\frac{\tau''}{\rho}}, \tag{7}$$

where τ'' refers to the bed-form contribution to the total shear stress acting on the bed. The parameter θ' is defined as

$$\theta' = \frac{\tau'}{\rho g(s-1)d}, \tag{8}$$

τ' being the portion of τ_0 due to skin friction. The postulated relationship, Equation (6), was obtained empirically, based on field data from rivers in the United States.

The method suggested by Engelund (1966) is an attempt to establish a nondimensional version of the relation of Figure 3, valid for both the upper and the lower flow regimes, but not for ripples or transition, as explained later. If τ_0 is replaced by the dimensionless version θ [Equation (5)], and if V is replaced by θ' [Equation (8)], the result will be as illustrated in Figure 4. In this figure all data from the Fort Collins experiments (except those termed "ripple" or "transition") are plotted and seem to define two curves for lower and upper regimes respectively.

If such a relationship is taken to be universally valid for sand-bed streams, it can be used to predict the stage-discharge relation of alluvial streams. Assume, for instance, that the lower branch may be given by the empirical relation

$$\theta' = 0.06 + 0.3\ \theta^{3/2}, \tag{9}$$

which is the curve shown in Figure 4. [Note that originally a somewhat different relation was suggested, namely $\theta' = 0.06 + 0.4\ \theta^2$. According to recent supplementary experiments, Equation (9) is preferable.]

It is remarkable that in this theory no effect of temperature is considered. Consequently, the ripple regime cannot be described adequately by this method. Furthermore, close to the transition between lower and upper flow regimes, the temperature has a definite effect on the bed configuration, and hence on the flow resistance, because the transition itself in certain

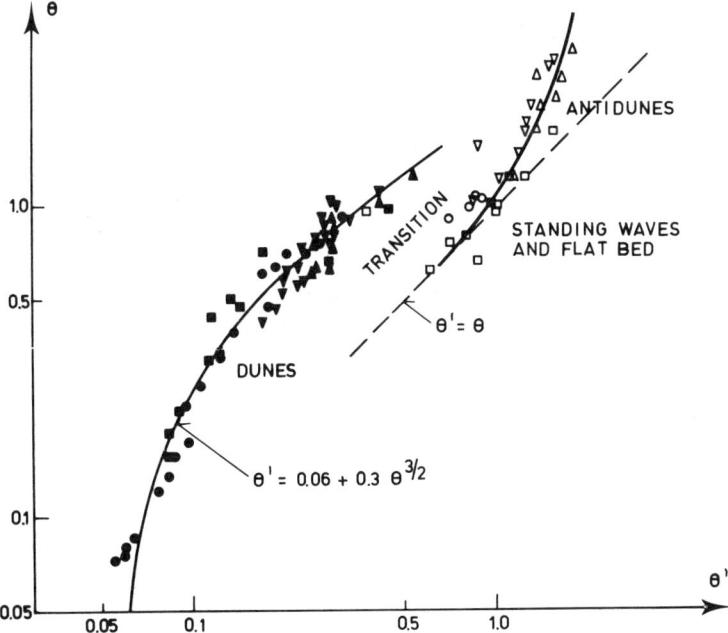

Figure 4 Relationship between θ and θ'. Open symbols correspond to upper flow regime.

cases may be rather sensitive to temperature changes, as will be mentioned in a later section.

While Einstein-Barbarossa's and Engelund's methods both have supplementary methods for calculating the rate of sediment transport, most alternative methods are specifically devoted to predicting the resistance.

Raudkivi (1967) suggested a graphical relation between mean velocity V and friction velocity U_* in the following dimensionless form

$$\frac{V}{\sqrt{U_*^2 - U_{*c}^2}} = F(\theta), \qquad (10)$$

where U_{*c} is the friction velocity at incipient sediment motion. Over a certain area the function is multiple-valued, which enables the method to predict multiple-valued depth-velocity relations, which may be important in the transition area.

The three methods mentioned here have been compared with an independent body of data by White et al. (1980). While these three methods are developed by "plausible reasoning" and subsequently tested by experimental and field data, White et al. went the other way round and considered an appreciable amount of data and tried, by dimensional analysis and plausible physical arguments, to find relationships involving several empirical functions determined by optimization.

Several comparisons between observed depth-velocity relations and predictions by different methods are presented in *Sedimentation Engineering* (Vanoni 1975).

THE SEDIMENT TRANSPORT

The total sediment load is traditionally divided into the following categories: 1. bed-material load, consisting of (a) bed load and (b) suspended load; 2. wash load.

The suspended load is the part of the total load that moves without continuous contact with the bed, being carried by the agitation of fluid turbulence. Besides this, most natural streams carry a certain amount of very fine particles (the so-called *wash load*) not represented in the bed. Today sediment-transport theories are mostly concerned with the bed-material load.

In Equation (11), q_s denotes the rate of bed-material transport volume of sediment material (without pores) per unit time and per unit width of the stream. q_B is the corresponding bed-load transport. It has been found convenient to use a dimensionless form:

$$\Phi_s = \frac{q_s}{\sqrt{(s-1)gd^3}}; \quad \Phi_B = \frac{q_B}{\sqrt{(s-1)gd^3}}. \qquad (11)$$

In the theory of sediment suspension developed by Rouse (1937) it is assumed that the natural tendency of heavy particles to settle is counteracted by turbulent diffusion. If the diffusivity is taken to be equal to the eddy viscosity, the following distribution of sediment concentration is obtained:

$$c = \text{const} \left(\frac{D-y}{y}\right)^z, \qquad (12)$$

in which z (occasionally called the Rouse number) is given by

$$z = \frac{w}{\kappa U_*}, \qquad (13)$$

where w is the settling velocity of the sediment and κ is von Kármán's constant.

This distribution has been compared with several sets of data from flumes and natural streams. The agreement with observation is fairly good but not perfect, and many attempts have been made to improve the basic theory; however, no generally accepted alternative is available to date. One obvious shortcoming of the theory is that it contains an unknown factor, so what it describes is the relative distribution and not the absolute values of c. What is needed next is a determination of the concentration at some reference level. Two different approaches exist.

Einstein's method (1950) is purely kinematical, as he defines the concentration at a distance $2d$ from the bed as being equal to the concentration of the bed load in the same zone. A method based on dynamical considerations has been suggested by Engelund & Fredsøe (1976). It takes into consideration the so-called dispersive stresses (introduced by Bagnold 1954) that are transmitted to the bed because of the interchange of momentum caused by collision of sediment particles. Both methods are found to give reasonably good results, but a closer examination and comparison has not been performed.

The total bed-material load is simply the sum of bed load and suspended load. The transport rate of the total bed material can, in principle, be obtained by adding the transport rates calculated for the bed load and suspension separately. However, many methods have been proposed for the easier procedure of estimating the total directly from the flow parameters. Again, for a complete review of all the methods that have been proposed, readers are referred to handbooks. In this review only a few newer methods under current discussion are mentioned.

The Engelund-Hansen formula (Engelund 1966, Engelund & Hansen 1967) is

$$f\Phi_s = 0.08 \, \theta^{5/2}, \qquad (14)$$

where f is the friction factor defined by Equation (2). If Equation (14) is combined with Equation (9), it is possible to develop a design chart that presents the stage-discharge relation and water discharge–sediment discharge relation in one graph. Equation (14) is applicable to both dunes and antidunes.

Ackers & White (1973) have developed empirical relations by comprehensive analysis of laboratory experiments. Instead of working with the Shields parameter they define a related quantity called the sediment mobility.

The general transport function resulting from their analysis states that the dimensionless transport rate is a function of the sediment mobility and a dimensionless grain diameter. The theory involves four empirical parameters, all given as empirical functions of the dimensionless grain diameter.

A somewhat similar analysis was made by Yang (1979), who proposed a transport relation where the total sediment concentration is given in terms of dimensionless quantities obtained from dimensional analysis and multiple regression analysis of laboratory data.

The success of such an approach depends on the relevance of the nondimensional groups involved and of the number of free coefficients to be determined by the regression analysis. Yang applied as many as six such coefficients and obtained good agreement when his formula was used as predictor. In fact, the three methods mentioned above were found to be of similar reliability.

Equation (14) was obtained by a general assumption concerning the parameters necessary for a description of the dune case and was, in fact, obtained as a by-product of the assumption leading to Equation (9).

Ackers & White's as well as Yang's method is based on plausible reasoning, dimensional analysis, and adaptation to a considerable body of data. Despite this, these methods can hardly claim to be definitely superior to the much simpler Equation (14). The reason for this may be that the two last methods seek to describe the entire lower flow regime (see Figure 3) including ripples as well as dunes. The often rather abrupt transition between the ripple and the dune is obviously a source of difficulty when a common description of these regimes is to be established.

Another point is that the use of laboratory data may introduce effects that are irrelevant to flow in rivers. In reality, most flume experiments are carried out in flumes that are too narrow to allow the true three-dimensional character of the dunes to develop, the Fort Collins experiments being about the only comprehensive exception. Today little is known about the effect of this artificial "two-dimensionality" of the bed configurations, although some few investigations have been made (Williams 1970). Furthermore, ripples are much more likely to occur in flumes than in rivers, and this may introduce yet another scale effect exaggerating the effect of viscosity.

In the case of a dune-covered bed it has not been possible to detect any influence of viscosity (or temperature), except at the transition to flat bed. This is in accordance with the parameters in the Engelund-Hansen method, whereas the nondimensional groups in Yang's and in Ackers & White's methods involve the viscosity of the fluid.

THE FORMATION OF BED FORMS

It has been a continuous challenge to scientists to determine why the bed forms grow and change from one form to another and yet remain stable for a given set of flow and sediment conditions. Two excellent review papers on the formation of bed forms are available (Kennedy 1969 and Reynolds 1976). Hence the present review emphasizes the results obtained during the last decade.

Most scientists today agree that the process of sand-wave formation is a problem of instability. Suppose that an originally plane bed is slightly perturbed so that the flow and the sediment transport are disturbed. Then there will be two main possibilities:

1. The changes in flow pattern and sediment transport will tend to attenuate the amplitude of the perturbation, so that the bed goes back to the original plane state. This means that the plane bed is stable.
2. The second possibility is that the flow causes the perturbation of the bed to increase in time, which corresponds to the unstable situation, ultimately leading to the formation of ripples, dunes, or antidunes.

To express this in mathematical form, we assume that the sand bed is given a small sinusoidal perturbation with the amplitude h_0, so that the surface of the sand bed is given by

$$h = h_0 \exp[ik(x - ct)], \tag{15}$$

where k is the wave number ($= 2\pi/L$, where L is the wavelength), x is a coordinate in the main flow direction, t is time, and c is the migration velocity of the sand wave, while i is the imaginary unit $\sqrt{-1}$.

Equation (15) may alternatively be written

$$h = h_0[\cos k(x - ct) + i \sin k(x - ct)] \tag{15a}$$

and only the real part of this expression is supposed to have physical significance.

As in most theories of hydrodynamic instability the idea is to assume the migration velocity c to be complex, i.e.

$$c = c_r + ic_i \tag{16}$$

so that the exponent in Equation (15) becomes

$$ik(x - ct) = ik(x - |c_r + ic_i|t) = ik(x - c_r t) + kc_i t .$$

Then Equation (15) becomes

$$h = h_0 \exp(kc_i t) \exp[ik(x - c_r t)]. \quad (17)$$

From this it is apparent that if c_i can be calculated, it may be used to determine whether the flow is stable or not. If c_i is positive, the factor

$$\exp(kc_i t)$$

indicates that the bed waves grow exponentially in time. On the other hand, if c_i is negative, the perturbation will attenuate and the bed becomes plane again.

We can proceed a little further by considering the well-known continuity equation for the sediment

$$\frac{\partial q_s}{\partial x} = -(1-n)\frac{\partial h}{\partial t}. \quad (18)$$

Here q_s is the rate of sediment transport and n is the porosity of the sand bed. It is easily seen that this equation can be satisfied by Equation (15) and the following expression for the transport rate:

$$q_s = q_{s0} + \Delta q_s \exp[ik(x - ct - \delta)], \quad (19)$$

where δ is a phase lag between the local rate of sediment transport and the bed form. q_{s0} is the average (undisturbed) transport rate. With Equations (15) and (19) the continuity equation is satisfied provided

$$\Delta q_s e^{-ik\delta} = (1-n)h_0 c$$

or

$$c = \frac{\Delta q_s}{(1-n)h_0}[\cos k\delta - i \sin k\delta]. \quad (20)$$

The imaginary part of c_i is found to be

$$c_i = \frac{\Delta q_s}{(1-n)h_0}(-\sin k\delta). \quad (21)$$

The Formation of Antidunes

Let us first consider the simple case in which the phase shift δk is positive and smaller than π, as will be the case when suspension is the dominating transport form. The sign of c_i is then determined by Δq_s, and a complete discussion comprises three different cases.

Case I is the subcritical flow, defined by the criterion for the Froude number Fr:

$$\text{Fr}^2 < \frac{\tanh(kD)}{kD} \quad (22)$$

(for long waves, $kD \gg 1$, this becomes equal to Fr < 1). In this case the depth and the bed form are 180° out of phase (see Figure 5). Consequently the sediment transport rate is maximum near the crest and minimum at the trough, apart from the lag $k\delta$, due to the time it takes for sediment to settle and rise in the nonuniform flow. Hence Δq_s is positive and—according to Equation (21)—c_i is negative, which means stability.

Case II relates to supercritical flow. When the Froude number Fr lies in the interval

$$\frac{\tanh(kD)}{kD} < \text{Fr}^2 < \frac{\coth(kD)}{kD}, \qquad (23)$$

it is known from traditional potential theory that the water-surface amplitude is larger than the bed-wave amplitude, as shown in Figure 5. Hence, sediment transport is minimum at the crest, which means that Δq_s is negative, i.e. c_i is positive, indicating instability.

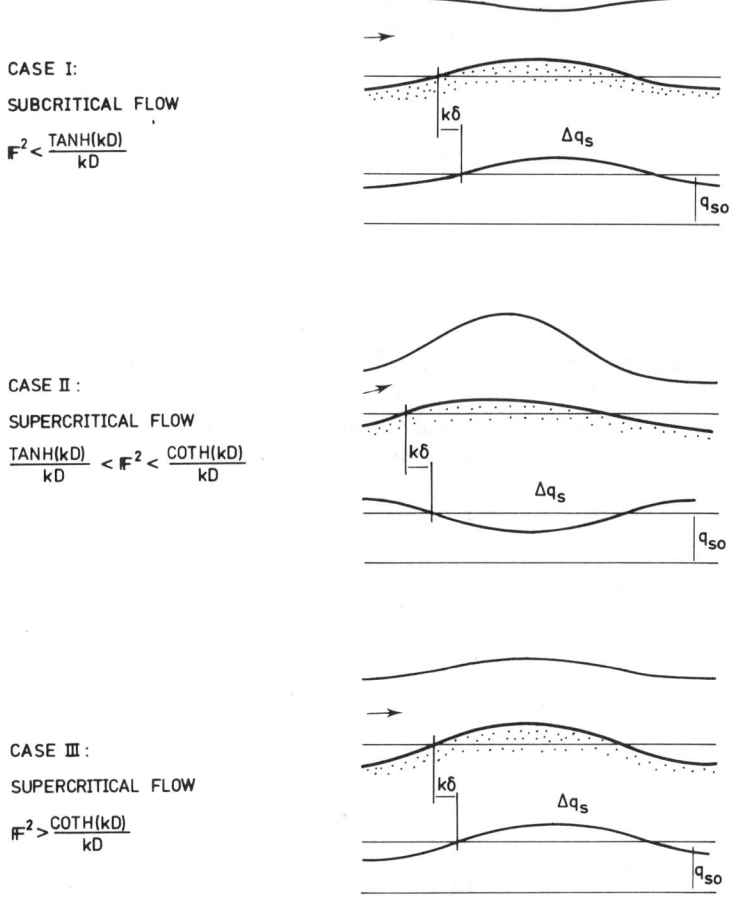

Figure 5 Variation in surface profiles and phase $k\delta$ with Froude number.

Case III comprises the largest values of Fr

$$\text{Fr}^2 > \frac{\coth(kD)}{kD}. \tag{24}$$

In this case the flow is still supercritical, but the surface amplitude is smaller than the bed-wave amplitude, so that again Δq_s is positive and the flow stable.

The result of this investigation is that the flow is unstable in case II, but stable in cases I and III.

From Equation (20), it is seen that the migration velocity c_r of the bed waves is negative in case II, corresponding to what we call antidunes. In Figure 6 are shown the areas corresponding to the three cases. The points correspond to all the antidune observations from Fort Collins (circles) and some data from J. F. Kennedy. The data confirm the instability in the area corresponding to case II, and the stability of case III.

Case I, which will now be discussed in greater detail, is much more complicated because the assumption of dominating suspension used in the above discussion is usually not fulfilled for subcritical flow.

The Formation of Dunes

Let us consider the subcritical flow situation ($\Delta q_s > 0$). By use of Equation (21) the stability condition can be expressed in terms of the phase shift δ. If δ is negative the flow is unstable, while $\delta > 0$ ensures stability.

The difficulties appear when δ has to be calculated from the laws of motion of (turbulent) fluid and sediment particles. The main conclusion that can be drawn from the available research is that δ depends on the

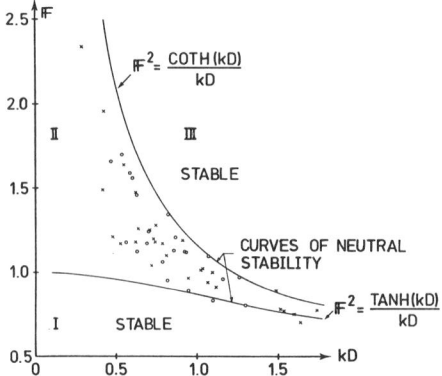

Figure 6 Stability boundaries for antidunes. After Engelund (1970).

following items, quoted in the order of decreasing importance (for sand-bed streams):

1. Fluid friction,
2. Rate of suspended sediment (versus bed load),
3. Gravity forces on moving bed load,
4. Inertia of sediment particles,
5. Percolation in the river bed.

The importance of these five factors is discussed below. When quantitatively acceptable estimates are applied, the theory is able to explain in detail the formation of sand waves also in the lower flow regime.

1. FLUID FRICTION The contribution from the fluid friction to δ is most important at low shear stresses, for which the sediment transport mainly occurs as bed load. Disregarding item 4 above, the bed load will respond immediately to changes in bed shear stress. In the subcritical area (Figure 5, case I) the flow upstream is converging, while the flow from crest to trough is diverging.

Now flow in converging channels gives larger fluid friction on the wall than uniform flow. Diverging flow, on the other hand, gives smaller wall friction than uniform flow; in fact, if the divergence angle is about 5°, friction drops to zero and separation comes up. As a consequence of this, bed friction is not exactly in phase with velocity (and bed form), but there exists a negative phase shift, or a negative contribution to δ. This explains why dunes develop at low flow stages, where the contribution to δ from fluid friction is dominant.

Dungan Smith (1970) calculated the flow over a wave bed, using an eddy-viscosity concept. Undulations on the water surface were disregarded. Dungan Smith's analysis yielded the result that the bed is always unstable. Further on, the growth rate of the perturbation increased with the wave number, so it was not possible to find a wave number corresponding to maximum growth rate, which would indicate the preferred wavelength.

In a paper in the same year by Engelund (1970) a more complete description of the flow over a wavy bed was attempted, taking into account undulations in the free water surface. Engelund used a simple eddy-viscosity concept to calculate the perturbed flow and found that the bed was unstable at low Froude numbers. The preferred wavelength determined by maximum growth rate was found to be rather too small. We discuss this question further under item 3.

2. SUSPENDED SEDIMENT At higher bed shear stresses the amount of suspended sediment increases, which gives a positive contribution to δ. In Engelund's paper (1970), this lag was found by calculating the nonuniform

sediment distribution using a continuity equation for the sediment, expressing equilibrium between convection, settling, and diffusion. Because it takes some time for the suspended sediment to settle after being picked up from the bed, the analysis indicates that maximum transport of suspended sediment occurs downstream of a maximum in bed shear stress, so that in the case of subcritical flow, the suspended sediment gives a positive contribution to δ. This mechanism was later used by Engelund & Fredsøe (1974) to explain the transition from dunes to plane bed in subcritical flow: for increasing shear stresses, where the ratio of suspended load to bed load increases, the positive contribution to δ from the suspended sediment ultimately surpasses the negative contribution from friction and causes stability of the bed.

From this theory it was also possible to explain that a decrease in temperature may cause transition from dunes to plane bed; in the case of a temperature drop, the ratio of suspended load to bed load is increased as the amount of suspended load increases due to the decrease in fall velocity, while the amount of bed load remains essentially unchanged. At sufficiently high bed shear stresses, this may cause the dunes to disappear. A similar method has been applied by Chen & Nordin (1976) in calculating the flow resistance in the river Missouri.

3. GRAVITY FORCES A phase shift between local bed shear stress and local bed-load transport exists because the gravity makes it more difficult to move the bed load uphill than downhill. This effect of local bed slope is quite significant for the result of the stability analysis at low bed shear stresses.

Hayashi (1970) introduced this effect into Kennedy's model (1963). In Kennedy's work, δ is not evaluated, so Hayashi's analysis was a step towards a better understanding of δ. Fredsøe (1974a) extended Engelund's analysis (1970) by incorporating the effect of gravity and showed that then it is possible to obtain a maximum growth rate for a wavelength that is more consistent with observation.

4. INERTIA OF SEDIMENT PARTICLES Parker (1975) introduced the effect of inertia of bed load into the stability theory and carried out a stability analysis with inertia giving the only contribution to δ. In subcritical flow, δ becomes positive, as maximum bed-load transport occurs downstream of maximum bed shear stress. In supercritical flow, δ becomes negative, and so the bed becomes unstable. Hence inertia of the bed load may cause the formation of antidunes for very small depths and coarse sediment, when bed-load transport is the dominant transport mechanism, even in supercritical flow. As pointed out by Fredsøe (1976), who carried out a complete stability analysis including the inertia of bed load, the contri-

bution from inertia is only of minor importance in sand-bed rivers, as the contribution from suspension dominates that of inertia.

5. PERCOLATION IN THE RIVER BED Because of the variation in pressure head along a wavy bed, groundwater flow occurs within the bed, and a small positive contribution to δ arises (see Ho & Gelhar 1973). To the authors' knowledge, no stability analysis has been carried out including this mechanism, but Ho & Gelhar's experiments indicate that this mechanism is usually unimportant as far as bed instability is concerned, as the effect arises at very large flow velocities combined with very coarse material.

Formation of Bed Waves in Closed Conduits

In a closed conduit the flow will always converge upstream of the top of the bed perturbation and diverge downstream of the top, so that Δq_s is always positive. This implies that at low flow velocities dunes occur as in open-channel flow, but that these dunes will disappear at higher bed shear stress, for which the contribution to δ from suspended sediment increases. At sufficiently high flow velocities, the bed will be flat because of the contribution to δ from suspended sediment.

Formation of Ripples

Until recently, stability analysis was concerned with the formation of dunes rather than ripples. Kelvin Richards (1980) extended the previous stability analysis by making a more advanced description of the flow over a wavy bed. Richards follows the line of Engelund (1970) and Engelund & Fredsøe (1974) but includes a more accurate description of the turbulent flow close to the bed by using a one-equation turbulence model. His stability analysis indicates that at low shear stress the bed is still unstable, and that there is not a single maximum growth rate, but rather two local maxima as is seen from Figure 7. One of these maxima depends strongly on the depth, while the other is essentially unaffected by changes in depth. According to Richards the latter fact is related to the instability of ripples, while the former yields the instability of dunes. The instability limits for dunes are consistent with what was found in earlier investigations.

Richards found that the preferred initial length of ripples was proportional to the roughness of the bed. As Richards' model assumes flow over a hydraulically rough bed, it would be of interest to carry out a similar analysis for a hydraulically smooth bed, ripples usually being associated with a smooth bed.

Alternative Theories

Other theories concerning the development of dunes in erodible beds relate dune formation to the properties of the eddies in the flow. In the past

Velikanov (1936) and later Yalin (1971) suggested that large eddies, rotating around a horizontal axis perpendicular to the flow direction, are able to form dunes whose length is about six times their depth. This is the correct order of magnitude in many cases, but not close to transition, where the dunes are much longer. Theories of this kind are at present not generally accepted (Kennedy 1971). They cannot explain changes in dune geometry due to temperature changes, and not even the ordinary transition from the lower to the upper flow regime.

Jackson (1976) relates the dune formation to the so-called burst phenomenon. The burst as described by Kline et al. (1967) is a three-dimensional pseudo-periodic process associated with the instability of the viscous sublayer, and is by Jackson and others postulated to be present also on rough and even dune-covered beds.

Whether this line of thinking is physically sound remains to be seen. At present the authors find it difficult to imagine that the strongly three-dimensional flow pattern associated with bursting is able to explain the two-dimensional nature of sand waves typical of the initial period of development.

SPECTRAL ANALYSIS OF RIPPLES AND DUNES

Statistical methods in the description of sand waves in alluvial rivers were introduced by Nordin & Algert (1966). The profiles determined by soundings of dune beds show that the dune shapes and sizes are rather irregular, so that an entire spectrum of wave lengths and wave heights is present.

Considering the elevation h of a dune bed as a stochastic function of the distance x along the channel, Nordin & Algert found that the process

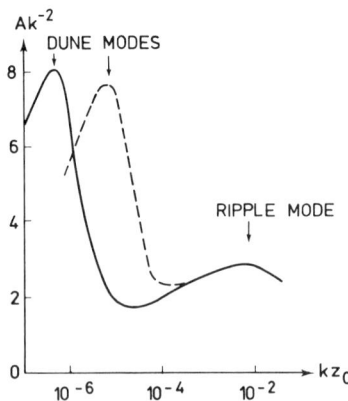

Figure 7 Variation in amplification factor A with wavenumber k. z_0 = bed roughness. ———, $D/z_0 = 10^5$; – – –, $D/z_0 = 10^4$. After Richards (1980).

was often (but not universally) well represented by a Markov second-order linear process.

By dimensional considerations Hino (1968) concluded that the wavenumber spectrum at high wave numbers should vary as the wave number k to the minus third power, while the frequency spectra should vary as the minus second power of the frequency.

The evolution of a sand bed from an originally flat state until it reaches its equilibrium configuration was investigated experimentally by Jain & Kennedy (1974). During the initial period two spectral peaks developed which were thought to correspond to two different physical mechanisms. One peak was supposed to be associated with the presence of a free surface and corresponds to an arrested surface wave. The second peak corresponds to the instability of the erodible bed as discussed above. Generally, the two mechanisms produce the two initial peaks, each with a different dominant wave number. The higher the wave number, the faster the wave will migrate. Hence the shorter waves will overtake the longer and slower moving ones and be absorbed by the latter. In this merging, bed-wave variance generated at higher wave numbers will be shifted to lower wave numbers, and this process will continue until the bed waves become so high that they reach a limiting steepness.

From consideration of this variance cascade Jain & Kennedy gave an independent derivation of Hino's "3rd power law." Nordin (1971) noticed that for both ripples and dunes the major part of the variance is contributed by the longer wavelength components, i.e. smaller values of k. Further he discovered that all spectra are remarkably similar in general shape so that a generally valid nondimensional form of the spectrum, including the flow parameters, may be obtained by dimensional considerations.

If G denotes the spectral density function for the process $h(x)$, Nordin suggests a relation of the type

$$G' = f(k'),$$

where

$$G' = g/V \text{ and } k' = kV/g.$$

Such dimensionless spectra are shown in Figure 8 including data for ripples as well as for dunes. For values of the dimensionless wave number higher than 0.03 the data follow a relation close to that predicted by Hino. For lower wave numbers, each set of data follows a separate curve without a consistent pattern.

However, for the hydraulic resistance of the flow channel, it is this lower wave-number range of the spectrum that is significant, just as in turbulence the lower frequencies describe the energy-containing eddies. For this reason it may be important to try to develop a dimensionless relation primarily to account for this significant range. An attempt in this direction was made by Engelund (1969), from the same ideas as Equation (9):

$$G(k) = \mathbf{h}^3 \phi(k\mathbf{h}/f)/f,$$

in which $\mathbf{h}$ is the rms value of bed elevation, f the friction factor, and ϕ a universal function (at least for the lower range of k). The validity of the relation is restricted to dune beds. Too little information is available today to decide whether it has a more universal validity in the lower range of k.

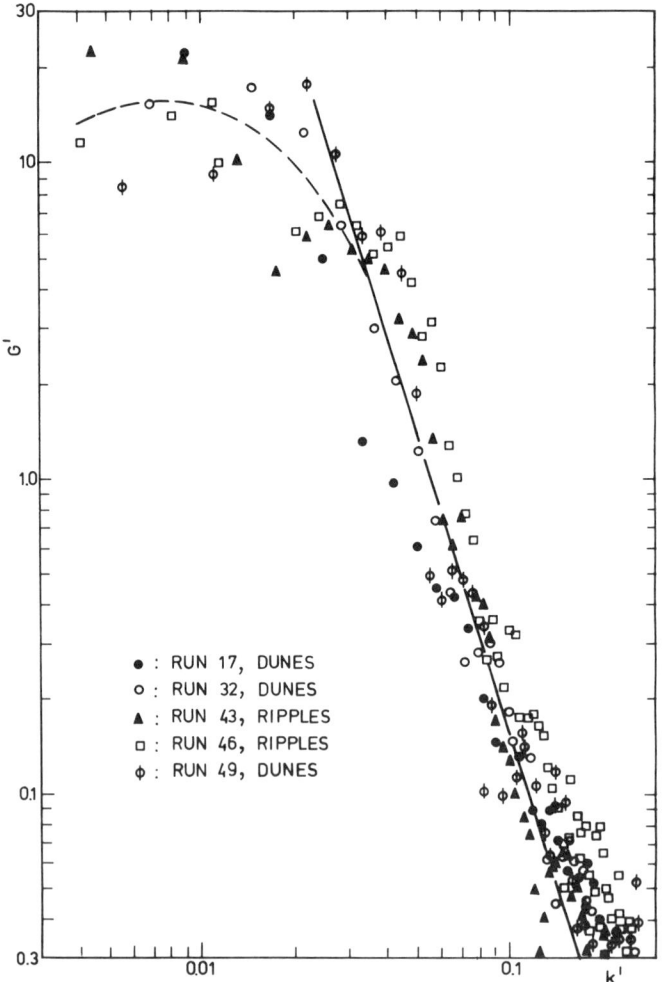

Figure 8 Dimensionless spectra for ripples and dunes. After Nordin (1971).

The physical interpretation of the spectral shape is still under discussion. Engelund & Fredsøe (1971) noticed that a Fourier analysis of a completely regular and periodic dune profile gives a discrete spectrum with the density decreasing as the wave number to the minus third power. This indicates that the part of the spectrum associated with high wave numbers or frequencies has no physical significance, but merely indicates the large number of Fourier components necessary to represent a break or discontinuity in the surface inclination.

A related point of view may lie behind the criticism of the spectral method put forward by Plate (1971). With regard to Jain & Kennedy's experiments he advocated the point of view that the second peak is not necessarily an effect of free surface flow only, but may be a second harmonic [like the one found theoretically by Fredsøe (1974a)] due to the non-sinusoidal shape of the sand waves. Further, Plate suggests "that more useful than spectral analysis is to isolate the average sediment wave shape from the data record, and to study its development in space and time." This seems to be a criticism of Fourier resolution using trigonometric functions. To the authors' knowledge, no alternative set of eigenfunctions has been successfully applied to the analysis of sand waves.

THE SHAPE OF FULLY DEVELOPED DUNES IN STEADY FLOW

The calculation of fully developed dunes, i.e. the shape, height, and length of dunes, is a rather complex problem, which, so far, has not been solved in a satisfactory way. The main difficulty is to describe the nonuniform flow, which changes from flow in a very strong adverse pressure gradient, just behind the top of the dunes, to flow in a slightly favorable pressure gradient along the mildly positive slope of the dune. The problem includes flow separation and development of a turbulent boundary layer. Several authors have suggested that this flow situation be compared with a flow past a rear-facing step. Dungan Smith (1970) made such a comparison and pointed out that a local maximum in the bed shear stress occurs downstream of the step: just after the point of reattachment, the boundary shear stress increases due to flow acceleration. Further downstream, the decrease in bed shear stress due to the increase in the boundary-layer thickness is larger than the increase due to flow acceleration, so the bed shear stress decreases slightly as shown in Figure 7 in the paper by Bradshaw & Wong (1972). Dungan Smith then related the length of a dune, built up downstream from a preceding dune, to this local maximum in bed shear stress.

Fredsøe (1980a,b) worked out a simple model to calculate the dune shape using an approximate curve fitted to Bradshaw & Wong's data. By relating the local dune height h to the local sediment transport through the equation of continuity, $q = ah(1 - n)$, (where a is the dune migration velocity and n the porosity), it is simple to calculate the shape of a dune formed behind a migrating step as demonstrated in Figure 3 (in 1980b).

The shape can be calculated if the total height H and the length L are known quantities. The height was calculated by a perturbation analysis that showed that only one dune height was stable. The length was found by stopping the dune development when flow separation occurs.

By distinguishing between bed load and suspended load, stating that only bed load settles on the dune front, Fredsøe (1980a) showed that the dune height will decrease at higher values of the bed shear stress, so that ultimately transition to plane bed will occur. By including a phase shift between the rate of suspended load and bed shear stress it is furthermore possible to explain the growth in the dune length at high transport rates and in this way to explain transition from dunes to plane bed. Thus it is possible to obtain a relation between total shear stress θ and skin friction θ' as shown in Figure 9, taken from Fredsøe (1980a).

OBLIQUE DUNES

The stability analysis discussed so far has been concerned with the simple idealized flow in straight rectangular channels. In the case of transverse

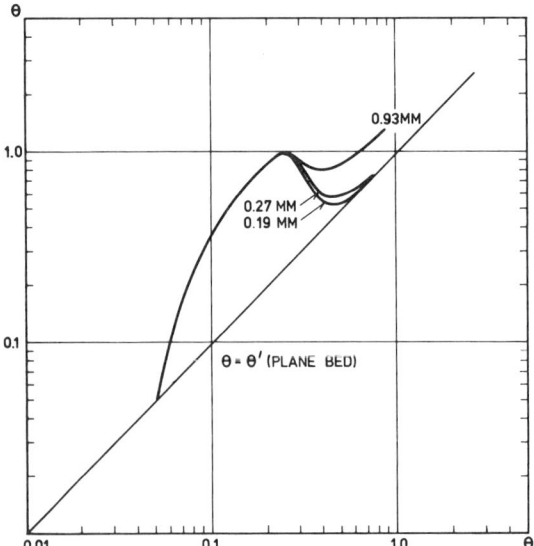

Figure 9 Theoretical prediction of relation between skin friction and total friction.

variation in flow depth or flow velocity the front of dunes will not be perpendicular to the direction of the flow, but will turn out to have some angle different from 90° relative to the flow direction. Barcilon & Lau (1973) considered the existence of transverse bars in the case of a transverse variation in flow depth. They carried out a stability analysis, following Kennedy's method by introducing δ as an unknown parameter. Engelund (1974) carried out a stability analysis with variable flow depth and flow velocity, using depth-averaged flow equations. His analysis yields the result that transverse variation in flow velocity does not lead to instability of the bed, while a variation in flow depth does: in this case, a perturbation of the bed with a front forming an angle with the flow direction will grow with time and form oblique dunes. This was later confirmed by a more complete three-dimensional analysis by Fredsøe (1974b,c), who also found that a transverse variation in flow velocity would lead to formation of oblique dunes, in agreement with observations. It is interesting to note that this kind of instability can be explained without introducing a phase shift δ between local sediment transport and bed level.

DUNE BEHAVIOR IN UNSTEADY FLOW

When the flow is unsteady, dunes will change their form with time, because the dune dimensions vary with the hydraulic conditions. In unsteady flow, the dimensions of the bed forms will normally not be related to the instantaneous hydraulic conditions, because it takes some time for the dunes to change their geometry. It is necessary to know this time lag in order to calculate the flow resistance and sediment transport in unsteady flow. The problem is not entirely deterministic, because of the stochastic nature of the dune behavior, such that small dunes disappear at the crest of larger dunes, and new dunes are formed in the river bed. Because of this, Allen (1976a,b) divided the change in configurations into two classes:

(*a*) changes in the properties of the individual dunes and
(*b*) replacement of the original dunes by others that were adjusted to the new hydraulic conditions.

Allen (1976b) proposed that the lifespan of individual dunes between creation and destruction (item *a* above) is the time during which the dune migrates a certain factor times its own length. This factor then has to be determined empirically. Fredsøe (1979, 1981) developed a mathematical model to determine the time-scale for item *a* above. This model considered the dune behavior from a totally deterministic point of view.

By assuming similarity in the bed shear stress distribution before and after a sudden change in water discharge, it is possible by application of the sediment-continuity equation to calculate the time scale for changes in dune height. The model does not allow one to calculate the change in dune length, a problem which may not be solved without taking into account stochastic elements of the dune behavior. However, the variation in dune height is normally the most important dimensional change, partly because it is more rapid and partly because the relative change in dune height is larger than that in dune length and has a more pronounced effect on flow resistance.

Literature Cited

Ackers, P., White, W. R. 1973. Sediment transport: new approach and analysis. *Am. Soc. Civ. Engrs., J. Hydraul. Div.* 99:HY11: 2041–60

Allen, J. R. L. 1976a. Computational models for dune time-lag: general ideas, difficulties and early results. *Sediment. Geol.* 15: 1–53

Allen, J. R. L. 1976b. Bed forms and unsteady processes: some concepts of classification and response illustrated by common one-way types. *Earth Surface Processes* 1: 361-74

Bagnold, R. A. 1954. Experiments on a gravity-free dispersion of large solid spheres in a Newtonian fluid under shear. *Proc. R. Soc. London Ser. A* 225:49-63

Barcilon, A. I., Lau, J. P. 1973. A model for formation of transverse bars. *J. Geophys. Res.* 78: 2656-64

Bradshaw, P., Wong, F. 1972. The reattachment and relaxation of a turbulent shear layer. *J. Fluid Mech.* 52: 113-36

Chen, H. Y., Nordin, C. F. 1976. Temperature effects in the transition from dunes to plane bed. *M.R.D. Sediment Series, no. 14.* US Army Engrs. District, Missouri River Div.

Einstein, H. A. 1950. The bed load function for sediment transportation in open channels. *Tech. Rep. 1026.* US Dept. Agric.

Einstein, H. A., Barbarossa, N. 1952. River channel roughness. *Trans. Am. Soc. Civ. Engrs.* 117: 1121–46

Engelund, F. 1966. Hydraulic resistance of alluvial streams. *Am. Soc. Civ. Engrs., J. Hydraul. Div.* 92:HY2: 315-26; Closure: 93:HY4: 287–96 (1967).

Engelund, F. 1969. On the possibility of formulating a universal spectrum for dunes. *Tech. Univ. Denmark, Hydraul. Lab., Basic Res., Prog. Rep.* 18: 1-4

Engelund, F. 1970. Instability of erodible beds. *J. Fluid Mech.* 42: 225-44

Engelund, F. 1974. The development of oblique dunes. Part 2. *Tech. Univ. Denmark, ISVA, Prog. Rep.* 32: 37-40

Engelund, F., Fredsøe, J. 1971. A mathematical model of flow over dunes. *Tech. Univ. Denmark, ISVA, Prog. Rep.* 22: 25-30

Engelund, F., Fredsøe, J. 1974. Transition from dunes to plane bed in alluvial streams. *Tech. Univ. Denmark, ISVA, Ser. Pap.,* 4, 56 pp.

Engelund, F., Fredsøe, J. 1976. A sediment transport for straight alluvial channels. *Nord. Hydrol.* 7: 293-306

Engelund, F., Hansen, E. 1967. *A Monograph on Sediment Transport in Alluvial Streams.* Copenhagen: Danish Technical Press

Fredsøe, J. 1974a. On the development of dunes in erodible channels. *J. Fluid Mech.* 64: 1–16

Fredsøe, J. 1974b. The development of oblique dunes. Part 3. *Tech. Univ. Denmark, ISVA, Prog. Rep.* 33: 15-22

Fredsøe, J. 1974c. The development of oblique dunes. Part 4. *Tech. Univ. Denmark, ISVA, Prog. Rep.* 34: 25-29

Fredsøe, J. 1976. Discussion of: Sediment inertia as cause of river anti-dunes. By G. Parker. *Am. Soc. Civ. Engrs., J. Hydraul. Div.* 102:HY1: 99-102

Fredsøe, J. 1979. Unsteady flow in straight alluvial streams: modification of individual dunes. *J. Fluid Mech.* 91:497-512

Fredsøe, J. 1980a. Dimensions of stationary dunes. Part 2. High transport rates. *Tech. Univ. Denmark, ISVA, Prog. Rep.* 52: 27-32

Fredsøe, J. 1980b. The form of dunes. *Proc. Int. Symp. River Sedimentation, Beijing,* Paper B-12, 13 pp.

Fredsøe, J. 1981. Unsteady flow in straight alluvial streams. Part 2. Transition from dunes to plane bed. *J. Fluid Mech.* 102: 431-53

Guy, H. P., Simons, D. B., Richardson, E. V. 1966. Summary of alluvial channel data from flume experiments, 1956-61. *US Geol. Surv., Prof. Pap. 462-I*

Hayashi, T. 1970. Formation of dunes and antidunes in open channels. *Am. Soc. Civ. Engrs., J. Hydraul. Div.* 96:HY2: 357-66

Hino, M. 1968. Equilibrium-range spectra of sand waves formed by flowing water. *J. Fluid Mech.* 34: 565-73

Ho, R. T., Gelhar, L. W. 1973. Turbulent flow with wavy permeable boundaries. *J. Fluid Mech.* 58: 403-15

Jackson, R. G. 1976. Sedimentological and fluid-dynamic implications of the turbulent bursting phenomenon in geophysical flows. *J. Fluid Mech.* 77: 531-60

Jain, S. C., Kennedy, J. F. 1974. The spectral evolution of sedimentary bed forms. *J. Fluid Mech.* 63: 301-14

Kennedy, J. F. 1963. The mechanics of dunes and antidunes in erodible-bed channels. *J. Fluid Mech.* 16: 521-44

Kennedy, J. F. 1969. The formation of sediment ripples, dunes and antidunes. *Ann. Rev. Fluid Mech.* 1: 147-68

Kennedy, J. F. 1971. Changes in alluvial beds composed of non-uniform material. [Déformation des lits alluvionnaires composés de matériaux à granulométrie étendue]. *Proc. 14th Cong. IAHR, General Rep.* 6: 241-52

Kline, S. J., Reynolds, W. C., Schraub, F. A., Runstadler, P. W. 1967. The structure of turbulent boundary layers. *J. Fluid Mech.* 30: 741-73

Nordin, C. F. 1971. Statistical properties of dune profiles. *US Geol. Surv., Prof. Pap.* 562-F

Nordin, C. F., Algert, J. H. 1966. Spectral analysis of sand waves. *Am. Soc. Civ. Engrs., J. Hydraul. Div.* 92:HY5: 95-114

Parker, G. 1975. Sediment inertia as cause of river antidunes. *Am. Soc. Civ. Engrs., J. Hydraul. Div.* 101:HY2: 211-21

Plate, E. 1971. Limitations of spectral analysis in the study of wind-generated water surface waves. *Proc. 1st Int. Symp. Stochastic Hydraul.*, Pittsburgh, pp. 522-39 (esp. p. 533)

Raudkivi, A. J. 1963. Study of sediment ripple formation. *Am. Soc. Civ. Engrs., J. Hydraul. Div.* 89:HY6: 15-33

Raudkivi, A. J. 1967. Analysis of resistance in fluvial channels. *Am. Soc. Civ. Engrs., J. Hydraul. Div.* 93:HY5: 73-84

Raudkivi, A. J. 1976. *Loose Boundary Hydraulics.* Oxford: Pergamon. 397 pp. (2nd ed.)

Reynolds, A. J. 1976. A decade's investigation of the stability of erodible stream beds. *Nord. Hydrol.* 7: 161-80

Richards, K. J. 1980. The formation of ripples and dunes on an erodible bed. *J. Fluid Mech.* 99: 597-618

Rouse, H. 1937. Modern conceptions of the mechanics of fluid turbulence. *Trans. Am. Soc. Civ. Engrs.* 102: 463-543

Simons, D. B., Richardson, E. V. 1961. Forms of bed roughness in alluvial channels. *Am. Soc. Civ. Engrs., J. Hydraul. Div.* 87:HY3: 87-105

Smith, J. D. 1970. Stability of a sand bed subjected to a shear flow of low Froude number. *J. Geophys. Res.* 75:5928-40

Vanoni, V. A., ed. 1975. *Sedimentation Engineering.* New York: Am. Soc. Civil Engrs. xvix + 745 pp.

Velikanov, M. A. 1936. Formation of sand ripples on the stream bottom. *Int. Assoc., Hydrol. Sci., Commiss. Potamology, Sect. 3, Rep. 13*

White, W. R., Paris, E., Bettess, R. 1980. The frictional characteristics of alluvial streams: a new approach. *Proc. Inst. Civil Engrs., Part 2* 69: 737–750

Williams, G. P. 1970. Flume width and water depth effects in sediment-transport experiments. *US Geol. Surv., Prof. Pap.* 562-H

Yalin, M. S. 1971. On the formation of dunes and meanders. *Proc. 14th Congr. IAHR, Paris* 3: 101-8

Yang, C. T. 1979. Unit stream power equations for total load. *J. Hydrol.* 40: 123-38.

STRONGLY NONLINEAR WAVES

L. W. Schwartz

Exxon Research and Engineering Company, Linden, New Jersey 07036

J. D. Fenton

School of Mathematics, University of New South Wales, Kensington, N.S.W., Australia 2033.

1. INTRODUCTION

As steep waves have recently come to be described with increasing accuracy, a number of unexpected physical and mathematical phenomena have been revealed. Until ten years ago it had been assumed that accurate solutions for high waves would hold few surprises. Examples of such suppositions are that deep-water solutions would converge for all waves short of the highest, that important integral quantities such as speed, energy, and momentum would increase with wave height until the highest is reached, that the solutions for periodic waves would be unique, and that if one solitary wave overtakes another any change of wave height would be a decrease. It is now known that all these suppositions are false, having been disproved in the last decade. The nonlinearity of the describing equations produces a complexity of solution structure that is only now beginning to be appreciated.

This review will deal with effectively exact solutions for nonlinear waves and the phenomena revealed by such solutions. The governing equation within the fluid is taken to be Laplace's equation, corresponding to irrotational flow of an incompressible fluid. Excluded are the physical effects of viscosity, density gradients, compressibility, and rotation. This model of the flow is the simplest, but one which is an excellent approximation in many cases of wave motion, and is the traditional avenue of approach to most problems of fluid flow. Throughout this review, however, the problems and solutions described are those where the complete nonlinear boundary conditions have been included. It has been the nonlinearity of these conditions which has made the accurate solution of water-wave problems so difficult.

The fluid is assumed to be inviscid and the flow irrotational, such that the velocity $\mathbf{q}$ may be expressed as the gradient of a potential ϕ, $\mathbf{q} = \nabla \phi$. If the fluid is assumed to be incompressible, such that $\nabla \cdot \mathbf{q} = 0$, the equation that holds throughout the fluid is Laplace's equation

$$\nabla^2 \phi = \phi_{xx} + \phi_{yy} + \phi_{zz} = 0, \qquad (1.1)$$

where the subscripts denote partial differentiation. The x and z coordinates are taken to be in a horizontal plane, the y axis vertically upwards. If the fluid is partly bounded by solid boundaries that are free to move such that $y = h(x, z, t)$ on the boundary, it can be shown that the condition that no fluid pass through the boundary is

$$h_t = \phi_y - \phi_x h_x - \phi_z h_z$$

on $y = h$. In many situations the boundary can be taken as stationary, $h_t = 0$, and horizontal, $h_x = h_z = 0$, so that the boundary condition becomes

$$\phi_y(x, h, z, t) = 0. \qquad (1.2)$$

The free-surface boundary conditions are to be satisfied on $y = \eta(x, z, t)$, which is also unknown. The kinematic requirement that a particle on the surface remain on it is expressed by

$$\frac{D}{Dt}(y - \eta) = -\eta_t - \phi_x \eta_x - \phi_z \eta_z + \phi_y = 0, \qquad (1.3)$$

on $y = \eta$. The dynamic boundary condition can be written

$$\phi_t + \tfrac{1}{2}(\phi_x^2 + \phi_y^2 + \phi_z^2) + g\eta + p_s/\rho$$
$$+ T(1/R_1 + 1/R_2) = C(t) \qquad (1.4)$$

on $y = \eta$, where g is gravitational acceleration, p_s is the pressure at the surface, ρ is fluid density, T is surface tension, R_1 and R_2 are principal radii of surface curvature, and $C(t)$ is a function only of time. In many situations, and throughout the rest of this review, air motion above the surface is neglected and the surface pressure taken to be a constant. The equations may be simplified in any or all of the cases of (a) steady flow, $\partial/\partial t \equiv 0$, (b) two-dimensional flow, $\partial/\partial z \equiv 0$, and (c) surface tension relatively unimportant, T set to zero. However, in all these physical simplifications, nonlinear terms remain in the free-surface boundary conditions.

2. THE CANONICAL PROBLEM : STEADY WAVES

The problem of a periodic train of waves propagating without change of form allows a considerable simplification by the addition of a suitable horizontal velocity to the reference frame, so that the fluid motion may be made steady and all time dependence and time derivatives vanish from

(1.1–1.4). By considering a coordinate frame in which one axis is parallel to the direction of propagation the problem is made two-dimensional. Surface tension will be neglected in this section, and its inclusion described in Section 3. The problem now formulated, the two-dimensional steady periodic gravity wave, is the simplest of all, and has often been the avenue by which solutions to more difficult and general problems have been approached. Despite its relative simplicity, it contains the full nonlinearity of the surface boundary conditions and has succumbed to accurate solution only in the last decade, during which several interesting phenomena have been discovered.

The existence of solutions to this problem has been studied by several authors. Krasovskii (1960) has devised the most significant proof. A recent discussion of his work has been given by Keady & Norbury (1978), who also established existence of a set of solutions of which his are a subset. Toland (1978) has shown that a solution exists for a wave of greatest height, and that this wave is the uniform limit of waves of almost extreme form.

A number of general theorems for the motions due to steady wave trains have been established, a summary of which is given in Wehausen & Laitone (1960, Section 32). These include theorems on the decrease of motion with depth, as well as relations between energy and momentum integrals. Longuet-Higgins (1974, 1975) has established a number of relations between the integral quantities of a wave train, and these have been generalized by Cokelet (1977a) to allow for the wave train moving at arbitrary speed relative to the frame of reference.

A steady wave train can be uniquely specified by three lengths—the peak-to-trough wave height H, the wavelength λ, and the mean depth D. From these, two independent dimensionless ratios can be formed, so that the steady-wave problem has a two-parameter family of solutions, although recently (see Section 2.2.1) it has been shown that for waves near the highest there may be more than one solution. Special limiting cases of the steady-wave problem are (a) $D/\lambda \rightarrow \infty$, λ and H finite, called the deep-water wave, and (b) $D/\lambda \rightarrow 0$, D and H finite, called the solitary wave.

2.1 Solution by Perturbation Expansion Methods

2.1.1 STOKES EXPANSIONS Stokes (1847) first used a systematic perturbation technique to solve the steady-wave problem. He assumed that the free-surface elevation may be represented by an infinite Fourier series

$$\eta(x) = a_1 \cos x + a_2 \cos 2x + \ldots \qquad (2.1a)$$

and that the velocity potential may be similarly represented by

$$\phi(x, y) = cx + b_1 \cosh(y + y_0) \sin x$$
$$+ b_2 \cosh 2(y + y_0) \sin 2x + \ldots \qquad (2.1b)$$

where the problem has been made dimensionless by referring lengths to $\lambda/2\pi$ and velocities to $c_0 = (g\lambda/2\pi)^{1/2}$, the speed of an infinitesimal wave in deep water. The coefficient c and the a_n and b_n are functions of the water depth. In some situations y_0 has been assumed to be D, while in inverse formulations of the problem (see Section 2.2.1) it is more convenient to let y_0, the height of the origin above the bottom, be $d = Q/c$ where Q is the area flow rate under the stationary wave. The leading coefficient c is the phase speed of the wave train in another frame through which the waves propagate such that the time-mean fluid velocity at all points is zero. The coefficients c, a_n, and b_n are assumed to be power-series expansions in a_1, such that the leading orders of a_n and $b_n \sim O(a_1^n)$. When these series are substituted into (2.1a and b), which are then substituted into the steady free-surface conditions, an ordered set of equations is obtained from which the coefficients in the power series can be found successively. The complexity of the manipulations makes a manual high-order calculation impractical. Fifth-order solutions have been obtained by De (1955), Chappelear (1961), who claimed mistakes in De's solution, and by Skjelbreia & Hendrickson (1961).

If the inverse problem is studied, where ϕ and ψ rather than x and y are the independent variables, the calculations can be greatly reduced. The free surface becomes a known boundary, $\psi = 0$.

Stokes (1880) computed the deep-water case to $O(a_1^5)$ and found results for finite depth to $O(a_1^3)$. In the most ambitious manual computations using the inverse method, Wilton (1914) carried the infinite depth calculation to $O(a_1^{10})$ (but has errors at the eighth order) and De (1955) has published a fifth-order solution for general depth.

In order to reveal details of highly nonlinear waves by the series method, solutions of much higher order must be obtained. Schwartz (1972, 1974) simplified the formalism of the inverse method by using complex functions. Each wave cycle in the physical $z = x + iy$ plane is mapped onto the interior of an annulus in the ζ plane, where $\zeta = \exp(-if/c)$ where f is the complex potential $f = \phi + i\psi$. The mapping function is an infinite polynomial in ζ with coefficients a_n, each being taken to be a power series in ϵ, a parameter that is zero for the undisturbed stream and assumed to increase monotonically with wave height. By substituting into the dynamic boundary condition, a set of recurrence relations is found, and the coefficients in the power series determined successively by computer. The system of equations is closed by defining ϵ. Choosing $\epsilon = a_1$ reproduces the procedure of Stokes (1880).

Stokes (1847) showed that the highest wave, assumed to be sharp-crested, must have an included angle of 120° at the crest. He conjectured that this limiting wave would correspond to the critical value of a_1 in the series expansion. In fact this is not the case, for a_1 is not a monotonic function of wave height but achieves its maximum value before the highest wave is attained, about 10% before for the deep-water wave. To surmount this difficulty Schwartz introduced the wave height as a new parameter, in fact, $\epsilon = H/2$. He found that all the coefficients in the expansions reached maxima before the highest wave is achieved. Thus when a_1 is used as the independent parameter, the maximum of a_1 as a function of H becomes a square-root singularity, which limits the convergence of Stokes' expansion in a_1. Schwartz subsequently used series-analysis techniques, including Padé approximants and Domb-Sykes plots, to estimate the limiting wave height in deep water. He found this to be $(H/\lambda)_{\max} = 0.1412$. From the accurate results of the enhanced series, Schwartz showed that the mass of the wave has a maximum in H. This has important implications for other integral quantities of the wave train such as energy and momentum.

Longuet-Higgins (1975) used Schwartz's program for the deep-water wave, recomputed to 32 decimal places by Cokelet, and re-expressed the series in terms of the parameter $\omega = 1 - (q_c q_t/cc_0)^2$, where q_c is the fluid speed at the crest and q_t that at the trough. The parameter ω has the useful property that its range is known ab initio, undisturbed flow corresponding to $\omega = 0$ and the limiting wave to $\omega = 1$. Longuet-Higgins found maxima in each of the integral quantities: wave speed, momentum, and potential and kinetic energy, *before* the highest wave was attained, as had been found by Longuet-Higgins & Fenton (1974) for the solitary wave. The variation of the integral quantities with wave steepness H/λ is shown in Figure 1 for the deep-water case. It is clear that the highest wave is not

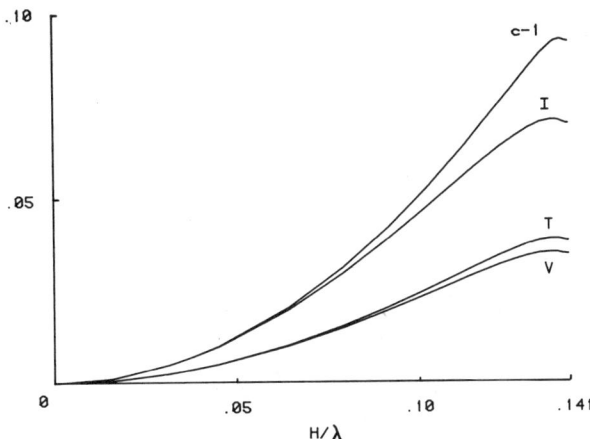

Figure 1 Dimensionless wave speed, impulse and kinetic and potential energies versus wave height for deep-water progressive gravity waves.

the fastest nor the most impulsive nor the most energetic. The physical implications of this are not well understood, but it may be responsible for the instability of the crest of high waves (see Section 4.1), for the multiplicity of solutions (see Section 2.2.1) and be relevant to the observed intermittency of spilling breakers (Longuet-Higgins & Turner 1974). As the maxima occur for high waves, when surface tension and viscosity would also be important, especially on a laboratory scale, experimental confirmation would be difficult.

Cokelet (1977a) used a method very similar to that of Schwartz, and for a wide range of finite water depths produced a number of accurate results for integral quantities of the wave train. Each showed a maximum before the highest wave was reached. Finite-depth results for wave speed are displayed on Figure 2, using data from Schwartz (1972), Cokelet (1977a), and Longuet-Higgins & Fenton (1974).

2.1.2. SHALLOW-WATER EXPANSIONS It has been shown by Ursell (1953) that the linear theory of periodic waves (the first term in Stokes' expansion) is valid only if the shallowness parameter $H\lambda^2/D^3$ as well as the wave steepness H/λ, is small. That is, the waves must not be too long relative to the water depth. This is shown by the results given by Schwartz (1974), where the radii of convergence of the Stokes expansions become smaller for longer waves. Analytical theories show the dependence on wavelength more explicitly; it can be shown that whereas the nominal Stokes expansion parameter is $a_1 k$, where k is the wavenumber $2\pi/\lambda$, the ratios of successive terms in the expansion actually behave like $a_1 k/\sinh^3 kD$. For shallow water

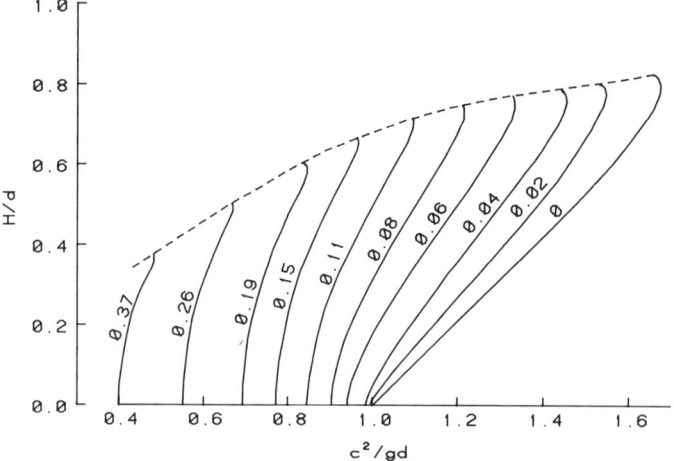

Figure 2 Wave speed versus amplitude and depth for progressive gravity waves. Numbers on curves are values of d/λ.--sharp-crested waves.

kD becomes small so that the effective expansion parameter varies like $a_1 k/(kD)^3$, proportional to Ursell's parameter.

A theory for waves in shallow water was put forward by Korteweg & de Vries (1895), who showed that to first order, the free surface of steady waves is of the form of a Jacobian elliptic cn function squared, giving rise to the term "cnoidal" waves which show the long flat troughs and short crests characteristic of waves in shallow water. Higher-order cnoidal wave theories have been obtained by Laitone (1960) and Chappelear (1962) to second and third order respectively. A systematic method using power-series expansions in terms of shallowness, $(D/\lambda)^2$, was used by Fenton (1979), who used computer manipulation of the long series to produce a ninth-order solution. It was shown that the most appropriate depth scale is h, the water depth under the troughs, and that the natural expansion parameter when the series are recast in terms of wave height is H/mh, where m is the parameter of the elliptic functions. For the long-wavelength limit of the solitary wave $m \to 1$; however, for shorter waves m becomes smaller and the expansion parameter larger, complementary to the manner in which the $\sinh^3 kD$ denominator in Stokes' expansion invalidates their application to shallow water. The cnoidal-wave solutions were found to be not accurate for very high waves, for reasons associated with the maxima of integral quantities as a function of wave height. While the cnoidal-wave results should be used instead of Stokes's wave solutions in shallow water, for physical applications where accurate solutions are necessary, both have been somewhat superseded by the numerical method of Rienecker & Fenton (1981), described in Section 2.2.2.

2.1.3 SOLITARY WAVE   The solitary wave is a steady wave of infinite wavelength. A recent specialized article on solitary waves has been written by Miles (1980), so that the treatment here will be brief. The only high-order series results are those presented by Fenton (1972) to ninth order, and by Longuet-Higgins & Fenton (1974) to fourteenth order. It was found that the series in terms of wave height H/h did not give accurate solutions for very high waves. The series was recast in terms of the parameter $\omega = 1 - q_c^2/gh$, where ω has the known range $(0,1)$. Using these series, Padé methods were implemented, and convergent results were obtained for all wave heights. It was found that the integral quantities of the wave all showed maxima as functions of wave height.

The other approach has been through an integral-equation formulation, the history of which is given by Miles (1980). Accuracy of the alternative equations and numerical methods has usually been measured by their ability to describe the wave of greatest height. Some of the more notable results for the maximum height $(H/h)_{\max}$ are 0.827 ± 0.008 (Yamada 1957b), 0.827 (Lenau 1966), 0.8262 (Yamada et al. 1968), and the result of 0.827 from Longuet-Higgins & Fenton (1974). All these results seem to support one another. However, Witting & Bergin, in an unpublished work mentioned by Witting (1975), obtained a value of 0.8332, precisely the result obtained by Fox (1977) in an unpublished dissertation. This agreement is rather striking, despite the use of some extrapolation in both cases. Finally, it should be noted that Witting (1975) has suggested that the method developed by Fenton (1972), on which the high-order series results are based, is defective in that the assumed expansions are incomplete.

2.2. *Solution by Numerical Methods*

Recent numerical solutions of the steady-wave problem are as accurate as the series results, are often easier to implement computationally, and generally do not need the convergence improvement techniques of the series. They do suffer from the usual disadvantage of numerical solutions in that they reveal less about the nature of the problem and its solution. In the case of high waves, however, so much numerical smoothing and extending of the series solutions is necessary that they too suffer from this disadvantage. Numerical methods can be divided into two categories depending on whether the authors choose to solve the problem in the physical (x,y) plane or the inverse (ϕ,ψ) plane.

2.2.1 INVERSE PLANE METHODS   In this case the problem is to be solved on a region which is known a priori. This huge advantage is offset by the fact that as the sharp-crested wave is approached, the singularities near the crest are stronger in the inverse plane. Also, the calculation of local quantities such as pressure and velocity as functions of position becomes a separate problem to be solved subsequently if the inverse plane is used.

Schwartz & Vanden-Broeck (1979) used an algorithm that is typical of inverse plane methods to solve the capillary-gravity wave problem on infinitely-deep fluid and that was generalized to the finite-depth case in Vanden-Broeck & Schwartz (1979). One wavelength of the flow was mapped onto an annulus, the dynamic boundary condition becoming a nonlinear differential equation for x and y on the unit circle. To satisfy Laplace's equation within the annulus, a Cauchy integral was written, valid on the unit circle and satisfying the bottom boundary condition identically. The equations were approximated by finite-difference expressions to give a system of nonlinear algebraic equations which were solved by Newton's method. The spacing of computational points at equal intervals of velocity potential was found to work well for capillary-gravity waves where the points tended to be clustered around the narrow troughs where velocities were greatest. For pure gravity waves, on the other hand, the sparse spacing occurred in the vicinity of the sharp crests and produced poor results for waves near the highest. This was overcome by a simple transformation which could be used to cluster points near the crest. The results given for some wave speeds and energies are probably the most accurate to date.

A remarkable result has recently been found by Chen & Saffman (1980a), providing convincing evidence that solutions for permanent gravity waves of finite amplitude are not unique when they are sufficiently high! They formulated the deep-water steady-wave problem as a nonlinear integro-differential equation, and approximated it by finite-difference methods to give a system of nonlinear algebraic equations which were solved by Newton's method. Having noted the existence of the "premature" maxima of the various integral properties of periodic waves, and its possible analogy with the analytic structure of the capillary-gravity wave problem, where multiple solutions were known to exist, they carefully monitored the determinant of the Jacobian matrix used in the Newton's method solution. Zeroes were found, identifying bifurcation points in the solutions, and the separate branches were followed. The new families of solutions, found to occur for very steep waves, $H/\lambda \approx 0.13$, corresponded to a doubling and tripling of the fundamental wavelength. In the doubling case, for example, the solution obtained was a train of steep waves in which alternate waves differed slightly in wave height.

2.2.2 PHYSICAL PLANE METHODS These have an important role to play in practical applications where, despite an approximate doubling of the number of unknowns because the free surface is also unknown, it is considerably easier to solve the problem from the beginning in the physical plane. Such solutions include the recent work of Rottman & Olfe (1979) and Rienecker & Fenton (1981). In the first of these, a boundary-integral technique was used to formulate an integro-differential equation, Newton iteration being used to find the vector of surface points. The method worked well for steep gravity waves in that the now well-known speed maximum was found. It is, however, not well suited to its nominal objective, the computation of capillary-gravity waves, since it fails when $\eta(x)$ becomes double-valued.

Rienecker & Fenton (1981) used a method in which the stream function is represented by a truncated Fourier series similar to that of Stokes in (2.1b). However, for a given wave the coefficients of the expansion are found by numerical means, obviating the introduction of general power series with their finite radius of convergence and breakdown in the inappropriate depth limit. The numerical method depends for its accuracy on the ability of a Fourier series to describe the wave train. This approach was originated by Chappelear (1961) and Dean (1965), but the method of Rienecker & Fenton is substantially different in that, for example, the solution method is simpler because Newton's method is used directly, the only approximation is in the truncation of the Fourier series, and the method recognizes that the waves may propagate at speeds determined by quantities such as mass flux. In comparing results for fluid velocity with experimental results, good agreement was found. Very close agreement for all waves including the highest was found between the results for phase speed and those reported by Cokelet (1977a) and Vanden-Broeck & Schwartz (1979).

2.3 The Highest and Almost-Highest Waves

2.3.1 THE HIGHEST WAVE A number of attempts have been made to solve the problem of the sharp-crested wave of greatest height, usually incorporating Stokes' discovery that it has an included crest angle of $120°$. For a sharp-crested wave with its apex at $z = x + iy = 0$, it follows from the Bernoulli condition that the complex potential $f = \phi + i\psi$ varies like $z^{3/2}$ locally. Thus the complex velocity df/dz varies like $f^{1/3}$ near the crest. If the computation is restricted to deep water, an expansion can be assumed (Michell 1893):

$$\frac{df}{dz} = c(1 - \zeta)^{1/3} (1 + b_1 \zeta + b_2 \zeta^2 + \ldots), \tag{2.2}$$

where $\zeta = \exp(-if/c)$. Michell substituted this into the dynamic boundary condition and determined the first few coefficients of the expansion to give a result for the limiting steepness of $(H/\lambda)_{\max} = 0.142$. The same method was used by Havelock (1919) who obtained a limiting steepness of 0.1418. He also calculated solutions for waves short of the highest by displacing the cube-root singularity above the crest. This technique is apparently defective, however, since Grant (1973) has shown that only square-root singularities are admissible in all cases but the limiting one. He showed that the singularity structure of the highest wave solution is much more complicated than had previously been assumed and that the sharp crest is not a regular singular point. The Stokes singularity is merely the first term in a local expansion about the crest in which irrational exponents occur.

Meanwhile Yamada (1957a) had assumed a solution equivalent to that of Michell, truncated after twelve terms, and obtained a limiting steepness of 0.1412. In a later paper, Yamada & Shiotani (1968), the method was extended to finite depth. McCowan (1894) and Lenau (1966) used comparable techniques for the solitary wave. Schwartz (1972, 1974) analyzed his high-order Stokes series and incorporated the inferred singularity structure in a recast form of the series. The highest wave was graphically indistinguishable from Yamada's and had the same steepness.

Michell's expansion (2.2) has been used again, but with a delightful difference, by Olfe & Rottman (1980). They observed that the nonlinear equation resulting from substitution of a one-term expansion into the dynamic boundary condition has multiple roots, one corresponding to Michell's solution but another real root almost eliminating the fundamental Fourier term so that the series is dominated by a higher harmonic—precisely the behavior found by Chen & Saffman (1980a) for irregular gravity waves. Subsequently, using (2.2) with up to 120 terms, Olfe & Rottman experimented with Newton's method and found other solutions in addition to Michell's, corresponding to every second, third, or fourth crest being sharp, the intermediate ones being lower and more rounded, as found by Chen & Saffman for waves lower than the highest.

Some simple and accurate irrational approximations, not part of systematic schemes, have been found by Longuet-Higgins (1973, 1974) for the highest steady, standing, and solitary waves.

2.3.2 ALMOST-HIGHEST WAVES A local expansion in the vicinity of the crest for waves just short of the highest was devised by Longuet-Higgins & Fox (1977). They found a class of self-similar flows with a length scale of $\ell = q_c^2/2g$, which have a smooth crest and whose free surface oscillates about the Stokes corner flow with a decaying amplitude like $(\ell/r)^{1/2}$, where r is the distance from the Stokes corner. The oscillations cause the maximum inclination of the surface to be greater than 30°, namely 30.37°, a result confirmed from extrapolated numerical results of Sasaki & Murakami (1973) and Byatt-Smith & Longuet-Higgins (1976). In a second paper, Longuet-Higgins & Fox (1978) matched the local-crest solution to a form of Michell's expansion for deep-water waves, valid far from the crest. A small parameter proportional to q_c was introduced, and a number of asymptotic expressions found for the height of the waves, the phase speed and other integral quantities. Unlike almost all other results for nonlinear waves, these are valid in the limit of the highest wave. These expressions have the unusual feature that they show an infinitude of local maxima and minima as the highest wave is approached. The global maximum and the first local minimum of speed and energy was found in the numerical solution of Schwartz & Vanden-Broeck (1979) for deep water and for a particular case of finite depth $D/\lambda = 0.110$. No doubt all finite-depth wave trains, and perhaps the solitary wave, show such behaviour.

Longuet-Higgins & Fox (1978) analytically continued their solution across the free surface in the crest neighborhood. They found a stagnation point above the crest corresponding to the square-root singularity found by Grant (1973) and Schwartz (1972, 1974). The apparent transition to the 2/3-power form for the sharp-crested wave was explained by Grant as the coalescence of several square-root singularities. Longuet-Higgins

(1979a) has subsequently developed a simpler approximation to the crest flow found by himself and Fox. This result, and other approximations, have been used to calculate fluid particle paths (Longuet-Higgins 1979b).

3. OTHER PERIODIC WAVES

There are two other important classes of spatially periodic surface waves that will be discussed here. The emphasis, as before, will be on strongly nonlinear effects.

3.1 *The Inclusion of Surface Tension*

In this case the waves considered are also two-dimensional, periodic, and the flow is steady, but the surface-tension term in (1.4) is retained. For two-dimensional waves, $1/R_2 = 0$, and $1/R_1 = \eta_{xx}/(1 + \eta_x^2)^{3/2}$ a highly nonlinear function of η. The parameter $\kappa = 4\pi^2 T/\rho g \lambda^2$ is used to measure the relative importance of surface tension and gravity. When κ is large, corresponding to short wavelengths, surface tension is the dominant restoring force. For longer waves gravity is most significant; however, capillarity becomes increasingly important as the wave steepens and the crest becomes more sharply rounded.

The problem of pure capillary waves in deep water has been resolved completely by Crapper (1957). In this most remarkable work, he obtained an exact closed-form solution for waves of arbitrary amplitude. More recently, closed-form solutions for finite depth were found by Kinnersley (1976), a result anticipated by Crapper. Unlike pure gravity waves, steep capillary waves are characterized by deep troughs and broad flat crests. The limiting wave occurs when the two sides of the trough meet, enclosing a pendant-shaped bubble. Steeper "waves" can be computed but they are physically impossible since the free surface crosses itself. Recently Vanden-Broeck & Keller (1980) discovered a new family of capillary waves by allowing a nonzero pressure within the enclosed pendant.

Wilton (1915) treated the combined capillary-gravity wave problem by using a Stokes expansion which he carried to fifth-order in amplitude, invoking Stokes' hypothesis that the nth Fourier coefficient is nth order. When the parameter $\kappa = 1/n$ for $n = 2,3, \ldots$, this expansion fails because certain series coefficients become infinite. For the particular case of $\kappa = 1/2$, Wilton was able to find two solutions, each of which he carried to third order, by revoking Stokes' hypothesis and re-ordering the terms in the series.

Recently the numerical methods described in Section 2.2.1 have been applied to the problem (Schwartz & Vanden-Broeck 1979, Chen & Saffman 1980b). Waves of maximum height can be computed without difficulty for most values of κ. In all cases for $\kappa > 0$, the highest waves are topologically limited, in that the surface encloses one or more bubbles, just as for pure capillary waves, and unlike pure gravity waves where the limiting wave is sharp-crested. The critical values of κ, where the Stokes-series solution fails, are indicative of multiple solutions. When $\frac{1}{n+1} < \kappa < \frac{1}{n}$, for moderate steepness the series yields a family of profiles with approximately n inflection points or "dimples." The numerical methods showed that each solution family can be analytically continued outside its "natural" domain. Thus there can be many different wave forms for a given wavelength. Certain of the multiple solutions arise via a finite-amplitude bifurcation from the regular wave train. As $\kappa \to 0$ the number of solution families increases. The structure of this highly singular limit, where surface tension should be important only in the neighborhood of the sharp crest, remains to be explored. A stability analysis for each of the several possible wave forms would be a useful contribution and might resolve the non-uniqueness in the ripple regime. Experimental results are not completely consistent, but Schooley (1960), for example, has published photographs of several multi-dimpled profiles. Analytical work, dealing with the genesis of multiple solutions, has been presented by Pierson & Fife (1961), Nayfeh (1970), and Chen & Saffman (1979).

Hogan (1979) obtained a number of relations between integral quantities of the steady wave train, which for Crapper's pure capillary wave reduced to very simple expressions. He then (Hogan 1980) extended the series technique of Schwartz (1972, 1974) to include surface tension for the deep-water case, and obtained members of the family of solutions corresponding to the gravity wave. For κ sufficiently small, each integral property was found to have a maximum as a function of wave height, but the effect of increasing κ was to move the maximum closer to the highest wave possible, showing how surface tension acts so as to make experimental verification of the maxima more difficult. Beyond a certain value of κ, the maximum wave height became limited by the wave geometry, and no maxima in the integral quantities were found before the limiting bubble was attained.

Figure 3 shows two steep wave profiles for the case $\kappa = 1/2$, taken from Schwartz & Vanden-Broeck (1979). The profile labelled 1 belongs to the family of Crapper's limiting wave and exhibits one trapped bubble per wave cycle; the other wave is from a different family which, while also symmetric, is limited by two bubbles per cycle.

3.2 Finite-Amplitude Standing Waves

A time-dependent relative of the steady gravity wave is the two-dimensional time and space periodic standing wave. Physically the problem is that of the periodic "sloshing" or "seiching" of water between two vertical side walls. To first order it corresponds to the reflection of a periodic wave train by a vertical barrier or the interaction of two oppositely propagating wave trains. Unlike the steady wave, no rigorous existence proof for finite amplitudes has been obtained.

In the spirit of Stokes, Rayleigh (1915) solved the problem by a third-order expansion in wave height. Penney & Price (1952) computed a fifth-order solution, and obtained a maximum steepness, for deep water, of 0.22. They also suggested that this highest wave has a 90° included angle at the crest, but the premises in their argument have been questioned by subsequent workers. Experiments of Taylor (1953) and Edge & Walters (1964) confirmed that the crest angle is close to 90°.

Recently Schwartz & Whitney (1977, 1981) have produced a 25th-order solution by a time-dependent conformal mapping method. They found that Penney & Price's procedure is defective in that it produces non-periodic secular time dependence if carried to higher order. This "resonance" may be suppressed by exploiting a degree of arbitrariness in certain of the series coefficients. Very steep waves were found to possess several inflection points near the crest, reminiscent of those for steady waves. The highest wave steepness was found to be about 0.208, with an included angle of about 90°. The oscillating water surface is never flat and has no nodes. The wave frequency, which in general decreases with increasing amplitude, appears to reach a minimum value just short of the limiting steepness.

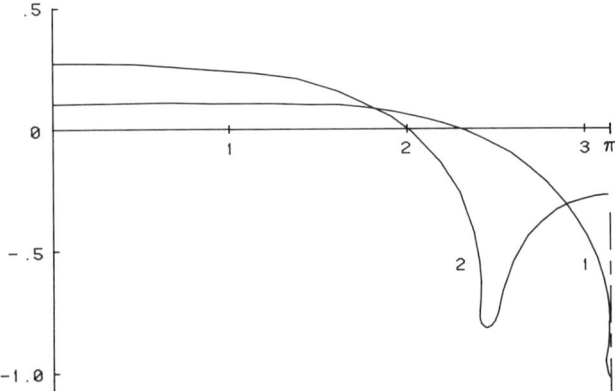

Figure 3 Two steep gravity-capillary wave profiles for $\kappa = 1/2$.

Unlike the deep-water case, shallow-water standing waves show an increase in frequency with amplitude, the critical value of the D/λ ratio at which the first nonlinear correction to the frequency changes sign being 0.17. This was obtained as part of the third-order series solution for finite depth obtained independently by Chabert-d'Hières (1960) and Tadjbakhsh & Keller (1960), and has been experimentally confirmed by Fultz (1962). An interesting feature of the series method is that certain values of D/λ must be excluded so as to suppress secular terms, which may indicate multiple solutions; however, the set of excluded depths becomes infinite as the order of the series solution becomes infinite (Concus 1964). Until this can be explained, the existence of standing-wave solutions for finite amplitude and depth remains in doubt.

A numerical method for finite depth, employing truncated Fourier series to represent space and time variation has been used by Vanden-Broeck & Schwartz (1981). While inappropriate for very steep waves because of computer storage limitations, the results confirmed the series results for moderate steepness and showed that, for some depths, the frequency is not a monotonic function of wave height.

Miche (1944) established that deep-water standing waves have a second-order contribution to the pressure field that is unattenuated with depth and varies with twice the frequency of the surface displacement. Longuet-Higgins (1950) proposed this effect as a likely cause of observed microseismic activity and explained the phenomenon very simply in terms of variation of the potential energy of the water mass. The phenomenon has been demonstrated experimentally by Cooper & Longuet-Higgins (1951) for low waves, but recent calculations by Schwartz (1980) indicate that the leading-order theory used by Longuet-Higgins may overpredict the pressures for very steep waves by as much as 40%.

4. UNSTEADY WAVES

The standing-wave problem described in the previous section is unsteady, but it is periodic in time and space and all flow quantities can be expanded in Fourier series in both variables. In this section, problems, methods and results are described in which the motion is more generally unsteady and where the effects of nonlinearities may bring about irreversible changes. Unlike the problems described in Sections 2 and 3, for which some of the most powerful methods were devised last century, almost all of the methods and results described here have appeared in the last decade.

4.1 *Stability of Steady Waves*

There has been much effort expended on the stability of weakly nonlinear waves, but very little on the stability of steep steady waves. The reason is simple enough—it was not until 1972 that accurate solutions for waves of arbitrary height became available. Longuet-Higgins (1978a, 1978b) examined the stability of steady waves in deep water for all heights, testing the linear stability of the nonlinear wave solutions to small harmonic perturbations. For low waves it was found that they were neutrally stable, but above a certain steepness the waves became unstable to subharmonic perturbations (disturbance wavelength greater than the fundamental). The growth rates obtained agreed quite well with experiment. For higher waves these modes became stable again, but for waves close to the highest, a very fast-growing superharmonic (small wavelength) instability was found. This was associated with the crest, and it seemed to owe its existence to the fact that the wave was steep enough that the first Fourier coefficient of the unperturbed wave had a maximum.

The results of this linear stability analysis were checked by Longuet-Higgins & Cokelet (1978), who followed the evolution of the perturbation by numerically solving the full nonlinear equations, and confirmed the accuracy of the linear analysis. It was found that the local superharmonic instability could quickly lead to overturning at the crest, providing the beginning of insight into the causes of that phenomenon. More remarkably, however, they found that the slow subharmonic instability, if followed for long enough, could also cause an overturning of the crest. Results from the linear theory have had some success in supporting a conjecture of Lake & Yuen (1978) that in very steep wind waves the modulation frequency of the wave amplitude may correspond to the fastest-growing subharmonic instability (Longuet-Higgins 1980d).

4.2 Breaking Waves

The plunging or overturning wave is the most important and dramatic of all breaker types and probably is the origin of the spilling breaker. It is one of the most difficult of all wave phenomena to analyze because of the rapidity with which it can occur, the large amplitudes and accelerations involved, and the contortions of the free surface. The origin of plunging in deep water remains unexplained; however, as described in Section 4.1, it has been shown that the wave crest is vulnerable to perturbations and that overturning of the crest is a common outcome even for waves that are considerably lower than the highest.

Some progress has been made in describing the overturning. Several methods are described in Cokelet (1977b), but the only method so far able to describe the overturning wave for most of its duration is that of Longuet-Higgins & Cokelet (1976, 1978), which will be outlined in Section 4.3. In the original 1976 paper they studied an idealized problem of a steady wave in deep water to which an asymmetric pressure distribution was applied for a finite time, after which the wave turned over. Cokelet (1977b) studied an initially sinusoidal wave train with an excess of energy and also found that the crest steepened and overturned. Subsequently Longuet-Higgins & Cokelet (1978) followed the motion of steady waves with a small perturbation and found the overturning as described in Section 4.1. A more detailed picture of the dynamics and kinematics inside a wave as it approaches breaking has been given by Peregrine, Cokelet & McIver (1980). The wave crest before breaking has been found to travel faster than the maximum phase speed for a wave of that length, while the whole front face of the wave had an acceleration several times greater than gravity!

Longuet-Higgins (1980e) has studied simple analytical models of the development of the overturning, and for the evolution of the tip of a plunging wave (Longuet-Higgins 1980b). In a very different study (Longuet-Higgins 1980c) he calculated the angular momentum of steady waves in deep water, and found that the relative persistence of wind-wave crests may be explained by the fact that for a force applied to the wave to have a minimum effect, it should act at about the wave-crest level; this is what actually occurs for both wind forces and drag forces due to whitecapping.

4.3 Solution of the Unsteady Equations

An early approach to the numerical solution of the full nonlinear equations (1.1–1.4) was through a marker-and-cell technique. The furthest development of this method is that of Chan & Street (1970) who used it to study the reflection of a solitary wave by a wall. This problem was also considered by Byatt-Smith (1971), who obtained a second-order analytical solution from a pair of nonlinear integro-differential equations. Other methods for the full equations include those of Brennen & Whitney (1970) (summarized in Brennen 1970), Whitney (1970), Multer (1973), and Chan (1975). None, however, seems to have been adopted and exploited further.

In 1976 a method was introduced which seems to have considerable potential. Longuet-Higgins & Cokelet (1976) studied the evolution of waves on water of infinite depth. If the initial free surface and ϕ on it are known, then a Cauchy-type integral equation may be formulated and approximated using discrete computational points on the surface, to give subsequently the velocity of each point normal to the surface as the solution of a matrix equation. The tangential velocity can be found by numerical differentiation. The surface particles are allowed to move a finite distance in a small time step, giving a new surface location. The new values of ϕ may be calculated from the dynamic free-surface condition and the process repeated, with a predictor-corrector method, for a large number of time steps. Longuet-Higgins & Cokelet found that a slow instability developed, but this could be countered by regular smoothing. This method has proved capable of describing the evolution of high waves, including the overturning of the crest (see Sections 4.1 and 4.2). Fenton & Mills (1976) showed how the method may be applied to water of finite depth with arbitrary boundary geometry, but were unable to produce solutions.

In two recent works, the free-surface boundary conditions have been given in different forms. Longuet-Higgins (1978a) expressed them in inverse form with dependent variables x and y as functions of ϕ and ψ. Another technique (Longuet-Higgins 1980a) has been developed in which the usual dynamic boundary condition with $p = 0$ is used, but the kinematic condition is written in terms of the material derivative of pressure, $Dp/Dt = 0$. This has been used (Longuet-Higgins 1980b, 1980e) to find exact solutions which mimic the overturning breaker.

A Fourier method has been developed by Fenton & Rienecker (1980, but described in greater detail in a manuscript submitted for publication). The method is applicable to irrotational flows over arbitrary bed topography and makes use of Fourier approximation throughout to represent horizontal variation. The truncation of Fourier series for ϕ and η, similar to (2.1), is the only approximation. The solution may be advanced in time by a leapfrog scheme, although it is necessary to solve a matrix equation at each time step. The method was found to be stable and accurate in describing solitary wave interactions, the use of Fourier series automating a number of numerical operations as well as facilitating deductions about stability or spectral growth. This makes it well suited to studies of finite-wave interactions, although it loses accuracy if at any stage a wave becomes sharp-crested, and it cannot describe overturning waves because it depends upon the surface elevation's being a single-valued function of x.

4.4 *Solitary Wave Interactions*

The problem of two interacting solitary waves has come to be considered a classical problem of nonlinear waves because it is completely specified by only two parameters (the incident wave heights), and because of the fundamental nature of the first-order equations which describe it, the existence of an exact solution to one of these equations, and the fact that the solution shows that the waves emerge unchanged from the interaction.

4.4.1 OVERTAKING SOLITARY WAVES To first order in wave height and shallowness, the interaction of one solitary wave overtaking another is described by the Korteweg–de Vries (K–de V) equation, for which an exact solution has been obtained (see Miles 1980). The only change after interaction contained in the solution is that the high and fast wave has received a finite phase shift forward and the low wave has been shifted backward. Weidman & Maxworthy (1978) conducted a number of experiments and found generally good agreement between experiment and K–de V theory, but with some consistent differences. Nothing was reported on the waves after interaction; this has been studied numerically using the full nonlinear equations by Fenton & Rienecker in the report mentioned in Section 4.3. For the one interaction studied they found that the waves after interaction, contrary to expectation, had a larger height difference than the incident waves—the high wave had grown slightly higher, at the expense of the lower wave. After interaction the waves propagated almost without further change, and there was no trace of a trailing wave train.

4.4.2. COLLIDING SOLITARY WAVES The other form of interaction, when the waves travel in opposite directions, has received much attention, as described in Miles (1980). In this case the interaction, although brief, is more nonlinear. Nevertheless, to second order in wave height, theory predicts that the waves emerge from the interaction unchanged, with a finite backward phase shift (Oikawa & Yajima 1973, Byatt-Smith 1971). In recent work, a third-order calculation has been made by Su & Mirie (1980). The waves after interaction were shown to have the same height as before, but the profiles are tilted backwards, and each sheds dispersive waves. Fenton & Rienecker (1980, and the above-mentioned report) studied a number of collisions with the numerical method outlined in Section 4.3. They found that both waves were actually degraded by the interaction, but that this change was slightly greater than that for overtaking. Accordingly, they recommend that the adjectives "weak" and "strong" not be used for the colliding and overtaking interactions, as the former has more effect on the waves than the latter. In contradiction of the third-order theory, they found strong evidence that the change of wave height is actually of third order, and that the waves after interaction were travelling faster than before, also noted experimentally by Maxworthy (1976). This change of speed has important implications for the measurements of phase changes due to the interaction, for it means that the change depends on the measuring location, which may explain some unusual features of Maxworthy's results for the phase change. In view of the ambiguity of the spatial phase shift, Fenton & Rienecker recommend use of the temporal shift at the wall, and show that this is considerably underestimated by second- and third-order theories.

Literature Cited

Brennen, C. 1970. Some numerical solutions of unsteady free surface wave problems using the Lagrangian description of the flow. *Proc. 2nd Int. Conf. Numer. Meth. in Fluid Dynam., Berkeley, Sept., 1970,* pp. 403–9

Brennen, C., Whitney, A. K. 1970. Unsteady, free surface flows; solutions employing the Lagrangian description of the motion. *8th Symp. Naval Hydrodynam., Pasadena, Aug., 1970,* pp. 117–45

Byatt-Smith, J. G. B. 1971. An integral equation for unsteady surface waves and a comment on the Boussinesq equation. *J. Fluid Mech.* 49:625–33

Byatt-Smith, J. G. B., Longuet-Higgins, M. S. 1976. On the speed and profile of steep solitary waves. *Proc. R. Soc. London Ser. A* 350:175–89

Chabert-d'Hières, G. 1960. Etude du clapotis. *Houille Bl.* 15:153–63

Chan, R. K. C. 1975. A generalized arbitrary Lagrangian-Eulerian method for incompressible flows with sharp interfaces. *J. Comput. Phys.* 17:311–31

Chan, R. K. C., Street, R. L. 1970. A computer study of finite-amplitude water waves. *J. Comput. Phys.* 6:68–94

Chappelear, J. E. 1961. Direct numerical calculation of wave properties. *J. Geophys. Res.* 66:501–8

Chappelear, J. E. 1962. Shallow-water waves. *J. Geophys. Res.* 67:4693–4704

Chen, B., Saffman, P. G. 1979. Steady gravity-capillary waves on deep water I. Weakly nonlinear waves. *Studies in Appl. Math.* 60:183–210

Chen, B., Saffman, P. G. 1980a. Numerical evidence for the existence of new types of gravity waves of permanent form on deep water. *Studies in Appl. Math.* 62:1–21

Chen, B., Saffman, P. G. 1980b. Steady gravity-capillary waves on deep water II. Numerical results for finite amplitude. *Studies in Appl. Math.* 62:95–111

Cokelet, E. D. 1977a. Steep gravity waves in water of arbitrary uniform depth. *Philos. Trans. R. Soc. London Ser. A* 286:183–230

Cokelet, E. D. 1977b. Breaking waves. *Nature* 267:769–74

Concus, P. 1964. Standing capillary-gravity waves of finite amplitude: Corrigendum. *J. Fluid Mech.* 19:264–66

Cooper, R. I. B., Longuet-Higgins, M. S. 1951. An experimental study of the pressure variations in standing water waves. *Proc. R. Soc. London Ser. A* 206:424–35

Crapper, G. D. 1957. An exact solution for progressive capillary waves of arbitrary amplitude. *J. Fluid Mech.* 2:532–40

De, S. C. 1955. Contributions to the theory of Stokes waves. *Proc. Cambridge Philos. Soc.* 51:713–36

Dean, R. G. 1965. Stream function representation of non-linear ocean waves. *J. Geophys. Res.* 70:4561–72

Edge, R. D., Walters, G. 1964. The period of standing gravity waves of largest amplitude on water. *J. Geophys. Res.* 69:1674–75

Fenton, J. D. 1972. A ninth-order solution for the solitary wave. *J. Fluid Mech.* 53:257–71

Fenton, J. D. 1979. A high-order cnoidal wave theory. *J. Fluid Mech.* 94:129–61

Fenton, J. D., Mills, D. A. 1976. Shoaling waves: numerical solution of exact equations. *Lecture Notes in Physics 64: Waves on water of variable depth*, pp. 94–101. Berlin: Springer

Fenton, J. D., Rienecker, M. M. 1980. Accurate numerical solutions for nonlinear waves. *Proc. 17th Int. Conf. Coastal Engrg., Sydney, March, 1980*

Fox, M. J. H. 1977. *Nonlinear effects in surface gravity waves on water*. PhD thesis. Cambridge Univ.

Fultz, D. 1962. An experimental note on finite-amplitude standing gravity waves. *J. Fluid Mech.* 13:193–212 (2 plates)

Grant, M. A. 1973. The singularity at the crest of a finite amplitude progressive Stokes wave. *J. Fluid Mech.* 59:257–62

Havelock, T. H. 1919. Periodic irrotational waves of finite height. *Proc. R. Soc. London Ser. A* 95:38–51

Hogan, S. J. 1979. Some effects of surface tension on steep water waves. *J. Fluid Mech.* 91:167–80

Hogan, S. J. 1980. Some effects of surface tension on steep water waves. Part 2. *J. Fluid Mech.* 96:417–45

Keady, G., Norbury, J. 1978. On the existence theory for irrotational water waves. *Math. Proc. Cambridge Philos. Soc.* 83:137–57

Kinnersley, W. 1976. Exact large amplitude capillary waves on sheets of fluid. *J. Fluid Mech.* 77:229–41

Korteweg, D. J., de Vries, G. 1895. On the change of form of long waves advancing in a rectangular canal and on a new type of long stationary wave. *Philos. Mag.* (5) 39:422–43

Krasovskii, Yu. P. 1960. The theory of steady-state waves of finite amplitude. *Dokl. Akad. Nauk SSSR* 130:1237–40; also *Zh. Vychisl. Mat. i Mat. Fiz.* 1:836–55(1961); transl. in *USSR Comp. Math. Math. Phys.* 1:996-1018

Laitone, E. V. 1960. The second approximation to cnoidal and solitary waves. *J. Fluid Mech.* 9:430–44

Lake, B. M., Yuen, H. C. 1978. A new model for nonlinear wind waves. Part 1. Physical model and experimental evidence. *J. Fluid Mech.* 88:33–62

Lenau, C. W. 1966. The solitary wave of maximum amplitude. *J. Fluid Mech.* 26:309–20

Longuet-Higgins, M. S. 1950. A theory of the origin of microseisms. *Philos. Trans. R. Soc. London Ser. A* 243:1–35

Longuet-Higgins, M. S. 1973. On the form of the highest progressive and standing waves in deep water. *Proc. R. Soc. London Ser. A* 331:445–56

Longuet-Higgins, M. S. 1974. On the mass, momentum, energy and circulation of a solitary wave. *Proc. R. Soc. London Ser. A* 337:1–13

Longuet-Higgins, M. S. 1975. Integral properties of periodic gravity waves of finite amplitude. *Proc. R. Soc. London Ser. A* 342:157–74

Longuet-Higgins, M. S. 1978a. The instabilities of gravity waves of finite amplitude in deep water. I. Superharmonics. *Proc. R. Soc. London Ser. A* 360:471–88

Longuet-Higgins, M. S. 1978b. The instabilities of gravity waves of finite amplitude in deep water. II. Subharmonics. *Proc. R. Soc. London Ser. A* 360:489–505

Longuet-Higgins, M. S. 1979a. The almost-highest wave: a simple approximation. *J. Fluid Mech.* 94:269–73

Longuet-Higgins, M. S. 1979b. The trajectories of particles in steep symmetric gravity waves. *J. Fluid Mech.* 94:497–517

Longuet-Higgins, M. S. 1980a. A technique for time-dependent free surface flows. *Proc. R. Soc. London Ser. A* 371:441–51

Longuet-Higgins, M. S. 1980b. On the forming of sharp corners at a free surface. *Proc. R. Soc. London Ser. A* 371:453–78

Longuet-Higgins, M. S. 1980c. Spin and angular momentum in gravity waves. *J. Fluid Mech.* 97:1–25

Longuet-Higgins, M. S. 1980d. Modulation of the amplitude of steep wind waves. *J. Fluid Mech.* 99:705–13

Longuet-Higgins, M. S. 1980e. The unsolved problem of breaking waves. *Proc. 17th Int. Conf. Coastal Engrg., Sydney, March, 1980*

Longuet-Higgins, M. S., Cokelet, E. D. 1976. The deformation of steep surface waves on water. I. A numerical method of computation. *Proc. R. Soc. London Ser. A* 350:1–26

Longuet-Higgins, M. S., Cokelet, E. D. 1978. The deformation of steep surface waves on water. II. Growth of normal-mode instabilities. *Proc. R. Soc. London Ser. A* 364:1–28

Longuet-Higgins, M. S., Fenton, J. D. 1974. On the mass, momentum, energy and circulation of a solitary wave. II. *Proc. R. Soc. London Ser. A* 340:471–93

Longuet-Higgins, M. S., Fox, M. J. H. 1977. Theory of the almost-highest wave: the inner solution. *J. Fluid Mech.* 80:721–41

Longuet-Higgins, M. S., Fox, M. J. H. 1978. Theory of the almost-highest wave. II. Matching and analytic extension. *J. Fluid Mech.* 85:769–86

Longuet-Higgins, M. S., Turner, J. S. 1974. An "entraining plume" model of a spilling breaker. *J. Fluid Mech.* 63:1–20

Maxworthy, T. 1976. Experiments on collisions between solitary waves. *J. Fluid Mech.* 76:177–85

McCowan, J. 1894. On the highest wave of permanent type. *Philos. Mag.* (5) 38:351–58

Miche, R. 1944. Mouvements ondulatoires de la mer en profondeur constante ou décroissante. *Ann. Ponts Chaussées* 114:25–61

Michell, J. H. 1893. The highest waves in water. *Philos. Mag.* (5) 36:430–37

Miles, J. W. 1980. Solitary waves. *Ann. Rev. Fluid Mech.* 12:11–43

Multer, R. H. 1973. Exact nonlinear model of wave generator. *J. Hydraul. Div. ASCE* 99:31–46

Nayfeh, A. H. 1970. Triple and quintuple-dimpled wave profiles in deep water. *Phys. Fluids* 13:545–50

Oikawa, M., Yajima, N. 1973. Interactions of solitary waves—a perturbation approach to nonlinear systems. *J. Phys. Soc. Jpn.* 34:1093–99

Olfe, D. B., Rottman, J. W. 1980. Some new highest-wave solutions for deep-water waves of permanent form. *J. Fluid Mech.* 100:801–10

Penney, W. G., Price, A. T. 1952. Finite periodic stationary gravity waves in a perfect fluid. *Philos. Trans. R. Soc. London Ser. A* 244:254–84

Peregrine, D. H., Cokelet, E. D., McIver, P. 1980. The fluid mechanics of waves approaching breaking. *Proc. 17th Conf. Coastal Engrg., Sydney, 1980*

Pierson, W. J., Fife, P. 1961. Some nonlinear properties of long-crested periodic waves with lengths near 2.44 centimeters. *J. Geophys. Res.* 66:163–79

Rayleigh, Lord 1915. Deep water waves, progressive or stationary, to the third order of approximation. *Proc. R. Soc. London Ser. A* 91:345–53

Rienecker, M. M., Fenton, J. D. 1981. A Fourier approximation method for steady water waves. *J. Fluid Mech.* 104:119–37

Rottman, J. W., Olfe, D. B. 1979. Numerical calculation of steady gravity-capillary waves using an integro-differential formulation. *J. Fluid Mech.* 94:777–93

Sasaki, T. K., Murakami, T. 1973. Irrotational, progressive surface gravity waves near the limiting height. *J. Oceanogr. Soc. Jpn.* 29:94–105

Schooley, A. H. 1960. Double, triple, and higher-order dimples in the profiles of wind-generated water waves in the capillary-gravity transition region. *J. Geophys. Res.* 65:4075–19

Schwartz, L. W. 1972. *Analytic continuation of Stokes' expansion for gravity waves*. PhD thesis. Stanford Univ.

Schwartz, L. W. 1974. Computer extension and analytic continuation of Stokes' expansion for gravity waves. *J. Fluid Mech.* 62:553–78

Schwartz, L. W. 1980. On the pressure field of nonlinear standing water waves. *Rep. JIAA TR-33*. Joint Inst. Aeronautics &

Acoustics, Stanford Univ.
Schwartz, L. W., Vanden-Broeck, J. M. 1979. Numerical solution of the exact equations for capillary-gravity waves. *J. Fluid Mech.* 95:119–39
Schwartz, L. W., Whitney, A. K. 1977. A high-order series solution for standing water waves. *Proc. 6th Australasian Hydraul. & Fluid Mech. Conf., Adelaide*, pp. 356–59
Schwartz, L. W., Whitney, A. K. 1981. A semi-analytic solution for nonlinear standing waves in deep water. *J. Fluid Mech.* 107:147–71
Skjelbreia, L., Hendrickson, J. 1961. Fifth order gravity wave theory. *Proc. 7th Conf. Coastal Engrg.* pp. 184–96
Stokes, G. G. 1847. On the theory of oscillatory waves. *Trans. Cambridge Philos. Soc.* 8:441–55, and in *Mathematical & Physical Papers* 1:197–229. Cambridge, Univ. Press
Stokes, G. G. 1880. Supplement to a paper on the theory of oscillatory waves. *Mathematical & Physical Papers* 1:314–26. Cambridge, England
Su, C. H., Mirie, R. M. 1980. On head-on collisions between two solitary waves. *J. Fluid Mech.* 98:509–25
Tadjbakhsh, I., Keller, J. B. 1960. Standing surface waves of finite amplitude. *J. Fluid Mech.* 8:442–51
Taylor, G. I. 1953. An experimental study of standing waves. *Proc. R. Soc. London Ser. A* 218:44–59
Toland, J. F. 1978. On the existence of a wave of greatest height and Stokes' conjecture. *Proc. R. Soc. London Ser. A* 363:469–85
Ursell, F. 1953. The long-wave paradox in the theory of gravity waves. *Proc. Cambridge Philos. Soc.* 49:685–94
Vanden-Broeck, J. M., Keller, J. B. 1980. A new family of capillary waves. *J. Fluid Mech.* 98:161–69
Vanden-Broeck, J. M., Schwartz, L. W. 1979. Numerical computation of steep gravity waves in shallow water. *Phys. Fluids* 22:1868–71
Vanden-Broeck, J. M., Schwartz, L. W. 1981. Numerical calculation of standing waves in water of arbitrary uniform depth. *Phys. Fluids.* 24:812–15
Wehausen, J. V., Laitone, E. V. 1960. Surface waves. In *Encyclopaedia of Physics* 9:446–778. ed, S. Flügge, Berlin: Springer
Weidman, P. D., Maxworthy, T. 1978. Experiments on strong interactions between solitary waves. *J. Fluid Mech.* 85:417–31
Whitney, A. K. 1970. The numerical solution of unsteady free surface flows by conformal mapping. *Proc. 2nd Int. Conf. Num. Meth. Fluid Dynam., Berkeley, Sept. 1970*, pp. 458–62
Wilton, J. R. 1914. On deep water waves. *Philos. Mag.* (6) 27:385–94
Wilton, J. R. 1915. On ripples. *Philos. Mag.* (6) 29:688–700
Witting, J. 1975. On the highest and other solitary waves. *SIAM J. Appl. Math.* 28:700–19
Yamada, H. 1957a. Highest waves of permanent type on the surface of deep water. *Rep. Res. Inst. Appl. Mech. Kyushu Univ.* 5, no. 18:37–52
Yamada, H. 1957b. On the highest solitary wave. *Rep. Res. Inst. Appl. Mech. Kyushu Univ.* 5, no. 18:53–67
Yamada, H., Kimura, G., Okabe, J. I. 1968. Precise determination of the solitary wave of extreme height on water of a uniform depth. *Rep. Res. Inst. Appl. Mech. Kyushu Univ.* 16:15–32
Yamada, H., Shiotani, T. 1968. On the highest water waves of permanent type. *Bull. Disaster Prev. Res. Inst. Kyoto Univ.* 18:1–22

TOPOLOGY OF THREE-DIMENSIONAL SEPARATED FLOWS[1]

Murray Tobak and David J. Peake

National Aeronautics and Space Administration, Ames Research Center, Moffett Field, California 94035

INTRODUCTION

Three-dimensional separated flow represents a domain of fluid mechanics of great practical interest that is, as yet, beyond the reach of definitive theoretical analysis or numerical computation. At present, our understanding of three-dimensional flow separation rests principally on observations drawn from experimental studies utilizing flow visualization techniques. Particularly useful in this regard has been the oil-streak technique for making visible the patterns of skin-friction lines on the surfaces of wind-tunnel models (Maltby 1962). It is a common observation among students of these patterns that a necessary condition for the occurrence of flow separation is the convergence of oil-streak lines onto a particular line. Whether this is also a sufficient condition is a matter of current debate. The requirement to make sense of these patterns within a governing hypothesis of sufficient precision to yield a convincing description of three-dimensional flow separation has inspired the efforts of a number of investigators. Of the numerous attempts, however, few of the contending arguments lend themselves to a precise mathematical formulation. Here, we shall single out for special attention the hypothesis proposed by Legendre (1956) as being one capable of providing a mathematical framework of considerable depth.

Legendre (1956) proposed that a pattern of streamlines immediately adjacent to the surface (in his terminology, "wall streamlines") be considered as trajectories having properties consistent with those of a continuous vector field, the principal one being that through any regular (nonsingular) point there must pass one and only one trajectory. On the

---

[1] The US Government has the right to retain a nonexclusive, royalty-free license in and to any copyright covering this paper.

basis of this postulate, it follows that the elementary singular points of the field can be categorized mathematically. Thus, the types of singular points, their number, and the rules governing the relations between them can be said to characterize the pattern. Flow separation in this view has been defined by the convergence of wall streamlines onto a particular wall streamline that originates from a singular point of particular type, the saddle point. We should note, however, that this view of flow separation is not universally accepted, and, indeed, situations exist where it appears that a more nuanced description of flow separation may be required.

Lighthill (1963), addressing himself specifically to viscous flows, clarified a number of important issues by tying the postulate of a continuous vector field to the pattern of skin-friction lines rather than to streamlines lying just above the surface. Parallel to Legendre's definition, convergence of skin-friction lines onto a particular skin-friction line originating from a saddle point was defined here as the necessary condition for flow separation. More recently, Hunt et al. (1978) have shown that the notions of elementary singular points and the rules that they obey can be easily extended to apply to the flow above the surface on planes of symmetry, on projections of conical flows (Smith 1969), on crossflow planes, etc. (see also Perry & Fairlie 1974). Further applications and extensions can be found in the various contributions of Legendre (1965, 1972, 1977), Oswatitsch (1980), and in the review articles by Tobak & Peake (1979) and Peake & Tobak (1980).

As Legendre (1977) himself has noted, his hypothesis was but a reinvention within a narrower framework of the extraordinarily fruitful line of research initiated by Poincaré (1928) under the title, "On the Curves Defined by Differential Equations." Yet another branch of the same line has been the research begun by Andronov and his colleagues (1971, 1973) on the qualitative theory of differential equations, within which the useful notions of "topological structure" and "structural stability" were introduced. Finally, from the same line stems the rapidly expanding field known as "bifurcation theory" (see the comprehensive review of Sattinger 1980). Applications to hydrodynamics are exemplified by the works of Joseph (1976) and Benjamin (1978). It has become clear that our understanding of three-dimensional separated flow may be deepened by placing Legendre's hypothesis within a framework broad enough to include the notions of topological structure, structural stability, and bifurcation. Bearing in mind that we still await a convincing description of three-dimensional flow separation, we may ask whether the broader framework will facilitate the emergence of such a description. In the following, we shall try to answer this question, limiting our attention to three-dimensional viscous flows that are steady in the mean.

THEORY

We consider steady viscous flow over a smooth three-dimensional body. The postulate that the skin-friction lines on the surface of the body form the trajectories of a continuous vector field is translated mathematically as follows: Let (ξ,η,ζ) be general curvilinear coordinates with (ξ,η) being orthogonal axes in the surface and ζ directed out of the surface normal to (ξ,η). Let the length parameters be $h_1(\xi,\eta)$, $h_2(\xi,\eta)$. Except at singular points, it follows from the adherence condition that, very close to the surface, the components of the velocity vector parallel to the surface (u_1,u_2) must grow from zero linearly with ζ. Hence, a particle on a streamline near the surface will have velocity components of the form

$$h_1(\xi,\eta)\frac{d\xi}{dt} = \zeta\frac{\partial u_1}{\partial \zeta}(\xi,\eta,0)$$
$$= -\zeta\omega_2(\xi,\eta) = \zeta P(\xi,\eta),$$
$$h_2(\xi,\eta)\frac{d\eta}{dt} = \zeta\frac{\partial u_2}{\partial \zeta}(\xi,\eta,0)$$
$$= \zeta\omega_1(\xi,\eta) = \zeta Q(\xi,\eta), \tag{1}$$

where (ω_1,ω_2) are the components of the surface vorticity vector. The specification of a steady flow is reflected by (u_1,u_2) being independent of time. With ζ treated as a parameter and P and Q functions only of the coordinates, Equations (1) are a pair of *autonomous* ordinary differential equations. Their form places them in the same category as the equations studied by Poincaré (1928) in his classical investigation of the curves defined by differential equations. Letting

$$\tau_{w_1} = \mu\frac{\partial u_1}{\partial \zeta}(\xi,\eta,0),$$
$$\tau_{w_2} = \mu\frac{\partial u_2}{\partial \zeta}(\xi,\eta,0) \tag{2}$$

be components of the skin friction parallel to ξ and η, respectively, we have for the equation governing the trajectories of the surface shear stress vector, from Equations (1),

$$\frac{h_1 d\xi}{\tau_{w_1}} = \frac{h_2 d\eta}{\tau_{w_2}}. \tag{3}$$

Alternatively, for the trajectories of the surface vorticity vector, which are orthogonal to those of the surface shear stress vector, the governing equation is

$$\frac{h_1 d\xi}{\omega_1} = \frac{h_2 d\eta}{\omega_2}. \tag{4}$$

Singular Points

Singular points in the pattern of skin-friction lines occur at isolated points on the surface where the skin friction (τ_{w_1}, τ_{w_2}) in Equation (3), or alternatively the surface vorticity (ω_1, ω_2) in Equation (4), becomes identically zero. Singular points are classifiable into two main types: nodes and saddle points. Nodes may be further subdivided into two subclasses: nodal points and foci (of attachment or separation).

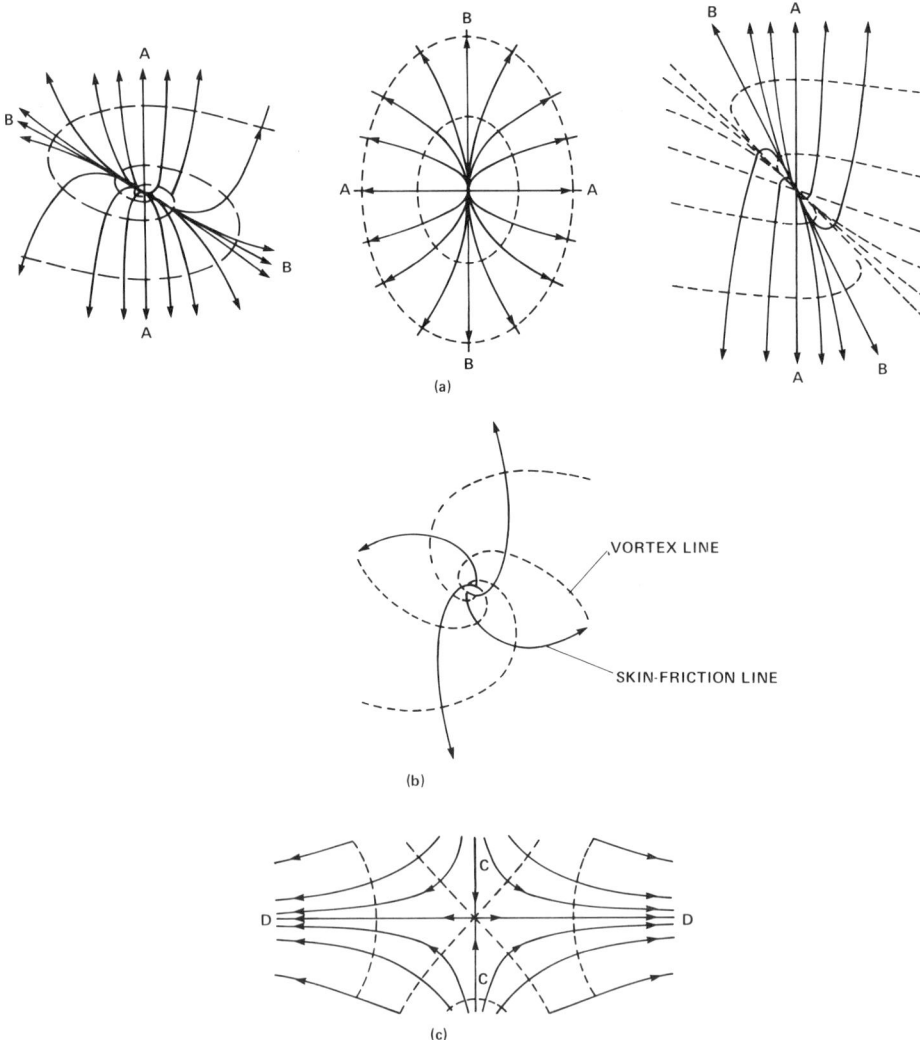

Figure 1 Singular points: (*a*) node; (*b*) focus; (*c*) saddle (Lighthill 1963).

A nodal point (Figure 1a) is the point common to an infinite number of skin-friction lines. At the point, all of the skin-friction lines except one (labeled AA in Figure 1a) are tangential to a single line BB. At a nodal point of attachment, all of the skin-friction lines are directed outward away from the node. At a nodal point of separation, all of the skin-friction lines are directed inward toward the node. In the exceptional case (e.g. in the presence of axisymmetry) the node degenerates into an isotropic node where every skin-friction line entering or leaving the singular point has a distinct tangent (Legendre 1977).

A focus (Figure 1b) differs from a nodal point on Figure 1a in that it has no common tangent line. An infinite number of skin-friction lines spiral around the singular point, either away from it (a focus of attachment) or into it (a focus of separation). Foci of attachment generally occur in the presence of rotation, either of the flow or of the surface, and will not figure in this study. In the exceptional case the trajectories of the focus form closed paths around the singular point. The focus is then called a center.

At a saddle point (Figure 1c), there are only two particular lines, CC and DD, that pass through the singular point. The directions on either side of the singular point are inward on one particular line and outward on the other particular line. All of the other skin-friction lines miss the singular point and take directions consistent with the directions of the adjacent particular lines. The particular lines act as barriers in the field of skin-friction lines, making one set of skin-friction lines inaccessible to an adjacent set.

For each of the patterns in Figures 1a–c, the surface vortex lines form a system of lines orthogonal at every point to the system of skin-friction lines. Thus, it is always possible in principle to describe the flow in the vicinity of a singular point alternatively in terms of a pattern of skin-friction lines or a pattern of surface vortex lines.

Davey (1961) and Lighthill (1963) have both noted that of all the possible patterns of skin-friction lines on the surface of a body, only those are admissible whose singular points obey a topological rule: the number of nodes (nodal points or foci or both) must exceed the number of saddle points by two. We shall demonstrate this rule and its recent extensions to the external flow field in a number of examples.

Topography of Skin-Friction Lines

The singular points, acting either in isolation or in combination, fulfill certain characteristic functions that largely determine the distribution of skin-friction lines on the surface. The nodal point of attachment is typically a stagnation point on a forward-facing surface, such as the nose of a body, where the external flow from far upstream attaches itself to the surface. The nodal point of attachment thereby acts as a source of skin-friction lines that emerge from the point and spread out over the surface. Conversely, the nodal point of separation is typically a point on a rearward-facing surface, and acts as a sink where the skin-friction lines that have circumscribed the body surface may vanish.

The saddle point acts typically to separate the skin-friction lines issuing from adjacent nodes, for example, adjacent nodal points of attachment. An example of this function is illustrated in Figure 2 (Lighthill 1963). Skin-friction lines emerging from the nodal points of attachment are prevented from crossing by the presence of a particular skin-friction line emerging from the saddle point. Lighthill (1963) has labeled the particular line a *line of separation* and has identified the existence of a saddle point from which the line emerges as the necessary condition for flow separation. As Figure 2 indicates, skin-friction lines from either side tend to converge on the line emerging from the saddle point. Unfortunately, the convergence of skin-friction lines on either side of a particular line occurs in other situations as well. It can happen, for example, that one skin-friction line out of the infinite set of lines emanating from a nodal point of attachment may ultimately become a line on which others of the set converge. All researchers agree that the existence of a particular skin-friction line on which other lines converge is a necessary condition for flow separation. The seeming nonuniqueness of the condition identifying the particular line has encouraged the appearance of alternative descriptions of flow separation that, in contrast to Lighthill's, do not insist on the presence of a saddle point as the origin of the line. Wang (1976), in particular, has argued that there are two types of flow separation: "open," in which the skin-friction line on which other lines converge does not emanate from a

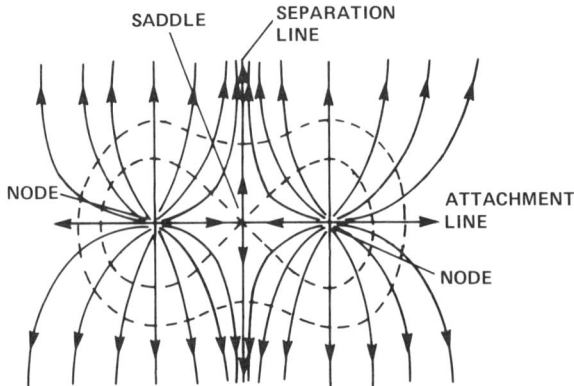

Figure 2 Adjacent nodes and saddle point (Lighthill 1963).

saddle point, and "closed," in which, as in Lighthill's definition, it does (see also Wang 1974, Han & Patel 1979). In what follows, we shall address the question of an appropriate description of flow separation by an appeal to the theory of structural stability and bifurcation. Like Wang, we shall find it necessary to distinguish between types of separation, but we shall adopt a terminology that is suggested by the theoretical framework. We shall say that a skin-friction line emerging from a saddle point is a *global* line of separation and leads to *global* flow separation. In the contrary case, where the skin-friction line on which other lines converge does not originate from a saddle point, we shall identify the line as being a *local* line of separation, leading to *local* flow separation. When no modifier is used, what is said will apply to either case. Thus (in either case), an additional indicator of the line of separation is the behavior of the surface vortex lines. In the vicinity of a line of separation the surface vortex lines become distorted, forming upstream-pointing loops with the peaks of the loops occurring on the line of separation.

The converse of the line of separation is the line of attachment. Two lines of attachment are illustrated in Figure 2, emanating from each of the nodal points of attachment. Skin-friction lines tend to diverge from lines of attachment. Just as with the line of separation, a graphic indicator of the presence of a line of attachment is the behavior of the surface vortex lines. Surface vortex lines form downstream-pointing loops in the vicinity of a line of attachment, with the peaks of the loops occurring on the line of attachment.

Streamlines passing very close to the surface, that is, those defined by Equations (1) are called *limiting streamlines*. In the vicinity of a line of separation, limiting streamlines must leave the surface rapidly, as a simple argument due to Lighthill (1963) explains. Referring to Equation (3), let us align (ξ, η) with external streamline coordinates so that τ_{w_1}, τ_{w_2} are the respective streamwise and crossflow skin-friction components. If n is the distance between two adjacent limiting streamlines (see Figure 3) and h is the height of a rectangular streamtube (being assumed small so that the local resultant velocity vectors are coplanar and form a linear profile), then the mass flux through the streamtube is

$$\dot{m} = \rho h n \bar{u}$$

where ρ is the density and $\bar{u}$ the mean velocity of the cross section. But the resultant skin friction at the wall is the resultant of τ_{w_1} and τ_{w_2}, or

$$\tau_w = \mu \left(\frac{\bar{u}}{h/2} \right)$$

so that

$$\bar{u} = \frac{\tau_w h}{2\mu}.$$

Hence,

$$\dot{m} = \frac{h^2 n \tau_w}{2\upsilon} = \text{constant},$$

yielding

$$h = C\left(\frac{\upsilon}{n\tau_w}\right)^{1/2}; \upsilon = \frac{\mu}{\rho}.$$

Thus, as the line of separation is approached, h, the height of the limiting streamline above the surface, increases rapidly. There are two reasons for this increase in h: first, whether the line of separation is global or local, the distance n between adjacent limiting streamlines falls rapidly as the limiting streamlines converge towards the line of separation; second, the resultant skin-friction τ_w drops toward a minimum as the line of separation is approached and, in the case of the global line of separation, actually approaches zero as the saddle point is approached.

Limiting streamlines rising on either side of the line of separation are prevented from crossing by the presence of a stream surface stemming from the line of separation itself. The existence of such a stream surface is characteristic of flow separation; how it originates determines whether the separation is of global or local form. In the former case, the presence of a saddle point as the origin of the global line of separation provides a

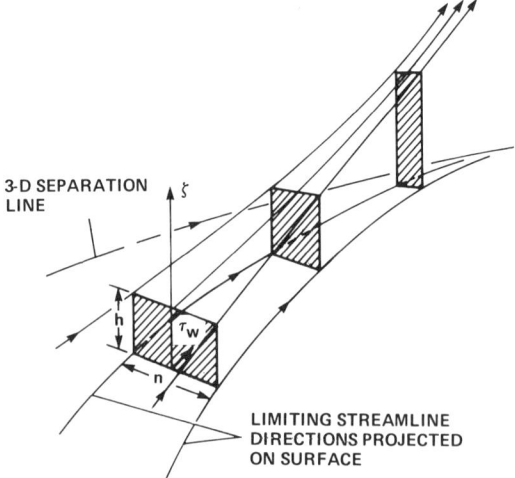

Figure 3 Limiting streamlines near three-dimensional (3D) separation line.

mechanism for the creation of a new stream surface that originates at the wall. Emanating from a saddle point and terminating at nodal points of separation (either nodes or foci), the global line of separation traces a smooth curve on the wall which forms the base of the stream surface, the streamlines of which have all entered the fluid through the saddle point. We shall call this new stream surface a *dividing surface*. The dividing surface extends the function of the global line of separation into the flow, acting as a barrier separating the set of limiting streamlines that have risen from the surface on one side of the global line of separation from the set arisen from the other side. On its passage downstream, the dividing surface rolls up to form the familiar coiled sheet around a central vortical core. Because it has a well-defined core, we use the popular terminology, calling the flow in the vicinity of the coiled-up dividing surface a *vortex*.

Now we consider the origin of the stream surface characteristic of *local* flow separation. We note that if a skin-friction line emanating from a nodal point of attachment ultimately becomes a local line of separation, then there will be a point on the line beyond which each of the orthogonal surface vortex lines crossing the line forms an upstream-pointing loop, signifying that the skin friction along the line has become locally minimum. A surface starting at this point and stemming from the skin-friction line downstream of the point can be constructed that will be the locus of a set of limiting streamlines originating from far upstream; this surface may also roll up on its development downstream.

This section concludes with a discussion of the remaining type of singular point, the focus (also called spiral node). The focus invariably appears on the surface in company with a saddle point. Together they allow a particular form of global flow separation. One leg of the (global) line of separation emanating from the saddle point winds into the focus to form the continuous curve on the surface from which the dividing surface stems. The focus on the wall extends into the fluid as a concentrated vortex filament, while the dividing surface rolls up with the same sense of rotation as the vortex filament. When the dividing surface extends downstream it quickly draws the vortex filament into its core. In effect, then, the extension into the fluid of the focus on the wall serves as the vortical core about which the dividing surface coils. This flow behavior was first hypothesized by Legendre (1965), who also noted (Legendre 1972) that an experimental confirmation existed in the results of earlier experiments carried out by Werlé (1962). Figure 4a shows Legendre's original sketch of the skin-friction lines; Figure 4b is a photograph illustrating the experimental confirmation. The dividing surface that coils around the extension of the focus (Figure 4c) will be termed here a "horn-type dividing surface." On the other hand, it can happen that the dividing surface to which the focus is

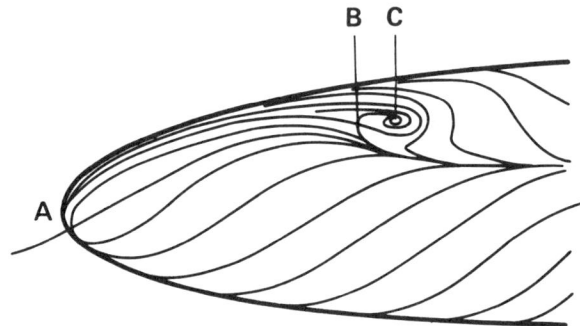

(a) A — NODAL ATTACHMENT POINT
B — SADDLE POINT
C — FOCUS OF SEPARATION

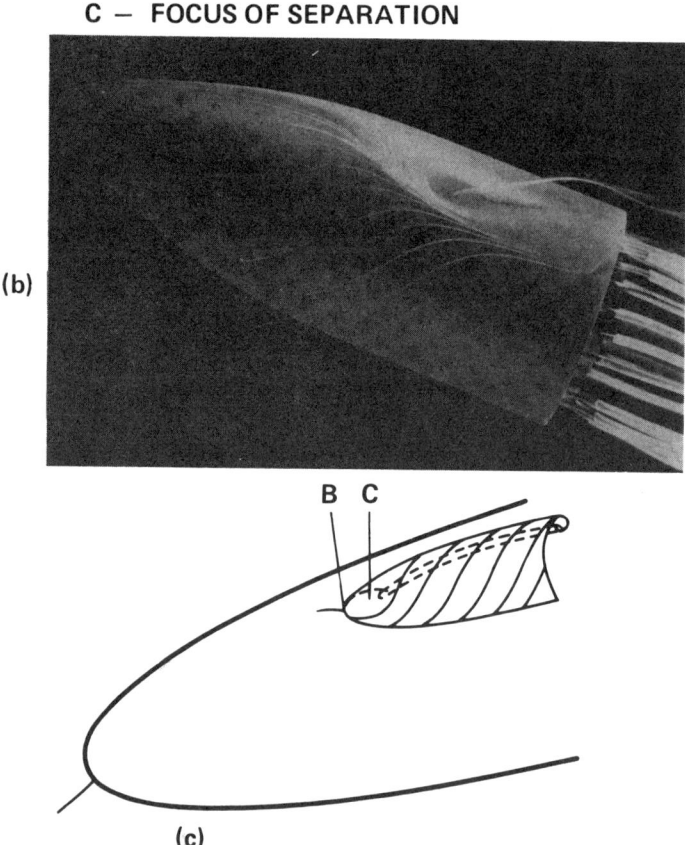

Figure 4 Focus of separation: (*a*) original sketch of skin-friction lines by Legendre (1965); (*b*) experiment of Werlé (1962) in water tunnel; (*c*) extension of focus, Legendre (1965).

connected does not extend downstream. In this case the vortex filament emanating from the focus remains distinct, and is seen as a separate entity on crossflow planes downstream of its origin on the surface. In an interesting additional interpretation of the focus, we begin by considering the pattern of lines orthogonal to that of the skin-friction lines; that is, the pattern of surface vortex lines. We see that what was a focus for the pattern of skin-friction lines becomes another focus of separation for the pattern of surface vortex lines, marking the apparent termination of a set of surface vortex lines. If we imagine that each of these surface vortex lines is the bound part of a horseshoe vortex, then the extension into the fluid of the focus on the wall as a concentrated vortex filament is seen to represent the combination into one filament of the horseshoe vortex legs from all of the bound vortices that have ended at the focus. One can envisage the possibility of incorporating this description of the flow in the vicinity of a focus into an appropriate inviscid flow model.

Forms of Dividing Surfaces

We have seen how the combination of a focus and a saddle point in the pattern of skin-friction lines allows a particular form of global flow separation characterized by a "horn-type dividing surface." The nodal points of attachment and separation may also combine with saddle points to allow additional forms of global flow separation, again characterized by their particular dividing surfaces. The characteristic dividing surface formed from the combination of a nodal point of attachment and a saddle point is illustrated in Figure 5a. This form of dividing surface typically occurs in the flow before an obstacle (see Figure 34 in Peake & Tobak 1980). In the example illustrated in Figure 5a it will be noted that the dividing surface admits of a point in the external flow at which the fluid velocity is identically zero. This is a three-dimensional singular point, which in Figure 5a acts as the origin of the streamline running through the vortical core of the rolled up dividing surface.

The characteristic dividing surface formed from the combination of a nodal point of separation and a saddle point is illustrated in Figure 5b. This form of dividing surface often occurs in nominally two-dimensional separated flows such as in the separated flow behind a backward-facing step (see Figure 24 in Tobak & Peake 1979) and the separated flow at a cylinder-flare junction (both two and three dimensional, compare Figures 47 and 48 in Peake & Tobak 1980). We note in both Figures 5a and 5b that the streamlines on the dividing surface have all entered the fluid through the saddle point in the pattern of skin-friction lines.

Topography of Streamlines in Two-Dimensional Sections of Three-Dimensional Flows

After an unaccountably long lapse of time, it has only recently become clear that the mathematical basis for the behavior of elementary singular points and the topological rules that they obey is general enough to support a much wider regime of application than had originally been realized. The results reported by Smith (1969, 1975), Perry & Fairlie (1974), and Hunt et al. (1978) have made it evident that the rules governing skin-friction line behavior are easily adapted and extended to yield similar rules governing behavior of the flow itself. In particular, Hunt et al. (1978) have noted that if $\mathbf{v} = [u(x,y,z_0), v(x,y,z_0), w(x,y,z_0)]$ is the mean velocity

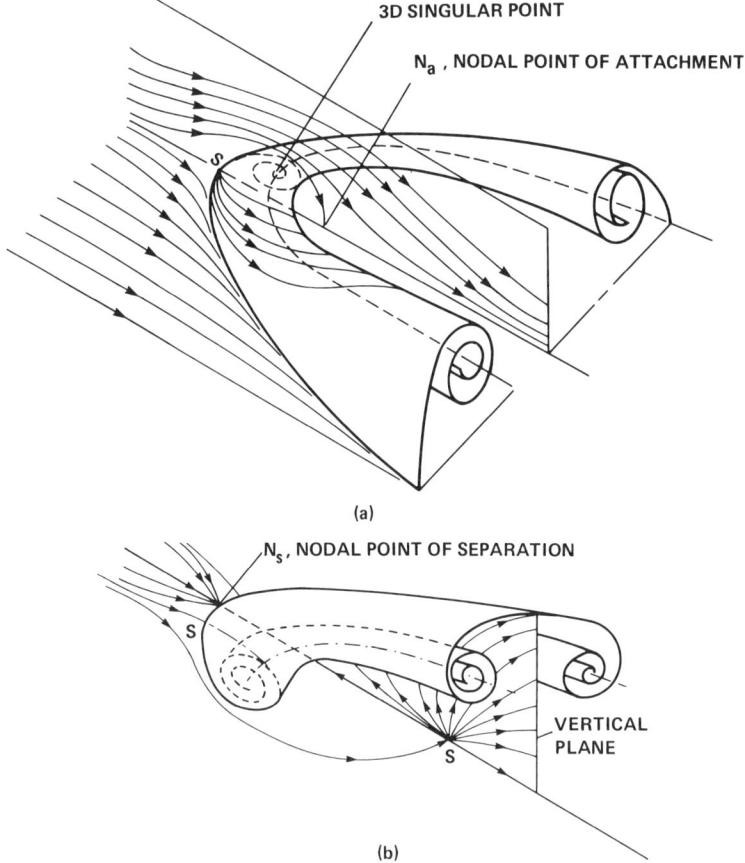

Figure 5 Dividing surfaces formed from combinations of (a) nodal point of attachment and saddle point; (b) nodal point of separation and saddle point.

whose u,v components are measured in a plane $z = z_0 =$ constant, above a surface situated at $y = Y(x;z_0)$ (see Figure 6), then the mean streamlines in the plane are solutions of the equation

$$\frac{dx}{u} = \frac{dy}{v} \qquad (5)$$

which is a direct counterpart of Equation (3) for skin-friction lines on the surface. Hunt et al. (1978) cautioned that for a general three-dimensional flow the streamlines defined by Equation (5) are no more than that—they are not necessarily the projections of the three-dimensional streamlines onto the plane $z = z_0$, nor are they necessarily particle paths even in a steady flow. Only for special planes—for example, a streamwise plane of symmetry (where $w(x,y,z_0) \equiv 0$)—are the streamlines defined by Equation (5) identifiable with particle path lines in the plane when the flow is steady, or with instantaneous streamlines when the flow is unsteady. In any case, since $[u(x,y),v(x,y)]$ is a continuous vector field $\mathbf{V}(x,y)$, with only a finite number of singular points in the interior of the flow at which $\mathbf{V} = 0$, it follows that nodes and saddles can be defined in the plane just as they were for skin-friction lines on the surface. Nodes and saddles within the flow, excluding the boundary $y = Y(x;z_0)$, are labeled N and S, respectively, and are shown in their typical form in Figure 6. The only new feature of the analysis that is required is the treatment of singular points on the boundary $y = Y(x;z_0)$. Since for a viscous flow, $\mathbf{V}$ is zero everywhere on the boundary, the boundary is itself a singular line in the plane $z = z_0$. Singular points on the line occur where the component of the surface vorticity vector normal to the plane $z = z_0$ is zero. Thus, for example, it is ensured that a singular point will occur on the boundary wherever it passes through a singular point in the pattern of skin-friction lines, since the surface vorticity is identically zero there. As introduced by Hunt et al. (1978), singular points on the boundary are defined as half-nodes N' and half-saddles S' (Figure 6). With this simple amendment to the types of singular points allowable, all of the previous notions and descriptions relevant to the analysis of skin-friction lines carry over to the analysis of the flow within the plane.

In a parallel vein, Hunt et al. (1978) have recognized that, just as the singular points in the pattern of skin-friction lines on the surface obey a topological rule, so must the singular points in any of the sectional views of three-dimensional flows obey topological rules. Although a very general rule applying to multiply connected bodies can be derived (Hunt et al. 1978) we list here for convenience only those special rules that will be useful in subsequent studies of the flow past wings, bodies, and obstacles. In the five topological rules listed below, we assume that the body is simply connected and immersed in a flow that is uniform far upstream.

1. Skin-friction lines on a three-dimensional body (Davey 1961, Lighthill 1963):

$$\sum_N - \sum_S = 2 . \tag{6}$$

2. Skin-friction lines on a three-dimensional body B connected simply (without gaps) to a plane wall P that either extends to infinity both upstream and downstream or is the surface of a torus:

$$\left(\sum_N - \sum_S \right)_{P+B} = 0 . \tag{7}$$

3. Streamlines on a two-dimensional plane cutting a three-dimensional body:

$$\left(\sum_N + \frac{1}{2} \sum_{N'} \right) - \left(\sum_S + \frac{1}{2} \sum_{S'} \right) = -1 . \tag{8}$$

4. Streamlines on a vertical plane cutting a surface that extends to infinity both upstream and downstream:

$$\left(\sum_N + \frac{1}{2} \sum_{N'} \right) - \left(\sum_S + \frac{1}{2} \sum_{S'} \right) = 0 . \tag{9}$$

5. Streamlines on the projection onto a spherical surface of a conical flow past a three-dimensional body (Smith 1969):

$$\left(\sum_N + \frac{1}{2} \sum_{N'} \right) - \left(\sum_S + \frac{1}{2} \sum_{S'} \right) = 0 . \tag{10}$$

Topological Structure, Structural Stability, and Bifurcation

The question of an adequate description of three-dimensional separated flow rises with particular sharpness when one asks how three-dimensional

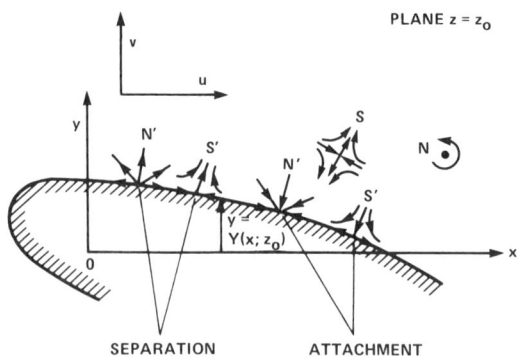

Figure 6 Singular points in cross section of flow (Hunt et al. 1978).

separated flow patterns originate and how they succeed each other as the relevant parameters of the problem (angle of attack, Reynolds number, Mach number, etc.) are varied. A satisfactory answer to the question may emerge out of the framework that we create in this section. We cast our formulation in physical terms although our definitions ought to be compatible with a more purely mathematical treatment based, for example, on whatever system of partial differential equations is judged to govern the fluid motion. In particular, we hinge our definitions of topological structure and structural stability directly to the properties of patterns of skin-friction lines, since this enables us to make maximum use of results from the principal source of experimental information on three-dimensional separated flow—flow-visualization experiments utilizing the oil-streak technique.

Adopting the terminology of Andronov et al. (1973), we say that a pattern of skin-friction lines on the surface of a body constitutes the *phase portrait* of the surface shear-stress vector. Two phase portrports have the same *topological structure* if a mapping from one phase portrait to the other preserves the paths of the phase portrait. It is useful to imagine having imprinted a phase portrait on a sheet of rubber that may be deformed in any way without folding or tearing. Every such deformation is a path-preserving mapping. A *topological property* is any characteristic of the phase portrait that remains invariant under all path-preserving mappings. The number and types of singular points, the existence of paths connecting the singular points, and the existence of closed paths are examples of topological properties. The set of all topological properties of the phase portrait describes the topological structure.

We define the *structural stability* of a phase portrait relative to a parameter λ as follows (see Andronov et al. 1971): A phase portrait is structurally stable at a given value of the parameter λ if the phase portrait resulting from an infinitesimal change in the parameter has the same topological structure as the initial one. The properties of structurally stable phase portraits can be elucidated via mathematical analysis (Andronov et al. 1971) although they depend to some extent on whether special conditions such as, for example, geometric symmetries, are to be considered typical (i.e. "generic"; see Benjamin 1978) or untypical ("nongeneric"). Here we wish to respect the conditions imposed by geometric symmetries whenever they exist. In this case structurally stable phase portraits of the surface shear-stress vector have two principal properties in common: (*a*) the singular points of the phase portrait are all *elementary* singular points; and (*b*) there are no saddle-point-to-saddle-point connections in the phase portrait. (We should note that condition (*b*) is a property only of the phase portrait representing the trajectories of the surface shear-stress vector. Saddle-point-to-saddle-point connections often occur on two-dimensional projections of the external flow—by external flow we mean the entire flow exterior to the surface—but these are artifacts of the particular projections and do not represent connections between actual (three-dimensional) singular points of the fluid velocity vector).

In speaking of the stability of the external flow, we find it necessary to distinguish between *structural stability* and *asymptotic stability* of the flow. The definition of structural stability follows from that introduced in reference to the phase portrait of the surface shear-stress vector. An external flow is called structurally stable relative to a parameter λ if a small change in the parameter does not alter the topological structure (e.g. the number and types of three-dimensional singular points) of the external three-dimensional velocity vector field. Asymptotic stability is discussed more fully later. Here, we note simply that a mean flow is called asymptotically stable if small perturbations from it (at fixed λ) decay to zero as time $t \to \infty$. In speaking of structural and asymptotic *instability*, we find it convenient to distinguish between *local* and *global* properties of the instabilities. We call an instability *global* if it permanently alters the topological structure of either the external three-dimensional velocity vector field or the phase portrait of the surface shear-stress vector. We call an instability *local* if it does not result in an alteration of the topological structure of either vector field. Thus, a structural instability is necessarily global while an asymptotic instability may be either local or global. On the other hand, an asymptotic instability necessarily implies nonuniqueness (mathematically speaking) in the solutions of the governing flow equations while a structural instability need not imply nonuniqueness.

The introduction of distinctions between local and global events helps to explain why we were led earlier to distinguish between local and global lines of separation in the pattern of skin-friction lines. If an (asymptotic) instability of the external flow does not alter the topological structure of the phase portrait representing the surface shear-stress vector, then the convergence of skin-friction lines onto one or several particular lines can only be a local event so far as the phase portrait is concerned; accordingly, we label the particular lines *local* lines of separation. On the other hand, if an (asymptotic or structural) instability of the external flow changes the topological structure of the phase portrait, resulting in the emergence of a saddle point in the pattern of skin-friction lines, then this is a global event so far as the phase portrait is concerned; accordingly, we label the skin-friction line emanating from the saddle point a *global* line of separation.

Asymptotic instability of the external flow leads to the notions of *bifurcation*, *symmetry-breaking*, and *dissipative structures* (Sattinger

1980, Nicolis & Prigogine 1977). Suppose that the fluid motions evolve according to time-dependent equations of the general form

$$u_t = G(u, \lambda),$$

where λ again is a parameter. Solutions of $G(u,\lambda) = 0$ represent *steady* mean flows of the kind we have been considering. As we have noted, a mean flow u_0 is an asymptotically stable flow if small perturbations from it decay to zero as $t \to \infty$. When the parameter λ is varied, one mean flow may persist [in the mathematical sense that it remains a valid solution of $G(u,\lambda) = 0$] but become unstable to small perturbations as λ crosses a critical value. At such a transition point, a new mean flow may *bifurcate* from the known flow. The behavior just described is conveniently portrayed on a bifurcation diagram, typical examples of which are illustrated in Figure 7. Flows that bifurcate from the known flow are represented by the ordinate ψ, which may be any quantity that characterizes the bifurcation flow alone. Stable flows are indicated by solid lines, unstable flows by dashed lines. Thus, over the range of λ where the known flow is stable, ψ is zero, and the stable known flow is represented along the abscissa by a solid line. The known flow becomes unstable for all values

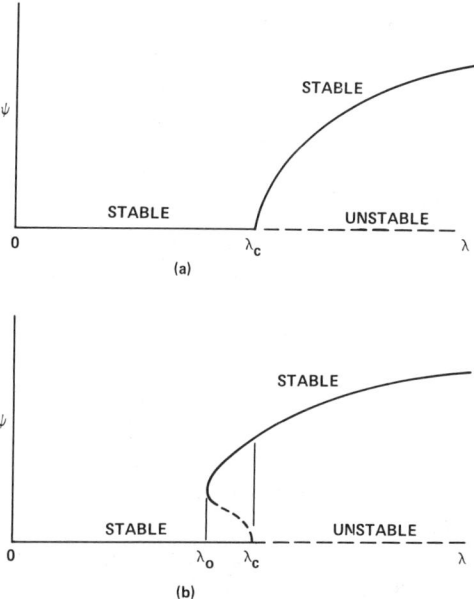

Figure 7 Examples of (*a*) supercritical bifurcation; (*b*) subcritical bifurcation.

of λ larger than λ_c, as the dashed line along the abscissa indicates. New mean flows bifurcate from $\lambda = \lambda_c$ either supercritically or subcritically.

At a supercritical bifurcation (Figure 7a), as the parameter λ is increased just beyond the critical point λ_c, the bifurcation flow that replaces the unstable known flow can differ only infinitesimally from it. The bifurcation flow *breaks the symmetry* of the known flow, adopting a form of lesser symmetry in which *dissipative structures* arise to absorb just the amount of excess available energy that the more symmetrical known flow no longer was able to absorb. Because the bifurcation flow initially departs only infinitesimally from the unstable known flow, the structural stability of the surface shear stress initially is unaffected. However, as λ continues to increase beyond λ_c, the bifurcation flow departs significantly from the unstable known flow and begins to affect the structural stability of the surface shear stress. *Ultimately* a value of λ is reached at which the surface shear stress becomes structurally unstable, evidenced either by one of the elementary singular points of its phase portrait becoming a singular point of (odd) multiple order or by the appearance of a new singular point of (even) multiple order. In either case, it is useful to consider the singular point of multiple order as being the coalescence of a number of elementary singular points, with the number divided among nodal and saddle points such as to continue to satisfy the first topological rule, Equation (6). An additional infinitesimal increase in the parameter λ results in the splitting of the singular point of multiple order into an equal number of elementary singular points. Thus there emerges a new structurally stable phase portrait of the surface shear-stress vector and a new external flow from which additional flows ultimately will bifurcate with further increases of the parameter.

At a subcritical bifurcation (Figure 7b), when the parameter is increased just beyond the critical point λ_c, there are no adjacent bifurcation flows that differ only infinitesimally from the unstable known flow. Here, there must be a finite jump to a new branch of flows that may represent a radical change in the topological structure of the external flow and perhaps in the phase portrait of the surface shear-stress vector as well. Further, with ψ on the new branch, when λ is decreased just below λ_c the flow does not return to the original stable known flow. Only when λ is decreased far enough below λ_c to pass λ_0 (Figure 7b) is the stable known flow recovered. Thus, subcritical bifurcation always implies that the bifurcation flows will exhibit *hysteresis* effects.

This completes a framework of terms and notions that should suffice to describe how the structural forms of three-dimensional separated flows originate and succeed each other. The following section is devoted to illustrations of the use of this framework in two examples involving supercritical and subcritical bifurcations.

EXAMPLES

Round-Nosed Body of Revolution at Angle of Attack

Let us first consider how a separated flow may originate on a slender round-nosed body of revolution as one of the main parameters of the problem, angle of attack, is increased from zero in increments. Focusing on the flow in the nose region alone, we adopt this example to illustrate a sequence of events in which *supercritical bifurcation* is the agent leading to the formation of large-scale dissipative structures.

At zero angle of attack (Figure 8a) the flow is everywhere attached. All skin-friction lines originate at the nodal point of attachment at the nose and, for a sufficiently smooth slender body, disappear into a nodal point of separation at the tail. The relevant topological rule, Equation (6), is satisfied in the simplest possible way ($N = 2, S = 0$).

At a very small angle of attack (Figure 8b) the topological structure of the pattern of skin-friction lines remains unaltered. All skin-friction lines again originate at a nodal point of attachment and disappear into a nodal point of separation. However, the favorable circumferential pressure gradient drives the skin-friction lines leeward where they tend to converge on the skin-friction line running along the leeward ray. Emanating from a node rather than a saddle point and being a line onto which other skin-friction lines converge, this particular line qualifies as a *local line of separation* according to our definition. The flow in the vicinity of the local line of separation provides a rather innocuous form of local flow separation, typical of the flows leaving surfaces near the symmetry planes of wakes.

As the angle of attack is increased further, a critical angle α_c is reached just beyond which the external flow becomes locally unstable. Coming into play here is the well-known susceptibility of *inflexional* boundary-layer velocity profiles to instability (Gregory et al. 1955, Stuart 1963, Tobak 1973). The inflexional profiles develop on crossflow planes that are slightly inclined from the plane normal to the external inviscid flow direction. Called a *crossflow* instability, the event is often a precursor of boundary-layer transition, typically occurring at Reynolds numbers just entering the transitional range (McDevitt & Mellenthin 1969, Adams 1971). Referring to the bifurcation diagrams of Figure 7 and identifying the parameter λ with angle of attack, we have that the instability occurs at the critical point α_c, where a *supercritical* bifurcation (Figure 7a) leads to a new stable mean flow. Within the local space influenced by the instability, the new mean flow contains an array of dissipative structures. The structures, illustrated schematically on Figure 8c, are initially of very small scale with spacing of the order of the boundary-layer thickness. Resembling an array

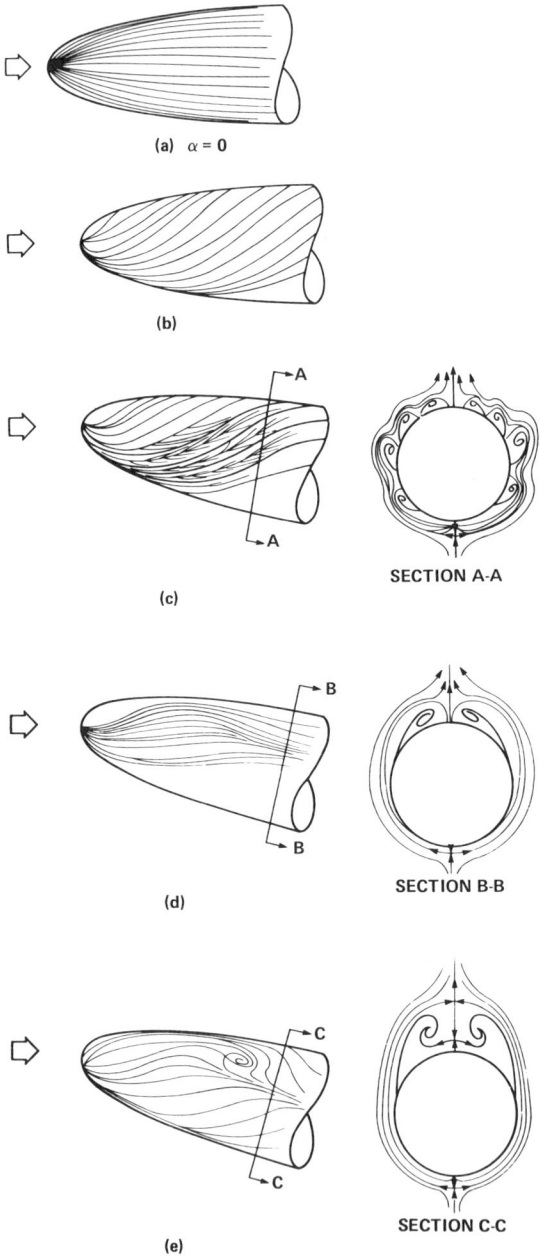

Figure 8 Sequence of flows leading to global three-dimensional flow separation on round-nosed body of revolution as angle of attack is increased.

of streamwise vortices having axes slightly skewed from the direction of the external flow, the structures will be called *vortical* structures. Although the representation of the structures on a crossflow plane in Figure 8c is intended to be merely schematic, nevertheless the sketch satisfies the topological rule for streamlines in a crossflow plane, Equation (8). As illustrated in the side view of Figure 8c, the array of vortical structures is reflected in the pattern of skin-friction lines by the appearance of a corresponding array of alternating lines of attachment and (local) separation. Because the bifurcation is supercritical, however, the vortical structures initially are of infinitesimal strength and cannot affect the topological structure of the pattern of skin-friction lines. Therefore, once again, these are *local* lines of separation, each of which leads to a locally separated flow that is initially of very small scale.

Although the vortical structures are initially all very small, they are not of equal strength, being immersed in a nonuniform crossflow. Viewed in a crossflow plane, the strength of the structures increases from zero starting from the windward ray, reaches a maximum near halfway around, and diminishes toward zero on the leeward ray. Recalling that the parameter ψ in Figure 7 was supposed to characterize the bifurcation flow, we find it convenient to let ψ designate the maximum crossflow velocity induced by the largest of the vortical structures. Thus, with further increase in angle of attack, ψ increases accordingly, as Figure 7a indicates. Physically, ψ increases because the dominant vortical structure captures the greater part of the oncoming flow feeding the structures, thereby growing while the nearby structures diminish and are drawn into the orbit of the dominant structure. Thus, as the angle of attack increases, the number of vortical structures near the dominant structure diminishes while the dominant structure grows rapidly. Meanwhile, with the increase in angle of attack, the flow in a region closer to the nose becomes subject to the crossflow instability and develops an array of small vortical structures similar to those that had developed further downstream at a lower angle of attack. The situation is illustrated on Figure 8d. We believe that this description is a true representation of the type of flow that Wang (1974, 1976) has characterized as an "open separation." We note that although the dominant vortical structure now appears to represent a full-fledged case of flow separation, nevertheless the surface shear-stress vector has remained structurally stable so that, in our terms, this is still a case of a *local* flow separation.

With further increase in the angle of attack, the crossflow instability in the region upstream of the dominant vortical structure prepares the way for the forward movement of the structure and its associated local line of separation. Eventually an angle of attack is reached at which the influence of the vortical structures is great enough to alter the structural stability of the surface shear-stress vector in the immediate vicinity of the nose. A new (unstable) singular point of second order appears at the origin of each of the local lines of separation. With a slight further increase in angle of attack, the unstable singular point splits into a pair of elementary singular points—a focus of separation and a saddle point. This combination produces the horn-type dividing surface described earlier (Figure 4) and illustrated again in Figure 8e (see also Figures 11 and 12 in Werlé 1979). We now have a *global* form of flow separation. A new stable mean flow has emerged from which additional flows ultimately will bifurcate with further increase of the angle of attack.

Asymmetric Vortex Breakdown on Slender Wing

In contrast to supercritical bifurcations, which are normally benign events, beginning as they must with the appearance of only infinitesimal dissipative structures, subcritical bifurcations may be drastic events, involving sudden and dramatic changes in flow structure. Although we are only beginning to appreciate the role of bifurcations in the study of separated flows, we can anticipate that sudden large-scale events, such as those involved in aircraft buffet and stall, will be describable in terms of subcritical bifurcations. Here we cite one example where it is already evident that a fluid dynamical phenomenon involving a subcritical bifurcation can significantly influence the aircraft's dynamical behavior. This is the case of asymmetric vortex breakdown which occurs with slender swept wings at high angles of attack.

We leave aside the vexing question of the mechanisms underlying vortex breakdown itself (see Hall 1972), as well as its topological structure, to focus on events subsequent to the breakdown of the wing's primary vortices. Lowson (1964) noted that when a slender delta wing was slowly pitched to a sufficiently large angle of attack with sideslip angle held fixed at zero, the breakdown of the pair of leading-edge vortices, which at lower angles had occurred symmetrically (i.e. side by side), became asymmetric, with the position of one vortex breakdown moving closer to the wing apex than the other. Which of the two possible asymmetric patterns was observed after any single pitchup was probabilistic, but once established, the relative positions of the two vortex breakdowns would persist over the wing even as the angle of attack was reduced to values at which the breakdowns had occurred initially downstream of the wing trailing edge. After identifying terms, we show that these observations are perfectly compatible with our previous description of a subcritical bifurcation (Figure 7b).

Let us denote by Δc the difference between the chordwise positions of the left-hand and right-hand vortex breakdowns and let Δc be positive when the left-hand breakdown position is the closer of the two to the wing apex. Referring now to the subcritical bifurcation diagram in Figure 7b, we identify the bifurcation parameter ψ with Δc and the parameter λ with angle of attack. We see that, in accordance with observations, there is a range of α, $\alpha < \alpha_c$, in which the vortex breakdown positions can coexist side by side, a stable state represented by $|\Delta c| = 0$. At the critical angle of attack α_c, the breakdowns can no longer sustain themselves side by side, so that for $\alpha > \alpha_c$, $|\Delta c| = 0$ is no longer a stable state. There being no adjacent bifurcation flows just beyond $\alpha = \alpha_c$, $|\Delta c|$ must jump to a distant branch of stable flows, which represents the sudden shift forward of one of the vortex breakdown positions. Further, with $|\Delta c|$ on the new branch, as the angle of attack is reduced $|\Delta c|$ does not return to zero at α_c but only after α has passed a smaller value α_0. All of this is in accordance with observations (Lowson 1964). At any angle of attack where $|\Delta c|$ can be nonzero under symmetric boundary conditions, the variation of Δc with sideslip or roll angle must necessarily be hysteretic. This also has been demonstrated experimentally (Elle 1961). Further, since Δc must be directly proportional to the rolling moment, the consequent hysteretic behavior of the rolling moment with sideslip or roll angle makes the aircraft susceptible to the dynamical phenomenon of wing-rock (Schiff et al. 1980).

SUMMARY

Holding strictly to the notion that patterns of skin-friction lines and external streamlines reflect the properties of continuous vector fields enables us to characterize the patterns on the surface and on particular projections of the flow (the crossflow plane, for example) by a restricted number of singular points (nodes, saddle points, and foci). It is useful to consider the restricted number of singular points and the topological rules that they obey as components of an organizing principle: a flow grammar whose finite number of elements can be combined in myriad ways to describe, understand, and connect together the properties common to all steady three-dimensional viscous flows. Introducing a distinction between local and global properties of the flow resolves an ambiguity in the proper definition of a three-dimensional separated flow. Adopting the notions of topological structure, structural stability, and bifurcation gives us a framework in which to describe how three-dimensional separated flows originate and how they succeed each other as the relevant parameters of the problem are varied.

ACKNOWLEDGMENTS

We are grateful to M. V. Morkovin for suggesting a useful thought experiment and to G. T. Chapman, whose contributions were instrumental framing an understanding of aerodynamic hysteresis in terms of bifurcation theory.

Literature Cited

Adams, J. C. Jr. 1971. Three-dimensional laminar boundary-layer analysis of upwash patterns and entrained vortex formation on sharp cones at angle of attack. *AEDC-TR-71-215*

Andronov, A. A., Leontovich, E. A., Gordon, I. I., Maier, A. G. 1971. *Theory of Bifurcations of Dynamic Systems on a Plane.* NASA TT F-556

Andronov, A. A., Leontovich, E. A., Gordon, I. I., Maier, A. G. 1973. *Qualitative Theory of Second-Order Dynamic Systems.* New York: Wiley

Benjamin, T. B. 1978. Bifurcation phenomena in steady flows of a viscous fluid. I, Theory. II, Experiments. *Proc. R. Soc. London Ser. A* 359:1–43

Davey, A. 1961. Boundary-layer flow at a saddle point of attachment. *J. Fluid Mech.* 10:593–610

Elle, B. J. 1961. An investigation at low speed of the flow near the apex of thin delta wings with sharp leading edges. *British ARC 19780 R & M 3176*

Gregory, N., Stuart, J. T., Walker, W. S. 1955. On the stability of three-dimensional boundary layers with application to the flow due to a rotating disc. *Philos. Trans. R. Soc. London Ser. A* 248:155–99

Hall, M. G. 1972. Vortex breakdown. *Ann. Rev. Fluid Mech.* 4:195–218

Han, T., Patel, V. C. 1979. Flow separation on a spheroid at incidence. *J. Fluid Mech.* 92:643–57

Hunt, J. C. R., Abell, C. J., Peterka, J. A., Woo, H. 1978. Kinematical studies of the flows around free or surface-mounted obstacles; applying topology to flow visualization. *J. Fluid Mech.* 86:179–200

Joseph, D. D. 1976. *Stability of Fluid Motions 1.* Berlin: Springer

Legendre, R. 1956. Séparation de l'écoulement laminaire tridimensionnel. *Rech. Aéro.* 54:3–8

Legendre, R. 1965. Lignes de courant d'un écoulement continu. *Rech. Aérosp.* 105:3–9

Legendre, R. 1972. La condition de Joukowski en écoulement tridimensionnel. *Rech. Aérosp.* 5:241–48

Legendre, R. 1977. Lignes de courant d'un écoulement permanent: décollement et séparation. *Rech. Aérosp.* 1977-6:327–55

Lighthill, M. J. 1963. Attachment and separation in three-dimensional flow. In *Laminar Boundary Layers*, ed. L. Rosenhead, II, 2.6:72–82. Oxford Univ. Press

Lowson, M. V. 1964. Some experiments with vortex breakdown. *J. R. Aero. Soc.* 68:343–46

Maltby, R. L. 1962. Flow visualization in wind tunnels using indicators. *AGARDograph No. 70*

McDevitt, J. B., Mellenthin, J. A. 1969. Upwash patterns on ablating and nonablating cones at hypersonic speeds. *NASA TN D-5346*

Nicolis, G., Prigogine, I. 1977. *Self-Organization in Nonequilibrium Systems.* New York: Wiley

Oswatitsch, K. 1980. The conditions for the separation of boundary layers. In *Contributions to the Development of Gasdynamics*, ed. W. Schneider, M. Platzer, pp. 6–18. Braunschweig/Wiesbaden: Vieweg

Peake, D. J., Tobak, M. 1980. Three-dimensional interactions and vortical flows with emphasis on high speeds. *AGARDograph No. 252*

Perry, A. E., Fairlie, B. D. 1974. Critical points in flow patterns. In *Advances in Geophysics*, 18B:299–315. New York: Academic

Poincaré, H. 1928. *Oeuvres de Henri Poincaré, Tome 1.* Paris: Gauthier-Villars

Sattinger, D. H. 1980. Bifurcation and symmetry breaking in applied mathematics. *Bull. (New Ser.) Am. Math. Soc.* 3:779–819

Schiff, L. B., Tobak, M., Malcolm, G. N. 1980. Mathematical modeling of the aerodynamics of high-angle-of-attack maneuvers. *AIAA Paper 80-1583-CP*

Smith, J. H. B. 1969. Remarks on the structure of conical flow. *RAE TR 69119*

Smith, J. H. B. 1975. A review of separation in steady, three-dimensional flow. *AGARD CP-168*

Stuart, J. T. 1963. Hydrodynamic stability. In *Laminar Boundary Layers*, ed. L. Rosenhead, IX:492–579. Oxford Univ. Press

Tobak, M. 1973. On local inflexional instability in boundary-layer flows. *Z. Angew. Math. Phys.* 24:330–54

Tobak, M., Peake, D. J. 1979. Topology of

two-dimensional and three-dimensional separated flows. *AIAA Paper 79-1480*

Wang, K. C. 1974. Boundary layer over a blunt body at high incidence with an open type of separation. *Proc. R. Soc. London Ser. A* 340:33–35

Wang, K. C. 1976. Separation of three-dimensional flow. In *Reviews in Viscous Flow, Proc. Lockheed-Georgia Co. Symp. LG 77ER0044*, pp. 341–414

Werlé, H. 1962. Separation on axisymmetrical bodies at low speed. *Rech. Aéro.* 90:3–14

Werlé, H. 1979. Tourbillons de corps fuselés aux incidences élevées. *L'Aero. L'Astro.*, no. 79, 1979–6

DYNAMICS OF GLACIERS AND LARGE ICE MASSES

Kolumban Hutter

Laboratory of Hydraulics, Hydrology and Glaciology, Federal Institute of Technology, Zürich, Switzerland

1. INTRODUCTION

In this article I discuss the fluid-mechanical aspects of that part of the world's ice which manifests itself as glaciers and ice sheets. Although different in their extent, glaciers and ice sheets have in common that they move under the influence of their own weight. Ice sheets spread in two spatial directions; glaciers are basically unidirectional and confined to a valley or a system of valleys (see Figures 1 and 2). On appropriate scales (intermediate, i.e. a few tens or hundreds of km, for ice sheets, global for glaciers) and for qualitative studies ice-sheet and glacier flow is planar. Plain strain is thus the restriction I concentrate on in this review.

Although glaciers are situated in the remoter parts of thinly populated areas their study is important, first, because their advance and retreat is an indicator of the variation of the climate; second, because they occasionally endanger valley inhabitants by ice avalanches and outbursts of ice-dammed lakes, and third, because they may someday be economically exploited as subglacial water catchments. The Arctic and Antarctic regions nowadays grow in significance mainly because they are exploited for their natural resources.

Figures 1 and 2 provide typical overviews of two Alpine glaciers, the *Grosser Aletschgletscher* and the *Fieschergletscher*. Both photographs illustrate the channelized nature of the flow. These glaciers are fed from the higher mountain regions by several "tributaries" and are in their lower portions confined to a single valley. The position of the snout varies according to the climate. The snout of the Fieschergletscher in Figure 2 is wedge-type, a form that is typical for a retreating glacier. By contrast, advancing glaciers have claw-like snouts (see Figure 10).

It is one purpose of glaciology to understand and describe how glaciers and ice sheets flow and in what sense their behavior can be related to that of the geophysical environment, the atmosphere, and the substratum the glacier is situated on. The ice motion is primarily due to the action of gravity, but the description of this motion is complicated by the fact that the thermal and mechanical aspects cannot be separated in general and, furthermore, the mechanical and thermal boundary conditions are often still not clearly understood. In addition, even if a sound theoretical description should exist, and numerical predictions of flow and temperature distributions could be made, such predictions would be complicated in reality by uncertainties in knowledge of basal topography, material properties, nourishment and wastage, geothermal and atmospheric heat, etc. Theoretical descriptions must bear in mind this limitation and are only useful as long as they can be subject to experimental verification.

Conceptually, glaciers and ice sheets are the same physical objects; they are usually grounded; frequently they extend into the ocean (or a lake). This floating portion is referred to as a shelf. Differences between glaciers and ice sheets can be attributed largely to size. Typical lengths and mean depths of glaciers are a few km and 100 to 300 m respectively. For ice sheets both length and depth are an order of magnitude larger; extents are of the order of up to thousands of km and depths vary from 100 m to more than 4000 m. Typical measured surface velocities on glaciers are of the order of 100 m per year, but these velocities can, on occasion, be very

Figure 1 View of the *Grosser Aletschgletscher*, the longest glacier in Switzerland. It is fed by several "tributaries" in the head region and then moves down a single valley. One can clearly see its sides and the moraines on the sides and in the middle. Courtesy of Dr. H. Röthlisberger, Laboratory of Hydraulics, Hydrology and Glaciology, ETH, Zürich.

much smaller or very much larger. In the central areas of Greenland, they are a few cm per year, in a glacier going through a surging phase they may be as large as several km per year [see Paterson (1969)]. The description of such a broad spectrum of velocities must depend upon a variety of parameters influencing the ice motion. It is theoretically desirable that one and only one mathematical model be capable of explaining all the pertinent velocity scales.

The ice motion is affected by the processes taking place within the ice mass, at the ice-rock or ice-water interface, and at the free surface between ice and atmosphere. Mathematical statements about these interactions yield the initial boundary-value problem governing motion and tempera-

Figure 2 A view of the *Fieschergletscher* in Switzerland, position of snout as of 14 July 1968. This photograph shows the typical channelized character of Alpine glaciers and it illustrates that the snout position may fluctuate with time. At the time of the shot the entire length was 16 km, but as can be seen from the photograph the foreground was once (40 years earlier) covered with ice. Courtesy M. Aellen, Laboratory of Hydraulics, Hydrology and Glaciology, ETH, Zürich.

ture evolution. Difficulties with the basic understanding of the glacier-flow problem arise on several levels. They are encountered in the formulation of the continuum approximation the physical phenomena should be based upon. Local phenomena both in space and time (e.g. stress concentrations, crevasse and crack formations, propagation of seismic waves) are ignored. But even when typical length and time scales are sufficiently large, the underlying physical model is a crude approximation to the true physical processes taking place. Even with the simplest physical model the emerging initial boundary-value problems led to complicated nonlinear equations. It is only within the last few years that rational solution procedures for the governing equations have been developed which make the dynamics of glaciers and large ice masses a rigorous branch of environmental fluid mechanics.

There are several books and earlier reviews on the subject. Lliboutry (1964, 1965) wrote the classic treatise in the French language giving thorough information on the state of the art at that time. Paterson (1969) gives a brief clearly written introduction to glacier and ice-sheet problems, very much suitable as a start. Shumskiy (1969) is similar in attempt but a bit more mathematical and more restrictive. Its advantage is again its brevity. Lliboutry (1971) reviews the physical aspects of glaciers. Colbeck (1980) contains review articles on several topics concerning snow and ice in the geophysical environment, two of which specifically deal with glaciers and ice sheets. The spirit of the book is physical and the information seems to be up to date. Hutter (1981c) analyzes the available literature and presents it as derived from basic principles.

2. FIELD EQUATIONS AND BOUNDARY CONDITIONS

2.1. Body Flow

Because the mechanical and thermal aspects of glacier and ice-sheet flows cannot be separated in general, a proper mathematical formulation must involve both mechanical and thermodynamic statements. The latter are complicated by the fact that glaciers and ice sheets are in general *polythermal,* i.e. they consist of two zones, "cold" and "temperate," in which the ice is respectively below and at the melting point. In the cold zone, heat generated by internal friction will affect the temperature distribution, and this, in turn, will influence the motion. In the temperate zone, on the other hand, frictional heat will melt some ice. Hence, whereas for cold glaciers a fluid model of a heat-conducting (non-Newtonian) viscous body may well be appropriate, such a model cannot be used for temperate ice, whose description should involve some notion of a binary mixture of ice with percolating or trapped water. In the cold zone, energy balance serves as an evolution equation for temperature; in the temperate zone it governs mass production of the constituent ice and water. The cold and temperate zones are separated by internal (non-material) surfaces that are capable of propagating at their own speed. Depending on the thermal conditions, these surfaces may be created or annihilated. Many Arctic and Antarctic ice sheets and glaciers are cold throughout, and most Alpine glaciers are thought to be temperate. Yet recent field investigations have provided convincing evidence that glaciers that were thought to be temperate often contain cold patches and that others, which are cold in their upper portions, may reach the melting point in their lower parts, or vice versa [see Vivian & Bocquet (1973), Müller (1976), Theakstone (1967)].

The formulation of the above-mentioned binary-mixture concept is still an open problem, but a reduced one-component model in which both temperature and water content are treated as state variables has been formulated (Fowler & Larson 1978). This model is a heat-conducting incompressible viscous fluid for which

$$\text{div } \mathbf{u} = 0,$$

$$\rho \dot{\mathbf{u}} = - \text{grad } p + \text{div } \mathbf{t}' + \mathbf{g}, \text{ and} \qquad (2.1)$$

$$\rho \dot{\epsilon} = \text{trace }(\mathbf{tD}) - \text{div } \mathbf{q}$$

are the local laws of balance of mass, linear momentum, and energy, respectively. In the above $\mathbf{u}$ is the velocity vector, ρ the density of ice, p the pressure, $\mathbf{t}'$ the (Cauchy) stress deviator, $\mathbf{g}$ the vector of external forces, ϵ the internal energy, $\mathbf{q}$ the heat flux, and $\mathbf{D}$ the stretching tensor, where

$$\mathbf{D} = \text{sym grad } \mathbf{u}. \qquad (2.2)$$

In *cold* ice, Equations (2.1) are complemented by constitutive relationships for the rate of internal heat, heat flux, and stress. These are usually postulated as

$$\rho \dot{\epsilon} = \rho c_p \dot{T}, \qquad \mathbf{q} = -\kappa \text{ grad } T. \qquad (2.3)$$

$$\mathbf{D} = A(T) f(t'_{II}) \mathbf{t}', \quad t'_{II} = \frac{1}{2} \text{ trace } \mathbf{t}'^2,$$

where c_p is the specific heat at constant pressure, κ the thermal conductivity, and T Kelvin temperature. Further, f is a creep-response function, assumed to depend on the second stress-deviator invariant t'_{II} and A, a rate factor that depends on the state variable (here temperature). There is considerable literature on the creep-response function f (see Glen 1955, Hawkes & Mellor 1972, Hobbs 1974, Mellor 1980, Michel 1979, Steinemann 1958). Classically, f is a power law $f(x^2) = |x|^{n-1}$, $n = 1.7 \to 4$, depending on stress range, and has in glaciology become known as Glen's flow law, but is better referred to as Norton's (1929) law in general. The modern trend, partly motivated by experimental evidence, partly by the mathematical difficulties connected with a creep-response function $f(x)$ with the property that $[f(x^2)x]' = 0$ at $x = 0$, tends towards a replacement of the simple power law. Such attempts are made by Meier (1960), Lliboutry (1969), Barnes et al. (1971), Colbeck & Evans (1973), Thompson (1979), Assur (1980), Hutter (1980a,b), and others. I shall leave f unspecified. Little is known about the rate factor A, except far from the melting point; the usual functional relationships are of the Arrhenius type (see Hobbs 1974),

$$A = \overline{A} \exp(-Q/k \cdot T), \qquad (2.4)$$

where Q is the activation energy (approximately 1 eV), k Boltzmann's constant, and $\overline{A}$ *a constant*.

We said that a proper description of temperate ice should use a binary-mixture concept. When postulating a one-component continuum one tacitly assumes the volume fraction of water to be small and to have negligible effect on the dynamics of the ice-flow processes. The balance laws (2.1) remain valid, but the constitutive relations change. First, $T = T_M$ where T_M is the melting temperature, now related to pressure; hence temperature is no longer a basic field variable and is replaced by a new state variable, the *moisture content w*, i.e. the volume content of water in the ice. Second, internal heating $\rho\dot{\epsilon}$ is given by the production of water mass and its associated heat production through latent heat: $\rho\dot{\epsilon} = \rho L \dot{w}$. Third, the rate factor A now depends on moisture rather than temperature. In summary, (2.3) are replaced by

$$\rho\dot{\epsilon} = \rho L \dot{w}, \qquad (2.5a)$$

$$\mathbf{q} = -\kappa \text{ grad } T_M \simeq 0, \qquad (2.5b)$$

$$\mathbf{D} = A(w) f(t'_{II}) \mathbf{t}', \qquad (2.5c)$$

$$(T_M - T_f) = -c_t(p - p^{\text{atm}}), \qquad (2.5d)$$

and have been complemented here by the (linearized) Clausius-Clapeyron equation. In these equations L is the latent heat of fusion per unit mass of ice, c_t is a constant, and T_f is the melting temperature at atmospheric pressure p^{atm}. A relationship $A(w)$ was first suggested by Lliboutry (1976), but nothing reliable is known quantitatively. To date, therefore, $A(w)$ is usually set constant.

Remarks (a) When the constitutive relationships (2.3) for cold ice and (2.5) for temperate ice are substituted, field equations are obtained for the unknown fields $\mathbf{u}$, p, T and $\mathbf{u}$, p, w, respectively. The structure of these equations is the same in both systems except for the energy equation, which reads

cold ice: $\rho c_p \dot{T} = \text{trace}(\mathbf{tD}) + \text{div}(\kappa \text{ grad } T)$, (2.6a)

temperate ice: $\rho L \dot{w} = \text{trace}(\mathbf{tD})$. (2.6b)

As an evolution equation for the state variable, the first of these equations is parabolic while the second is hyperbolic. Fowler & Larson (1978) have shown that this hyperbolicity excludes, for instance, the existence of temperate ice surrounded by cold ice. For this to be physically possible, a parabolic (or elliptic) equation is required for w. Adding the term div(δ grad w) with δ = constant, a diffusivity, on the right-hand side of (2.6b), will do and can be motivated by Fick's law, but needs a mixture concept to motivate it. (*b*) The constitutive relationships adopted above are generally accepted as appropriate in glaciology, but are subject to revision and improvement. First, assuming ice to be incompressible yet still using the pressure-melting relationship (2.5d) is inconsistent because the Clausius-Clapeyron equation requires compressibility for its existence (see Hutter 1981c). Equations (2.1)–(2.5) should thus be regarded as derived from a formulation of a compressible fluid as the isochoric limit is approached. Second, constitutive postulates for stress of the above form are motivated by creep tests in uniaxial compression or simple shear and by establishing a relationship between stress and strain rate at minimum creep rate (see Glen 1974). The only possible three-dimensional extension of these simple stress tests is $\mathbf{D} = A f(t'_{II}) \mathbf{t}'$, as originally proposed by Nye (1953), but, as shown by Morland (1979) and well known to continuum mechanicians, no f can be found in general that would fit given creep-curve data. For that purpose the constitutive relation of stress must be the more general isotropic functional relation (see Glen 1958),

$$\mathbf{D} = -\frac{2}{3} g(t'_{II}, t'_{III}, T) t'_{II} \mathbf{1} + f(-)\mathbf{t}' + g(-)\mathbf{t}'^2,$$ (2.7)

or its inverse (Morland 1979), or even more general relations involving rate-dependent terms (Morland & Spring 1981), in which **1** is the identity tensor and t'_{II} and t'_{III} denote the second and third invariants of the stress deviator $\mathbf{t}'$. I restrict myself in this review to the simpler form, as virtually no ice-flow problems have so far been analyzed with the more general constitutive relationship (2.7).

Equation (2.7) is a stress-deviator-stretching relationship for an (isotropic) fluid. For the polycrystalline ice of glaciers and ice sheets this is a fair approximation; because of high pressure and shearing at large depths ice may recrystallize thus causing stress-induced anisotropies. Very little is known about the process of recrystallization under these circumstances [see Lile (1978) for a laboratory experiment] and present-day glaciology is far from incorporating such processes into the continuum description.

2.2. Ice-Rock Interface

To the nonlinearities induced by non-Newtonian behavior, further complexity, not encountered in other branches of fluid mechanics, is added at the ice-rock interface. Depending on whether the glacier is cold or temperate, different boundary conditions apply at the base. Three types of mechanical boundary conditions have been proposed and each is used under different conditions. The *no-slip* condition applies at the cold portions of the bed; adhesive forces are so large as to prevent the ice from sliding over the rock, while basal stresses are moderate enough not to cause rupture. At the temperate portion of the base the heat generated by deformation causes melting; a layer of water separating ice sole and ice bed is formed. Often this layer is just a few microns thick and acts as a lubricant with negligible film thickness. Its effect is to change the boundary condition from no-slip to *perfect sliding*. When sufficient water is available, ice sole and rock bed may separate because the interstitial water pressure exceeds the ice-overburden pressure, causing the glacier to become partly afloat. Water-filled cavities are formed, reducing thereby the frictional resistance. Evidence for the existence of such cavities in times when they are not water filled is shown in Figures 3 and 4. Sliding *with* and *without cavity formation* characterize, therefore, two different types of boundary conditions at the temperate portion of the bed. Strictly speaking, there is a further possibility when the pressure at the base reaches the triple point

Figure 3 A subglacial caverne in the *Aletschgletscher* formed by the intraglacial water runoff. Courtesy of Dr. H. Röthlisberger, Laboratory of Hydraulics, Hydrology and Glaciology, ETH, Zürich.

so that ice, water, and vapor may coexist. This case is referred to as *sliding with cavitation* and occurs at pressures of 0.006 bars and an ice temperature of 0.0075°C (see Dorsey 1940).

On a global scale velocities and stresses in a glacier or ice sheet cannot depend on the fine details of the topography of the bed. These are confined to a boundary layer whose thickness is usually small compared with the glacier thickness (see Figure 5). The global (*outer*) flow may be determined accurately enough by considering flow past a smoothed-out surface. The outer flow then "feels" the inner flow as a viscous drag, and the inner flow "feels" the outer flow as a uniform shearing flow at infinity. Matching the two leads mathematically to a functional relationship between shear traction t_b and the velocity component u_b tangential to the smoothed bed,

$$u_b = F(t_b),\qquad(2.8)$$

in which the functional form of F depends on the roughness of the true bed, the creep-response function f, and the sliding conditions (with or without cavity formation). There is considerable literature on basal sliding, but hardly any of this work goes beyond the application of arguments of dimensional analysis unless restriction is made to Newtonian behavior (see Nye 1969b, 1970, Kamb 1970, Morland 1976a,b, Fowler 1977, 1979b, 1980b, Lliboutry 1979, Weertmann 1979, the latter two authors giving interesting contrasting reviews). To date, insufficient knowledge is avail-

Figure 4 A subglacial vault in a sliding portion of the steep hanging *Gietrogletscher*, date 30 August 1967. Courtesy of Dr. H. Röthlisberger, Laboratory of Hydraulics, Hydrology and Glaciology, ETH, Zürich.

able, but for sliding without cavity formation and plane strain a power law

$$u_b = Ct_b^m \quad \text{(no cavity formation)} \tag{2.9}$$

is suggested, where C and m depend on bed roughness and the creep-response function f. To be precise, (2.9) holds only when Glen's flow law $f(x^2) = |x^{n-1}|$ applies, and regelation arguments indicate that $m = \frac{1}{2}(n + 1)$ (Weertman 1957a) or $m = n$, as shown by a more rigorous argument (Fowler 1980b). For a general creep-response function f the functional form of $\mathbf{F}$ is unknown, but as remarked by Fowler (1980c) ". . . we can expect the sliding law $\mathbf{t}_b = \mathbf{F}^{-1}(\mathbf{u}_b)$ to be a well-behaved, increasing function of $\mathbf{u}_b$ as long as the ice maintains effective contact (separated only by a thin water film) with the bedrock: This concurs with the intuitive notion that the frictional resistance increases as the velocity increases. . . ." This situation persists as long as the base is lubricated, no reattachment occurs, and no cavities are formed. In this case the ice flow is not affected by the regelation film at least as far ". . . as the ice sole boundary is concerned; in other words, the ice flow/bedrock temperature sliding problem (as formulated by Nye 1969b, for example) uncouples from the determination of the regelation film thickness" (after Fowler 1980c, who derives an equation for it in 1980b). Film thicknesses are *a posteriori* data in this case, not restricted in any way; they may, in par-

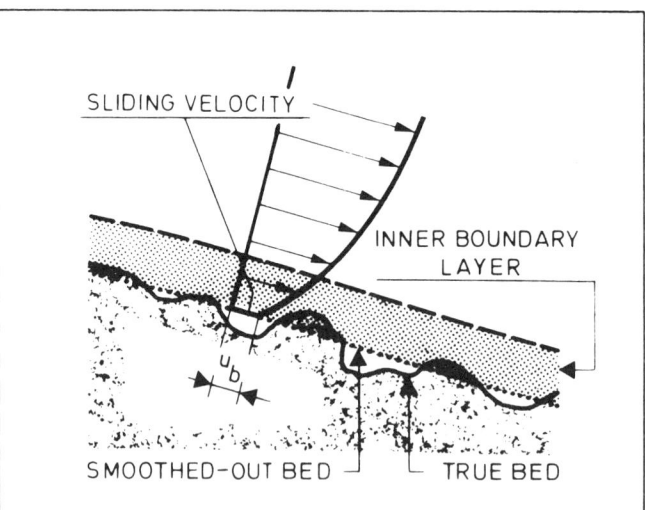

Figure 5 Ice flow far from the base cannot depend on the fine details of the topography of the base. Global flow can be described by introducing a smoothed-out bed and applying a viscous sliding law relating the sliding velocity u_b tangential to the smoothed-out bed to the shear traction t_b acting on the same smoothed-out bed. The undulating characteristics of the true bed about the smoothed-out bed are contained as roughness coefficient in the sliding law.

ticular, become negative (which is obviously contradictory) and may also grow indefinitely. In both cases the problem of ice flow/bedrock temperature sliding and the problem of the lubricating film cannot uncouple and both should be solved interactively. These inconsistencies were observed by Nye (1973), have been further followed up by Morris (1976, 1979), and are succinctly discussed by Fowler (1980c), but mathematical analyses treating the problem rationally are still lacking.

Nonetheless a thorough mathematical treatment of basal sliding with cavitation or cavity formation is urgent because extensive and less mathematical but nevertheless inspiring studies with great physical insight (Lliboutry 1968, 1976, 1978, 1979) have convincingly demonstrated that the sliding function $\mathbf{F}$ in (2.8) may become *multi-valued*. In other words for a given shear traction t_b two sliding velocities may exist and thus give rise to a bifurcation mechanism. The active phase of glacier surges, i.e. a sudden unexpected multiplication of flow velocities, would correspond to the high-velocity branch whereas the quiescent phase would be described by the low-velocity branch. In Hutter (1981b), I have explained how a multi-valued sliding function $\mathbf{F}$ gives rise to the cusp catastrophe (Poston & Stewart 1978). Multi-valuedness of $\mathbf{F}$ is, however, not necessary for $\mathbf{F}$ to cease being well behaved. It suffices that $\mathbf{F}'(x)$ become large for a certain domain of x so that sliding velocities change by several orders of magnitude when x changes only slightly (see Figure 6).

Further difficulty is added at the transition line between cold and temperate ice. On the cold side no-slip requires $\mathbf{F} \equiv \mathbf{0}$. On the temperate side

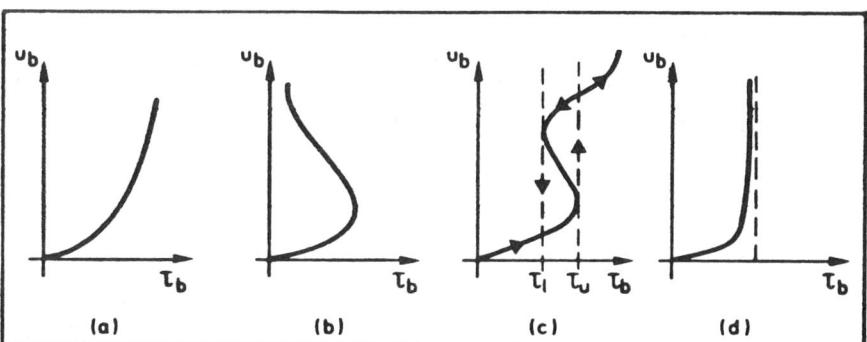

Figure 6 Four possible functional relationships describing the sliding law. In case (*a*) basal shear and sliding velocity are uniquely related. Sliding without cavity formation is firmly established to obey such functional characteristics. The cases (*b*) and (*c*) are somewhat speculative but allow for a bifurcation of a given flow configuration. In case (*d*) the sliding function remains single-valued but large derivatives of F cause sliding velocities to grow substantially when basal shear changes are only small. Cases (*c*), (*d*) give a possible explanation of sudden catastrophic glacier advance, as can be observed in seasonal waves and surges.

$F \neq 0$, giving rise to discontinuities in boundary data and associated stress concentrations. Fowler (1977) argues that for the outer flow $F \neq 0$ should be smoothly continued into the cold portion of the boundary to match $F \equiv 0$ at a point where a threshold temperature is attained. Below this temperature no-slip applies; above it sliding is continuously developed, reaching a maximal value at melting. The effects of a sliding function F that is continuous across the "cold-temperate" transition line are presently not clearly understood and results are somewhat controversial. Fowler & Larson (1978) make F temperature-dependent to account for the continuous transition and claim to prove that no stress and velocity singularities develop. Hutter & Olunloyo (1980, 1981) and Olunloyo & Hutter (1981, in preparation) on the other hand, show that for a Newtonian fluid and a sliding law (2.9) with position-dependent C, stress singularities will arise even for a C-function that is continuous but not differentiable at the "cold-temperate" transition line. Hence, while Fowler & Larson's sliding function F is made continuous by introducing a temperature dependency, that of Hutter & Olunloyo achieves the same by simply assuming F to be position-dependent. Granted that each author's claims are based on correct calculations, the different results imply that the form of F will crucially determine the stress and velocity distribution in the transition zone between cold and temperate ice. To date it seems plausible that F, which applies to the outer flow, should be such that the outer flow does not experience a singularity of any kind. For the boundary-layer flow, stress singularities must be expected, however, as follows from a flat-plate boundary-layer analysis of a power-law fluid (see Nachman & Callegari 1980).

The preceding discussion has concentrated on the traction condition. Clearly, if difficulties arise with it, similar complications should also be expected with the energy flux q through the basal boundary. The current regelation literature does not deal with this problem, and it has never been mentioned to my knowledge. The classical regelation theory (Nye 1967) contradicts observations (Drake & Shreve 1973) and needs rectification; and in a proper treatment of glacier sliding, energy flux through the base should be as carefully treated as is momentum flux. To date, the thermal boundary condition has been presented only for no-slip and for sliding without cavity formation. If we denote by q^{geoth} the geothermal heat and by $\mathbf{n}$ the unit normal vector pointing out of the ice, the thermal boundary condition reads

$$q^{geoth} - \mathbf{q} \cdot \mathbf{n} = \mathbf{t}_b \cdot \mathbf{u}_b. \tag{2.10}$$

This equation expresses that the heat added to the lubrication layer from above and below is used up within the layer by friction. The term $\mathbf{t}_b \cdot \mathbf{u}_b$ is the power of working of the shear traction $\mathbf{t}_b$ of the outer flow at the smoothed-out bed and $\mathbf{u}_b$ is the outer-flow sliding velocity. In general,

$t_b \cdot u_b$ should probably be replaced by a function $G(t_b \cdot u_b)$, with the property $G(0) = 0$, the form of which will depend on the bed roughness. In summary, the boundary conditions at the smoothed-out bed read

$$\left. \begin{array}{c} \mathbf{u}_b = \mathbf{F}(\mathbf{t}_b) \, , \\[2mm] u_n = u_{\text{drain}} = \dfrac{G(\mathbf{t}_b \cdot \mathbf{u}_b)}{\rho L} \, , \\[2mm] q^{\text{geoth}} - \mathbf{q} \cdot \mathbf{n} = G(\mathbf{t}_b \cdot \mathbf{u}_b) \, , \\[2mm] T = T_M \, , \text{ for temperate ice only} \, . \end{array} \right\} \begin{array}{c} \text{at the} \\ \text{smoothed-} \\ \text{out base} \end{array} \quad (2.11)$$

The second equation relates the velocity component u_n normal to the bed to a suction velocity u_{drain} that is usually set to zero, since $|u_n| \ll \|\mathbf{u}_b\|$. To relate it to the frictional heat generated in the regelation film means that all meltwater is instantly squeezed out. With appropriate interpretations for $\mathbf{F}$ and G the first two equations of (2.11) hold for cold as well as temperate ice. In temperate ice $\mathbf{q} \cdot \mathbf{n}$ is usually negligible compared with the geothermal heat so that $u_n = q^{\text{geoth}}/\rho L$ in this case. This approximation makes determination of the function G superfluous and explains why the problem of evaluating G has never arisen in the glaciological literature.

2.3. *Free Surface*

Boundary conditions at the free surface are better understood than those at the base; there is a kinematic, a dynamic, and a thermal boundary condition. The latter two are straightforward as they simply express continuity of the stresses and temperature (or occasionally heat flux). Hence,

$$\mathbf{t} \cdot \mathbf{n} = -p^{*\text{atm}} \mathbf{n} \, ,$$
$$T = T_S \, [\text{or } \mathbf{q} \cdot \mathbf{n} = q^{*\text{atm}}] \, , \quad (2.12a)$$

in which T_S denotes surface temperature, $q^{*\text{atm}}$ the given atmospheric energy input, and $p^{*\text{atm}}$ the atmospheric pressure. Shear traction due to wind is always negligible. To derive the kinematic surface condition let $S(s^\alpha, v, t) = 0$ ($\alpha = 1, 2$) be a parameterization of the free-surface geometry with coordinate lines s^α within and v orthogonal to the free surface; s^α and v are themselves functions of time. If the free surface were material, $dS/dt \equiv 0$ would form the kinematic surface condition. However, because of accumulation (nourishment due to snowing) and ablation (wastage due to surface melting) the free surface is non-material; whence,

$$\frac{dS}{dt} = a_\perp^* (s^\alpha, v, t) \, . \quad (2.12b)$$

Here $a^*_\perp$ is the accumulation/ablation function, supposedly known as a function of position. Realistically, it may depend on depth. It has the dimension of a velocity and is the rate of change of ice thickness measured perpendicular to the instantaneous surface. An order of magnitude is a few cm per year to m per year. Despite this small value $a^*_\perp$ is an important variable dominating the glacier and ice-sheet geometry (see below). When $S = \overline{S}(s^\alpha, t) - v$, then $dS/dt = \partial\overline{S}/\partial t + \partial\overline{S}/\partial s^\alpha \cdot \dot{s}^\alpha - \dot{v} = a^*_\perp$. This form of the kinematic surface condition shows that, in steady state and when $\dot{s}^\alpha = 0$ (no tangential velocities), $\dot{v} = -a^*_\perp$. Variations of physical quantities within the surface are usually *slow*, and transient effects are often ignored so that most applied glaciologists use this simplified surface condition. On the other hand, when Cartesian coordinates (x,y,z) are used, Equation (2.12b) is often written as

$$\frac{\partial \hat{y}}{\partial t} + \frac{\partial \hat{y}}{\partial x}u + \frac{\partial \hat{y}}{\partial z}w - v = a^*(x, y, z, t),$$

where $S(x,y,z,t) = \hat{y}(x,z,t) - y \equiv 0$ and (u,v,w) are the velocity components of surface particles in the (x,y,z)-directions; a^* now measures accumulation/ablation per unit area in (x,z)-planes. Physically, the regimes of positive and negative accumulation are separate, as $a^* > 0$ can only be obtained from precipitation measurements whereas $a^* < 0$ may follow from an energy budget at the free surface, since $\rho L a^* = \mathbf{q} \cdot \mathbf{n} - q^{\text{atm}}$ in this case, where $\mathbf{q} \cdot \mathbf{n}$ is usually negligible.

It should be noticed that the use of the kinematic surface condition in glaciology is very recent (Fowler & Larson 1978, Morland & Johnson 1980, Hutter 1980a, 1981a,c). Nonetheless the above presentation is still defective, as the free surface is a non-material surface of discontinuity for which jump discontinuities in the normal tractions $\mathbf{t} \cdot \mathbf{n}$ and in the heat flux arise. These jumps have always been assumed to vanish. In fact the idea of formulating the kinematic and dynamic surface conditions as jump conditions for the field variables, as is common for propagation of shock waves, has never arisen in the glaciological literature. Hutter (1981d) has recently done this.

2.4. "Cold-Temperate" Transition Surface and Ice-Water Interface

A complete description of polythermal glaciers must also include transition conditions at inner surfaces where cold ice reaches the melting point. At these surfaces it is natural to suppose that velocity, temperature, moisture content, stresses, and heat flux are continuous, but simultaneously evolution equations for the surface motion should be formulated. If $S_M(\mathbf{x}, t) = 0$ denotes this surface (M for melting point) then

$$\left.\begin{array}{c} \mathbf{u}, T, w, \\ \\ \mathbf{t}, \mathbf{q} \end{array}\right\} \text{ are continuous on } S_M(\mathbf{x}, t) = 0 \quad (2.13)$$

and, in particular,

$$\left.\begin{array}{c} w \equiv 0, \text{ and} \\ T - T_M = T - T_f \\ + c_t(p - p^{atm}) \equiv 0 \end{array}\right\} \text{ on } S_M(\mathbf{x}, t) = 0. \quad (2.14)$$

Moisture content vanishes and temperature equals melting temperature. The latter in turn is given by the Clausius-Clapeyron equation. Let D/Dt denote total time derivative following (virtual) surface particles. Then, since (2.14) are valid for all time,

$$\frac{DS_M}{Dt} \equiv 0, \frac{Dw}{Dt} \equiv 0, \frac{D(T - T_M)}{Dt} = 0, \text{ on } S_M(\mathbf{x}, t) = 0 \quad (2.15)$$

are three further conditions for the surface geometry and the velocity of propagation $\mathbf{u}_M$ of S_M ($\mathbf{u} \neq \mathbf{u}_M$). The conditions (2.13) and (2.14) were first given by Fowler & Larson (1978) and have subsequently been complemented with (2.15) to a full set of evolution equations for S_M by Hutter (1981c). To date, no problem has been attacked for polythermal ice in which the "cold-temperate" transition surface would have been determined.

The preceding transition conditions agree with those in the mentioned literature. However, they are *not* reasonable. In still ongoing research Dr. Straughan and I are able to show in a special but important case that with $\mathbf{q}$ being continuous on $S_M = 0$ a cold-temperate transition surface would not exist. In order simply to allow for the existence of such surfaces at least the heat flow across the surface must jump. The requirement that $\mathbf{q}$ is continuous in (2.13) must, consequently, be replaced by

$$\mathbf{q} \cdot \mathbf{n}|_{\text{temp}} - \mathbf{q} \cdot \mathbf{n}|_{\text{cold}} = \lambda,$$

where λ is a *surface energy production*. In Hutter (1981c) and in the study of Straughan and Hutter λ is treated as a constitutive quantity, which must be determined by suitable experiments on propagating cold-temperate transition surfaces. In the more careful study, modeling temperate ice as a binary mixture, Hutter (1981d) proves that with $w = 0$ on $S_M = 0$ this is the only possibility, if a determinate system of equations should evolve.

The state of the art is not much different with regard to the boundary conditions at an ice-water interface. The parameter governing the geometry of the ice-water interface is the melting and freezing rate, which must be related to the energy budget at the interface. If $S_w(\mathbf{x}, t) \equiv 0$ is the equation of the interface, then $dS_w/dt = a_w^\perp$ is the kinematic surface equation. Continuity of traction, on the other hand, requires $\mathbf{t} \cdot \mathbf{n} = -p_w \cdot \mathbf{n}$, where p_w is the pressure exerted by the water; shear tractions induced by the water boundary layer are generally and justly neglected (see Hutter 1981c). The thermal boundary condition is best derived by considering the layer of water frozen to the shelf from below per unit time (see Figure 7). The energy released by freezing within this layer, $a_w^\perp \rho L$, must equal the heat removed from the layer by heat flux into the ice and the ocean, $q_w^* - \mathbf{q} \cdot \mathbf{n}$, where $\mathbf{q}$ is the heat-flux vector on the ice side whereas q_w^* is its normal component on the water side. The quantity q_w^* is treated here as a known function. Relations for it would follow from basic studies of heat transfer in the oceanic boundary layer. Detailed analysis not being available, the best we can do is to write $q_w^* = h(T - T_w)$, where h is the heat-transfer coefficient that is generally given as a relation between Nusselt, Prandtl, and Reynolds number (see Schlichting 1979). In summary we may thus write

$$\left. \begin{aligned} \frac{dS_w}{dt} &= a_w^\perp , & &\text{(2.16a)} \\ \mathbf{t} \cdot \mathbf{n} &= -p_w^* \mathbf{n} , & &\text{(2.16b)} \\ a_w^\perp &= h(T_w^* - T) - \mathbf{q} \cdot \mathbf{n} , & &\text{(2.16c)} \end{aligned} \right\} \text{ on } S_w(\mathbf{x}, t) = 0 .$$

Depending on the value of T on $S_w = 0$ Equation (2.16c) assumes three different forms. For freezing, $a_w^\perp > 0$, $T = T_w^*$, and the heat-transfer

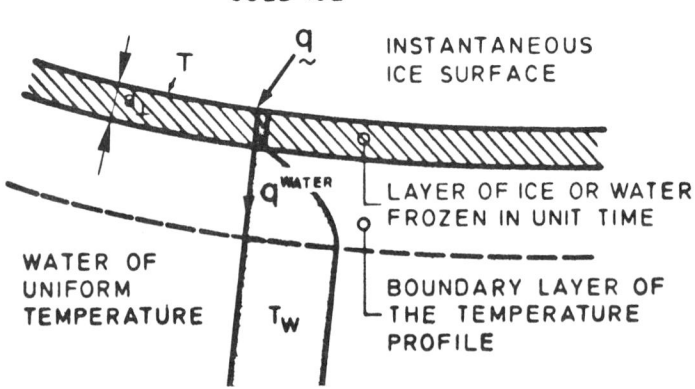

Figure 7 Illustration of the derivation of the thermal boundary condition at the ice-water interface.

on the right-hand side of (2.16) vanishes; on the other hand, when ice is melting, $a_w^\perp < 0$, then $T = T_M$ and all terms in (2.16c) survive. For $T_w < T < T_M$ neither melting nor freezing occurs, whence $a_w^\perp = 0$. This shows that either one knows T at the interface and must then determine $a_w^\perp$, or else $a_w^\perp(=0)$ is known and T must be determined.

Boundary conditions (2.16) are used in analyses of ice-shelf responses but have only scarcely been used. Hutter & Williams (1980) following work of Weertman (1957b) relate ice-shelf thickness to thermal conditions of the environment and F. M. Williams (personal communication) gives the corresponding stability analysis. Another interesting analysis of ice shelves, based however, on other boundary conditions, is given by Shumskiy & Krass (1976).

A final remark concerns the boundary condition for moisture content. Above, w was prescribed only on the "cold-temperate" transition surface. This is in line with the hyperbolicity of (2.6b) and the further assumption that ice flow is from the cold to the temperate ice. When moisture is governed by a parabolic equation, boundary conditions must be prescribed on all boundary surfaces. At the base this would give rise to the prescription of moisture flux (drainage of water through the ground) and a similar statement would govern the moisture flux at the free surface. Using a mixture concept for temperate ice Hutter (1981d) was able to derive these conditions from the general jump conditions, which must hold at surfaces of discontinuities. Replacing (2.6b) by the parabolic equation

$$\rho L \dot{w} = - \operatorname{div} \mathbf{j} + \operatorname{trace} \mathbf{t'D}, \quad \mathbf{j} = -v \operatorname{grad} w, \tag{2.17}$$

Hutter's study shows that the boundary conditions are

$$\mathbf{j} \cdot \mathbf{n} + \rho L w a_\perp^* = j_{\text{ext}}^*, \quad \text{on } S(\mathbf{x}, t) = 0,$$

$$\mathbf{q} \cdot \mathbf{n}|_{\text{temp}} - \mathbf{q} \cdot \mathbf{n}|_{\text{cold}} = \lambda, \quad \text{on } S_M(\mathbf{x}, t) = 0,$$

$$\mathbf{j} \cdot \mathbf{n} - \rho L(1 - w)a_w^\perp = 0, \quad \text{on } S_w(\mathbf{x}, t) = 0,$$

$$\mathbf{j} \cdot \mathbf{n} = j_{\text{drain}}^* \quad \text{on } S_B(\mathbf{x}) = 0. \tag{2.18}$$

Consequently, on the free surface, $S = 0$, moisture content and its flux are related to the ablation rate and the external water flux j_{ext}^* into the ice. On the cold-temperate transition surface, $S_M = 0$, the jump in heat flux is given by λ, a quantity describing the melting rate. On the ice-water interface, $S_w = 0$, moisture content and its flux are related to the melting rate, and on the base, $S_B = 0$, moisture flux equals drainage of water through the base. A description of polythermal ice including water motion through the ice necessitates knowledge of the diffusivity v (Darcy's constant) and the constant λ, values for both of which have not been determined yet. Once these are known, solution of a boundary-value problem expediates prescription of $a_\perp^*$, j_{ext}^*, and j_{drain}^*.

3. PLANE FLOW–NONDIMENSIONAL EQUATIONS

3.1. *Scaling of the Equations*

In the remainder of this review I focus attention mainly on plane flow. Cartesian coordinates (x, y) will be used; x is downglacier and is inclined to the horizontal by the angle γ; y is perpendicular to x and roughly transverse to the main flow direction. The gravity constant will be denoted by g.

The non-dimensionalization involves the finding of typical dimensionless quantities to serve as scaling factors for the various dimensional variables in the problem. Lengths are made dimensionless with a typical thickness D of the glacier, stresses with the overburden pressure $\rho g D$ at zero inclination, and velocities and accumulation rate with a typical longitudinal surface velocity U. Denoting by T_f and T_0 the melting temperature at atmospheric pressure and a typical ice temperature range within the glacier, we introduce dimensionless variables as follows:

$$(x, y) = D(\hat{x}, \hat{y}), \quad t = \frac{D}{U}\hat{t},$$

$$(t_{xx}, t_{yy}, t_{xy}, p) = \rho g D (\sigma_x, \sigma_y, \tau, \hat{p}),$$

$$(u, v, a) = U(\hat{u}, \hat{v}, \hat{a}), \quad T = T_f + T_0 \theta, \quad (3.1)$$

Here Greek and circumflexed quantities are nondimensional; $\hat{y}$, σ_x, σ_y, $\hat{p}$, and $\hat{u}$ are of order unity or smaller, but $\hat{x}$ and $\hat{t}$ may become large, and $\hat{v}$ and $\hat{a}$ are small. This observation is important as it will suggest the stretching transformation introduced below. If we delete circumflexes, the field equations for cold ice assume the form

$$\frac{\partial u}{\partial x} + \frac{\partial v}{\partial y} = 0, \quad (3.2a)$$

$$\mathbb{F}\frac{du}{dt} = \frac{\partial \sigma_x}{\partial x} + \frac{\partial \tau}{\partial y} + \sin\gamma, \quad (3.2b)$$

$$\mathbb{F}\frac{du}{dt} = \frac{\partial \tau}{\partial x} + \frac{\partial \sigma_y}{\partial y} - \cos\gamma, \quad (3.2c)$$

$$\frac{\partial v}{\partial x} + \frac{\partial u}{\partial y} = 2\,\mathbb{G}\,\exp(\mathbb{A}\Theta)\,f(\tau_{\mathrm{II}})\,\tau, \quad (3.2e)$$

$$\frac{\partial \theta}{\partial t} + \frac{\partial \theta}{\partial x}u + \frac{\partial \theta}{\partial y}v = \mathbb{D}\left(\frac{\partial^2 \theta}{\partial x^2} + \frac{\partial^2 \theta}{\partial y^2}\right)$$

$$+ \mathbb{E}\left\{\frac{1}{2}(\sigma_x - \sigma_y)\left(\frac{\partial u}{\partial y} - \frac{\partial v}{\partial x}\right) + \tau\left(\frac{\partial u}{\partial y} + \frac{\partial v}{\partial x}\right)\right\}, \quad (3.2f)$$

where

$$\tau_{\text{II}} = \frac{1}{4}(\sigma_x - \sigma_y)^2 + \tau^2, \; \Theta = \frac{1+\mathbb{Z}}{1+\mathbb{Z}\theta}\theta \qquad (3.3)$$

are the dimensionless second stress-deviator invariant and a temperature measure, called coldness;

$$f(\tau_{\text{II}}) = \frac{f(\rho^2 g^2 D^2 \tau_{\text{II}})}{f(\rho^2 g^2 D^2)} \qquad (3.4)$$

is the dimensionless creep-rate function, and σ is total stress. All variables are dimensionless and hollow quantities are characteristic parameters defined as

$$\mathbb{F} = \frac{U^2}{gD} \lesssim 10^{-14},$$

$$\mathbb{G} = \frac{D}{U} A \exp\left(-\frac{Q}{RT_f}\right) \rho g D f(\rho^2 g^2 D^2) \simeq 10^2$$

[and smaller for steep glaciers $\gtrsim O(1)$]

$$\mathbb{A} = \frac{Q}{RT_f} \frac{\mathbb{Z}}{1+\mathbb{Z}} \simeq 2,$$

$$\mathbb{Z} = \frac{T_0}{T_f} \sim (0.05 - 0.1),$$

$$\mathbb{D} = \frac{\kappa}{\rho c D U} \gtrsim 10^{-3},$$

$$\mathbb{E} = \frac{gD}{cT_0} \lesssim 10^{-1}, \quad \text{and}$$

$$\mathbb{C} = \frac{C(\rho g D)^m}{U} \lesssim 10^0 \quad \text{(see below)}. \qquad (3.5)$$

$\mathbb{F}$, $\mathbb{D}$, and $\mathbb{E}$ are Froude, thermal-diffusion, and energy-dissipation numbers. $\mathbb{A}$, $\mathbb{G}$, and $\mathbb{Z}$ characterize the constitutive behavior. Orders of magnitudes for these are obtained by substituting numerical values of the phenomenological constants for ice (see Paterson 1969) and selecting characteristic values for U and D. When $U = 10^2$ m per year and $D = 100$–500 m, orders of magnitudes are as listed in (3.5). [A careful discussion of the nondimensionalization is given by Hutter (1981c)]. Accordingly, $\mathbb{F}$ is extremely small. Thus "Stokesian" flow conditions apply and accelerations may be ignored. $\mathbb{A}$ is of order unity and $\mathbb{D} < \mathbb{E}$, but both are small; (3.2f) then shows that shear heating (the term involving $\mathbb{E}$) is more important than heat conduction. Neglecting the latter causes difficulties, however, as the equation would change from parabolic to hyperbolic. Matched asymptotic expansions will necessarily be involved (Grigoryan et al. 1976, Shumskiy & Krass 1976). $\mathbb{G}$ is usually large; but solutions in the asymptotic limit $\mathbb{G} \to \infty$ are too special, so that with respect to $\mathbb{G}$ a singular perturbation must be performed. For many problems $\mathbb{G}$ is, however, regarded as finite and incorported in the velocities by rescaling velocities and time according to $(u, v, t) \to (u/\mathbb{G}, v/\mathbb{G}, \mathbb{G}t)$. This corresponds to formally setting $\mathbb{G} = 1$ in (3.2); scaled velocities are now no longer $O(1)$ but $O(\mathbb{G}^{-1})$ (see Hutter 1981c, Hutter et al. 1981).

In the following analysis I restrict attention to grounded glaciers that are either cold or temperate throughout. In the dimensionless coordinate system (x, y) the boundary conditions, therefore, read:

At the base:

$$\left. \begin{array}{l} u_b = \mathcal{F}(\tau_b) \ [= \mathbb{C} \, \tau_b^m] \, , \\[4pt] \qquad \mathcal{F}(\tau_b) = \dfrac{F(\rho \, g \, D \, \tau_b)}{U} \, , \\[6pt] u_n \simeq 0 \, , \\[4pt] -\dfrac{\partial \theta}{\partial n} = q^{\text{geoth}} \, , \ \text{cold ice} \, , \\[8pt] T = T_M \, , \ \text{temperate ice} \, , \end{array} \right\} \ \text{on } y = y_B(x) \, , \qquad (3.6)$$

at the free surface:

$$\left. \begin{array}{l} \boldsymbol{\sigma} \cdot \mathbf{n} = -p^{*\text{atm}} \, , \\[4pt] \theta = \theta_S \, , \ \text{cold ice} \, , \\[4pt] \theta = \theta_M \, , \ \text{temperate ice} \, , \\[4pt] \dfrac{\partial y_S}{\partial t} + \dfrac{\partial y_S}{\partial x} u - v = a^* \, , \end{array} \right\} \ \text{at } y = y_S(x, t) \, , \quad \begin{array}{l}(3.7\text{a}) \\[4pt] (3.7\text{b}) \\[4pt] (3.7\text{c}) \\[4pt] (3.7\text{d})\end{array}$$

in which $\partial/\partial n$ is the derivative perpendicular to the respective surface.

Equations (3.2) have been written down for cold ice, but with little change they also apply for an approximate description of temperate ice. In view of the insufficient knowledge of the rate factor $A(w)$ in the constitutive relation for stresses, Equation (2.5c), A is often assumed as independent of the moisture content w. In (3.2) this corresponds to setting $\mathbb{A} = 0$. The mechanical field equations thus uncouple from the evolution equation of the state variable (temperature or moisture content). Since, according to (3.6) and (3.7), the mechanical boundary conditions are also uncoupled from those of the state variable the latter can be regarded as an a posteriori quantity.

3.2. Simple Solutions

Before presenting the stretchings referred to above we seek exact solutions of (3.2), (3.6), and (3.7) in the limit $\mathbb{F} = 0$. This is the Stokesian approximation and extremely accurate so that leading-order solutions need not be perturbed. The strictly parallel-sided infinitely long ice slab is a configuration with some practical application. Steady conditions are discussed below and time-dependent processes are also briefly treated.

ALL FIELDS ARE x-INDEPENDENT If stresses, velocities, and temperature are independent of the longitudinal coordinate, $\partial/\partial x = 0$, then the first five equations of (3.2), together with the corresponding mechanical boundary conditions (3.6) and (3.7) at the free surface $y = H$ and the base $y = 0$, can be integrated to yield

$$\tau = \sin \gamma \, (H - y), \tag{3.8a}$$

$$\sigma_x = \sigma_y = - \cos \gamma \, (H - y) - p^{\text{atm}}, \tag{3.8b}$$

$$u = \mathcal{F} \, (\sin \gamma \, H)$$
$$+ 2 \, \mathbb{G} \int_0^y \exp(\mathbb{A} \, \Theta) \, f \, (\sin \gamma^2 \, (\xi - H)^2) \tag{3.8c}$$
$$\sin \gamma \, (H - \xi) \, \partial \xi,$$

$$v = 0,$$

whereas for the temperature θ the following two-point boundary-value problem is obtained:

$$\left.\begin{aligned}
& \theta'' + \frac{2 \, \mathbb{E}}{\mathbb{D}} \exp(\mathbb{A} \, \Theta) \, f \, (\sin \gamma^2 \, (y - H)^2) \\
& \sin \gamma \, (y - H)^2 = 0, \\
& \theta = \theta_S, \text{ at } y = H, \frac{d\theta}{dy} = - q^{\text{geoth}} \text{ at } y = 0,
\end{aligned}\right\} \text{cold ice} \tag{3.9}$$

in which primes denote derivatives with respect to y. (3.8) is compatible with a steady flat surface at $y = H$ only when accumulation vanishes. For temperate ice one has $\mathbb{A} = 0$, so (3.8) uncouples from (3.9); for cold ice (3.9) must be solved first and the resulting transverse temperature distribution be used to evalute the coldness before the quadrature to obtain longitudinal velocities can be performed.

Remarks (a) The stress distribution (3.8a,b) is materially independent and simply follows by exploring equilibrium and associated boundary conditions. (b) Shear traction at the base is, from (3.8a), $\tau_b = H \sin\gamma \sim H\gamma$ for small γ, thus proportional to glacier thickness and surface inclination γ (if the latter differs from that of the bed). This formula is one of the "glaciologist's commandments." (c) As a rule of thumb, ice can be regarded as an ideally plastic body with yield stress 10^5 Pa. Setting basal shear stress equal to this value and measuring surface inclination will yield thickness. Qualitatively, results are reasonable but need improvement. (*d*) Longitudinal velocity consists of two contributions, one due to sliding the other due to gliding. For cold ice $\mathcal{F} = 0$; otherwise, often both contributions are of comparable magnitude. In a surging stage sliding overrides gliding so that the motion is almost rigid. (*e*) The nonlinear two-point boundary-value problem (3.9) for temperature was first solved by Yuen & Schubert (1979). They found that for most cases strain heating has a marginal effect on the temperature distribution. Temperature is then mostly linearly distributed; this contradicts observation, see Figure 8 below. When strain heating is accounted for, using H (the dimensionless thickness) as input and integrating (3.8c), (3.9) yields the surface velocity $u(H)$ as a multi-valued function of H. In such a range, to a given value of H, there correspond at least two values for $u(H)$. Physically, to a given depth and temperature distribution there can only be one single solution. This suggests that one solution branch may become unstable, while the other branch is stable. Such a bifurcation of the velocity distribution is referred to as a thermal runaway instability. Yuen & Schubert looked at the stability of the multi-valued steady-state solutions by performing a small-perturbation time-dependent harmonic analysis, but no unstable branches were detected. This points at an imperfection in the basic formulation and, indeed, the authors argue that a finite-amplitude analysis may eventually yield an unstable branch. Fowler (1980a), on the other hand, has criticized the author's approach because accumulation (which is zero here) rather than glacier thickness is a true input parameter. Fowler's argument is that when accumulation is an input parameter the multi-valuedness may disappear and the detected stability of Yuen & Schubert is no longer a surprise. (*f*) As follows from (3.7d) steady straight surfaces require $a^* = 0$ when $v = 0$ or else there must be transverse motion. This motion gives rise to convective heat transport, which changes temperature distributions drastically. This was recognized by Robin (1955).

EXTENDING AND COMPRESSING FLOW, VELOCITIES x-DEPENDENT Solutions for a strictly parallel-sided slab in which stresses and temperature are x-independent, but velocities are not, go back to Nye (1957). Stresses are again given by (3.8a,b), but temperature and velocities can be shown to be determinable from the equations (see Hutter 1981c):

$$u = \frac{a^*}{H} x + h(y), \qquad v = -\frac{a^*}{H} y, \qquad (3.10a)$$

$$(3.10b)$$

$$h(y) = 2 \, \mathbb{G} \int_0^y f(\tau_{II}) \exp(\mathbb{A} \, \Theta) \sin \gamma (H - \xi) \, d\xi, \qquad (3.10c)$$

$$\left(\frac{a^*}{H}\right)^2 + \exp(2 \, \mathbb{A} \, \Theta) f^2(\tau_{II}) [\sin \gamma^2 (H-y)^2 - \tau_{II}] = 0, \qquad (3.10d)$$

$$D\,\theta'' - v\,\theta' + 2\,\mathbb{E}\exp(\mathbb{A}\,\Theta)\,f(\tau_{11})\,\tau_{11} = 0,\quad\quad\quad (3.10e)$$

$$\theta = \theta_S,\text{ at }y = H,\,\theta' = -q^{\text{geoth}},\text{ at }y = 0,\quad\quad\quad (3.10f)$$

with the consistency condition that a^* is constant. Velocity components vary linearly with x and y, respectively, with constant stretching a^*/H. The flow is therefore extending (compressing) when $a^* > 0$ ($a^* < 0$). In the accumulating area, ice is squeezed; in the ablating area it is stretched. Qualitatively this is corroborated by field observations. Transverse variations of longitudinal velocities are similar to those obtained before, but now depend on a^* through (3.10d). Similarly, accumulation rate has entered the temperature problem through both a y-dependent convective term and strain heating, as τ_{11} depends on a^*.

Remarks (*a*) The velocity field (3.10a) contradicts boundary conditions at the base in general since $u(0) = a^* x/H = \mathcal{F}(\sin\gamma H)$, which is impossible, because $\mathcal{F}$ is x-independent. Nye's solution of extending and compressing flow is inconsistent with the assumption of a stress-dependent sliding law. For over 20 years glaciologists have not been bothered by this. (*b*) At an ice-water interface drag exerted on the ice by the water is zero or negligible. Hence, there is no basal friction and the inconsistency due to the sliding law can not arise. For an unconfined ice shelf with $\gamma = 0$, a rationale identical to the above one yields a stress, velocity, and temperature problem that is free of internal inconsistencies. The mechanical part of this problem was presented by Weertman (1957b); the corresponding thermal problems are approximately treated by Shumskiy & Krass (1976) and Hutter & Williams (1980). (*c*) If we neglect strain heating, the temperature problem is linear and can be integrated straightforwardly. We can then express temperature in terms of exponential functions

$$\theta = \theta_S + q^{\text{geoth}}\int_y^H \exp\left(-\text{sgn}(a^*)\frac{1}{2}\sqrt{\frac{|a^*|}{H\mathbb{D}}}\,\xi^2\right)d\xi.\quad\quad\quad (3.11)$$

For $a^* > 0$ this solution is due to Robin (1955); the solution for $a^* < 0$ is given by Lliboutry (1963). Temperature distributions based on (3.11) are monotone as shown in Figure 8*b,c*, in fair agreement with field observations. Results that lead to better agreement with field data were obtained by adding to v a suction velocity to account for basal melting so that $v = -a^*y/H + v_{\text{drain}}$ (Weertman 1968; see Figure 8*b*). Temperature inversions as shown in Figure 8*d* are still not predictable. Such inversions can be attributed to longitudinal heat advection (see Robin 1955, Weertman 1968) and non-steady behavior, all effects ignored in (3.10). (*d*) Strain heating has been approximately incorporated into the temperature problem by Clarke et al. (1977) by substituting reasonable guesses for τ_{11}. When $\tau_{11} = \tau^2$ a slab with dominant shearing is treated, while $\tau_{11} =$

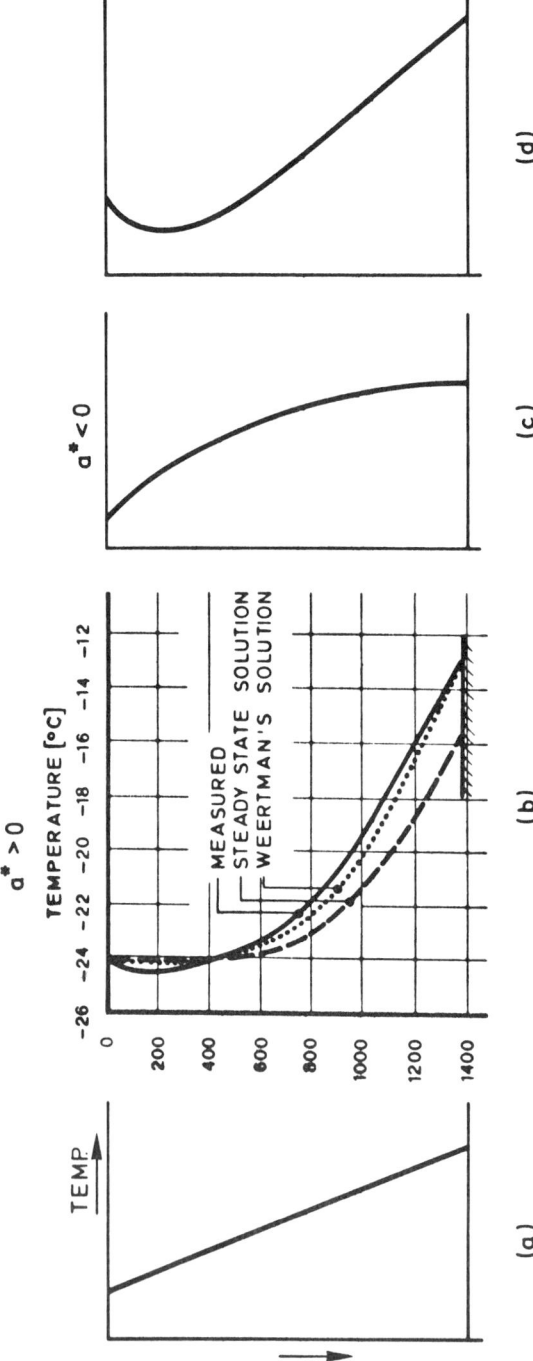

Figure 8 (*a*) In a strictly parallel-sided cold ice slab subject to zero accumulation, the temperature is nearly linearly distributed with $T_u < T_L < T_M$, where T_M is the melting temperature described by the Clausius-Clapeyron equation. (*b*) Transverse convective heat transport changes the temperature distribution drastically; compare with (*a*). When $a^* > 0$ then temperature growth with depth is overlinear. Theoretical curves are monotonic. (*c*) Same as for (*b*), but $a^* < 0$. Temperature growth is now underlinear with depth and also monotonic. (*d*) Typical observed temperature curve with an inversion. A measured curve of this type is shown also in (*b*).

constant approximates a region of dominant longitudinal stretching. With such approximations the linear distribution of transverse velocities v can no longer be regarded as an accurate result but as an approximation, which however proves to be a reasonable one (see Morland & Johnson, 1980). The analysis of Clarke et al. extends Yuen & Schubert's treatment of thermal instability to flows with temperature distributions as shown in Figures 8b,c and indicates that thermal conditions may indeed exist which render one branch of the steady-flow configurations unstable.

FURTHER RESULTS FOR PARALLEL-SIDED ICE SLABS In order to remove the inconsistency referred to in Remark (a) and, moreover, to obtain information regarding small deviations from the strictly parallel-sided slab, all fields must vary with x. Calculations now become much more complicated and cannot even be hinted at in this review. One central question is how undulations of the base with amplitudes that are small compared with the glacier thickness affect the steady-state surface topography. Such a nearly parallel-sided slab allows investigation of longitudinal stretching effects and of the influence of the basal geometry on that of the surface. Corresponding research was initiated by Budd (1968, 1970, 1971), but alternative calculations by Hutter et al. (1981) imply that Budd's approximation is not generally valid and even partly erroneous. The results of Hutter et al. indicate that in a steady state, bottom-protuberance transfer to the surface depends chiefly on the sliding law and on protuberance wavelength, yet it is also influenced by the inclination angle of the ice slope. There are situations in which transfer of bottom undulations to the surface is larger the larger the protuberance wavelength, whereas for other conditions transfer is optimal for a finite wavelength of about 3 to 5 times the glacier thickness (see Figure 9). Preferred wavelengths of this order have, for instance, been observed in the Wilkes Ice Cap in Antarctica by Budd (1966). Further results concern the stress distribution, which can no longer be given by Equations (3.8) but must reflect a constitutive dependence induced by the undulating boundaries. In Hutter et al. (1981) these corrections of formulas (3.8) are mainly analyzed for temperate ice slabs. In Hutter & Spring (1979) corresponding results are presented for cold ice. Even though it is slightly oversimplified, it can nevertheless be said that the temperature distribution has a sizeable effect on the stress correction. This correction is much more pronounced than the corresponding effect manifested in the velocity distribution or in the transfer function of bed undulations to the top surface. In other words stresses react critically to temperature variations. The approach of the analysis of Hutter et al. is in the spirit of small-perturbation airfoil theory—the undulation amplitude acts as perturbation parameter and boundary conditions at the true bed and true surface are replaced by approximate conditions on the corresponding flat mean surfaces—and it fails when protuberance wavelengths are in the order of the slab thickness and when inclination angles γ become small. This indicates that Equations (3.2) need to be rescaled.

Another problem in which small-amplitude perturbation theory has been used is a linear stability analysis of surface perturbations that are superposed on the steady motion given by (3.8). Such an analysis was given by Thompson (1979). Its purpose was to search for conditions of instability as possible explanations for surging motion. One speculation is that a glacier surge might be a phenomenon akin to the laminar-turbulent transition of ordinary hydrodynamics. Yet, whereas laminar-turbulent transition is caused by advection, these terms are negligibly small in glaciers, so possible instabilities must be due to rheological nonlinearities, transverse

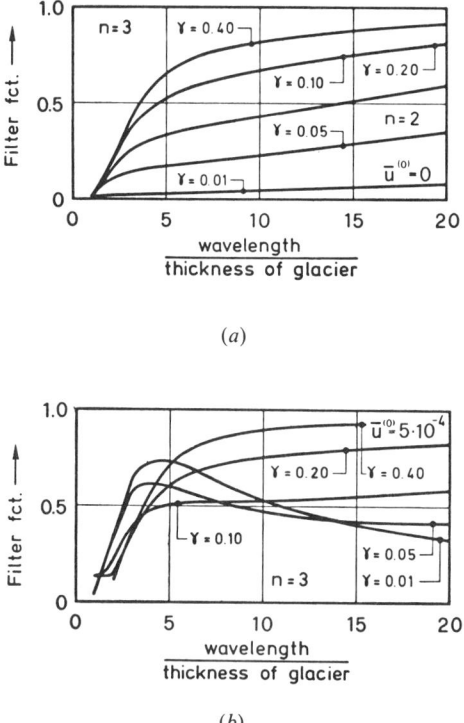

Figure 9 The effect of bottom undulations on the steady-state flow of parallel-sided slab manifests itself in surface undulations and longitudinal stress variations. (*a*) The ratio of surface amplitude to bottom amplitude is a measure of how much of the transfer bottom protuberances is filtered out at the top. Here it is plotted against dimensionless protuberance wavelength and parameterized for various inclinations γ of the ice slope. No sliding at bottom and uniform temperature. (*b*) Same as (*a*) but allowing for sliding without cavity formation.

temperature variations, longitudinal stretching gradients, etc. No instabilities were detected, however.

Whereas Thompson in his study aimed at flow-induced instabilities of temperate ice, Yuen & Schubert (1979) looked at the stability of a cold ice slab in the range of thermal conditions where multi-valued steady flow conditions exist. Their linear stability analysis did not allow detection of instabilities, and the authors muse that a nonlinear large-amplitude analysis might change this. On the other hand, Clarke et al. (1977) in their approximate treatment of thermal conditions in cold ice do find unstable steady-state conditions. As mentioned before, these analyses are criticized by Fowler (1980a). Rectifications of the problem are given by Fowler & Larson (1980a,c). However, these rectifications are only partly satisfactory. Fowler argues that Yuen & Schubert and Clarke et al. keep the glacier thickness constant, whereas it should be determined in due course with the solution of a problem, in which accumulation serves as input parameter. Fowler and Larson demonstrate that for their choice of accumulation $u(H)$ is single-valued in this case. This does not exclude the possibility of multi-valuedness for other situations. Furthermore, slab problems are realistic models for creep flow in other applications, and for these the above flow stability problem remains also unsolved.

Further studies with parallel-sided ice slabs concern the sliding problem and local stress analyses. Mathematically, such problems are usually so complex that Newtonian behavior is assumed, which formally makes the analysis linear and thus amenable to solution. Qualitatively good physical insight can be obtained thereby but quantitatively the results must be used with some care. For instance, the linear analyses of the sliding problem by Nye (1969b, 1970), Kamb (1970), and Morland (1976a,b) show that the sliding velocity is composed of two parts: one contribution to sliding can be traced to regelation in the absence of viscous deformation, the other is attributable to creep flow around the protuberances with no regelation. The linear analyses indicate that both are equally responsible for the total sliding velocity, so the two contributions are additive. A careful nonlinear analysis (Fowler 1980b), on the other hand, shows that the contribution due to regelation is usually negligible. As a consequence the sliding law without cavity formation yields for the exponent m of (2.8) the value $m = n$ rather than $m = \frac{1}{2}(n + 1)$, which obtains when regelation velocities are important.

3.3. Stretching Transformation

Slab problems are central-zone analyses, not completed to the margins. Validity of solutions is necessarily local. Lengths of glaciers are finite, and when processes with scales of these dimensions are considered, variations of glacier depth can no longer be treated as higher-order corrections.

Simplifications of the governing equations must be invoked; these are suggested by the following facts: First, $\mathbb{F}$ is small, hence, as was assumed already above, $\mathbb{F} = 0$. Second, $\mathbb{G}$ may be large so that solutions in the asymptotic limit $\mathbb{G} \to \infty$ could be searched for. An easy calculation involving (3.2), (3.6), and (3.7) then shows that for a flat bed with inclination γ the only permissible solution is a static ice reservoir with horizontal surface resting against an embankment (Morland & Johnson 1980). Third, surface and bottom geometry in glaciers varies slowly in general, suggesting a stretching of coordinates.

When velocities and accumulation rate are scaled with a representative *longitudinal* surface velocity, dimensionless longitudinal velocities are of order unity, but dimensionless *transverse* velocity components and accumulation rate are small. *Incorporating slow variation along x*, the following stretching transformations are therefore suggested:

$$\xi = \epsilon x, \; \bar{t} = \epsilon t, \; v = \epsilon V, \; a^* = \epsilon \bar{a}^*, \; 0 < \epsilon \ll 1, \tag{3.12}$$

with yet unspecified parameter ϵ. When (3.12) is incorporated into (3.2), the equilibrium equations and stress-deviator-stretching relationships become

$$\epsilon \frac{\partial \sigma_x}{\partial \xi} + \frac{\partial \tau}{\partial y} + \sin \gamma = 0, \tag{3.13a}$$

$$\epsilon \frac{\partial \tau}{\partial \xi} + \frac{\partial \sigma_y}{\partial y} - \cos \gamma = 0, \tag{3.13b}$$

$$\epsilon \frac{\partial u}{\partial \xi} = \frac{1}{2} \mathbb{G} \exp(\mathbb{A}\Theta) \, f(\tau_{II}) (\sigma_x - \sigma_y), \tag{3.13c}$$

$$\frac{\partial u}{\partial y} + \epsilon^2 \frac{\partial V}{\partial \xi} = 2 \mathbb{G} \exp(\mathbb{A}\Theta) \, f(\tau_{II}) \, \tau, \tag{3.13d}$$

whereas the temperature problem assumes the form

$$\frac{\partial \theta}{\partial \bar{t}} + u \frac{\partial \theta}{\partial \xi} + V \frac{\partial \theta}{\partial y} = \frac{\mathbb{D}}{\epsilon} \frac{\partial^2 \theta}{\partial y^2}$$

$$+ \frac{\mathbb{E}}{2\epsilon} f^2(\tau_{II}) \exp(2\mathbb{A}\Theta) \, \tau_{II} + \epsilon \mathbb{D} \frac{\partial^2 \theta}{\partial \xi^2},$$

$$\theta = \theta_S, \quad \text{at } y = y_S(\xi, \bar{t}), \tag{3.14}$$

$$\frac{\partial \theta}{\partial y} - \epsilon^2 \frac{\partial \theta}{\partial \xi} \frac{dy_B}{d\xi} = -q^{\text{geoth}} \left[1 + \frac{\epsilon^2}{2} \left(\frac{dy_B}{d\xi} \right)^2 \right],$$

at $y = y_B(\xi)$.

Two scaling factors have now entered the equations. By balancing the physically important terms one can be determined as a function of the other, and when this is done perturbation solutions can be constructed.

Two cases can obviously be distinguished: (a) $\sin \gamma \simeq O(1)$, i.e. steep glaciers, and (b) $\sin \gamma = O(\epsilon)$ or smaller, i.e. ice sheets and glaciers on a horizontal or weakly sloping bed.

BED INCLINATION $O(1)$ To lowest order, transverse stress gradients balance gravity, while longitudinal stress gradients are small. This means that the last two terms on the left-hand side of (3.13a,b) balance each other, and integration is immediate. The corresponding solution for the leading-order stresses agrees with (3.8a) with H replaced by $y_S(\xi,t)$. Hence, $\tau = O(1)$, so that (3.13d) implies, since $|\partial u/\partial y| = O(1)$, that $\mathbb{G} = O(1)$. I shall set $\mathbb{G} = 1$. Equation (3.13c) then implies $\sigma_x = \sigma_y$, in agreement with (3.8). Now that functional relationships for the stresses have been found these can be substituted into (3.13d), revealing an integrable relationship for the leading-order longitudinal velocity u. Use of the continuity equation $\partial u/\partial \xi + \partial V/\partial y = 0$ and a further integration then yields an expression for the leading-order transverse velocity. The integrations subject to the corresponding boundary conditions are easy to perform and the results read

$$\tau = \sin \gamma \, (y_S - y), \qquad (3.15a)$$

$$\sigma_x = \sigma_y = -\cos \gamma \, (y_S - y) - p^{\text{atm}}, \qquad (3.15b)$$

$$u = \mathcal{F} \left[\sin \gamma \, (y_S - y_B) \right]$$

$$+ 2 \int_{y_B}^{y} \exp(\mathbb{A}\Theta) f \left(\sin \gamma^2 (y_S - \bar{y})^2 \right) \sin \gamma \, (y_S - \bar{y}) \, d\bar{y}, \qquad (3.15c)$$

with an expression for V that is too complex to write down here (see Hutter 1981a,c). Notice that since τ does not change sign and because $\mathcal{F}(\tau) \gtreqless 0$ if $\tau \gtreqless 0$ the sliding velocity cannot change sign. Thus *there is no ice divide*. For that to be possible a different scaling is needed. In form the results (3.15) agree with (3.8a,b,c), the stress and velocity distribution of the strictly parallel-sided ice slab; but in (3.15) the surface geometry is not yet determined and neither is the temperature field θ. But on the assumption that surface profiles and temperature distributions are measured or estimated, the stress and velocity fields can be calculated by mere quadratures. Mathematically these results suggest the following solution procedure for the thermomechanical problem. For temperature, an iterative procedure is chosen whereas for stresses and velocities a perturbation scheme using ϵ as perturbation parameter is selected. This allows determination of perturbation series for stresses and velocities independently of temperature. Formulas involve quadratures and can be exploited as soon as the surface geometry is known. Once stress and velocity series are carried up to a certain order the temperature field can be iterated on in a second step by solving the two-point boundary-value problem (3.14). A further calculation of the stress and velocity fields based on the new temperature iterate will improve on these fields, etc. This process must at each iteration step also involve determination of the surface geometry. It is believed that with the exception of thermal runaway instabilities this iteration procedure is sound and converges rapidly.

Remarks (a) The above solution procedure has the advantage of being applicable equally to cold and temperate ice, as the stress problem is solved first and the temperature problem in a second step which for temperate ice can simply be deleted. (b) Because for finite bed inclination $\mathbb{G} = O(1)$ the small parameter ϵ is not related to $\mathbb{G}$ and must be chosen independently from the assumption of slow variation of the fields. Hutter (1981a) suggests choosing ϵ as a ratio of a representative thickness to a length over which fields vary by $O(1)$. A similar choice is also suggested by Fowler & Larson (1978). The perturbation scheme based on $\mathbb{G} = O(1)$ and $\epsilon \ll 1$ is in this form referred to by these authors as *the shallow-ice approximation;* it is the same as the shallow-water approximation. (c) To lowest order, the temperature problem is meaningless, but when $O(\epsilon^1)$ terms are included the temperature problem (3.14) includes heat conduction, strain heating, and transverse and longitudinal advection. Only at the $O(\epsilon^1)$ level do longitudinal advective terms enter, indicating that temperature inversions (see Figure 8) can only be obtained in steady state, when the stress-velocity problem is carried up to and including first-order terms. This has been done by Hutter (1981a,c), although the lowest-order problem is independently treated also by Fowler (1977, 1980c), Fowler & Larson (1978), and Morland & Johnson (1980, 1982). (d) Equation (3.15a) relates shear stress to depth and mean inclination angle γ. Extending the perturbation solution to higher-order terms in ϵ shows that when terms of $O(\epsilon)$ are included

$$\begin{aligned}\tau &= \left(\sin \gamma - \epsilon \cos \gamma \frac{\partial y_S}{\partial \xi}\right)(y_S - y) \\ &= \left(\sin \gamma - \cos \gamma \frac{\partial y_S}{\partial x}\right)(y_S - y)\end{aligned} \quad (3.16)$$

is obtained. For small inclination angles the first term is the surface inclination angle relative to the horizontal, α_S, say; hence in this case $\tau = \alpha_S(y_S - y)$, which is the classical formula well known to glaciologists. Pursuing the calculations to $O(\epsilon^2)$ allows us to improve on this formula (Hutter 1981a). The corresponding effects are known as longitudinal strain-rate effects on basal shear. Their determination has a long history (Lliboutry 1965, Robin 1967, Collins 1968, Nye 1969a) culminating in the simple perturbation scheme of the shallow-ice approximation. (*e*) Mathematically, the perturbation method suggested by the shallow-ice approximation is not without difficulties, difficulties that were already encountered in previous regular perturbation schemes but were left unmentioned there. They arise with a creep response function $f(x)$ which vanishes at $x = 0$ as is the case, for instance, for a power law. It corresponds to infinite apparent viscosity at zero stress. Since f enters denominators at higher stress and velocity formulas, the corresponding quantities may become unbounded whenever $f = 0$. This happens at the free surface and makes perturbation solutions invalid there. The problem can be rectified by replacing the simple power law by a "finite-viscosity law." One possibility is to replace $f(x^2) = |x|^{n-1}, n > 1$, by $f(x^2) = (|x|^{n-1} + k)/(1 + k)$ with k small. This eliminates the singularities mentioned above, but does not prevent higher-order stresses and/or velocities from becoming large. Asymptotic methods are therefore needed with respect to the parameter k. Matched asymptotic expansions with a boundary layer at the free surface are (most likely) inappropriate as no boundary conditions have to be reactivated. Multiple variable scales are suggested, yet the problem is still not resolved and warrants further study.

There still remains the presentation of a prediction equation for $y_S(\xi, \bar{\tau})$. An equation follows from the kinematic surface condition written in the stretched variables,

$$\frac{\partial y_S}{\partial \bar{\tau}} + \frac{\partial y_S}{\partial \xi} u - V = \bar{a}^*, \text{ at } y = y_S(\xi, \bar{\tau}). \tag{3.17}$$

By integrating the continuity equation $\partial u / \partial \xi + \partial V / \partial y = 0$ between $y = y_B$ and $y = y_S$ and incorporating the bottom sliding condition, Equation (3.17) can easily be shown to transform to (see Fowler 1980c, Hutter 1981c)

$$\frac{\partial y_S}{\partial \bar{\tau}} + \frac{\partial}{\partial \xi} \int_{y_B}^{y_S} u(\xi, y, \bar{\tau}) \, dy = \bar{a}^*, \text{ at } y = y_S(\xi, \bar{\tau}), \tag{3.18}$$

which has the form of the kinematic wave equation (see Whitham 1974). The integral in the second term is usually denoted by $Q(y_S, \xi, t)$ and characterizes the flux. In the classical kinematic wave theory its functional form is a constitutive statement. On such information the kinematic wave theory as developed by Lighthill & Whitham (1955) was applied to a linearized description of the response of a glacier to nourishment and wastage by Nye (1960, 1963a,b). He found that surface bulges travel downglacier at speeds that are roughly four times the particle speed, and this result was corroborated in a nonlinear analysis by Fowler & Larson (1980a). The stretchings (3.12) have led us into a much more privileged situation than Nye, for u in (3.18) is known in terms of a perturbation series in ϵ [of which the lowest-order term is given in (3.15c)].

Equation (3.18) serves as the governing equation for surface geometry for time-dependent and time-independent problems. Both have only recently been analyzed. Morland & Johnson (1980, 1982) treat the problem of steady-state geometry. Fowler (1980c) and Fowler & Larson (1980b) analyze waves on glaciers. Here we focus attention on temperate ice and leading-order, i.e. $O(\epsilon^0)$, terms only. Substituting (3.15c) into (3.18) yields (Hutter 1981c)

$$\frac{\partial H}{\partial \bar{t}} + \left\{ \left[\mathcal{F}'(\sin \gamma\, H) \sin \gamma\, H + \mathcal{F}(\sin \gamma\, H) \right] \right.$$

$$\left. + 2 \sin \gamma\, H^2\, f(\sin \gamma^2\, H^2) \right\} \frac{\partial H}{\partial \xi} = \bar{a}^*, \qquad (3.19)$$

or (3.18) with Q, $\quad Q = \mathcal{F}(\sin \gamma\, H)\, H$

$$+ 2 \int_0^H f(\sin \gamma^2\, y^2) \sin \gamma y^2\, dy,$$

where $H = y_S - y_B$. Notice that $f(0) = 0$ but $\mathcal{F}(0)$ may have any value.

Steady-state surface profile When $\partial H/\partial \bar{t} = 0$ steady-state surface profiles are obtained. They follow from (3.19) by a forward integration in ξ subject to the boundary condition $D = 0$ at $\xi = \xi_M^-$, where ξ_M^- denotes the left margin. The extent of the glacier then follows from the second boundary condition $H = 0$, at $\xi = \xi_M^+ > \xi_M^-$, where ξ_M^+ is defined as the position where H vanishes for the first time. This integration process would be routine were it not for the reasons stated in the following remarks.

Remarks (a) For a sliding law with $\mathcal{F}(0) = 0$ (the usual case!) the term in curly bracket in (3.19) vanishes with H, implying that the profile equation (3.19) may be singular. The behavior near the margin of the solution of the steady-state version of (3.19) can be analyzed by looking at power-series solutions of the form $H = k_d(\xi - \xi_M)^\delta$ valid when $\xi \to \xi_M$. Substituting such an "ansatz" into the differential equation will yield values for k_d and δ which depend on the near-margin behavior of the sliding function $\mathcal{F}$ and of the accumulation rate function $\bar{a}^*$. The result is that, in general, $\delta < 1$, so that $H' \to \infty$ as $\xi \to \xi_M$. This invalidates the assumption of slow variation and implies that solutions of (3.19) can, in this case, only hold in an outer region far from the margin, which must be matched with a margin solution of the full equations. Figure 10 shows a closeup of an actively advancing glacier, indicating that the small-slope assumption may indeed be violated in the snout region. (b) By postulating a sliding function $\mathcal{F}$ that does not necessarily vanish with H, (3.19) can be regularized and integration commenced at the margin. For instance, when Weertman-type sliding applies, $\mathcal{F} = \mathbb{C}(\xi) \sin \gamma^m H^m$, we must have $\mathbb{C} \to \mathbb{C}_M |\xi - \xi_M|^{-m}$, $\mathbb{C}_M$ = constant, as $\xi \to \xi_M$. With this choice a margin solution for H has the form $H = k_d |\xi - \xi_M|$ where $k_d \neq 0$, so $H' \neq \infty$ at $\xi = \xi_M$; to be sure,

$$H \simeq \text{sgn}(\bar{a}_M^*) \left\{ \frac{|\bar{a}_M^*|}{(m+1)\, \mathbb{C}_M \sin^m \gamma} \right\}^{\frac{1}{m+1}} |\xi - \xi_M|, \quad \bar{a}^* \neq 0, \qquad (3.20)$$

where $\bar{a}_M^* = \bar{a}^*(\xi_M)$. Integration can now be commenced at the margin, but for this purpose $\bar{a}_M^*$ and $\mathbb{C}_M$ must be known. (c) Physically $\mathbb{C} = 0$ means no-slip and $\mathbb{C} \to \infty$ perfect sliding. Perfect sliding of the ice at the margin is therefore necessary to make solutions (3.19) valid there. Values for $\mathbb{C}_M$ can be found by measuring snout ablation and snout velocities, for (3.15c) implies a relation between u, $\bar{a}_M^*$, and $\mathbb{C}_M$. For practitioners this provides a clue as to the order of magnitude for $\mathbb{C}_M$ (Hutter 1981c). (d) Equation (3.19) is obtained from an $O(\epsilon^0)$-analysis. When $O(\epsilon)$ terms are included, a diffusion term, i.e. a term involving $\partial^2 H/\partial \xi^2$, arises (see Fowler & Larson 1980b, Hutter 1981c). Such terms are known to smooth out processes and will, perhaps, make corresponding solutions for H uniformly valid. This

Figure 10 Snout of the advancing *Trientgletscher* on 11 June 1971. Since 1958 this glacier was constantly advancing except in 1960/61 when it was retreating. The photograph shows the typical claw-like form of the snout of actively advancing glaciers, with locally steep slopes and with many crevasses in the snout region. Courtesy of Prof. P. Kasser, Laboratory of Hydraulics, Hydrology and Glaciology, ETH, Zürich.

does not happen, however, unless $\mathcal{F}(0) \neq 0$ at the margin, as before (Hutter 1981c).

Waves on glaciers Waves on glaciers manifest themselves in three typical forms, namely, as surface waves, seasonal waves, and surges. Surface waves are undulations of the glacier surface that travel downglacier at speeds of (typically) three to four times the surface-particle speed. Early analyses are by Blümcke & Finsterwalder (1905) and Finsterwalder (1907), more recent ones by Weertman (1958), Nye (1960, 1963a,b), Lick (1970), and Lliboutry (1971). This work is all based on the kinematic wave theory; its basic equation is (3.18), but, unlike what was demonstrated above, the flux term Q [the second term on the left-hand side of (3.18)] is not obtained from a consistent application of momentum balance. These early authors specify the flux in an ad hoc manner, and some linearize the governing equations (Nye). A careful recent study of surface waves was made by Fowler & Larson (1978, 1980b). Seasonal waves manifest themselves as fluctuations of surface velocities rather than surface bulges, which travel at speeds in the range of 20 to 150 times the surface velocity. The first thorough analysis of seasonal waves was done by Fowler (1980c). Surges, finally, are a relaxation-type oscillation of a glacier consisting of a long (30 to 100 years) passive phase, during which time dependent processes are manifested by surface waves, and a phase of short duration (typically one to two years) which has seasonal-wave character. Meier & Post (1960) give a good qualitative description of surges and ad hoc models have been presented by Budd (1975), Budd & McInnes (1974), Robin (1969), and Robin & Weertman (1973).

One understanding of surface and seasonal waves on glaciers is that the former arise when the sliding function $\mathcal{F}$ is well behaved (which is the case for sliding without cavity formation) while seasonal waves are triggered by a sliding function that may become multi-valued or whose derivative may be large in a small range of its argument. This view is held by Lliboutry and Fowler, and it implies that Equation (3.19) is appropriate for surface waves but needs to be rescaled for seasonal waves. Both types of waves are expertly discussed by Fowler & Larson (1980b) and Fowler (1980c).

For surface waves (3.19) applies, and can be integrated as soon as boundary and initial conditions are prescribed. The boundary condition is not simply $H = 0$ at $\xi = \xi_M^-$, since ξ_M varies with time; it can be *derived* from a ξ-integral of (3.19), namely

$$\frac{\partial}{\partial t} \int_{\xi_M^-}^{\xi} H(\sigma, \bar{t}) \, d\sigma + Q(\xi, \bar{t}) = \int_{\xi_M^-}^{\xi} \bar{a}^* \, d\sigma . \qquad (3.21)$$

Evaluating this at $\xi = \xi_M^-$ yields

$$\left. \begin{array}{ll} H \dot{\xi}_M^- - Q = 0, \dot{\xi}_M^- > 0 \\ H \phantom{\dot{\xi}_M^- - Q} = 0, \dot{\xi}_M^- < 0 \end{array} \right\} \text{ at } \xi = \xi_M^- \quad (3.22)$$

with Q as in (3.19). In a steady state (3.22) reduces to $H(\xi_M^-) = 0$. The initial condition, on the other hand, reads

$$H(\xi, 0) = H_0(\xi), \text{ for } \xi_M^-(0) \leq \xi \leq \xi_M^+(0). \quad (3.23)$$

Equations (3.19), (3.22), and (3.23) constitute the complete initial-boundary-value problem for surface-wave phenomena. For $\mathcal{F} \equiv 0$ and $f(x^2) = |x|^{n-1}$ the problem was solved by Fowler & Larson (1980b), using the method of characteristics. Further details including a more general sliding function $\mathcal{F}$ are given by Fowler (1980c). Their calculations are summarized in the following remarks.

Remarks (*a*) The surface wave speed is given by the term in curly brackets in (3.19) or by dQ/dH, while surface-particle speeds follow from (3.15c). Using a Weertman-type sliding law and a power-flow law the two surface speeds have been correlated and bounds been found for the surface-wave speed. From such a correlation it follows that when $\mathcal{F}$ is well behaved, wave speeds are likely to be three to four times the surface-particle speeds.
(*b*) Integration of (3.15), (3.22), (3.23) along characteristics shows that smooth solutions do exist up-glacier, i.e. from the head downwards, but that generally shocks are formed before the snout is reached. The shock fronts $\xi_S(t)$ travel at the speed

$$\frac{d\xi_S}{dt} = \frac{[Q]}{[H]},$$

where $[Q]$ and $[H]$ are the jumps of Q and H across the shock, i.e. $[Q] = Q_1 - Q_2$, before and behind the shock. Stable shocks are those where H decreases across a shock as ξ increases. The analysis then parallels usual procedures of weak solutions of hyperbolic equations that result from conservation laws (Dafermos 1974, Courant & Friedrichs 1948).
(*c*) Carrying the perturbation solution to $O(\epsilon)$ alters Equation (3.19) to a *convection-diffusion equation,* i.e. a term in ϵ involving $\partial^2 H/\partial \xi^2$ enters. It must, therefore, be expected that the shock solutions shall be smoothed out and made continuous. Difficulties arise because the nonlinear diffusivities vanish at the margins, making the second-order differential equation singular. As a result shocks close to the snout may not be smoothed out. Moreover, addition of second spatial derivatives supposedly necessitates a further boundary condition, but in this case it is not clear what constitutes a proper set of spatial boundary conditions. A steady-state

margin analysis (Hutter 1981c) with power-series solutions suggests that no new boundary condition must be provided, but it also indicates that margin slopes are finite only when $\mathcal{F}(0) \neq 0$, at the margin. This may have led Fowler & Larson (1980b) to conjecture that no further boundary condition is needed to solve the convection-diffusion equation with singular coefficient. (*d*) Things become even more complicated when perturbation schemes are pushed to $O(\epsilon^2)$ and higher-order terms. Indeed, a study of surface elevation in a parallel-sided slab by Hutter (1980a) indicates that dispersion, i.e. a third-order derivative, arises, and numerical estimates for the dispersion coefficient seem to point to the importance of these terms. The occurrence of higher and higher derivatives as the perturbation is continued shows that the question of boundary conditions is unsettled and warrants a thorough mathematical study.

SMALL BED INCLINATION We return now to (3.12) and (3.13) and assume small bed inclinations, $\gamma \leqslant O(\epsilon)$. It will be shown that the scalings introduced below allow prediction of profile geometry *including ice divides*. Since $\gamma \leqslant O(\epsilon)$, transverse shear-stress gradients must balance longitudinal stress gradients, whence, $\tau = O(\epsilon)$ as seen from (3.13a). With this result, (3.13d) implies $\mathfrak{G} f(\tau_{II}) \tau = O(1)$, or because $f(\tau_{II}) \sim f(\tau^2)$,

$$\mathfrak{G} \tau^\alpha = O(1) \to \mathfrak{G} = O(\tau^{-\alpha}) = \epsilon^{-\alpha}, \tag{3.24}$$

where $\alpha = n$ for a power law and $\alpha = 1$ for a finite-viscosity law. I call this the MJ-scaling (Morland & Johnson, 1982). With this value of $\mathfrak{G}$, Equations (3.13b,c) together with the corresponding boundary condition imply to leading order

$$\sigma_x = \sigma_y = -\cos\gamma\,(y_S - y) - p^{\text{atm}}. \tag{3.25}$$

Shear stresses follow by integrating (3.13a); when $\sin\gamma \sim \gamma \sim \epsilon\,\gamma_0$, $\gamma_0 = O(1)$; this yields to leading order

$$\tau = \epsilon\,\zeta^2 \left(\gamma_0 - \frac{\partial y_S}{\partial \xi} \right)(y_S - y),$$

$$\zeta = \text{sign}\left(\gamma_0 - \frac{\partial y_S}{\partial \xi} \right). \tag{3.26}$$

Since $\epsilon(\gamma_0 - \partial y_S/\partial\xi) \sim \alpha_S$, where α_S is the surface inclination relative to the horizontal, the "classical shear-stress formula" is obtained, $\tau_b = \alpha_s H$. Equation (3.26) also indicates that $\tau = 0$ at positions $\xi = \xi_d$, where surface slopes are horizontal, $\gamma_0 = \partial y_S/\partial\xi$. With $\mathcal{F}(0) = 0$, which is reasonable for interior points, this implies that sliding velocities vanish. It will be shown later on that $u = 0$ for all y, so that $\xi = \xi_d$ is an ice divide. In other words, ice divides are where surfaces slope horizontally. Substituting (3.26) into (3.13) yields

$$\frac{\partial u}{\partial y} = 2\,\epsilon^{-\alpha} f(\tau^2)\,\tau$$

as a differential equation for the leading-order longitudinal velocity component, in which the right-hand side is $O(1)$ despite the explicit appearance of the small parameter ϵ. Integration subject to the basal boundary condition then yields

$$u = \zeta \mathcal{F}[\tau(y_B)] + 2\epsilon^{-\alpha} \int_{y_B}^{y} f[\tau^2(\bar{y})] \tau(\bar{y}) \, d\bar{y}. \tag{3.27}$$

At interior points we have with $\mathcal{F}(0) = 0$ and $\tau = 0$, also $u = 0$, confirming that $\xi = \xi_d$ is indeed an ice divide. Finally, the equation describing the surface follows from (3.18) by substituting (3.27). This yields

$$\frac{\partial y_S}{\partial t} + \frac{\partial Q}{\partial \xi} = \bar{a}^*,$$

$$Q = \zeta f[\tau(y_B)](y_S - y_B) \tag{3.28}$$
$$+ 2\epsilon^{-\alpha} \int_{y_B}^{y_S} (y_S - y) f[\tau^2(y)] \tau(y) \, dy,$$

in which τ is given by (3.26). Boundary and initial conditions are derived in a similar fashion as before and will not be rederived. Equation (3.28) has not been investigated so far except in steady state and for special choices of $\mathcal{F}$ and f, see Morland & Johnson (1982). Their results are summarized in the following remarks.

Remarks (*a*) Because $y_S - y$ and $\partial y_S/\partial \xi$ enter the leading-order shear-stress formula, (3.27) will yield an equation for y_S rather than $y_S - y_B$. Thus, the bed profile is not simply superimposed on the geometry of the flat bed as was the case for steep glaciers. (*b*) For the same reason, the surface-wave equation (3.22) is second order in ξ and thus of convection-diffusion type; it may be singular at the margin, giving rise to problems like those already discussed. In fact the steady-state equation shows singular behavior at the margin as $\partial y_S/\partial \xi$ may be infinitely large unless the sliding function $\mathcal{F}(0) \neq 0$ at the margin. A careful analysis of the margin behavior *under these circumstances* again shows that margin velocity, surface slopes, and accumulation are related by (Hutter 1981c)

$$u_M \left(\frac{\partial y_S}{\partial \xi} - \frac{\partial y_B}{\partial \xi} \right) = \bar{a}_M^*.$$

In other words, measuring margin velocities u_M and accumulation rate $\bar{a}^*_M$ will yield margin slopes needed for the integration of the second-order differential equation. This result is interesting because the number of boundary conditions to be prescribed at the margin is larger than for the corresponding $O(1)$-slope analysis. It is clear that careful study of the problem is urgent. (*c*) For this case integrations have explicitly been performed by Morland & Johnson (1982). They show that for an accumulation function that depends on altitude, with integration started at the upper margin, a surface results that is flat and horizontal far from the margin. Such a solution corresponds to an ice reservoir on a sloping embankment. Integration should thus always be started at the lower ablating margin. As integration proceeds away from it, the thickness increases, reaching a maximum and sloping down towards the rising bed beyond this maximum. For most cases analyzed by Morland & Johnson the solution for y_S ceases to be valid near the upper margin, since H' is large again. Hence, even though finite surface slopes at the lower margin are enforced this does not suffice to make the solution uniformly valid unless the ice-sheet geometry is symmetric about a vertical line. (d) Comparison of flat-bed ice-sheet analyses based on this theory with more ad hoc treatments of Bodvardsson (1955), Nye (1959), and Vialov (1958) (see also Paterson 1980) reveals that the latter require the prescription of the ratio of maximal ice-sheet thickness to semi-length of the ice sheet, while this need not be done with Morland & Johnson's more accurate theory. When maximal ice-sheet depths of the two formulations are made to coincide, it is found that Nye's semi-lengths are much smaller than those predicted by Morland & Johnson. Hence, agreement of the early models with observed surface topographies is accidental and, indeed, generally enforced.

Johnson (1981) has extended this steady-state plane-deformation analysis to axisymmetric geometry, but no new qualitative features are observed. In addition to this, surface waves have never been looked at so far in this analysis of small bed inclination. On the other hand, it is straightforward to apply the perturbation scheme followed in this approximation of small-bed inclination to three-dimensional ice sheet flow. This was done by Hutter (1981c,d) who derives the lowest ordering governing equations. Qualitatively, results are similar to (3.25)–(3.28), but now, two horizontal coordinates arise. The interesting feature of this three-dimensional analysis is that to lowest order and for a horizontal mean bed ($\gamma = 0$),

$$\tau_\xi = -\frac{\partial y_S}{\partial \xi}(y_S - y),$$

where ξ is now *any* horizontal direction. The classical shear stress formula therefore applies for each component. It follows that τ_ξ vanishes whenever $\partial y_S/\partial \xi = 0$. In other words, horizontal shear stress is zero along lines of equal height and maximal in the direction of steepest descent. This result can be used to prove that forward flow at a surface point is in the direction of steepest descent of the free surface and flow does not change direction with depth. To the exclusion of margin effects an approximate flow field of an ice sheet can thus be constructed by completing a topographic map by the lines of steepest descent, and velocities can be found when (3.26) and (3.27) are calculated for these directions.

4. LOCAL FLOW PROCESSES AND THREE-DIMENSIONAL EFFECTS

In the preceding discussion, glacier and ice-sheet flow was considered *in the large*. In other words, the behavior was viewed as a response to the global geophysical environment. In many problems of applied glaciology, practical problems stay in the foreground and, indeed, glaciers in Switzerland have been observed from the beginning of glaciology with the object of preventing catastrophes. Problems in this category are of local nature and often three-dimensional. Further, glaciers are channelized systems in which plane motion can be regarded as being of qualitative nature only. Effects of boundedness of the valley by flanks have been studied by Nye (1965) and Reynaud (1973). They consider free-surface channel flow of ice through semi-ellipsoidal, rectangular, and parabolic channels and find that the maximal basal shear stress can be written as $\tau = \phi \sin \gamma \, H$ where $\phi < 1$ depends on the aspect ratio "width/depth." For a semicircular channel $\phi = 1/2$. ϕ is regarded as *the* fudge-factor by glaciologists to make simple concepts, valid under plane flow, applicable to channelized flow. Such calculations have been made, for instance, by Bindschadler (1978) and Raymond (1980). Local flow problems are geometrically complex and thus often defy an analytic treatment. Numerical methods are thus suggested. In these the geometry of the structure being analyzed is usually already known and attention is focused on the stress-velocity problem at known temperature distribution. The boundary-value problems are then elliptic though nonlinear, so that the use of the finite-element method is suggested. Numerous solution procedures are suggested, and a large number of codes are available, which only have to be adjusted to a particular constitutive class of ice. For instance, codes that solve the linear elasticity equations apply equally to linear Stokes flow, and nonlinear constitutive behavior may simply be modeled by a stress-increment approach. This approach was used by Iken (1977) in an analysis of the movement of a large ice mass before breaking off. Its aim was to see whether the time development of the velocity of a typical material point could by any means provide information as to the time of breaking off. A similar problem arises in the formation of ice avalanches. In another numerical analysis, an ice slope on an undulating base was looked at in the process of forming cavities. This problem involves a simultaneous adjustment of boundary geometry (Iken 1981). The results indicate that steady-state sliding, both with and without cavity formation, differs substantially from sliding during cavity formation. This points out a possible necessity for revision of the usual steady-state approach to sliding and opens an entirely new field for extremely difficult theoretical research.

Often finite-element codes use a direct approach to the constitutive nonlinearities. For the non-Newtonian fluid that polycrystalline ice represents, dual varational principles exist (see Johnson 1960, 1961, Oden & Reddy 1976) which allow construction of upper- and lower-bound solutions; the corresponding finite-element formulations have, however, found only limited applications in glaciology (see, for example, Raymond 1980).

5. CLOSING REMARKS

Many subtopics have not been touched upon in this brief overview, including intraglacial water flow, effects of deforming beds, and effects of till, glacier erosion, and abrasion, and their mathematical description. Undoubtedly intensive research is still needed until a thorough and proper understanding of the physics and fluid mechanics of glaciers and ice sheets is obtained. Much of the complexity arises because of the interdisciplinary nature of the field, and communication among the various specialists is often aggravated by "language barriers" that exist between different scientific disciplines. On the other hand, I hope that I have demonstrated that the fluid dynamics of glaciers and ice sheets is a challenging subject that has a sound basis and to which a rational approach is possible. Many prominent scientists have contributed to the field through often rather diverse approaches; my attempt here is to demonstrate that all these can be unified and derived from rather simple but consistent concepts.

ACKNOWLEDGMENTS

The foundations for this manuscript were laid in a series of graduate seminars in Applied Mathematics at the University of Arizona, Tucson. I thank Professor R. A. Seebass for inviting me to give these lectures during my visiting appointment in 1979/80. I further thank Dr. L. W. Morland, University of East Anglia, and Dr. A. C. Fowler, University of Dublin, for sending me preprints of their papers. As is apparent from the text, much of the development of the rigorous aspects of glacier and ice sheet dynamics is due to these authors. This review would have been written differently had I not known their yet unpublished work. I hope that I have correctly interpreted their work, which in form deviates from the above presentation. Finally I acknowledge the help received from Ch. Bucher and F. Langenegger who drew the final graphs and typed the various versions of the manuscript.

Literature Cited

Assur, A. 1980. Some promising trends in ice mechanics. In *Physics and Mechanics of Ice*, ed. P. Tryde, pp. 1–12. IUTAM-Symposium, Copenhagen, 1979. Berlin/Heidelberg/New York: Springer

Barnes, P., Tabor, D., Walker, J. C. F. 1971. The friction and creep of polycrystalline ice. *Proc. R. Soc. London. Ser. A* 125:670–93

Bindschadler, R. A. 1978. *A time-dependent model of temperate glacier flow and its application to predict changes in the surge type variegated glacier during its quiescent phase.* PhD thesis. Univ. Washington, Seattle, Washington

Blümke, A., Finsterwalder, S. 1905. Zeitliche Aenderungen in der Geschwindigkeit der Gletscherbewegung. *Sitzungs-*

berichte der Mathematischphysikalischen Klasse der K. Bayerischen Akademie der Wissenschaften zu München. 35:109–31

Bodvardsson, G. 1955. On the flow of ice sheets and glaciers. *Jökull.* 5:1–8

Budd, W. F. 1966. Glaciological studies in the region of Wilkes, Eastern Antarctica, 1961. *Australian National Antarctic Research Expeditions (ANARE) Scientific Reports, Ser A (IV). Glaciology* (88)

Budd, W. F. 1968. The longitudinal velocity profile of large ice masses. *International Union of Geodesy and Geophysics/International Association of Scientific Hydrology, General Assembly of Bern, 25 Sept.–7 Oct. 1967. (Commission of Snow and Ice). Reports and discussions. IASH-Publ.* 79:58–77

Budd, W. F. 1970. Ice flow over bedrock perturbations. *J. Glaciology* 9 (55):29–47

Budd, W. F. 1971. Stress variations with ice flow over undulations. *J. Glaciology* 10 (59):177–95

Budd, W. F. 1975. A first simple model for periodically self-surging glaciers. *J. Glaciology* 14 (70):3–21

Budd, W. F., McInnes, B. J. 1974. Modelling periodically surging glaciers. *Science* 186:925–27

Clarke, G. K. C., Nitsan, U., Paterson, W. S. B. 1977. Strain heating and creep instability in glaciers and ice sheets. *Rev. Geophys. Space Phys.* 15:235–47

Colbeck, S. C., Evans, R. J. 1973. A flow law for temperate glaciers. *J. Glaciology* 12 (64):71–86

Colbeck, S. C., ed. 1980. *Dynamics of Snow and Ice Masses.* New York/London/Toronto: Academic 468 pp.

Collins, I. F. 1968. On the use of the equilibrium equations and flow law in relating surface and bed topography of glaciers and ice sheets. *J. Glaciology* 7 (50):199–204

Courant, R., Friedrichs, K. O. 1948. *Supersonic Flow and Shock Waves.* New York: Interscience. Reprinted as Vol. 21, *Appl. Math. Sci.* New York: Springer 464 pp.

Dafermos, C. 1974. Quasilinear hyperbolic systems that result from conservation laws. In *Nonlinear Waves,* ed. S. Leibovich, A. R. Seebass, pp. 82–102 Ithaca/London: Cornell Univ. Press

Dorsey, N. E. 1940. *Properties of Ordinary Water Substance.* New York: Reinhold

Drake, L. D., Shreve, R. L. 1973. Pressure melting and regelation of ice by round wires. *Proc. R. Soc. London. Ser. A* 332:51–83

Finsterwalder, S. 1907. Die Theorie der Gletscherschwankungen. *Z. Gletscherkunde* 2:81–103

Fowler, A. C. 1977. *Glacier dynamics.* PhD thesis. Univ. Oxford, Corpus Christi College, England

Fowler, A. C. 1979a. The use of a rational model in the mathematical analysis of a polythermal glacier. *J. Glaciology* 24 (90):443–56

Fowler, A. C. 1979b. A mathematical approach to the theory of glacier sliding. *J. Glaciology* 23 (89):131–41

Fowler, A. C. 1980a. The existence of multiple steady states in the flow of large ice masses. *J. Glaciology* 24 (91):183–84

Fowler, A. C. 1980b. A theoretical treatment of the sliding of glaciers in the absence of cavitation. *Philos. Trans. R. Soc. Ser. A* 298 (1445):637–81, 1981

Fowler, A. C. 1980c. Waves on glaciers. Preprint

Fowler, A. C., Larson, D. A. 1978. On the flow of polythermal glaciers I. Model and preliminary analysis. *Proc. R. Soc. London. Ser. A* 363:217–42

Fowler, A. C., Larson, D. A. 1980a. The uniqueness of steady state flows of glaciers and ice sheets. *Geophys. J. R. Astron. Soc.* 63:333–45

Fowler, A. C., Larson, D. A. 1980b. On the flow of polythermal glaciers II. Surface wave analysis. *Proc. R. Soc. London. Ser. A* 370:155–71

Fowler, A. C., Larson, D. A. 1980c. Thermal stability properties of a model of glacier flow. *Geophys. J. R. Astron. Soc.* 63:347–59

Glen, J. W. 1955. The creep of polycrystalline ice. *Proc. R. Soc. London. Ser. A* 228:519–38

Glen, J. W. 1958. The flow law of ice: A discussion of the assumptions made in glacier theory, their experimental foundations and consequences. *International Union of Geodesy and Geophysics/International Association of Scientific Hydrology, Symposium of Chamonix, 16–24 Sept. 1958. Physics of the movement of the ice. IASH-Publ.* 47:171–83

Glen, J. W. 1974. Physics and mechanics of ice as a material. *US Army Cold Regions Research and Engineering Laboratory, Cold Regions Science and Engineering Monographs.* II-C 2a,b

Grigoryan, S. S., Krass, M. S., Shumskiy, P. A. 1976. Mathematical model of a three dimensional non-isothermal glacier. *J. Glaciology* 17 (77):401–17

Hawkes, I., Mellor, M. 1972. Deformation and fracture of ice under uniaxial stress. *J. Glaciology* 11 (61):103–31

Hobbs, P. V. 1974. *Ice Physics.* Oxford: Clarendon

Hutter, K. 1980a. Time dependent surface elevation of an ice slope. *J. Glaciology* 25 (92):247–66

Hutter, K. 1980b. Rate process theory and

the creep law of ice. *Cold Regions Science and Technology* 3:335-36
Hutter, K. 1981a. The effect of longitudinal strain on the shear stress of an ice sheet. In defense of using stretched coordinates. *J. Glaciology.* 25(95):39-56
Hutter, K. 1981b. Glacier flow. *Am. Sci.* In press
Hutter, K. 1981c. *Theoretical Glaciology.* Vol 1. Dordrecht: Reidel. In press
Hutter, K. 1981d. Mathematical foundation of flow of glaciers and large ice masses. *Lecture Notes for Banach International Center of Mathematical Sciences, Polish Academy of Sciences, Warsaw.* In press
Hutter, K., Olunloyo, V. O. S. 1980. On the distribution of stress and velocity in an ice strip, which is partly sliding over and partly adhering to its bed using a Newtonian viscous approximation. *Proc. R. Soc. London Ser. A* 373:385-403
Hutter, K., Olunloyo, V. O. S. 1981. Basal stress concentrations due to abrupt changes in boundary conditions. A cause for high till concentration at the bottom of a glacier. *Ann. Glaciology* 2:29-33
Hutter, K., Spring, U. 1979. Distribution of stress and velocity in an ice slope with spatially material response. *Festschrift Professor Kasser. Mitteilung der Versuchsanstalt für Wasserbau, Hydrologie und Glaziologie, ETH Zürich* 41:99-122
Hutter, K., Williams, F. M. 1980. Theory of floating ice sheets. *Proc. IUTAM-Symposium on the Mechanics and Physics of Ice*, ed. P. Tryde, pp. 147-62. Berlin/Heidelberg/New York: Springer
Hutter, K., Legerer, F., Spring, U. 1981. First order stresses and deformations in glaciers and ice sheets. *J. Glaciology* 27(96):227-70
Iken, A. 1977. Movement of a large ice mass before breaking off. *J. Glaciology* 19 (81):595-605
Iken, A. 1981. The effect of the subglacial water pressure on the sliding velocity of a glacier in an idealized numerical model. *J. Glaciology* In press
Johnson, M. W. Jr. 1960. Some variational theorems for non-Newtonian flow. *Phys. Fluids* 3:871-78
Johnson, M. W. Jr. 1961. On variational principles for non-Newtonian fluids. *Trans. Soc. Rheology* 5:9-21
Johnson, I. R. 1981. The steady profile of an axisymmetric ice sheet. *J. Glaciology.* 27(95):25-38
Kamb, W. B. 1970. Sliding motion of glaciers: Theory and observation. *Rev. Geophys. Space Phys.* 8:673-728
Lick, W. 1970. The propagation of disturbances on glaciers. *J. Geophys. Res.* 75:2189-97
Lighthill, M. J., Whitham, G. B. 1955. On kinematic waves: I. Flood movement in long waves. II. Theory of traffic flow along crowded roads. *Proc. R. Soc. London Ser. A* 229:281-345
Lile, R. C. 1978. The effect of anisotropy on the creep of polycrystalline ice. *J. Glaciology* 21 (85):475-83
Lliboutry, L. A. 1963. Le régime thermique de la base des calottes polaires. *Union Géodésique et Géophysique Internationale/Association Internationale d'Hydrologie Scientifique, Assemblée Générale de Berkeley, 19-31 Aug. 1963. (Commission des Neiges et des Glaces) IASH-Publ.* 61:232-41
Lliboutry, L. A. 1964. *Traité de Glaciologie.* Tome I. Paris: Masson
Lliboutry, L. A. 1965. *Traité de Glaciologie.* Tome II. Paris: Masson
Lliboutry, L. A. 1968. General theory of subglacial cavitation and sliding of temperate glaciers. *J. Glaciology* 7 (49):21-58
Lliboutry, L. A. 1969. The dynamics of temperate glaciers from the detailed viewpoint. *J. Glaciology* 8 (53):185-205
Lliboutry, L. A. 1971. The glacier theory. In *Advances in Hydroscience*, ed. Ven te chow. 7:81-167
Lliboutry, L. A. 1976. Physical processes in temperate glaciers. *J. Glaciology* 16 (74):151-58
Lliboutry, L. A. 1978. Glissement d'un glacier sur un plan parsemé d'obstacles hémisphériques. *Ann. Geophys.* 34:147-62
Lliboutry, L. A. 1979. Local friction laws for glaciers: a critical review and new openings. *J. Glaciology* 23 (89):67-95
Meier, M. F. 1960. Mode of flow of Saskachewan Glacier, Alberta, Canada. *US Geol. Surv. Prof. Pap. 351.* 70 pp.
Meier, M. F., Post, A. 1960. What are glacier surges? *Can. J. Earth Sci.* 6:807-16
Mellor, M. 1980. Mechanical properties of polycrystalline ice. In *Physics and Mechanics of Ice;* ed. P. Tryde, pp. 217-45. IUTAM-Symposium Copenhagen, 1979. Berlin/Heidelberg/New York: Springer
Michel, B. 1979. *Ice Mechanics.* Les presses de l'Université Laval. 499 pp.
Morland, L. W. 1976a. Glacier sliding down an inclined wavy bed. *J. Glaciology* 17 (77):447-62
Morland, L. W. 1976b. Glacier sliding down an inclined wavy bed. *J. Glaciology* 17 (77):463-77
Morland, L. W. 1979. Constitutive laws for ice. *Cold Regions Science and Technology* 1:101-8
Morland, L. W., Johnson, I. R. 1980. Steady motion of ice sheets. *J. Glaciology* 25 (92):229-46
Morland, L. W., Johnson, I. R. 1982. Effects of bed inclination and topography on steady ice sheets. *J. Glaciology.* In press

Morland, L. W., Spring, U. 1981. Viscoelastic fluid law for ice. *J. Cold Region Science and Technology.* 4(3):255–68

Morris, E. M. 1976. An experimental study of the motion of ice past obstacles by the process of regelation. *J. Glaciology* 17 (75):79–98

Morris, E. M. 1979. The flow of ice, treated as a Newtonian viscous liquid, around a cylindrical obstacle near the bed of a glacier. *J. Glaciology* 23 (89):117–29

Müller, F. 1976. On the thermal regime of a high-arctic valley glacier. *J. Glaciology* 16 (74):119–29

Nachman, A., Callegari, A. 1980. A nonlinear singular boundary value problem in the theory of pseudo-plastic fluids, SIAM. *J. Appl. Math.* 38:275–81

Norton, F. H. 1929. *The Creep of Steel at High Temperature.* New York: McGraw Hill

Nye, J. F. 1953. The flow law of ice from measurements in glacier tunnels, laboratory experiments and the Jungfraufirn experiment. *Proc. R. Soc. London Ser. A* 219:477–89

Nye, J. F. 1957. The distribution of stress and velocity in glaciers and ice sheets. *Proc. R. Soc. London Ser. A* 239:113–33

Nye, J. F. 1959. The motion of ice sheets and glaciers. *J. Glaciology* 3 (26):493–506

Nye, J. F. 1960. The response of glaciers and ice sheets to seasonal and climatic changes. *Proc. R. Soc. London Ser. A* 256:559–84

Nye, J. F. 1963a. On the theory of advance and retreat of glaciers. *Geophys. J. R. Astron. Soc.* 7:431–56

Nye, J. F. 1963b. The response of a glacier to changes in the rate of nourishment and wastage. *Proc. R. Soc. London Ser. A* 275:87–112

Nye, J. F. 1965. The flow of a glacier in a channel of rectangular, elliptic or parabolic cross-section. *J. Glaciology* 5 (41):661–90

Nye, J. F. 1967. Theory of regelation. *Philos. Mag.* 16:1249–66

Nye, J. F. 1969a. The effect of longitudinal stress on the shear stress at the base of an ice sheet. *J. Glaciology* 8 (53):207–13

Nye, J. F. 1969b. A calculation on the sliding of ice over a wavy surface using a Newtonian viscous approximation. *Proc. R. Soc. London Ser. A* 311:445–67

Nye, J. F. 1970. Glacier sliding without cavitation in a linear viscous approximation. *Proc. R. Soc. London Ser. A* 315:381–403

Nye, J. F. 1973. The motion of ice past obstacles. In *Physics and Chemistry of Ice*: papers presented at the Symposium on the Physics and Chemistry of Ice, Ottawa, Canada, 14–18 August 1972, ed. E. Whalley, S. J. Jones, L. W. Gold, pp. 387–94. Ottawa:Royal Society of Canada

Oden, J. T., Reddy, J. N. 1976. *Variational Methods in Theoretical Mechanics.* Universitext. Berlin/Heidelberg/New York: Springer. 302 pp.

Paterson, W. S. B. 1969. *The Physics of Glaciers.* Oxford/New York/Toronto: Pergamon

Paterson, W. S. B. 1980. The sheets and ice shelfs. In *Dynamics of Snow and Ice Masses*, ed. S. C. Colbeck, pp. 1–78. New York/London/Toronto: Academic

Poston, T., Stewart, I. 1978. *Catastrophe Theory and its Applications.* London/San Francisco/Melbourne: Pitman

Raymond, C. F. 1980. Temperate valley glaciers. In *Dynamics of Snow and Ice Masses*, ed. S. C. Colbeck, pp. 79–139. New York/London/Toronto: Academic

Reynaud, L. 1973. Flow of a valley glacier with a solid friction law. *J. Glaciology* 12 (65):251–58

Robin, G. de Q. 1955. Ice movement and temperature distribution in glaciers and ice sheets. *J. Glaciology* 2 (18):523–32

Robin, G. de Q. 1967. Surface topography of ice sheets. *Nature* 215:1029–32

Robin, G. de Q. 1969. Initiation of glacier surges. *Can. J. Earth Sci.* 6:919–28

Robin, G. de Q., Weertman, J. 1973. Cycling surging of glaciers. *J. Glaciology* 12 (64):3–18

Schlichting, H. 1979. *Boundary-Layer Theory.* New York: McGraw-Hill. 817 pp.

Shumskiy, P. A. 1969. *Dinamicheskaya Glastiologiya I.* Moscow: Itogi Nanki. English transl. 1978. *Dynamic Glaciology.* Amerind Publ.

Shumskiy, P. A., Krass, M. S. 1976. Mathematical models of ice shelves. *J. Glaciology* 17 (77):419–32

Steinemann, S. 1958. Experimentelle Untersuchungen zur Plastizität von Eis. Beiträge zur Geologie der Schweiz. *Geotechnische Serie Hydrologie* 10:72S

Theakstone, W. H. 1967. Basal sliding and movement near the margin on the glacier Østerdalsisen. *J. Glaciology* 6 (48):805–16

Thompson, D. E. 1979. Stability of glaciers and ice sheets against flow perturbations. *J. Glaciology* 24 (90):427–41

Vialov, S. S. 1958. Regularities of ice deformation. *International Union of Geodesy and Geophysics/International Association of Scientific Hydrology, Symposium of Chamonix, 16–24 Sept. 1958. Physics of the movement of ice. IASH-Publ.* 47:383–91

Vivian, R. A., Bocquet, G. 1973. Subglacial cavitation phenomena under the Glacier d'Argentière; Mont Blanc, France. *J. Glaciology* 12 (66):439–51

Weertman, J. 1957a. On the sliding of glaciers. *J. Glaciology* 3 (21):33–38

Weertman, J. 1957b. Deformation of floating ice shelves. *J. Glaciology* 3 (21):38–42

Weertman, J. 1958. Traveling waves on glaciers. *International Union of Geodesy and Geophysics/International Association of Scientific Hydrology, Symposium of Chamonix, 16–24 Sept. 1958. Physics of the movement of ice. IASH-Publ.* 47:162–68

Weertman, J. 1968. Comparison between measured and theoretical temperature profile on the Camp Century, Greenland, borehole. *J. Geophys. Res.* 73:2691–2700

Weertman, J. 1979. The unsolved general glacier sliding problem. *J. Glaciology* 23 (89):97–115

Whitham, G. B. 1974. *Linear and Non-Linear Waves*. New York: Wiley. 636 pp.

Yuen, D. A., Schubert, G. 1979. The role of shear heating in the dynamics of large ice masses. *J. Glaciology* 21:195–212

THE MATHEMATICAL THEORY OF FRONTOGENESIS

B. J. Hoskins

Department of Meteorology, University of Reading, Reading, England

1. INTRODUCTION

For over one hundred years it has been known that atmospheric depressions have cold frontal regions in which typically the temperature falls rapidly, the wind changes, and precipitation occurs. Sixty years ago Norwegian meteorologists shaped their polar front model in which depressions were considered to grow on a preexisting sloping thermal discontinuity; as the depressions grew it was considered that they distorted the discontinuity into a cold front and also a warm front. The dynamical view changed thirty years ago with the discovery by Charney (1947) and Eady (1949) that realistic depression structures may be obtained as the normal modes of the baroclinic instability of a smooth rather than discontinuous thermal contrast. Fronts then had to be viewed as being formed in a growing nonlinear baroclinic wave, though this secondary role in no way diminished their importance in practical meteorology. The theoretical interest shifted towards the mechanism for generating fronts, namely frontogenesis.

When upper air data became available in the 1950s it was apparent that strong frontal regions occurred in the upper troposphere as well as near the surface. Many studies (e.g. Reed & Danielsen 1959) suggested strongly that these upper air fronts are associated with the descent of a tongue of stratospheric air into the troposphere.

The existence of fronts in the ocean has been known for many years and when vertical sections were obtained (e.g. Voorhis & Hersey 1964 and Katz 1969) they suggested many structural similarities with atmospheric fronts. Further, the experimental work of Fultz (1952) and Faller (1956) has shown that fronts can be reproduced in laboratory experiments using differentially heated rotating water. Thus we conclude that details of latent-heat release and other diabatic heating, and of the boundary layer and turbulent friction, are probably not crucial in the process of frontogenesis though they will clearly have important modifying effects in any particular case, and will be crucial in any steady-state front. The adiabatic, frictionless equations will be the basis of the work reviewed in this paper.

It is probably a mistake to attempt a rigorous definition of a front. However, in order to make it clear that the subject of this review is not the sea-breeze front, inter-tropical convergence zone, etc., some definition is required. A front is considered to be a region whose length scale is comparable with the radius of deformation (NH/f where N is a buoyancy frequency, H a relevant height scale, and f the Coriolis parameter) in one direction but much less than this in the cross direction, and across which there are significant changes in buoyancy and velocity with gradients tending to become very large in a finite time. The rotation of the fluid is crucial to the existence of the front.

2. ADVECTION OF A PASSIVE SCALAR

Although much of the interesting theory concerning frontogenesis is concerned with nonlinear feedback, it is helpful to begin by considering the advection of a passive scalar b, later taken to be the buoyancy, by a horizontal two-dimensional, nondivergent flow $(u,v,0)$:

$$Db/Dt = b_t + ub_x + vb_y = 0 . \quad (1)$$

A usual way of discussing the rate of change of the gradient of b moving with the fluid (e.g. Mudrick 1974) is by introducing the deformation D with the stretching oriented along the dilatation axis. Then the so-called frontogenesis function (Miller 1948) is

$$\frac{D}{Dt} |\nabla b|^2 = D |\nabla b|^2 \cos 2\phi , \quad (2)$$

where ϕ is the angle between the dilatation axis and the b contour. If this angle is of magnitude $\pi/4$ or less, gradients in b grow following a fluid particle.

More information can often be obtained from a slightly different approach. Suppose that we are interested in the x gradient of b. Then (1) gives

$$\frac{D}{Dt} b_x = Q_1 , \quad (3)$$

where

$$Q_1 = -u_x b_x - v_x b_y = -\partial (v , b)/\partial (x , y) . \quad (4)$$

The x gradient of b on a fluid particle increases in time if the y component of wind increases along a b contour, moving with larger b on the right.

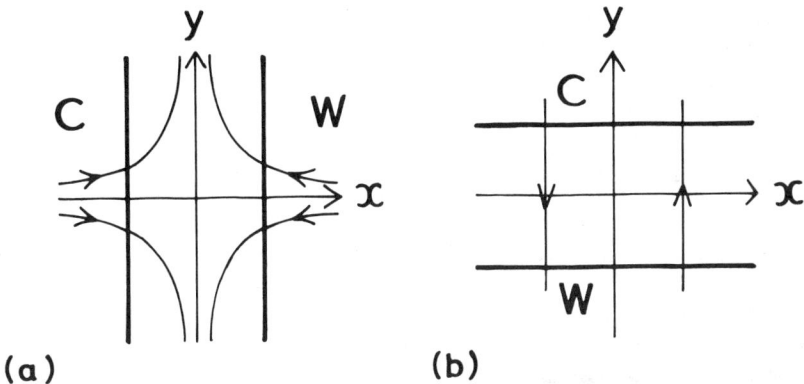

Figure 1 Mechanisms for forming an x gradient in buoyancy. Buoyancy contours are indicated by thick lines and light (warm) and heavy (cold) fluid are indicated by W and C respectively. Streamlines are shown by thin lines with arrows. (*a*) relies on horizontal convergence and (*b*) on horizontal shear.

Two simple cases have been discussed in many frontogenesis papers. Bergeron (1928) postulated that frontogenesis could occur in a simple deformation field $u = -\alpha x$, $v = \alpha y$. The case of b independent of y is shown in Figure 1*a*. The other component of the Jacobian is highlighted in the shear model investigated numerically by Williams (1967). A simple example is indicated in Figure 1*b*.

The vector form of (3) may be written

$$\frac{D}{Dt}\nabla b = \mathbf{Q},\qquad(5)$$

where $\mathbf{Q} = -\left(\mathbf{k}\wedge\frac{\partial \mathbf{v}}{\partial s}\right)\left|\frac{\partial b}{\partial n}\right|$, s and n being coordinates tangential and normal to b contours with n increasing in the direction of decreasing b. From (5) the "frontogenesis function" given in (2) has the alternative form $2\mathbf{Q}\cdot\nabla b$.

3. QUASI-GEOSTROPHIC FRONTOGENESIS

For a first look at the interaction of the changing buoyancy[1] and velocity fields we consider the small Rossby number approximation of quasi-geostrophic theory (for a detailed discussion of this theory, see, for example, Pedlosky 1979). The y momentum and buoyancy equations for a quasi-

---

[1] In meteorological parlance, following Hoskins & Bretherton (1972), b can be taken to be g times potential temperature divided by a standard temperature.

geostrophic Boussinesq fluid in a coordinate system rotating with angular velocity $f/2$ about a vertical axis may be written

$$D_g v_g + f u_a = 0 , \tag{6}$$

and

$$D_g b + N^2 w = 0 , \tag{7}$$

where the geostrophic wind is $(u_g, v_g) = (-f^{-1}\phi_y, f^{-1}\phi_x)$, $u_a = u - u_g$ is the ageostrophic flow in the x direction, $b = \phi_z$, D_g is the time derivative moving with the geostrophic velocity, and $N(z)$ is the buoyancy frequency. v_g and b are linked by the "thermal wind" relation

$$f v_{g_z} = b_x . \tag{8}$$

Rather as in the previous section, an x derivative of (7) gives

$$D_g b_x = Q_1 - N^2 w_x , \tag{9}$$

where Q_1 is given by (4) evaluated using geostrophic velocities. A z derivative of (6) shows, after a little manipulation, that we also have

$$D_g f v_{g_z} = -Q_1 - f^2 u_{a_z} . \tag{10}$$

Equations (9) and (10) describe how the geostrophic velocity field acting through the term Q_1 attempts to destroy thermal wind balance (8) by changing the two parts by equal but opposite amounts. The role of the ageostrophic motion (u_a, w) is to maintain thermal wind balance. Subtracting (9) and (10) gives an equation for the ageostrophic motion in the x,z plane:

$$N^2 w_x - f^2 u_{a_z} = -2Q_1 . \tag{11}$$

Geostrophic motion is nondivergent so that, if the y component of ageostrophic motion is independent of y, we may introduce a stream function ψ for the ageostrophic motion such that $(u_a, w) = (-\psi_z, \psi_x)$. From (11) the equation for ψ is

$$N^2 \psi_{xx} + f^2 \psi_{zz} = -2Q_1 . \tag{12}$$

Using these ideas, we are able to obtain valuable insights into the dynamics of frontogenesis. Consider the frontogenetic situation shown in

Figure 2. There is a positive x gradient in b in balance with a positive vertical shear in v_g. Suppose that the large-scale geostrophic motion is tending to increase b_x everywhere, i.e. $Q_1 > 0$. In Figure 2 this is accomplished by large-scale convergence. As discussed above, the geostrophic motion must imply a tendency to decrease v_{g_z} through the appearance of $-Q_1$ in (10). Thermal wind balance is maintained by the ageostrophic motion with w_x positive and u_{a_z} negative. The upward motion at positive x and downward motion at negative x act to decrease and increase b, respectively, thereby decreasing b_x. The negative u_a at upper levels and positive u_a at lower levels act to increase and decrease v_g, respectively, thereby increasing v_{g_z}.

The stream-function equation (12) and the arguments from it are essentially simpler quasi-geostrophic forms of those given by Sawyer (1956) and Eliassen (1959, 1962). The more complete theory will be discussed in the next section.

Stone (1966), Williams & Plotkin (1968), and Williams (1968) used quasi-geostrophic theory to investigate the simple deformation model of frontogenesis. Their solutions exhibited the growth in gradients especially near boundaries. Away from horizontal boundaries the ageostrophic motion stopped the horizontal length scale of the b field from decreasing below the Rossby radius of deformation NH/f (H is the depth of fluid). On the boundaries w is zero and the scale decreases as dictated by the imposed field. The largest gradients in b occur in a vertical region and positive and negative vorticity exhibit equal growth. Many of these unrealistic features are easily traced to the omission of certain feedbacks in

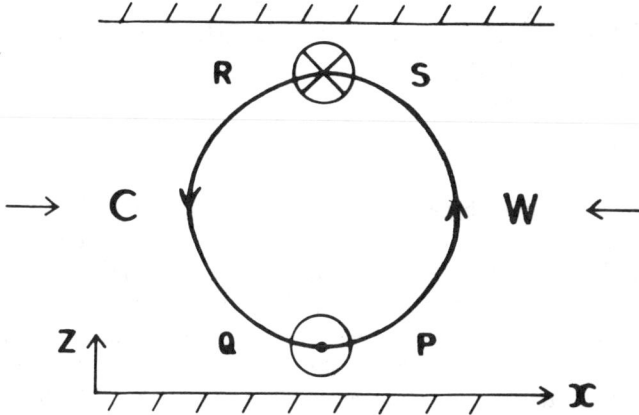

Figure 2 A frontogenetic situation in which a buoyancy gradient from heavy (*C*) to light (*W*) in thermal wind balance with a flow into the section at the top (⊗) and out of it at the bottom (⊙) is being increased by a large scale convergence in x indicated by arrows. The resulting ageostrophic cross-frontal circulation is shown by an arrowed streamline.

quasi-geostrophic theory. However, the qualitative effect of some of these feedbacks can be deduced from the quasi-geostrophic results. For example, consider the region P at the ground on the "warm" side of the temperature contrast. The ageostrophic velocity is clearly convergent ($u_{a_x} < 0$) there and would, if included, imply the generation of larger gradients in b. Also at P the positive vorticity generation is underestimated because the nonlinear term (ξw_z) in the vorticity (ξ) equation

$$D\xi/Dt = (f + \xi)w_z \qquad (13)$$

is not included. These arguments also apply to point R on the cold side at upper levels. However, at points Q and S the ageostrophic divergence would imply weaker gradients in b, and the neglected nonlinear term would imply smaller negative vorticity.

Quasi-geostrophic theory thus correctly suggests, though its solutions do not include, the formation of strong surface fronts with positive vorticity on the warm side of the buoyancy contrast, and that the region of maximum gradients slopes in the vertical from P to R. It highlights the role of horizontal boundaries in nonlinear frontogenesis. In the free atmosphere the ageostrophic circulation acts to inhibit the formation of large gradients. The final point to be made is that, unless the ageostrophic convergence at P increases as the local gradients increase, the vorticity [from (11)] and the gradients in b can only increase exponentially with time. Quasi-geostrophic theory does not even suggest the formation of frontal discontinuities in a finite time.

4. TWO-DIMENSIONAL SEMI-GEOSTROPHIC FRONTOGENESIS

Introduction

To produce a frontogenesis theory that is less restrictive than quasi-geostrophic theory and that includes the ageostrophic effects highlighted in the last section, we perform a scale analysis in a coordinate system in which the front is stationary and oriented in the y direction. Let ℓ, U and L, V be respectively length and velocity scales across and along the front. Then $\ell \ll L$ and observations show that $V \gg U$. Assuming that $D/Dt \sim U/\ell$, the ratios of the accelerations and the Coriolis forces in the x and y directions are

$$(Du/Dt)/fv \sim (U/V)^2\, (V/f\ell) \text{ and } (Dv/Dt)/fu \sim V/f\ell\ .$$

In the long-front (y) direction the acceleration is of first-order importance if the front is strong enough that $V/f\ell \sim 1$. However, the cross-front acceleration, Du/Dt, should be smaller than the Coriolis force in the x direction so that this force must be balanced by the pressure gradient. A more detailed scale analysis is given in Hoskins & Bretherton (1972). The implication is that v may be approximated by its geostrophic value. Then the y momentum and buoyancy equations may be written

$$Dv_g/Dt + fu_a = 0 , \qquad (14)$$

and

$$Db/Dt = 0 . \qquad (15)$$

Comparing with the quasi-geostrophic versions (6) and (7), the crucial extra ingredient is the advection by the full ageostrophic motion in the (x,z) plane, so that D/Dt is a full three-dimensional material time derivative.

This approximation, which is a two-dimensional version of the geostrophic momentum approximation introduced by Eliassen (1948), was first used by Sawyer (1956) and Eliassen (1959, 1962) to derive a more complicated version of (12). Performing an analysis identical with that used in deriving (12) but starting from (14) and (15) gives the cross-front stream-function equation:

$$N^2\psi_{xx} - 2S^2\psi_{xz} + F^2\psi_{zz} = -2Q_1 , \qquad (16)$$

where $N^2 = b_z$, $S^2 = b_x = fv_z$, and $F^2 = f(f + v_x)$. The operator on the left-hand side is elliptic provided the potential vorticity $q = F^2N^2 - S^4$ is positive or, equivalently, the fluid is symmetrically stable (Hoskins 1974). It can be shown from (14) and (15) plus the mass conservation and hydrostatic relations that q is conserved following a fluid particle. The forcing term in (16) is identical with that in the quasi-geostrophic version (12).

The qualitative ideas on vertical circulations in a region of frontogenesis gained from quasi-geostrophic theory and illustrated in Figure 2 need only a little modification. As indicated in Figure 3a thermal wind balance, from (14) and (15), may be shown to be maintained by gradients in w along buoyancy surfaces and by gradients in u_a along "absolute momentum" lines $X = x + v_g/f =$ constant. Now it is easily seen that

$$q = f^2 \partial(X,b)/\partial(x,z) . \qquad (17)$$

Since q is conserved on a fluid particle, as gradients in X and b increase, i.e. N^2, S^2, and F^2 increase, so the angles between X and b lines decrease. The ellipticity of Equation (16) becomes weaker and the boundary condition $\psi = 0$ at $z = 0$ becomes more difficult to satisfy.

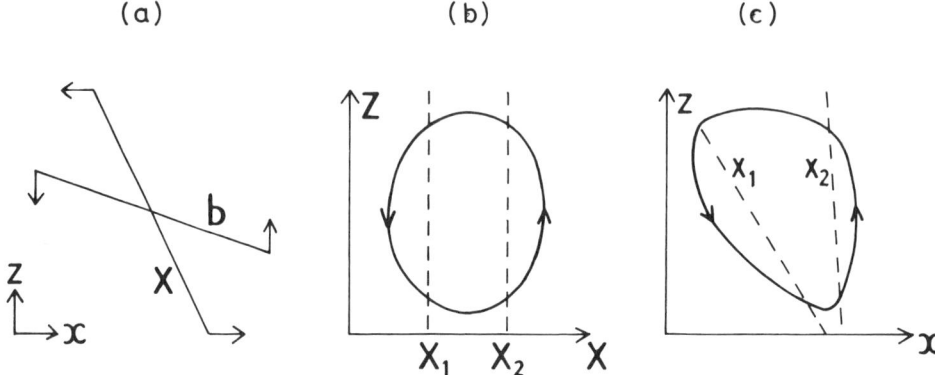

Figure 3 (a) Ageostrophic motions (indicated by arrows) which tend to restore thermal wind balance when Q_1 is positive. By analogy with (1), (3), and (4), the gradient in w along the b contour decreases b_x and the gradient in u_a along the $X = x + v_g/f$ contour increases v_z. (b) The circulation around a closed contour in the XZ plane in a region where Q_1 is positive. (c) The same circulation in the xz plane. The dashed lines are lines of constant X which are close together near the surface in a region of large vorticity.

Mathematically it is simplest to follow Eliassen (1962) and use X as independent variable instead of x which then turns (16) into

$$f^{-2}(q\,\psi_X)_X + f^2\psi_{ZZ} = -2Q_1/J. \tag{18}$$

Here $Z = z$, and $J = \partial X/\partial x = 1 + f^{-1}\,\partial v_g/\partial x$ is the Jacobian of the transformation. For a closed circulation around a region where $Q_1 > 0$, the circulation must have the sense shown in Figure 3b. Transformation back to x,z coordinates tilts the ellipses along lines $X = $ constant and produces more intense flow in regions in which J is large, i.e. the relative vorticity $\xi = \partial v_g/\partial x$ is significant in comparison with f.[2] This is sketched in Figure 3c.

We now have the extra effect that quasi-geostrophic theory did not even suggest. As the frontal gradients increase in region P in Figure 2, so the ageostrophic convergence there increases. To be specific, suppose that near P, $\partial w/\partial z = \gamma\xi$ where $\gamma \sim .2$ would be not unreasonable. For a fluid particle remaining at P, infinite vorticity would be produced in a time $\ln 2/\gamma f$ after the relative vorticity reached f. Typical numbers give this time to be ~ 10 hours. In reality, of course, the approximations of no frictional processes and geostrophic long-front velocity would break down somewhat before this time.

As mentioned in the introduction, the other region of strong atmospheric frontogenesis is near the tropopause. The quasi-geostrophic arguments

---

[2] Note that $\partial u/\partial y$ is neglected in ξ consistent with the approximations made.

presented in Section 3 suggested frontogenesis on the cold side of the thermal contrast but assumed that the tropopause was rigid. To obtain a better indication of the process of upper-air frontogenesis it is necessary to consider atmospheric structure in the region of the tropopause. In terms of potential vorticity, the tropopause marks a transition from the relatively low values in the troposphere to high values in the stratosphere. Near the tropopause in a region of tropospheric temperature gradients the thermal structure is generally as shown in Figure 4. The tropopause slopes and there is a tendency for a reversed horizontal thermal gradient about it. If Q_1 is positive then the cross-frontal circulation must look approximately as indicated. The high potential vorticity in the stratosphere implies generally small vertical motion there. The effect of the circulation is to steepen the tropopause and also to make it descend near R. Large velocity and temperature gradients may be expected in the region R but a tendency to produce infinite gradients in a finite time is unlikely.

The cross-frontal circulation equation either in the form (16) or (18) has been the subject of numerous investigations. Sawyer (1956) first derived the equation in a study of frontogenesis in a prescribed deformation field. He exhibited numerical solutions for particular geostrophic fields and also assessed possible modifications due to latent-heat release. Eliassen (1959) concentrated on the forcing of (16) by a latent-heat release term $-H_x$ where H is a term added to the right-hand side of (15) and also on an Ekman pumping boundary condition. Eliassen (1962) exhibited for the first time the full forcing term Q_1 and also the transformation to the normal form (18). More recently, Shapiro (1981) has solved the equation for

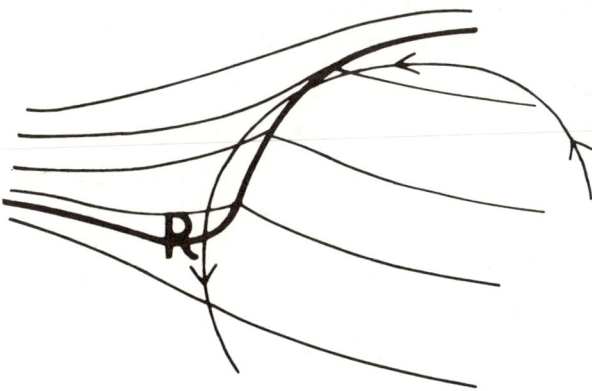

Figure 4 The ageostrophic circulation in the region of the tropopause (thick line) assuming Q_1 is positive. Temperature contours are marked by continuous lines. In three-dimensional models, as discussed below, it is often found that Q_1 is negative near R, which forces the descent to be concentrated and to the right of R.

geostrophic fields taken from observed and simulated frontogenesis cases. He has also incorporated a parameterization of clear-air turbulence into the forcing term.

Time-Dependent Models

GENERAL COMMENTS Simple time-dependent models of frontogenesis have also been formulated in terms of potential vorticity q, buoyancy b, and coordinate X. We recall that q is a conservative quantity that is proportional to the Jacobian of X and b with respect to x and z as shown in (17). From its definition, $X = x + \phi_x/f^2$ and also $b = \phi_z$. As Gill (1981) pointed out, the thermal wind relation can be written

$$\partial(b, z)/\partial(X, x) = -f^2. \tag{19}$$

Thus the potential vorticity definition and the thermal wind relation both reduce to linear form if X and z or x and b are taken as independent variables. Buoyancy (or isentropic) coordinates x and b may often be useful for describing interior motions (Shapiro 1975) but where bounding surfaces are at constant z it is more convenient to use X and z. The boundary condition on such a surface is then that b is conserved in horizontal motion. To see how X changes on a fluid particle we use the y equation of motion (14) to obtain

$$DX/Dt = u_g. \tag{20}$$

Thus in X space the horizontal motion is geostrophic.

For the case where the frontogenesis is forced by a simple deformation $u = -\alpha x, v = \alpha y$, it is most convenient, following Hoskins & Bretherton (1972), to return to the full equations and set

$$u = -\alpha x + u', v = \alpha y' + v',$$
$$\phi = f\alpha xy - 1/2\, \alpha^2(x^2 + y^2) + \phi', b = b'.$$

Scale analysis then suggests that v' is approximately the geostrophic value ϕ_x'/f. It is then consistent to consider a model in which all the primed variables are independent of y. Defining $X = x + v'/f$ all the above relations are valid for this problem except that (20) is replaced by

$$DX/Dt = -\alpha X. \tag{21}$$

Thus $X = X_0 e^{-\alpha t}$ on a fluid particle.

ZERO POTENTIAL VORTICITY DEFORMATION MODEL The simplest time-dependent frontogenesis model that yields realistic results is obtained by considering the simple deformation acting on a fluid of zero potential vorticity and bounded by surfaces at $z = \pm H/2$. From (17), at any instant b must be a function of X only. Since b is conserved and the behavior of X is given by (21), we find that

$$b = F(X\, e^{\alpha t}). \tag{22}$$

From the thermal wind equation (19),

$$(\partial z/\partial x)|_{b=\text{const}} = -f^2/b_X. \tag{23}$$

Thus lines on which b and X are constant are straight and of known slope at time t if F is known.

Mass conservation gives that at $z = 0$ the line must be at the same position as it would have been if it had moved with the deformation velocity alone, i.e. $X = x$. Thus on an X,b line,

$$x = X - f^{-2}\, z\, b_X. \tag{24}$$

Therefore

$$v' = f(X - x) = f^{-1} z b_X. \tag{25}$$

The vertical component of absolute vorticity is

$$\zeta = f\, \partial X/\partial x = f(1 - f^{-2}\, z\, b_{XX})^{-1}. \tag{26}$$

Specification of the geostrophic fields at an initial instant gives the function F in (22). At later times the solution is given by (22), (24), and (25). Time t has become a simple parameter in the solution.

A specific solution is shown in Hoskins & Bretherton (1972). Here we will note that the solution quite generally supports the inferences made above. In particular, in the lower half of the fluid the frontogenesis proceeds most rapidly at the lower boundary on the warm side where θ_{XX} is most negative. There infinite gradients are predicted in a finite time. In fact, it is easily shown that on the lower boundary the parameter γ giving the ratio between the ageostrophic convergence and the vorticity is $2\alpha/f$, so that the time scale for infinite vorticity is α^{-1}.

Finally, we note that this solution gives a good indication of the range of validity of our geostrophic approximation and neglect of mixing processes. The former is found to be consistent if $\zeta/f \ll (2\alpha/f)^{-2} \sim 25$. For zero potential vorticity, the Richardson number is equal to f/ζ so that one may expect mixing processes to be important before $\zeta = 10f$.

OTHER SIMPLE MODELS Multiplying (19) by the Jacobian of X and x with respect to X and Z shows that thermal wind balance takes the same form in X,Z coordinates as in real space. The potential function

$$\Phi = \phi + 1/2\, v_g^2 \tag{27}$$

then gives

$$f v_g = \Phi_X, \quad b = \Phi_Z. \tag{28}$$

The Jacobian of the transformation to X coordinates is

$$J = \zeta/f = (1 - \Phi_{XX}/f^2)^{-1} \qquad (29)$$

and (17) gives

$$f^{-2} \Phi_{XX} + f^2 q^{-1} \Phi_{ZZ} = 1 . \qquad (30)$$

The next simplest frontogenesis model after the deformation model with a zero potential vorticity fluid is that with a uniform potential vorticity fluid. On rigid horizontal boundaries, b is again given for all time by (22). Thus the total geostrophic solution at any subsequent time is given by a solution of the simple second-order elliptic equation (30) with Φ_Z specified on horizontal boundaries and suitable conditions on Φ at large $|X|$ away from the region of interest. Such solutions have been discussed in detail in Hoskins (1971), and compared with observed frontogenesis by Blumen (1980). There are no substantial differences from the zero potential vorticity solution. Again, infinite gradients are predicted in a finite time $O(\alpha^{-1})$. Determination of the ageostrophic circulation is not necessary for solution of the time-dependent problem but is a simple matter from (18) with the right-hand side $- 2 \alpha b_X$.

A uniform potential vorticity model in which the frontogenesis is initiated by the horizontal shear mechanism of Figure 1b is given by the two-dimensional Eady problem (Eady 1949). The basic state is a uniform negative meridional buoyancy gradient $\bar{b}_y$ in thermal wind balance with u_g increasing linearly with height. Perturbations of the form (a cosh kZ cos $kX + b$ sinh kZ sin kX) trivially satisfy the interior equation (30) and substitution in the buoyancy-conservation equation on the two horizontal boundaries shows that exponential growth is possible. The growing baroclinic wave again contains frontal gradients on the boundaries which tend to become infinite in a finite time $O(N/\bar{b}_y)$. This problem was first solved numerically by Williams (1967) without the approximation of the geostrophy of v. The later analytical solution (Hoskins & Bretherton 1972) using this approximation showed good agreement.

It is worth noting that both the uniform potential vorticity solutions for v_g and b are mathematically identical with quasi-geostrophic solutions except that these values are obtained in the distorted X space. The simple interpretation from (14) is that to accelerate to a geostrophic velocity V in the y direction, a particle must receive an ageostrophic displacement $- V/f$ in the x direction. This displacement is not included in quasi-geostrophic theory. It implies a strengthening of positive relative vorticity and weakening of negative vorticity such that relative vorticities of $-f$, 0, $f/2$, and f in a quasi-geostrophic solution become $-f/2$, 0, f, and ∞.

The next level of complexity is achieved by proceeding to a model in which the deformation acts on a fluid with two regions of uniform potential vorticity separated by an interface. In the two regions (30) holds and on rigid boundaries (22) again gives Φ_Z. On the interior boundary, Φ_Z is also given by (22) but the position of the boundary is determined such that Φ_X is continuous. The solution may be determined numerically using iterative techniques (Hoskins & Bretherton 1972). In the atmospheric case the interest is in representing frontogenesis near the tropopause, which is an interface between high potential vorticity stratospheric air and low potential vorticity tropospheric air. Solutions exhibited in Hoskins (1971, 1972) show the descent of a narrow tongue of stratospheric air and the formation of large gradients at the base of the tongue.

MacVean & Woods (1980) have studied an oceanic example of a low-stratification "well-mixed" layer above a higher potential vorticity layer and shown the frontogenesis and distortion of the interface. One extra ingredient highlighted by their paper is that oceanic buoyancy depends on temperature and salinity. Although the theory suggests that the horizontal length scale of the buoyancy field in the interior does not fall below the radius of deformation, a sharp front in temperature and salinity is possible. This is illustrated in Figure 5. There is, however, no tendency to form infinite interior gradients in a finite time.

5. THREE-DIMENSIONAL SEMI-GEOSTROPHIC THEORY

Introduction

The two-dimensional frontogenesis models show the essential nature of the dynamics but only within the imposed large-scale geostrophic frame-

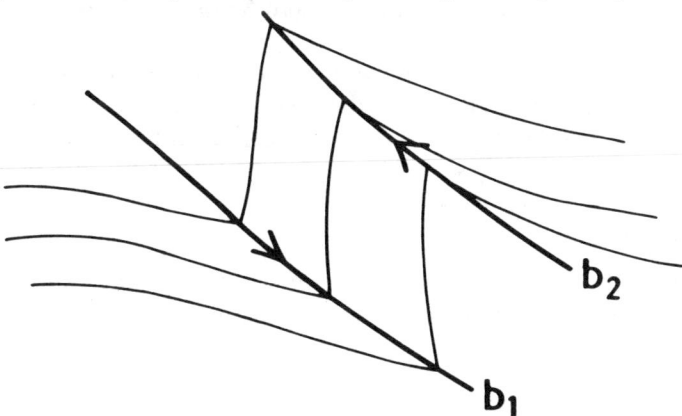

Figure 5 The interior situation during frontogenesis when buoyancy depends on more than one conservative quantity, e.g. temperature and salinity. The ageostrophic motion tends to make the total velocity lie along buoyancy contours b_1, b_2 so that increased gradients are not formed. However, gradients of any other conservative quantity (light contours) increase indefinitely.

work. The next stage is to study the formation of fronts in growing three-dimensional baroclinic waves. The numerical integration of the primitive equations by Arakawa (1962), Edelmann (1963), Eliassen & Raustein (1970), Mudrick (1974), and others indicates that such growing waves do indeed form surface fronts though the intensity of the process is limited by numerical resolution. Shapiro (1975) and Hoskins & Draghici (1977) have used isentropic coordinates to simulate upper-tropospheric frontogenesis. However, in order to extend the understanding of the processes involved from two to three dimensions we shall consider a three-dimensional theory which in a two-dimensional situation reduces to the previous theory.

Following Eliassen (1948), Fjortoft (1962), and Hoskins (1975) it is assumed that the time scale for change in velocity following a fluid particle is much larger than $f^{-1} \sim 3\ h$. Now the exact, horizontal, Boussinesq vector momentum equation may be written

$$\mathbf{v} = \mathbf{v}_g + \mathbf{k} \wedge f^{-1} D\mathbf{v}/Dt \ . \tag{31}$$

Using this expression to substitute for $\mathbf{v}$ in the last term gives

$$\mathbf{v} = \mathbf{v}_g + \mathbf{k} \wedge f^{-1} D\mathbf{v}_g/Dt - f^{-2} D^2\mathbf{v}/Dt^2 \ . \tag{32}$$

Consistent with the above assumption, the last term is neglected. This is called the geostrophic momentum approximation because the momentum equation becomes

$$D\mathbf{v}_g/Dt + f\mathbf{k} \wedge \mathbf{v} + \nabla\phi = 0 \ . \tag{33}$$

The ageostrophic motion is retained in the advection but not in the momentum. The advection of buoyancy is also by the full (geostrophic plus ageostrophic) flow.

Like the full Boussinesq equations, the modified set conserves buoyancy, has a full energy equation, a three-dimensional vorticity equation, and an Ertel potential vorticity conservation equation. As in the two-dimensional case, the equations are most easily considered by making a transformation to "geostrophic" coordinates:

$$X = x + v_g/f, Y = y - u_g/f \ .$$

Some interesting comments on this coordinate transformation have been made by Blumen (1981). As described in Hoskins (1975) a potential function

$$\Phi = \phi + 1/2(u_g^2 + v_g^2) \tag{34}$$

has derivatives in the transformed space giving the geostrophic velocities and buoyancy. The conservation of buoyancy and potential vorticity are

$$Db/Dt = 0 \tag{35}$$

and $Dq/Dt = 0$, \hfill (36)

where the expression for the full material time derivative in geostrophic coordinates is $D/Dt = \partial/\partial T - f^{-1}\Phi_Y \partial/\partial X + f^{-1}\Phi_X \partial/\partial Y + w\, \partial/\partial Z$.

Making one further approximation consistent with the original assumption, Φ and the potential vorticity q are related by the elliptic equation

$$f^{-2}(\Phi_{XX} + \Phi_{YY}) + f^2 q^{-1} \Phi_{ZZ} = 1 . \tag{37}$$

Given suitable lateral boundary conditions, numerical solution is possible using (36) to predict q, (35) to predict $b = \Phi_Z$ on horizontal boundaries, and (37) to obtain values of Φ everywhere. The vertical velocity, w, which is required in (36), may be obtained by consistency of the predictions of (35) and (36) in the interior of the fluid. Alternatively, as suggested in Hoskins & Draghici (1977), it can be obtained from an elliptic equation that is the analogue of the cross-frontal circulation equation (18).

This set of equations, known as the semi-geostrophic equations, is clearly an extension to three dimensions of the two-dimensional theory given in particular by (27)–(30). On the large scale they are certainly as accurate as the quasi-geostrophic equations but, unlike them, are able to represent the vertical advection of a region of large gradients in potential vorticity, e.g. the tropopause, and the dynamics of frontogenesis. McWilliams & Gent (1980) have discussed in great detail a variety of theories whose accuracy is greater than that of quasi-geostrophic theory. There is little present knowledge about the practical usefulness of the other possibilities raised.

Uniform Potential Vorticity Models

If the potential vorticity is initially uniform, then it remains uniform for all time. Thus (36) is trivial and w is not required. Time dependence enters only through the buoyancy equation (35) with $w = 0$ on horizontal boundaries. Solution proceeds by time marching the boundary equations for Φ_Z and solving the interior equation (37) at each time step.

The source of the energy for baroclinic instability is the potential energy associated with the horizontal temperature contrast and its attendant vertical shear in the horizontal wind. Various basic states $\Phi(Y,Z)$ have been used in different papers. The Eady model has a uniform buoyancy gradient:

$$\overline{\Phi} = N^2 Z^2/2 - fUYZ/H .$$

Other flows in Hoskins & West (1979) are modifications of this but always such that

$$f^{-2}\overline{\Phi}_{YY} + N^{-2}\overline{\Phi}_{ZZ} = 1 ,$$

where $N^2 = q/f^2$ is constant. The fluid is considered to be doubly periodic in the horizontal. The next step is to study the linear stability of the flow and then to initiate an integration of the semi-geostrophic equations using as initial conditions $\overline{\Phi}$ plus a small-amplitude unstable normal mode.

The Eady mode independent of Y is a solution of the nonlinear equations and has been discussed above and illustrated in Hoskins & Bretherton (1972) and Hoskins & West (1979). Eady modes with a finite periodicity in Y have been discussed in Hoskins (1976). These solutions did not produce very realistic frontogenesis. In Hoskins & West (1979) the first flows considered were only minor modifications to the Eady flow in which buoyancy gradients were concentrated more in the center of the region of interest. The most unstable modes were modifications of the two-dimensional Eady modes. These solutions exhibited more realistic frontogenesis.

Here we shall comment only on two particular solutions in Hoskins & West (1979). The first corresponds to a basic flow that is zero at $z = 0$ rising to $U(1 + \cos \ell Y)/2$ at the top. At day 6.3 of the integration the situation at the surface is as shown in Figure 6. A strong surface cold front has formed, there being a tendency to form infinite gradients in velocity and buoyancy in a finite time. A weaker warm front has just started to appear to the northeast of the low. Both fronts are on the "warm side" of the contrast consistent with the two-dimensional theory. As discussed in much more detail in Hoskins & West (1979) the crucial factors to consider are

1. the frontogenetic nature of the large-scale geostrophic motion as shown by the deformation or the vector **Q** [see(5)],
2. the trajectories of fluid particles relative to the system.

The cold front region, particularly towards the low pressure, is a frontogenetic region. There the **Q** vectors point almost directly towards the warm air (Hoskins & Pedder 1980). Fluid particles in a small-amplitude mode move westwards through the system. As the amplitude increases so particles slow down equatorward of the low and move along the cold frontal region in the manner indicated. The warm frontal region is also frontogenetic but particles move quickly through it and around the low. The warm front only really develops when the velocities in the eddy are large enough that fluid particles begin to move northwards and eastwards relative to the system as shown.

The second case to be discussed is for a basic flow $-.15\, U \cos \ell y$ at the lower surface rising to $.85\, U$ at the top. In this case the most unstable linear mode tilts in the horizontal as if advected by the low-level flow. At day 5.5 the surface development is as shown in Figure 7. There is this time no real cold front (except perhaps the rather unrealistic region on the poleward side of the low) but a strong warm front in a different position from that of the weak warm front in the previous example. It can be shown from the **Q** vectors that the previous cold frontal position is no longer frontogenetic but that the new warm frontal position is. The trajectories of fluid particles are such that they move along the developing warm front towards the low-pressure center.

These two examples indicate the possibility in a developing baroclinic wave of either a primary cold front and a secondary warm front (type A) or a primary warm front (type B). The nonlinear frontogenetic tendency to form infinite gradients in a finite time is just as described in the two-dimensional theory. The primary cold front and warm front have very different characteristics. In particular the former has a relatively steep slope and is associated with significant positive vorticity through the depth whilst the latter is relatively shallow in slope and in the depth of positive vorticity. It is shown in Hoskins & Heckley (1981) that these quite realistic differences are associated with the forward tilt with height of the temper-

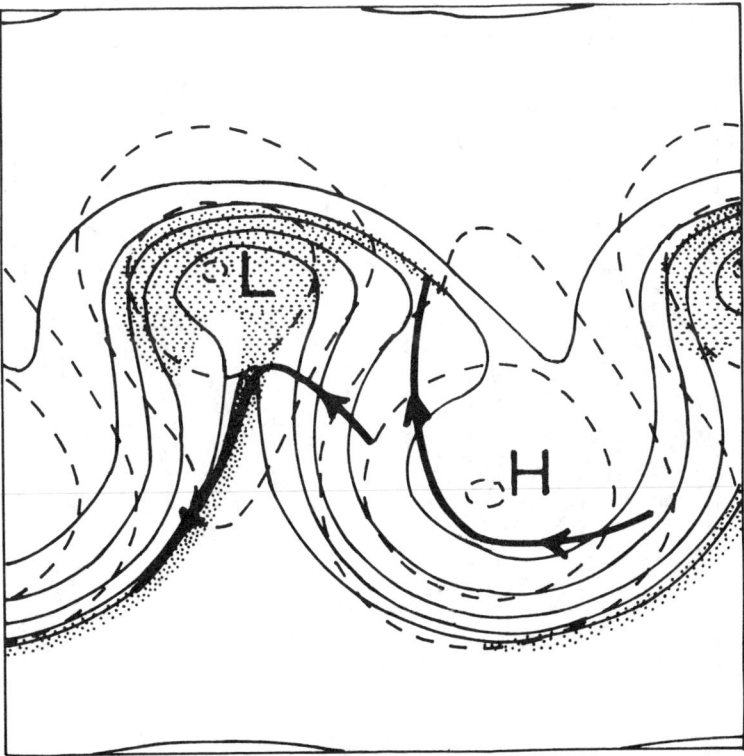

Figure 6 Surface map at day 6.3 for the first case discussed in the text. Temperature contours every 4 K are continuous lines, ϕ contours dashed lines, and the region of relative vorticity larger than $f/2$ is shaded. The relative vorticity in the cold front region has a maximum of $5f$. The bold lines indicate two trajectories relative to the system from day 3.

ature wave in a growing disturbance. The theoretical work of Gidel (1978) should also be mentioned. He attributed some of the differences between warm and cold fronts to the opposite sign of the thermal gradient along the two sorts of front.

Non-Uniform Potential Vorticity Models

Full semi-geostrophic integrations have been described in Heckley (1980). In particular some models have included an explicit stratosphere. The surface frontogenesis is not altered but upper troposphere frontogenesis can be simulated. One point of agreement with Shapiro (1981) and numerous observational studies is that just in the region of the upper-air front the **Q** vectors point towards colder air suggesting the possibility of an indirect circulation with cold air rising and warm air descending. However, in the context of the generally direct forcing in the wave, the negative geostrophic frontogenesis function locally means that descent is strongest on the warm side of the descending stratospheric tongue (see Figure 4).

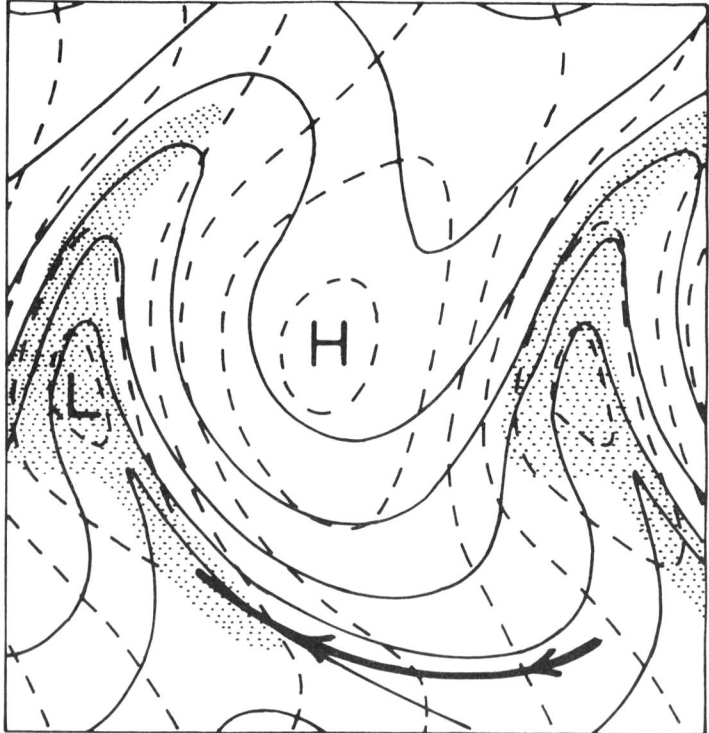

Figure 7 Surface map at day 5.5 for the second case discussed in the text. Conventions are as in Figure 6 except that the temperature contour interval is 8 K.

6. SUMMARY

The analysis described in this paper has indicated why contrasts in velocity and buoyancy in a rapidly rotating fluid tend to be concentrated in frontal regions. The sharpness of these regions depends on the strength of the mixing processes which eventually must balance the frontogenetic mechanism. In frontogenesis all scales are linked directly as discussed by Blumen (1978a,b). If mixing is weak enough that sharp fronts may form, an energy spectrum behaving like a $-8/3$ power of the wavenumber is predicted by semi-geostrophic theory (Andrews & Hoskins 1978).

For a more realistic comparison with frontogenesis in the Earth's atmosphere the modifying effects of various additional physical processes may be considered. Compressibility effects may be added to the theory in a simple manner (Hoskins & Bretherton 1972) and produce only qualitative changes. The feedback on frontogenesis of latent-heat release in the rising warm air may be expected to quicken the process (Orlanski & Ross 1978). Bennetts & Hoskins (1979) suggested that it could also make the effective potential vorticity negative, thereby rendering (16) hyperbolic and the fluid symmetrically unstable. This could be the reason for the frequently banded structure of the precipitation in frontal regions. The interactions of frontal regions with the surface boundary layer and with topographic features are important and not well understood. One example of the latter has been highlighted recently by Baines (1981). As mentioned above, Shapiro (1981) has discussed the effects of turbulent fluxes in the free atmosphere. Williams (1974) exhibited a numerical model in which mixing processes eventually balanced the frontogenetic tendency to produce a steady-state front.

Another important area is that of the stability of fronts. The forecasting of the development of waves on atmospheric fronts is based on poorly understood empirical rules. Some theoretical work has been performed by Eliasen (1960), Orlanksi (1968), Duffy (1976), and Grotjhan (1979).

As discussed in this paper, much progress has been made on the understanding of why fronts form, but our knowledge of the interplay of all the processes occurring in frontal regions is still in its infancy.

ACKNOWLEDGMENTS

The author is very grateful to Dr. David Andrews and Dr. Malcolm MacVean for their speedy and thorough reviews of the first version of this paper.

Literature Cited

Andrews, D. G., Hoskins, B. J. 1978. Energy spectra predicted by semi-geostrophic theories of frontogenesis. *J. Atmos. Sci.* 35: 509–12

Arakawa, A. 1962. Non-geostrophic effects in the baroclinic prognostic equations. *Proc. Int. Symp. Numerical Weather Prediction, Tokyo*, pp. 161–75

Baines, P. G. 1981. The dynamics of the Southerly Buster. *Aust. Meteorol. Mag.* In press

Bennetts, D. A., Hoskins, B. J. 1979. Conditional symmetric instability—a possible explanation for frontal rainbands. *Q. J. R. Meteorol. Soc.* 105:945-62

Bergeron, T. 1928. Uber die dreidimensional verknüpfende Wetteranalyse I. *Geofys. Publikasjoner* 5(No. 6):1-111

Blumen, W. 1978a. Uniform potential vorticity flow: Part I. Theory of wave interactions and two dimensional turbulence. *J. Atmos. Sci.* 35:774-83

Blumen, W. 1978b. Uniform potential vorticity flow: Part II. A model of wave interactions. *J. Atmos. Sci.* 35:784-89

Blumen, W. 1980. A comparison between the Hoskins-Bretherton model of frontogenesis and the analysis of an intense surface frontal zone. *J. Atmos. Sci.* 37:64-77

Blumen, W. 1981. The geostrophic coordinate transformation. *J. Atmos. Sci.* 38:1100-5

Charney, J. G. 1947. The dynamics of long waves in a baroclinic westerly current. *J. Meteorol.* 4:135-63

Duffy, D. G. 1976. The application of the semi-geostrophic equations to the frontal instability problem. *J. Atmos. Sci.* 33:2322-37

Eady, E. T. 1949. Long waves and cyclone waves. *Tellus* 1:33-52

Edelmann, W. 1963. On the behaviour of disturbances in a baroclinic channel. Summary Rep. No. 2, *Research in Objective Weather Forecasting*, Part F. Contract AF61(052)-373. Deut. Wetter., Offenbach. 35 pp.

Eliasen, E. 1960. On the initial development of frontal waves. *Publ. Danske Met. Inst. 13*. 107 pp.

Eliassen, A. 1948. The quasi-static equations of motion. *Geofys. Publikasjoner 17*, No. 3

Eliassen, A. 1959. On the formation of fronts in the atmosphere. *The Atmosphere and the Sea in Motion*, pp. 277-87. NY: Rockefeller Institute Press

Eliassen, A. 1962. On the vertical circulation in frontal zones. *Geofys. Publikasjoner* 24 (No. 4):147-60

Eliassen, A., Raustein, E. 1970. A numerical integration experiment with a six-level atmospheric model with isentropic information surfaces. *Meteorol. Ann.* 5:429-49

Faller, A. J. 1956. A demonstration of fronts and frontal waves in atmospheric models. *J. Meteorol.* 13:1-4

Fjortoft, R. 1962. On the integration of a system of geostrophically balanced prognostic equations. *Proc. Int. Symp. Numerical Weather Prediction, Tokyo*, pp. 153-59

Fultz, D. 1952. On the possibility of experimental models of the polar front wave. *J. Meteorol.* 9:379-84

Gidel, L. T. 1978. Simulation of the differences and similarities of warm and cold surface frontogenesis. *J. Geophys. Res.* 83:915-28

Gill, A. E. 1981. Homogeneous intrusion in a rotating stratified fluid. *J. Fluid Mech.* 103:275-95

Grotjhan, R. 1979. Cyclone development along weak thermal fronts. *J. Atmos. Sci.* 36:2049-74

Heckley, W. A. 1980. *Frontogenesis*. PhD thesis. Univ. Reading. 209 pp.

Hoskins, B. J. 1971. Atmospheric frontogenesis: some solutions. *Q. J. R. Meteorol. Soc.* 97:139-53

Hoskins, B. J. 1972. Non-Boussinesq effects and further development in a model of upper tropospheric frontogenesis. *Q. J. R. Meteorol. Soc.* 98:532-41

Hoskins, B. J. 1974. The role of potential vorticity in symmetric stability and instability. *Q. J. R. Meteorol. Soc.* 100:480-82

Hoskins, B. J. 1975. The geostrophic momentum approximation and the semigeostrophic equations. *J. Atmos. Sci.* 32:233-42

Hoskins, B. J. 1976. Baroclinic waves and frontogenesis. Part I: Introduction and Eady waves. *Q. J. R. Meteorol. Soc.* 102:103-22

Hoskins, B. J., Bretherton, F. P. 1972. Atmospheric frontogenesis models: mathematical formulation and solution. *J. Atmos. Sci.* 29:11-37

Hoskins, B. J., Draghici, I. 1977. The forcing of ageostrophic motion according to the semi-geostrophic equations and in an isentropic coordinate model. *J. Atmos. Sci.* 34:1859-67

Hoskins, B. J., Heckley, W. A. 1981. Cold and warm fronts in baroclinic waves. *Q. J. R. Meteorol. Soc.* 107:79-90

Hoskins, B. J., Pedder, M. A. 1980. The diagnosis of middle latitude synoptic development. *Q. J. R. Meteorol. Soc.* 106:707-19

Hoskins, B. J., West, N. V. 1979. Baroclinic waves and frontogenesis. Part II: Uniform potential vorticity jet flows—cold and warm fronts. *J. Atmos. Sci.* 36:1663-80

Katz, E. J. 1969: Further study of a front in the Sargasso Sea. *Tellus* 21:259-69

MacVean, M. K., Woods, J. D. 1980. Redistribution of scalars during upper ocean frontogenesis: a numerical model. *Q. J. R. Meteorol. Soc.* 106:293-312

McWilliams, J. C., Gent, P. R. 1980. Intermediate models of planetary circulations in the atmosphere and ocean. *J. Atmos. Sci.* 37:1657–78

Miller, J. E. 1948. On the concept of frontogenesis. *J. Meteorol.* 5:169–71

Mudrick, S. E. 1974. A numerical study of frontogenesis. *J. Atmos. Sci.* 31:869–92

Orlanski, I. 1968. Instability of frontal waves. *J. Atmos. Sci.* 25:178–200

Orlanski, I., Ross, B. B. 1978. The circulation associated with a cold front. Part II: Moist case. *J. Atmos. Sci.* 35:445–65

Pedlosky, J. 1979. *Geophysical Fluid Dynamics*. NY: Springer. 624 pp.

Reed, R. J., Danielsen, E. F. 1959. Fronts in the vicinity of the tropopause. *Arch. Meteorol. Geophys. Bioklim.* A11:1–17

Sawyer, J. S. 1956. The vertical circulation at meteorological fronts and its relation to frontogenesis. *Proc. R. Soc. London Ser. A* 234:346–62

Shapiro, M. A. 1975. Simulation of upper-level frontogenesis with a 20-level isentropic coordinate primitive equation model. *Mon. Weather Rev.* 103:591–604

Shapiro, M. A. 1981. Frontogenesis and geostrophically forced secondary circulations in the vicinity of jet stream-frontal zone systems. *J. Atmos. Sci.* 38:954–73

Stone, P. H. 1966. Frontogenesis by horizontal wind deformation fields. *J. Atmos. Sci.* 23:455–65

Voorhis, A. D., Hersey, J. B. 1964. Oceanic thermal fronts in the Sargasso Sea. *J. Geophys. Res.* 69:3809–14

Williams, R. T. 1967. Atmospheric frontogenesis: a numerical experiment. *J. Atmos. Sci.* 24:627–41

Williams, R. T. 1968. A note on quasi-geostrophic frontogenesis. *J. Atmos. Sci.* 25:1157–59

Williams, R. T. 1974. Numerical simulation of steady-state fronts. *J. Atmos. Sci.* 31:1286–96

Williams, R. T., Plotkin, J. 1968. Quasi-geostrophic frontogenesis. *J. Atmos. Sci.* 25:201–6

DYNAMICS OF LAKES, RESERVOIRS, AND COOLING PONDS

Jörg Imberger
Department of Civil Engineering, University of Western Australia, Nedlands 6009, Western Australia

Paul F. Hamblin
Canada Center for Inland Waters, Burlington, Ontario L7R 4A6, Canada

1. INTRODUCTION

In this article we review recent observational and theoretical studies of the water motion and the distribution of heat and material substances within a contained body of water. A lake is defined as a naturally occurring body of water, with both the inflow and the outflow regulated by the natural riverine system. On the other hand, a reservoir is a man-made impoundment that has a controlled throughflow and very often a subsurface outlet. With the exception of the Australian billabongs, reservoirs have in general a larger throughflow, a considerably shorter average residence time, and usually a severely fluctuating water level. These differences induce a considerably different ordering of the important physical mechanisms, which may, in turn, result in a quite different water quality. Water quality is defined here quite generally as the concentration set of all dissolved and particulate substances as well as living organisms. We shall treat cooling ponds together with side arms and embayments in Section 6.

The scales of motion in a lake encompass the turbulent field, an internal and surface wave field, a gyre field (non-overturning motions), and finally a wind- or buoyancy-driven mean motion. In general, the turbulent components result from actively overturning motions and extend from millimeters to meters, the internal waves from meters to kilometers, and the gyres are embedded in the mean motion with a scale length very much dependent on the shape, size, and location of a particular lake. The corresponding time scales of the motion range from a fraction of a second to the seasonal time scale. The time scales governing the water quality extend from milliseconds (chemical reactions) to many years (sediment-water interactions). The multiplicity of temporal and spatial scales interacting in a complex basin geometry with a wide range of imposed forcing functions gives rise to a large number of physical phenomena in lakes and reservoirs.

Two very fine and reasonably current reviews of lake hydrodynamics already exist (Mortimer 1974, Csanady 1975), both emphasizing the dynamics of waves in large lakes. For this reason we place greater emphasis here on the mixing mechanisms and gravitational adjustments found in medium and small lakes and reservoirs. Since the above reviews were written a great many reviews have been published emphasizing individual mechanisms relevant to physical limnology. Internal waves are highlighted by Garrett & Munk (1979) and LeBlond & Mysak (1978); mixed-layer dynamics is the subject of the book edited by Kraus (1977); microstructure in mixing forms the topic of the reviews by Gregg & Briscoe (1979) and Garrett (1979b); thermohaline fine structure is fully treated by Fedorov (1978); progress in the understanding of frontogenesis is reviewed in the book edited by Bowman & Esaias (1978); mixing in lakes and reservoirs forms a chapter in the recent book by Fischer et al. (1979) and is also discussed in the article by Sherman et al. (1978); the dynamics of outflows is summarized by Imberger (1980); the problems encountered in numerical modeling of motions in large lakes was reviewed by Lick (1976); water quality in small impoundments is reviewed in great detail by Orlob (1977) and by Chen & Orlob (1978); and, finally, the research concerning water-quality criteria is reviewed by Steele & Stefan (1979). The formation of ice, numerical-simulation techniques, observational methods, and the construction of water quality models are not discussed in this review.

2. EXTERNAL FORCES AND LAKE MODELING

The forcing on a lake may be represented schematically as shown in Figure 1 (p. 161), where the symbols used in this review are also defined. Together, the forcing must determine not only the mean motions, but also the internal wave spectra and the turbulent fluctuations in the lake. Hence, specification of the appropriate nondimensional groups, derivable from these inputs only, must fix the flow and the mixing regimes in the lake. Further, since most fluid motions are governed by a simple balance of two or at most three forces we are able to proceed through the various classes of motion one by one.

Wind-Forced Motions

BAROTROPIC SEICHING   A wind stress on the water surface initiates a tilting motion of that surface leading to what are called storm surges. The water accelerates under the action of the applied force $O(\tau LB)$ until a reverse pressure force $O(\rho_0 g H^2 B)$ due to the sloping water surface is built up. Balancing these two forces leads to the nondimensional group

$$W_s = gH/u_*^2 \cdot H/L. \tag{1}$$

INTERNAL SEICHING   In a lake where the water has a definite two-layer character, the wind-induced flow in the upper layer is balanced by a counterflow in the lower layer causing the pycnocline to tilt. Once again, the tilting will continue until the applied wind force is balanced by the pressure force, but now the pressure force arises from the tilting interface and g must be replaced by $g' = \Delta_0 g$ in Equation (1). This leads to the nondimensional parameter $W = g'd/u_*^2 \cdot d/L$, first introduced by Thompson & Imberger (1980), who called it the Wedderburn number. If W is large, only weak motions are induced at the interface, but as W decreases the interface will tilt more and more, until at $W \sim 1$ the interface will surface at the upwind end.

COASTAL JETS   In a large lake, the earth's rotation enters the above simple force balance (Csanady 1975). Once again, consider a wind aligned along the length of the lake. If the lake is quite long then the upper layer will accelerate such that $u \sim u_*^2 t/d$, until the long waves originating at the two ends have propagated back into the center of the lake. As Spigel & Imberger (1980) have shown, this will take a time $O(L/\sqrt{g'd})$ which may be long compared with the inertial period $2\pi f^{-1}$. In such cases the transverse flow, induced by the Coriolis force $O(\rho_0 fudBL)$ will build up a transverse pressure force, which will have a value $O(\rho_0 g'd^2 L)$ at the initiation of upwelling. The ratio of the two forces yields once again a parameter $W_T = g'd/u_*^2 \cdot d/B \cdot 1/Tf$, where $T = $ minimum $(T_W, 2\pi f^{-1}, T_i/4)$. If T_W is the minimum, then obviously the cessation of the wind stops any build up of the motion, the time $T_i/4$ is the cut-off if the long-waves from the ends are first to build up the longitudinal pressure gradient and $2\pi f^{-1}$ is relevant for large lakes where the earth's rotation turns the velocity vector. In a particular lake Tf is a fixed number and so $W_T \approx W$.

For barotropic motions g' must be replaced by g and d by H, and for a continuously stratified lake g' should be replaced by $H(\overline{N})^2$, where $\overline{N}$ is a mean buoyancy frequency.

TOPOGRAPHIC GYRES   Csanady (1975) has classified all motions that are the result of a balance between wind stress and bottom friction as topographic gyres. In a homogeneous shallow lake the topography thus completely determines the configuration of the gyre. Neglecting rotation and stratification, the balance at steady state will be a local balance between the applied stress $O(\rho_0 u_*^2)$, the bottom resistance $\rho_0 U_*^2$ where U_* is the bottom-shear velocity and the reverse pressure force. Since at steady state $u \sim u_* \sim U_*$ the ratio of the bottom to the top stress is just C_D, the bottom drag coefficient.

Free-Wave Motion

Many different types and modes of basin-scale motions are possible in a lake that is stratified. Csanady (1975) and Mortimer (1974) give a full account of the mathematical complexities of these motions. The surface modes depend on g and f while the baroclinic modes are determined by g' or N and f. Long waves can be defined by a phase speed c_n, given by the usual long wave approximation (see Fischer et al. 1979). The character of a wave, whether it be a free internal mode, a Poincaré wave, or a Kelvin wave, is given by the ratio,

$$P = c_n/Bf. \tag{2}$$

The wind-disturbance pattern obviously determines the amplitude of each wave component, but for $P < 1$ Kelvin waves are excited and circumnavigate the lake boundaries, for $P \sim 1$ Poincaré waves fill the whole basin (Mortimer 1971), and for $P > 1$ the internal waves propagate across the basin in a fraction of the time the earth takes to rotate the basin and are therefore simple internal seiches unaffected by the rotation. It is perhaps interesting to remember that for a continuously stratified fluid c_n decreases with the vertical wave number and so for the jet structures initiated by selective withdrawal, intrusions, and double diffusion, rotation will nearly always limit the transverse horizontal extent of the jet structure.

Velocity Concentrations

INSTABILITIES, FLOW CONCENTRATIONS, FLOW BLOCKING   Often in lakes the flow velocity is increased by some type of energy focusing. This could be due to an instability, a contraction of the flow, or a vertical sill preventing horizontal movement. In all these cases inertia may enter the force balance and so to assess this likelihood we must estimate the Froude number (internal and surface).

$$F_i = u/\sqrt{g'd}, \tag{3}$$

$$F = u/\sqrt{gH}, \tag{4}$$

and second, the Rossby number

$$R_0 = u/fL. \tag{5}$$

In spirit these numbers are similar to P, except that the information is carried here by u rather than c_n.

For small Froude numbers the flow will be smooth with both the upstream and downstream levels acting as controls. Upstream control is called blocking and for small Froude numbers the fluid moves in horizontal planes around obstructions. Supercritical flow has sufficient energy, on the other hand, to overcome these obstructions. The Rossby number influences inertially dominated flows such as coastal jets. However, most motions in lakes affected by the earth's rotation are dominated by a geostrophic balance, with inertia playing only a minor role.

FLOW SEPARATION   The vertical vorticity generated at the boundaries of the lake by the longshore pressure gradient often causes separation at headlands or spits. This is a very common occurrence, but so far seems to have been neglected by researchers. The relevant ratio is obtained by comparing the advection time of the vorticity $O(\ell u^{-1})$ with the diffusion time $O(\ell^2 A_x^{-1})$. This yields the turbulent Reynolds number

$$\text{Re} = \ell u / A_x . \tag{6}$$

The process is, however, complicated by the sloping bottom, which causes a stretching of the vertical vortex lines as they propagate past the spit.

Inflow

PLUNGE POINT   A river entering a lake has normally a different density from that of the water at the surface of the lake. If it is heavier, then the water will plunge beneath the reservoir at the point where the buoyancy force equals the inertia force. If it is lighter, it will overflow at the same point. A simple buoyancy-inertia balance leads to the ratio

$$F_i = Q_i^2 / g' \tilde{H}^5 , \tag{7}$$

where $\tilde{H}$ is the channel depth at the entrance to the main basin. If this internal Froude number is very small then the plunge point will be located far up the river bed where it is shallower. Increasing the Froude number causes entrainment leading to a balance similar to that of the mixed layer.

UNDERFLOW   For definiteness we consider only the underflowing river case for a very mild slope angle α. The downflow will be governed by a balance of the gravitational force down the slope $O(\Delta_i \rho_0 g \tilde{H}^2 \tilde{L} \alpha)$ and the bed friction force $O(C_D \rho_0 u^2 \tilde{L} \tilde{H})$. The ratio yields the modified Froude number

$$F_m = C_D Q_i^2 / \alpha g' \tilde{H}^5 \tag{8}$$

If F_m is large, then the flow velocity is large and the entrainment velocity u_e at the interface will dominate the momentum balance. Since the entrainment can either be due to shear production at the interface or surface production of turbulent kinetic energy at the bed,

$$R_i^* = g'h/U_*^2 \tag{9}$$

also enters the dynamics.

INTRUSIVE FLOW   The underflow entrains lake water and at a certain depth reaches neutral buoyancy. At this depth the underflow leaves the river bed and propagates horizontally into the lake. For an inertia-buoyancy intrusion the thickness will be $O(Q_i^2/N^2\tilde{B}^2)^{1/4}$ for a two-dimensional flow and $O(Q_i^2/N^2)^{1/6}$ for a three-dimensional flow (see Imberger 1980). For the two-dimensional flow this leads to the thickness-to-depth ratio squared:

$$F_Q = Q_i/N\tilde{B}H^2. \tag{10}$$

For an intrusion governed by a balance between viscosity and buoyancy, Imberger (1980) showed that the ratio of the layer thickness to the depth is given by

$$\frac{\delta}{H} = (A_z/NH^2)^{1/3} \cdot (L/H)^{1/3}. \tag{11}$$

To determine which type of layer forms, we must compute the ratio

$$\tilde{R} = \frac{Q_i}{\tilde{B}} (A_z^2 L^2 N)^{-1/3}. \tag{12}$$

In a large lake, the Coriolis force $O(\rho_0 f Q_i)$ may be sufficient to compress the inflowing fluid layer δ, which has a velocity $u = O(N\delta)$ against the lake bank. For such a flow we must balance the transverse pressure force $O(\rho_0 N^2 \delta^3)$ against this Coriolis force and conserve mass $[\delta b u = O(Q_i)]$ so that the horizontal extent at the layer,

$$b = O\left(\frac{Q_i N}{f^2}\right)^{1/3}. \tag{13}$$

The ratio b/B thus becomes equal to P [Equation (2)]. For a surface intrusion a similar balance leads to the parameter

$$P_i = \left(\frac{Q_i g'}{f^3 B^4}\right)^{1/4}. \tag{14}$$

Outflow

As shown in Imberger et al. (1976), the outflow dynamics are governed by the same scaling laws as the inflow intrusion and the ratios $F_Q, \tilde{G}_r, \tilde{R}$ apply.

Lake Heating

The heat enters the lake at the surface, some being absorbed immediately in the first few centimeters while the short-wave radiation may penetrate to great depth. Any modeling must take care that the ratio ad is carefully matched, where a is the overall absorption inverse length scale. If ad is small, then the heat will penetrate to great depth, and little net buoyancy is added to the water column. If ad is large, then the water will quickly stabilize during a heating phase.

Turbulence in the Mixed Layer; Mixed-Layer Deepening

The ratio W determines the slope of the pycnocline and, therefore, it is naturally the dominant ratio determining the behavior of mixed-layer deepening. If W is small, we may expect severe upwelling, active billowing, and rapid deepening. This deepening may, however, be offset by a positive buoyancy flux at the surface due to the heat input. Defining the heat input per unit time

$$H^* = \int_0^\infty I(x, y, t; \lambda)d\lambda + \text{other inputs} , \qquad (15)$$

where λ is the radiation wavelength, one finds that for ad large the ratio of the buoyancy flux to the stirring energy becomes

$$M = \alpha g d H^* / c_p \rho_0 u_*^3 . \qquad (16)$$

This is the ratio of the depth d to the familiar Monin-Obukhov length.

Mixing in the Hypolimnion

Once again a global parameterization must be able to describe at least the relevant dependent parameters. However, so far, as will be seen in Section 4, our understanding does not allow us to go beyond saying that W, H/L, and M must somehow combine to determine the mixing in the interior of the fluid and at the boundaries. When W is small, mixing most likely will be vigorous. When double diffusion prevails, then in addition to the above numbers, the ratio $\beta \Delta S / \alpha \Delta T$, as well as the Rayleigh number, influences the vertical and horizontal transport, where $\beta \Delta S$ and $\alpha \Delta T$ are the buoyancy jumps due to the salinity and temperature jumps.

Dynamics of Side Arms and Cooling Ponds

An embayment in a lake is usually formed from a drowned river valley. Consider, first, the case where the river is flowing with a discharge Q_i and a density anomaly $\Delta_i \rho_0$. For very weak flows, the value of the internal Froude number sets the plunge point. As the discharge is increased, billowing commences at the nose, causing entrainment and a compensating flow in the stationary layer. The entrainment causes an effective stress, $\rho_0 \overline{u'w'}$ at the interface. For an underflow where bottom friction dominates, this will be $O(\rho_0 U_i^2)$ and for a layer in which shear production dominates it will be $O(\rho_0 u^2)$. The ratio of this interfacial stress force to the maximum pressure force (see section on forced internal seiching) leads once again to a modified Wedderburn number:

$$\tilde{W} = (g'\tilde{H})/(Q^2/\tilde{B}^2\tilde{H}^2) \cdot \tilde{H}/\tilde{L} . \tag{17}$$

This is basically the inverse of the pond number postulated by Jirka & Watanabe (1980) in their cooling-pond work. Small $\tilde{W}$ leads to an active vertical-mixing situation with a longitudinal gradient. On the other hand, for large $\tilde{W}$ the embayment becomes stratified.

Once the flow ceases, the adjustment of the longitudinal gradient $\Delta_i\rho_0/\tilde{L}$ takes place by natural convection. A simple balance (see Imberger 1974) of the horizontal pressure gradient $O[(g\Delta_i\rho_o \tilde{H}^2\tilde{B})/L]$ and the friction force induced by the motion $O[\rho_o A_z u \tilde{B}/\tilde{H}]$ leads to the estimate

$$u \sim g\Delta_i \tilde{H}^3 / A_z \tilde{L} . \tag{18}$$

At the end of the embayment the flow must turn and so inertia forces are set up. Because the appropriate length scale in the end region is $\tilde{H}$, a ratio of the inertia forces to the viscous forces (the term retained) is the Reynolds number,

$$\text{Re} = g'_i \tilde{H}^4 / A_z^2 \tilde{L} . \tag{19}$$

If it is remembered that $A_Z \sim u'H$, where u' is the turbulent velocity scale, then

$$\text{Re} = g'_i \tilde{H}/u'^2 \cdot \tilde{H}/\tilde{L} = W . \tag{20}$$

So we see that the dissipation of the longitudinal gradient is also governed by W, but with $Q/\tilde{B}\tilde{H}$ replaced by u'. For Re small, the flow is parallel and the density gradient linear. On the other hand, if Re is large, closed recirculation forms. More complicated structures form if the basin geometry is explicitly accounted for.

3. SURFACE MIXED LAYER

The so-called mixed layer is that region of water near the surface that is characterized by a uniform temperature and salinity. The epilimnion, as the mixed layer is called in limnology, is confined to the surface and is usually separated from the deeper waters by a rapidly varying density layer called the metalimnion. The stability of this metalimnion has led many past workers (for a summary see Kraus 1977) to think of the epilimnion as a slab of mixed water moving in a uniform coherent fashion (Figure 1); the surface mixing derives its energy from the surface-wind stress and a cooling-surface buoyancy flux.

Very near the surface, wave action and surface stress lead to a "law of the wall" boundary layer as recently shown by Dillon et al. (1981). Below this, the mean shear develops similarly to the bottom boundary layer, forming an Ekman layer (see Davis et al. 1981 and Svensson 1979) to a depth where the pycnocline prevents any further penetration. The bulk of the mixed layer is thus made up of a homogeneous boundary layer in which the shear is concentrated in the surface "law of the wall" layer, but which still has appreciable shear in the remainder (enough to cause appreciable production of turbulent kinetic energy).

The metalimnion is where the density changes rapidly, where the turbulent fluxes of momentum, heat, and mass are damped by the increased buoyancy and where the velocity changes from the high mixed-layer value

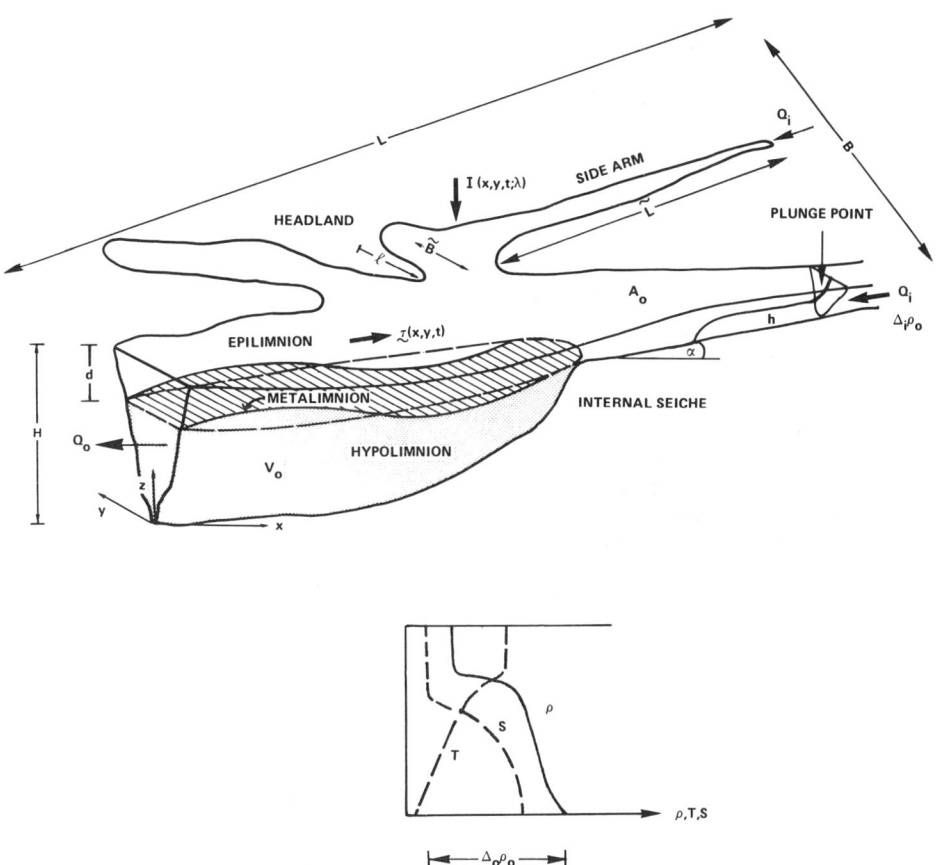

Figure 1 Schematic of reservoir, defining symbols used in the text.

to the very small velocity in the hypolimnion. It is, therefore, a region of high shear and thus also production of turbulent kinetic energy.

On the other hand, when strong cooling takes place at the surface, the induced natural convection penetrates the surface layers and once again sets up a uniform mixed layer (Denton 1978).

The energetics of this surface layer are thus sustained by a so-called stirring source $O(\rho_0 u_*^3)$ originating in the "law of the wall" layer, a shear production term $O(\rho_0 u^2 dd/dt)$ derived from the metalimnion shear, a shear production input $O(\rho_0 u_*^2 u)$ from the shear in the main mixed layer, and, lastly, a buoyancy-flux source $O[(\alpha g d H^*)/c_p]$. These inputs are redistributed within the layer by turbulent diffusion and Langmuir circulation but are mostly dissipated near the source of production (Dillon & Powell 1979). Some of the energy remains to increase the potential energy of the system and so cause a deepening of the mixed layer.

Under weak wind conditions, or in the initial stages, the stirring and natural convection sources dominate the process (Spigel & Imberger 1980). However, for strong winds, shear production is the main source of energy once the surface velocities have become appreciable (Thorpe et al. 1977). The shear production mechanism, however, ceases once the interface has set up and the baroclinic circulation dominates the flow. This one-dimensional description must be augmented by what Blanton (1973) calls "leakage around the edge." Thompson & Imberger (1980) have shown with numerical simulations that once the Wedderburn number W is lower than about 3–4, quite appreciable upwelling is caused upwind in a small lake or near the sides in a large lake (see Lee & Mysak 1979). The deepening then takes place by a progressive movement of the upwelling front downwind. As the wind shifts, or relaxes, a gravitational adjustment of the upwelled water causes large intrusions into the metalimnion which eventually degenerate into second-mode solitons moving within the pycnocline (Maxworthy 1980). Uneven winds (Farmer 1978) or a periodic wind forcing (Thorpe 1974) leads to similar surges in and on the metalimnion.

The evolution of the mixed layer is different from that in the atmosphere since the time scale (d/u_*) is comparable to the wind-forcing time scale T_W, whereas in the atmosphere the adjustment is much shorter, leading to a rapid development of equilibrium layers. The mixed layer in a lake is rarely at equilibrium and evolves by the propagation of fronts from the surface down into the water in a direct response to wind increases (Rayner 1980 and Price 1978). The seasonal mixed layer may thus conveniently be thought of as an accumulation of individual diurnal mixed-layer events.

Modeling of the mixed layer has taken two very definite paths, although in principle they account for the same energy sources and sinks: (a) integral methods (Kraus 1977, Imberger & Patterson 1980), and (b) closure schemes (Mellor & Durbin 1975 and Svensson 1978).

In the past, the integral methods have neglected the layer-shear-production term $O(\rho_0 u_*^2 u)$ because all assumed a constant velocity in the epilimnion, and thus many workers claim that the closure schemes are a better representation of the dynamics. However, in principle there is no reason why such a term could not be added to the integral methods and then the two approaches would have the same number of empirical "tuning" constants. Both assume essentially a local balance in the turbulent kinetic-energy equation. The integral methods average over the billows and it is easy to superimpose a thickening of the metalimnion to account for the billowing (Imberger & Patterson 1980). On the other hand, large turbulent structures are difficult to resolve in a closure scheme and it is not surprising that Svensson (1978) found considerable divergence of the results from his closure schemes from observations during periods of active billowing. Further, in principle it is feasible to add other deepening algorithms to the integral formulation, in order to account for differential deepening, boundary intrusions, and surges by averaging over these several events and including separate parameterizations. By comparison, however, the closure-scheme formulation would require resolution of the gravitational adjustments (these have a vertical scale of only about .5 m); this would require a prohibitive amount of computer time.

Integral methods as first proposed by Niiler (1975) and summarized by Sherman et al. (1978) are very economical to run and have proven very successful both in the ocean (Davis et al. 1981) and in lakes (J. Patterson et al., in preparation). The efficiencies of energy conversion have been empirically determined and the values reliably reproduce deepening in both large and small lakes.

Dillon & Caldwell (1978) have provided documentation of a severe overturning event in an ocean thermocline. Their simple scaling analysis implies that the depth of the thermocline is mostly due to a few isolated severe deepening events. This explains the success of the models using only shear production as the energy source (Price 1978 and Price et al. 1978). In lakes the situation is somewhat different. Wind mixing, as distinct from surface cooling, is still dominated by shear production as explained by Spigel & Imberger (1980), but the combination of stirring and a negative buoyancy flux determine the mixed-layer depth in many small lakes (Dillon & Powell 1979, Stefan & Ford 1975, and Imberger & Patterson 1980) simply because lakes are more sheltered from the wind than the open ocean.

The simulation of the mixed layer in Kootenay Lake ($L \sim 100$ km) by J. Patterson et al. (in preparation) using an integral model that accounts for stirring, buoyancy flux, and shear production, reproduces the temperature and depth accurately. However, even though the simulation contained an algorithm to thicken the metalimnion by billowing, the metalimnion was consistently too thin. On the other hand, Rayner (1980) simulated diurnal mixing events in the Wellington Reservoir with much the same model and found good agreement for the metalimnion thickness of diurnal events. Internal seiching would lead to an apparent thicker metalimnion in Kootenay Lake since the profiles were averaged over the basin, but even correcting for this, the metalimnion simulated by the model was too thin. This is further strong evidence that the generation and propagation of intrusions may be important in mixed-layer deepening in lakes and reservoirs.

Special problems are encountered when the mixed layer temperature is near 4°C, and as Farmer (1981) shows these may lead to subtle inversions.

4. HYPOLIMNETIC MIXING

The horizontal and vertical mixing of mass, heat, and momentum in the weakly stratified hypolimnion is one of the most important processes requiring study since it directly affects the health of the lake (Imberger 1977). Great advances have been made in the understanding of mixing in the ocean (see Gregg & Briscoe 1979 and Garrett 1979b for comprehensive reviews) and it is now possible to identify, at least conceptually, energy-flux paths in the ocean. The same concentration of effort has not taken place in limnology. In the following discussion we attempt to apply the knowledge gained from oceanic conditions to the hypolimnion of a lake.

During quiet periods in the summer heating cycle the hypolimnion as a whole will normally be quiescent with only internal waves moving through the water. Usually a great deal of fossil structure remains and this may act to distort the internal-wave field. At the start of an external forcing event (storm, inflow, outflow, differential heat input) basin scale motions will be excited. Often smaller-scale internal waves are also generated in the vicinity of the region of excitation and these propagate into the main basin. This leads to the establishment of a fairly continuous broad spectrum of internal waves that quickly fill the whole lake basin. As such, these wave motions do not usually induce sufficient shear for mixing (Wunderlich & Fan 1971). However, interaction with a lake boundary or internal interactions may lead to the appearance of isolated sporadic turbulent patches (Gibson 1980). Further, inflows and intrusions initiated by the large-scale forcing may lead to a rearrangement of the temperature-dissolved substance field such that double-diffusive instabilities release potential energy of one of the constituents and so cause mixing (Turner 1973).

Internal and boundary mixing usually involves some critical-flow condition at which an active event erupts, leading to an isolated and often sporadic overturn of the stable stratification. This introduces vertical scales of motion ranging from the Osmidov (1965) scale $(\epsilon/N^3)^{1/2}$ to the Kolmogorov dissipation scale $(v^3/\epsilon)^{1/4}$. The turbulent kinetic energy contained in this mixing patch is drained in the first few minutes by local dissipation and then progressively by the internal waves radiating energy away from the disturbance, resulting in an increasingly fossil step-like density structure. Steps ranging from centimeters to the size of the patch itself become embedded in the mixed area (Gregg & Broscoe 1979). The size of the patch is set by the Osmidov scale for isolated energy inputs (Kelvin-Helmholtz billows) but may also have a rather larger scale if the instability is brought about by an intrusion or has its own characteristic larger scale as would be the case for a wave–mean shear interaction. The potential energy locked in this finescale and microstructure remains until it is mixed by a subsequent event or dissipated by weak gravitational adjustment and molecular diffusion. Thus, it may take days or even years (Armi & D'Asaro 1980) for the Cox number, $\overline{(\nabla T'^2)}/(\partial T/\partial z)^2$, to decrease from its equilibrium value $O(\epsilon/DN^2)$ even though the structure is completely fossil and the diffusion is by molecular processes.

No direct measurements of the local value of the vertical diffusion coefficient K_z or the vertical vorticity diffusion coefficient A_z have been made either in the ocean or in lakes. We must therefore rely completely on estimates of the basin area average $\overline{K}_z$ or estimates from the spreading of large dye clouds. Many estimates exist for this, all based on global thermal or tracer budgets. Imboden & Emerson (1977), using radon and phosphorus, arrived at a number of 0.2–$0.05 \times 10^{-4}\,\text{m}^2\,\text{s}^{-1}$ in the Greifen See, the larger values applying to the bottom boundary layer. Kullenberg el al. (1974) used the growth of large dye clouds in Lake Ontario to obtain a value of 1–$50 \times 10^{-6}\,\text{m}^2\,\text{s}^{-1}$. Weiss et al. (1979) computed 2–$10 \times 10^{-5}\,\text{m}^2\,\text{s}^{-1}$ for Lake Constance and Robarts & Ward (1978) obtained $2 \times 10^{-5}\,\text{m}^2\,\text{s}^{-1}$ in Lake McIlwaine. Values from Castle Lake by Jassby & Powell (1975) of only $2 \times 10^{-6}\,\text{m}^2\,\text{s}^{-1}$, reflect the small size of the lake. Many other investigations could be cited but all would confirm that the lake average vertical transport lies between the molecular value and about $10^{-4}\,\text{m}^2\,\text{s}^{-1}$, depending on the lake, shape and size, its location, and the estimation technique. Ward (1977) has summarized some of these data and suggested an empirical formula:

$$\overline{K}_z = 4.9 \times 10^{-12} g A_0^{1/2}/N, \tag{21}$$

where A_0 is the surface area of the lake and N is the local buoyancy frequency. On the other hand, Murthy (1976) has summarized the empirical state for horizontal diffusion and suggested the equation

$$\overline{K}_x = 1.8 \times 10^{-4} \sigma_x^{1.33}, \tag{22}$$

where σ_x is the standard deviation in meters of the cloud in the direction of flow. Transverse values were generally somewhat less. The coefficients were derived from measurements in Lake Ontario. Neither Equation (21) nor (22) involved the forcing energy and so must obviously be regarded as the time-averaged values with the coefficients applying only to the geographic locations where the data was collected.

Little information exists regarding the vertical diffusion of momentum, although Ivey & Imberger (1978) computed a ratio $\overline{A}_z/\overline{K}_z$ of about 20 from measurements of the withdrawal-layer thickness.

So far we have outlined the general energy flow leading to the mixing in the hypolimnion and have given values of the mean diffusion coefficient measured in lakes. In an effort to reconcile the two, we must look more closely at the mechanisms actually responsible for the mixing.

Boundary Mixing

In basins with mildly sloping sides, basin-scale motions will induce quite severe shear adjacent to the bottom boundary (Fischer et al. 1979). These velocities could be steady if the motion arose from topographic gyres, oscillating cross shore if simple internal-wave sloshing were set up or long shore if the internal waves were affected by the earth's rotation. Imberger et al. (1978) and Armi (1978) postulated that such bottom shear induces a turbulent bottom boundary layer in which the transport of mass is enhanced. Following Armi (1978) an equivalent vertical diffusion coefficient $\overline{K}_z$ may be defined,

$$\overline{K}_z = \left(\frac{Ph}{A_T}\right) K_u, \qquad (23)$$

where P is the perimeter of the lake at that depth, h is the thickness of the turbulent boundary layer, A_T is the plan area of the lake at that depth, and K_u is the turbulent cross-shore eddy coefficient applicable to the turbulent boundary layer. Equation (23) is intuitively pleasing but so far no data have been obtained to substantiate the existence of such a layer, nor is it possible to define h or K_u unambiguously. The measurements of Tyler & Buckney (1974) indicate a very stable boundary layer of dissolved manganese and iron in some of the Tasmanian lakes. This, in fact, would indicate that the bottom boundary layer does not always have to be turbulent even when there is internal-wave activity in the lake. Presumably such a layer does become active (Caldwell & Chriss 1979) when the velocities become appreciable. Garrett (1979a) suggests an estimate of $0.15hu_*$ for K_u, this being the transverse diffusion coefficient in a wide-open channel (see Fischer et al. 1979). This choice would be appropriate for a long-shore-current situation. The value of h could be taken as the "law of the wall," $0.03u_*/f$ (Smith & Long 1976), or even as the rotating velocity-defect thickness given by the Ekman layer $0.4u_*/f$. For a strongly stratified lake the bottom mixed layer would be more appropriate (Armi & D'Asaro 1980). For an oscillating boundary layer, on the other hand, as would form in small lakes Fischer et al. (1979) suggest a value of $h = a\beta e/133$, where a is the amplitude of motion, e is the roughness length, and β is the inverse of the Stokes thickness. An equally suitable estimate for such oscillating layers is given by Grant & Madsen (1979) who suggest a layer thickness equal to $2kU_*/\omega$ where k is the von Kármán constant, U_* is the maximum shear velocity, and ω is the oscillation frequency. Substituting $k = 0.41$, $U_* = 3 \times 10^{-3}\,\text{ms}^{-1}$, $\omega = 7.3 \times 10^{-5}\,\text{s}^{-1}$ leads to a layer thickness of nearly 32 m, a magnitude where in most lakes the stratification would strongly influence the thickness. The equation suggested by Fischer et al. (1979) leads to a more realistic layer size of about 3–5 m so that stratification can be ignored. Assuming a conical-shaped basin and substituting the above values leads to a value of $\overline{K}_z = 1.8 \times 10^{-4}\text{–}2.8 \times 10^{-6}\,\text{m}^2\,\text{s}^{-1}$ depending on which estimate of h is used. As is seen from the above, this number straddles all known measurements, so one may conclude that boundary mixing is an important contributor to the vertical transfer of mass and heat no matter which boundary-layer thickness is used. Garrett (1979a) suggested yet another approach. He estimated the working in the boundary layer and compared this with the potential energy stored in the lake at that level. Assuming an efficiency η for the conversion of kinetic energy to potential energy led him to postulate the expression

$$\overline{K}_z = (4\eta C_D u_b^3)/\alpha L N^2 , \qquad (24)$$

where C_D is the bottom drag coefficient, u_b is the bottom velocity, α is the lake slope, L is the lake length, and N the buoyancy frequency. The same argument was also applied earlier by Imberger et al. (1978) who arrived at the result:

$$\overline{K}_z = 0.048 H^2/T_M S , \qquad (25)$$

where T_M is equal to the potential energy stored in the lake, divided by the power input from the wind and the river inflow and S is proportional to N^2. Garrett (1979a) suggested an efficiency of conversion η of about 1%. This compares favorably with the efficiencies implied by the coefficient 0.048 in Equation (25), which yields an efficiency ranging from 0.1% to 5% depending on the shape of the lake (Fischer et al. 1979).

The energy arguments thus lead to a constant-flux model quite different from that proposed by Armi (1978) and neither agree in their dependence on N with the summary formula provided by Ward (1977). Further, it is noteworthy that these boundary-mixing models all require a horizontal adjustment in the density profile that is rapid compared with the vertical flux in the turbulent boundary layer. Ivey (1980) carried out a laboratory experiment with a vertically oscillating grid at one end of a stratified tank. He observed a rapid internal adjustment, not by an intrusion mechanism but rather by basin-scale shear-wave propagation. The dependence of $\overline{K}_z$ on the stratification was $N^{-1/2}$, considerably weaker than that determined by Ward (1977) or suggested by Garrett (1979a), but stronger than that estimated by Armi (1978). Perhaps internal-wave breaking accounts for the difference.

Internal Mixing

A great many mechanisms have been identified that are capable of inducing a local overturning event in a stratified shear flow (Fischer et al. 1979). These events are characterized by mixing patches with a scale of up to 20 m (Gregg & Briscoe 1979) but which rarely contain individual overturnings larger than the Osmidov scale which is about 3 m (Thorpe 1977). The Cox number C takes on an initial value given by $C_0 = O[\epsilon/N^2 D]$ where D is the molecular diffusivity of heat. Most of the production is immediately dissipated, but some is utilized to raise the potential energy. If N is low then the maximum turbulence level is determined by homogeneous isotropic turbulence theory and the spin-up of the turbulence will be limited by the unsteady inertia terms. Thus, if the forcing becomes too severe the Cox-number formulation, as expressed above, breaks down and we may expect a cutoff in the diffusion coefficient calculated by Weinstock (1978). Under most circumstances the energy levels are below this critical state and the energy source is quickly used up. Fossilization begins at the smaller wave numbers and progressively moves to higher wave numbers, with a corresponding decrease in the activity number (Gibson 1980). This relaxation continues until the high-wave-number part of the spectrum recedes into the characteristic internal-wave spectrum and the structure may be called fossil. The microstructure spectrum is then solely due to internal-wave straining. In the absence of internal waves the activity number decreases to zero, but of course the Cox number will remain finite. This state represents a passive step-like structure slowly evolving by gravitational adjustments, reducing the initial Cox number over a much longer time scale. Initial large Cox numbers are likely to lead to fossilization more quickly because the high-energy-density sources required to bring the Cox number to large values are fairly rare events within a reservoir, and also the dissipation in such an event is proportionally higher (Gartrell 1979). Caldwell et al. (1980) and Dillon & Caldwell (1980) compared the dissipation computed from the temperature-gradient spectra cutoff with that estimated from the Cox-number measurements and concluded that the turbulence was in equilibrium for Cox numbers between 0 and 2500.

The shapes of the spectra also clearly showed the transition from the internal-wave dependence to the Batchelor spectra for this range of Cox numbers. This would indicate that the internal-wave-number spectrum has a cutoff just sufficient to cause an energy dissipation appropriate to the energy input. However, it is worth noting that measurement of the Cox number alone does not show the degree of fossilization, and caution must be exercised in using it to compute the vertical-diffusion coefficient. If the turbulence is in equilibrium then the vertical-diffusion coefficient may be estimated from the Weinstock relationship:

$$K_z = DC_0 = 0.8\epsilon/N^2 . \qquad (26)$$

This relationship has been justified theoretically by Weinstock (1978) who assumes that the mixing is done by the inertial subrange between the Osmidov length and the viscous cutoff. Weinstock estimated the coefficient to be about 0.8. However, measurements summarized by Gregg & Briscoe (1979) indicate a value closer to 0.2. Direct measurement of both the Cox number (by many investigators) and the dissipation (by Osborn 1980) has yielded values of K_z ranging from $10^{-5} \, \text{m}^2 \, \text{s}^{-1}$ in the hypolimnion to $10^{-4} \, \text{m}^2 \, \text{s}^{-1}$ in regions of very active shear (Thorpe 1977), values comparable to those estimated from the boundary-mixing models.

At Cox numbers greater than 2500 the above methodology does not appear to work and most of the structure is apparently fossilized to some degree. Application of the Weinstock relationship would thus consistently yield values of the diffusion coefficient that are too large. Verification of this conclusion, however, requires the simultaneous measurement of the distribution of the Cox number and the local dissipation. Such measurements do not appear to have been carried out to date.

The model proposed by Imberger et al. (1978) can be recast to conform with Equation (26) provided it is assumed that all processes in the lake are in equilibrium with the lake forcing. With this assumption $\epsilon \sim \bar{\epsilon} = P/M$, where P is the power input from the wind and the inflow and M is the total mass of water in the lake. Noting that the stability S may be written in the form $S = HN^2/g\Delta_0$, Equation (25) becomes

$$\overline{K}_z = 0.048 \, (HMg'/E) \, \bar{\epsilon}/N^2 , \qquad (27)$$

where E is the potential energy locked in the stratification, g' equals $g\Delta_0$, and Δ_0 is the nondimensional density difference between the top and bottom of the lake. The quantity in brackets may be estimated by assuming that ρ is equal to $\rho_0 + \Delta_0\rho_0(1 - z/H)$ and A the area of the lake at depth z is given by $A_0(z/H)^n$. This leads to a value of $(n+3)(n+2)/(n+1)$ for the value of the factor contained in the brackets in Equation (27). Most lakes have a shape such that $n \sim 3$ leading to the expression $\overline{K}_z = 0.4\bar{\epsilon}/N^2$. Assuming an average factor of 0.5 in Equation (26), and balancing the flux through the active patches with that calculated as a mean over the total area A, leads to a simple ratio of $1.2\overline{K}_z/K_z$ for the mass fraction of the lake involved in the active mixing processes.

If the lake is in equilibrium in the above sense, then $\epsilon \sim Nu'2$, where u' is the scale for the velocity fluctuations at the Osmidov scale. Numerous estimates may be inserted for u', but if we assume that $u' \sim u_{max}$ as defined by Spigel & Imberger (1980), as would apply for lake billowing, then the flux Richardson number becomes equivalent to the inverse of the square of the Wedderburn number. Other estimates of u' give different relationships between the Cox and the Wedderburn numbers. Hence, we see that the Wedderburn number not only determines the mixed layer dynamics but also fixes the mixing intensity in the active patches and thus the magnitude of the Cox number. Lastly, following Gibson (1980) it is possible to define a lake average activity number $\bar{\epsilon}^{1/2}/D^{1/2}NC^{1/2}$. If this is larger than one then the mixing is in the process of intensifying under the applied forcing. If on the other hand, the number is less than one then fossilization has commenced and the vertical transport of mass will be decreasing.

These deductions suggest an experiment where the microstructure of temperature field and finescale temperature and salinity is documented in combination with measurements of the mean flux and direct measurements of the local dissipation. The signature of the T-S (temperature-salinity) relationship would allow the differentiation of the internal waves and mixing from microstructure induced by intrusions and double diffusion. The two estimates of K_z obtained from the Cox number and the dissipation would allow a direct check on the degree of fossilization and the relationship between $\overline{K}_z$ and K_z would determine the patchiness of the turbulence provided the site was unaffected by the direct influence of the boundaries. These properties should then be determined by the value of W.

Double Diffusion

This process was reviewed in Turner (1973) and more recently by Sherman et al. (1978). Quite large vertical fluxes may be sustained by opposing gradients of temperature and a dissolved substance such as salt. The induced microstructure shows a characteristic regular stepiness illustrated beautifully by the measurements of Newman (1976) in Lake Kivu where the double-diffusion process was triggered by a warm salt seep. This is not an isolated feature: opposing gradients of substances with different diffusivities may be expected to be a feature of all meromictic lakes. The origin of meromictic lakes is varied. Tyler & Buckney (1974) showed evidence of how low-oxygen contents brought about by severe temperature stratification may in turn lead to high concentrations of manganese and iron in the bottom waters thus setting up temporary meromixis. Stewart & Hollan (1975) illustrate the history of Ulmener Maar and Hemmelsdorfer See indicating states of meromixis. Many examples exist where the lake becomes meromictic due to dissolved matter or saline underflows. Wiegand & Carmack (1980), J. Patterson et al. (in preparation), and Carmack & Killworth (1979) all have demonstrated strong temperature and salinity inversions during temporary meromixis. Under such circumstances it is likely that double diffusion becomes the dominant vertical-transport mechanism. Furthermore, since the dissolved substances are more often than not introduced at the lake boundaries, horizontal gradients are usually set up. The formation of horizontal intrusions similar to that found by Turner (1978) and Ruddick & Turner (1979) may thus be expected in meromictic lakes. Such intrusions would lead to active horizontal mixing by the Taylor shear mechanism (Fischer et al. 1979). To see this, suppose we model the intrusion by a combination of Couette and Poiseuille flow; then $K_x \sim U^2 h^2 / 100 K_z$, where U is the speed of the intrusion, h the vertical scale of the intrusion, and K_z the effective vertical diffusivity driven by salt fingering. So far only preliminary scales exist for h and K_z (Ruddick & Turner 1979) and none for U. However, comparisons of these scalings and field observations is encouraging and it is reasonable to choose for the ocean $h \sim 10$ m, $U \sim 10^{-1}$ m s^{-1} and $K_z \sim 10^{-4}$ m^2 s^{-1}. This would lead to a value for K_x of about 10^2 m^2 s^{-1}, even larger than that observed by Joyce (1977) who measured values closer to 10 m^2 s^{-1}. In the Wellington such intrusions have been identified on the echo sounder of the research vessel Djinnang (see Figure 2). Work is in progress but a tentative set of values is $h \sim 0.75$ m, $U \sim 0.05$ m s^{-1}, $K_z \sim 10^{-6}$ m^2 s^{-1}, leading to a horizontal-diffusion coefficient of about 14 m^2 s^{-1}. Such active mixing may have great importance in the redistribution of dissolved substances originating from the sediments of the lake.

Intrusions

The picture that evolves from the previous three sections is of a stratified hypolimnion characterized by random-mixing events, which locally raise the potential energy and so induce gravitational intrusions and interleaving movements along isopycnals. This activity is augmented by intrusions originating from the inflow, outflow, and any unequal heat capture at the surface. Thus, the hypolimnion may be thought of as a stratified body of water with isopycnal mixing being sustained by the vertical diffusivity and the vertical shear of the intrusions (Taylor mechanism). Little or no documentation exists to substantiate this point of view, but Fischer et al. (1979) used this model to derive an expression for the horizontal-diffusion coefficient which is in close agreement with the observations of Murthy

(1976) in Lake Ontario. It is surprising that more effort has not been directed to this field as it is of great importance to the eutrophication process of a lake. Further, as Fedorov (1978) shows, intrusions themselves may often induce vertical mixing. This may come about either due to a build-up of shear, or by the interaction of internal waves and the intrusional structure (Delisi & Orlanski 1975) or by the intrusion rearranging the temperature and salinity field so that double diffusion may result.

In summary, four major classes of mixing mechanisms have been identified as contributing towards both vertical and horizontal mixing in the hypolimnion of a lake. It is extremely difficult, when making measurements of the microstructure, to differentiate among these mechanisms and to determine which is responsible and to what degree. However, as is so clearly explained by Fedorov (1978), much can be learned by a careful analysis of the T-S relationship. Based on this, a methodology for the

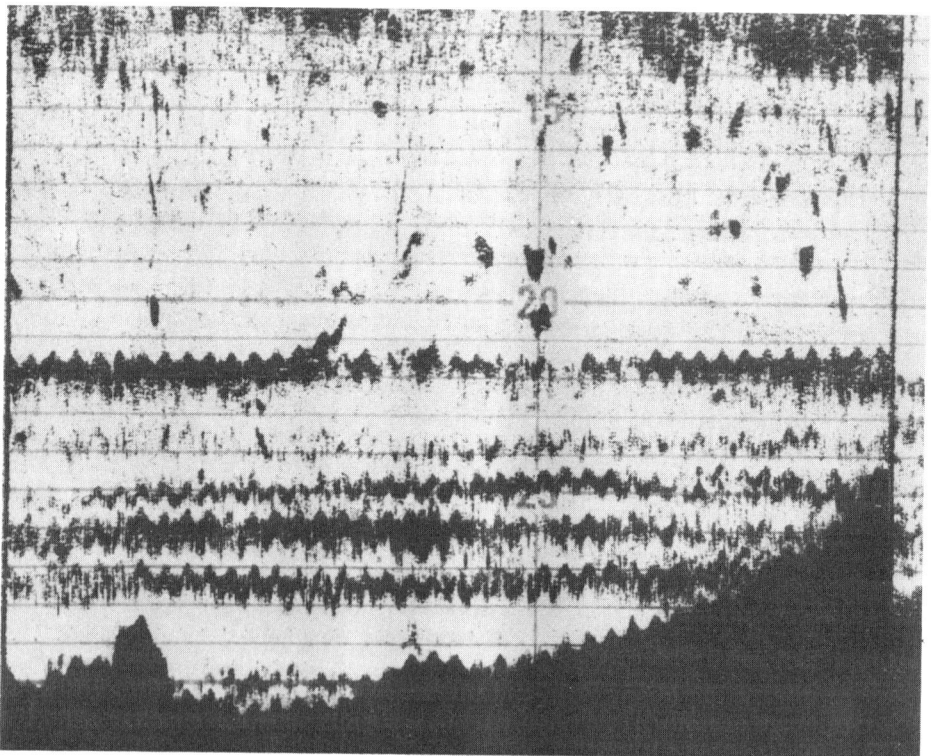

Figure 2 Echo sounding of fossil intrusions. Vertical scale is 0.5 m per division and the extent of the photo is about 200 m. Black at bottom is lake floor and regular waviness is an interference pattern characteristic of echo sounder.

extraction of internal waves signatures has been developed by Johnson et al. (1978) and refined by Toole (1980). This technique allows the removal of the internal-wave contribution leaving a T-S relationship originating from either double diffusion or isopycnal intrusions. An exceptionally clear documentation of this methodology may be found in the article by Middleton & Foster (1980). In this connection, the work of Brubaker (1979) should also be mentioned. He obtained measurements of finescale and microstructure in Lake Tahoe and showed that the characteristics of the displacement spectra could be explained by internal waves moving through a body of water with a weakly varying buoyancy frequency. Records taken over a day or less indicated that the variability from one record to the next admitted a scaling consistent with internal-wave straining. On the other hand, the same analysis revealed that for periods longer than a few days, intrusions or long-period basin-scale waves had distorted the temperature structure.

5. INFLOWS AND OUTFLOWS

The flow configuration of a river entering a lake may conveniently be subdivided into basically four distinct regimes.

1. The river water is warmer than the surface waters of the lake and the water propagates down the river channel until buoyancy causes an overflow into the main basin.
2. The river water is a great deal colder than the bottom lake water causing a complete underflow with the lake being filled from beneath.
3. The inflow density is intermediate between these two extremes, leading to a partial underflow and a subsequent horizontal intrusion.
4. The lake water is above (or below) 4°C and the inflow river water is below (or above) 4°C leading, upon contact, to a plunging underflow by the cabeling mechanism.

These four states are documented for the Thompson River inflow into Kamloops Lake by Carmack et al. (1979), Lake deGrey by Ford & Johnson (1980), Lake Kinneret by Serruya (1974), Lake Powell by Johnson & Merritt (1979), and the Wellington Reservoir by Hebbert et al. (1979).

A fifth regime dominated by double diffusion was demonstrated by Buchan (1975) in the laboratory. A solution of sugar water was allowed to flow into a reservoir stratified with salt. The heavy inflow initially plunged down the slope much like an ordinary inflow until it reached its level of neutral buoyancy. Then instead of forming an intrusion, an instability developed at the lower depth which quickly propagated upstream along the interface to the surface plunge point. The growth of the instability continued and multiple intrusions with a spacing equal to that predicted by Ruddick & Turner (1979) slowly began to penetrate horizontally into the reservoir over its complete depth.

The dynamics of the plunge line and the underflow are described in detail in Fischer et al. (1979) for the case of weak inflows into a continuously stratified lake with no rotation. The mixing at the plunge line at small W may be estimated from the cooling-pond work by Jirka & Watanabe (1980). The entrainment at the underflowing interface was shown by Fischer et al. (1979) to gain its energy from the bottom-generated turbulence for gentle valley slopes and the efficiency of the energy conversion was nearly ten times higher than that found at the base of the mixed layer. It was suggested that the different mean-velocity distribution in the underflow gave rise to a more active diffusion of the turbulent kinetic energy generated at the boundary and transported to the interface. The calculations of Svensson (1978) also resulted in an entrainment efficiency larger for underflows than at the base of the mixed layer. Overflows and underflows where the shear production dominates are treated by Huppert & Simpson (1980) who also extend the original analysis of Koh (1976). These authors documented the evolution of the flow from its initial entrance, inertia-buoyancy-forced balance, to the final viscosity-dominated regime.

Surface inflows in a large lake where rotation is important (P small) had been analyzed by Hamblin & Carmack (1978), whose integral analysis reproduced most of the features observed by Carmack et al. (1979) in Kamloops Lake. For P below a certain critical value Stern (1980) has shown that the surface intrusion becomes unstable and geostrophic eddies are formed at the boundary of the intrusion and ejected into the main body of the fluid.

Once the fluid reaches the depth of neutral buoyancy, the mean characteristics of the internal intrusion into a continuously stratified lake are completely determined by the parameter $\tilde{R}$. However, for an unsteady inflow shear waves are generated and they may interact with the intrusion to cause long-waves or even pinching off of the intrusion (Lyne 1976). Holyer & Huppert (1980) investigated the counterpart for an intrusion entering a lake at the pycnocline. These authors illustrate that, distinct from a boundary or surface intrusion, long-waves may be formed at the intrusion interface and propagate away from the nose of the intrusion. T. W. Kao (personal communication) has observed the growth of these waves to an extent that the intrusion degenerates into a series of slugs, each moving with a solitary-wave speed and each containing a parcel of the original intrusion fluid.

We are thus in a position to predict qualitatively many of the observed characteristics of the river-water inflows. However, quantification of the entrainment, once the inflow forms an intrusion or once the rotational instability takes hold, is at present not possible.

The dynamics of the outflow has recently been reviewed by Imberger (1980). He concluded that much work remained to be done for the case of unsteady outflows, outflows to a point sink, and outflows for P small.

6. INTRUSIONS AND GRAVITATIONAL ADJUSTMENTS

In order to derive the basic scales of motion, consider first large-scale adjustments of the density field. Basin-scale horizontal gradients may be established in an embayment, cooling pond, or river side arm in basically three ways:

(a) Uniform incoming radiation will raise the temperature of the water column from the old equilibrium temperature $T_e^{(1)}$ at a rate inversely proportional to the depth of the water. The shallow end heats rapidly until the temperature reaches the new equilibrium temperature $T_e^{(2)}$, whereupon the incoming radiation is balanced by the heat loss at the surface. The deeper end will heat more slowly and it will take longer to come to equilibrium at the temperature $T_e^{(2)}$. During a cooling cycle the gradient is reversed. The equilibrium-density variation, in the absence of motion, would thus be approximated by

$$\rho = \Delta\rho_0 \left[1 - \exp(-\beta t/h(x)c_p\rho_0)\right], \qquad (28)$$

where $\Delta = \alpha(T_e^{(2)} - T_e^{(1)})$ and β is the heat-transfer coefficient. The gradient initiates a longitudinal gravitational motion, with a length scale $\tilde{L}$, which transfers heat from the hot end to the cold central lake basin, but at a rate small compared with the radiation input.

(b) A heat exchanger from a power station raises the end temperature to, say, T_s. Natural convection will spread the temperature field horizontally, the horizontal length scale $\tilde{L}_e$ being determined by the surface heat losses given by $\beta(T_s - T_e)$.

(c) Heat and mass are discharged at one end either by an inflowing river or from a power station. In this case both the surface density ρ_s and the volume discharge Q are fixed and the circulation may be forced convection.

A solution for case (a) may be found from the model proposed by Imberger (1974). He assumed a linear variation of T with no heat exchange through the top and bottom boundaries. If we identify the gradient $(T_e^{(1)} - T_e^{(2)})/\tilde{L}$ with the linear variation and assume a perfect balance between incoming radiation and surface losses then a parallel-flow solution with vertical diffusion of vorticity balancing baroclinic generation is admitted, so that the velocity

$$u \sim \frac{g\Delta \tilde{H}^3}{A_z \tilde{L}}, \tag{29}$$

where $\tilde{H}$ is the depth of the water body, $\tilde{L}$ is the length, A_z the vertical eddy-diffusion coefficient and $\Delta \rho_0 = (\rho_e^{(1)} - \rho_e^{(2)})$, the difference between the two end densities. In this flow the horizontal convection of the background density field must be balanced by the vertical diffusion of mass (heat). This leads to a vertical density differential $\delta \rho_0$ such that

$$\frac{\delta}{\Delta} \sim \frac{g\Delta \tilde{H}^5}{A_z K_z \tilde{L}} = \tilde{G}r A^2, \tag{30}$$

where $\tilde{G}r$ is the Grashof number and $A = \tilde{H}/\tilde{L}$ is the aspect ratio.

The time required to establish this flow from rest is simply $O(\tilde{H}^2/A_z)$, which typically can be as short as an hour or so, implying that the circulation pattern will keep pace with the diurnal heating and cooling.

Inertia forces enter the force balance for $\tilde{G}r \sim Pr^{-1} A^{-2}$, a condition often encountered in side arms and cooling ponds. The simple creeping-flow solution must in such cases be replaced by a full solution of the boundary-layer equations. Such a solution does not appear to have been found.

Let us now turn to the heat-exchanger problem, where the surface-heat losses dominate the flow and where T_e is no longer representative of the mean temperature in the water column; the heat exchanger raises the overall temperature as well as the differential until the input heat is balanced by the losses. On the other hand, T_e is fixed by the local meteorological conditions. If one introduces a water-column mean density $\bar{\rho}$, a surface-density scale ρ_s, a density difference $\delta = (\bar{\rho} - \rho_s)/\rho_0$, and an "overheat" density scale $\Delta_e = (\rho_e - \rho_s)/\rho_0$, then the surface heat exchange balance requires:

$$\frac{u\delta \tilde{H}_e}{\tilde{L}_e} \sim \beta \Delta_e, \tag{31}$$

since $\Delta \sim \delta$ for this case. The length $\tilde{L}_e$ is the effective scale of the hot overflow and $\tilde{H}_e$ is its effective depth.

The vertical gradients of temperature set up by the surface losses must be sufficient to diffuse the heat through a distance $\tilde{H}_e$, so that

$$u \sim \frac{K_z \tilde{L}_e}{\tilde{H}_e^2}. \tag{32}$$

Comparison of Equations (32) and (29) yields the scale for δ,

$$\delta \sim \frac{A_z K_z \tilde{L}_e^2}{g\tilde{H}_e^5}, \tag{33}$$

and substitution of this estimate in Equation (31) leads to the estimate

$$\tilde{L}_e \sim \left[\frac{g\tilde{H}_e^6 \beta(\rho_s - \rho_e)}{K_z^2 A_z \rho_0}\right]^{1/2} \tag{34}$$

Thus, for a given $\tilde{H}$ we must compare $\tilde{L}_e$ with $\tilde{L}$. If $\tilde{L} > \tilde{L}_e$ then only part of the water body will be at an overheat and the gravitational circulation is confined to a length $\tilde{L}_e$. If for a given $\tilde{H}$, the length $\tilde{L}$ is less than $O(g\tilde{H}^6 \beta(\rho_s - \rho_e)/K_z^2 A_z \rho_0)^{1/2}$ then the circulation will adjust to its own depth $\tilde{H}_e < \tilde{H}$, the density difference $\delta\rho_0$ will decrease, the heat exchange will decrease, and the water will stratify. The above estimates imply that in any solution of the problem the inertia forces must be included for all flow conditions.

The remaining case, where a flow Q is imposed in addition to imposing ρ_s, is easily recovered from the above scaling by requiring

$$u \sim \frac{Q}{\tilde{H}_e}. \tag{35}$$

Equating the two velocity scales leads to the relationship,

$$\frac{\tilde{H}_e}{\tilde{H}} = \frac{Q^2/\tilde{H}^2;}{g\tilde{H}(\rho_s - \rho_e)/\rho_0} \cdot \frac{A_z \tilde{L}}{Q\tilde{H}}. \tag{36}$$

Jirka & Watanabe (1980) assumed $A_z \sim \tilde{H}U_* \sim \tilde{H}u \sim Q$ so that $\tilde{H}_e/\tilde{H}$ becomes their Pond number (inverse of the Wedderburn number). The significance of the Pond number is thus clear. A small Pond number leads to a small $\tilde{H}_e/\tilde{H}$ and a stratified pond. Conversely, for $\tilde{H}_e/\tilde{H}$ large the throughflow causes sufficient turbulence to thoroughly mix the water over the depth.

The transition between a naturally convecting pond and one undergoing forced convection occurs at $Q \sim \tilde{L}\beta$. It has thus been possible to classify all regimes in a side arm and to show that there is a natural connection between the shallow-cavity flow studies by Imberger (1974) and the cooling-pond solutions put forth by Jirka & Watanabe (1980). However, it does not appear to be generally appreciated that a naturally convecting water body can possess both a stratified and an unstratified mode.

Much of the above scaling should be refined to include the effects of buoyancy on K_z and A_z, as these influences will sharpen the transitions. Basin shape has also been ignored, but apart from the results of Imberger (1976) all theoretical results appear to be confined to rectangular cavities. Jirka & Harleman (1979) review the empirical literature.

Very often wind will move cold or warm water masses into the proximity of a side arm forming a front at the mouth. The gradient at the front thus propagates into the side arm leaving behind it a horizontal gradient. Such initial-value problems are very complex with a great diversity of flows (Patterson & Imberger 1980) depending on the density gradient and the depth.

A further complication arises when the nonlinear nature of the equation of state is included. Cabeling then lends to the formation of regions of enhanced temperature gradient known as a thermal bar. Hamblin (1981, personal communication) has shown that at the density extremum horizontal motion is cut off, thereby reducing the horizontal heat transport by convection. Heat conduction must balance horizontal convection away from the 4°C isotherm with augmented gradients. This balance leads to a $Gr^{-1/4}\tilde{H}$ scaling of the frontal thickness which is in agreement with the data of Carmack (1979) in Kamloops Lake. The influence of rotation is discussed by Mortimer (1977).

At smaller scales there are the intrusions formed in the metalimnion once upwelling relaxes. No actual documentation of this process exists, but it may be expected that the vertical scale of the initial mixed region at the boundary will be large compared to the thickness of the metalimnion. The collapse may thus be similar to the laboratory experiments of Kao & Pao (1979) and Maxworthy (1980) whose results show the evolution from the initial state to the formation of solitons. The solitons pump the mixed upwelled water in a peristaltic fashion along the metalimnion contributing to a thicker metalimnion. Gravitational motions may also be expected along a boundary (Phillips 1970, Ivey 1980, and Turner & Chen 1974).

Finally, in the hypolimnion, as mentioned in Section 4, overturning motions and double-diffusive instabilities lead to collapsing intrusions. Lyne (1976) has analyzed the time scales for such a collapse but so far little attention has been given to the internal circulation and mass transfer at the boundaries of individual intrusions. Since the initiation time scale for the internal motion is $O(\delta^2/A_z)$, where δ is the intrusion thickness, it is apparent that an active internal circulation may be expected by the time the intrusion is itself governed by a viscous buoyancy balance. This circulation will, in turn, affect the mass transfer across the boundaries, and therefore the intrusion life. As the aspect ratio is small it may be expected that the cooling-pond circulation [case (b)] would serve as a useful model.

7. BASIN-SCALE WAVES AND MOTIONS

Steady Flows in a Rotational System

Considerable attention has been devoted to the theoretical understanding of the steady gyral motions observed in lakes. In a homogeneous fluid circulation is induced by stretching of vortex columns over bottom topography by wind and pressure forces. As an example of a numerical technique based on this principle Hamblin (1976) has developed a finite-element treatment suitable for multiply connected basins. Vorticity input directly by the wind field by variation in boundary-layer characteristics (Emery & Csanady 1973) and by varying internal-layer thicknesses (Bennett 1975) have been employed in accounting for the observed circulation. An additional mechanism for driving the internal circulation in lakes, namely the rectification of internal waves by nonlinearity, has been explored by Wunsch (1973) and by Ou & Bennett (1979).

Barotropic Waves

Waves in a homogeneous fluid have been studied extensively and, in general, their basic properties reasonably well understood. Embarking upon our discussion at the shortest scales, spectra of wind-generated surface waves under deep-water conditions have been observed by numerous workers (e.g. Liu 1971) and are well represented over limited ranges of fetch by various empirical relations such as the Jonswap formulas. Forecasting of wave energy at a given point in a lake is less certain, with the dissipation of wave energy in shallow water being a major unknown factor. Operational wave-forecasting methods have been developed for large lakes employing input from numerical weather models (Thompson 1978).

Long standing gravity waves in lakes, known as seiches, have received considerable observational and computational investigation. Accurate numerical representation of structure configuration, bathymetry, and the earth's rotation by a variety of methods have permitted the computation of free periods to within several percent of observed periods. As an example, Raggio and Hutter (1981, personal communication) have computed the free oscillations of the Lake of Lugano. The generation of surface seiches by wind has been studied in a natural basin by Whitaker et al. (1975) and by barometric pressure gradients by Hamblin & Hollan (1978). Only in very deep lakes are barometric pressure gradients important in the excitation of surface seiches. The closely related problems of the generation of astronomical tides in closed basins have been treated by Hamblin et al. (1977). Careful comparison of tidal excursions of water level with theory has permitted some inference on the magnitude of earth tides in lake basins.

Another class of long-waves admissible theoretically in rotating basins of variable depth, namely topographic Rossby waves, is difficult to establish experimentally because of the very long period of oscillation demonstrated to exist in lake basins. Saylor et al. (1980) have proposed a rotational wave model of sub-basin in a lake to account for persistent current fluctuations of 100 hr in period in the open lake.

Baroclinic Waves

When the Wedderburn number is less than two, Spigel & Imberger (1980) have shown that in narrow channels long-period internal waves are highly transient, but when W is in the range from two to $L/(2H)$ long-period standing waves are a prominent feature. Finally when W exceeds $L/(2H)$ basin-scale internal waves have small amplitudes which persist for a long time. Lazier & Sandstrom (1978) have developed a theory for the wind-forced basin-scale internal waves in a viscous fluid stratified at a constant rate. In elongated basins internal waves are influenced by the earth's rotations. Numerical modeling of these motions or Kelvin waves, based upon the assumption of constant depth and horizontal similarity of density structure (i.e. Kanari 1975), has indicated some correspondence to observed periods. Nonlinear effects ($R_0 \sim 1$) in Kelvin waves have been studied by Bennett (1973). Basin-scale standing internal waves with the principal displacement aligned across the lake are termed Poincaré waves when influenced by the earth's rotation. At present, observational studies have not established the basin-wide coherence required for the Poincaré wave model and thus the inertial wave interpretation has been favored (Marmorino 1978).

Internal waves in lakes and reservoirs also occur on scales less than basin scales. Shorter-period internal waves are observed in lakes and range in frequencies from inertial to buoyancy. Energy levels of high-frequency internal waves are closely related to sources of shear in lakes; the case of wind shear has been studied by Boyce (1977) and shear due to river inflow and shear due to basin-wide Kelvin wave shear has been investigated by Hamblin (1977). The role played by short-period internal waves in the transport of momentum and the mixing induced by breaking waves is not understood but is likely to be vital to mixing processes in deep lakes and reservoirs as discussed in Section 4.

8. TIME-DEPENDENT WIND-DRIVEN CIRCULATION

This topic has been a central theme of lake studies of the past decade, the studies being so numerous that it is impossible to describe more than several. On account of the close connection of circulation problems with the free surface we include storm-surge studies under this topic.

In an exceptionally fine analysis of the problem of the wind generation of lake circulation, Birchfield & Hickie (1977) have for the first time been able to show how the relative contributions of surface seiches and topographic waves modulate the seiche response by a slow rotation in the cyclonic sense of the pattern of coastal jets and return flow. Hopefully their analysis will be extended in the future to include more realistic geometry, friction, and stratification.

An evaluation of current capability of numerical models to account for the observed flows in a large lake has been undertaken by Allender (1977). A statistical analysis of model simulations of lake currents for periods up to one month showed that model and observed currents do not compare well and that the lack of agreement probably results from poorly resolved wind data and grid sizes used in the models. It is likely that inadequate knowledge of other model parameters such as the eddy coefficients also contributed to the degradation of model performance.

Although the agreement between hindcasts and observations of water-level fluctuations known as storm surges is much closer than in the case of lake currents, numerous studies nevertheless indicate the need for increased accuracy in the specification of the wind field. For example, Schwab (1978) arrived at this conclusion in an examination of a number of occurrences of storm surges in Lake Erie. In this paper, as well as in Hamblin (1976), the problem of operational storm-surge prediction is simplified by the introduction of the water-level response function for a point of interest in the lake. The storm-surge hindcast is calculated by the convolution of the time history of the wind forcing with the response function.

As common as wind-induced upwelling is and as important to the productivity of a lake, the phenomenon has received scant attention until recently. Although upwelling in lakes is a highly transient phenomenon, considerable progress has been made by considering the final equilibrium configuration of the upwelled fluid. Csanady (1977) has developed expressions relating the magnitude of the upwelling of the layer depths to the wind impulse in a constant-depth inviscid and two-layer fluid that are in quantitative agreement with observations of temperature transects in Lake Ontario. Other analytical work on upwelling in lakes includes the application of small-amplitude theory to the problem of transient response in a rotating two-layer channel. Lee & Mysak (1979) found that both longitudinal and transverse wind forcing are responsible for upwelling in a rotating two-layer channel of constant depth, but that, in general, the time scale associated with longitudinal forcing is longer than that associated with transverse forcing.

It is experienced observationally that when upwelling is sufficiently strong so that the thermocline intersects the surface it retains its sharp frontal nature. Conversely at the downwind side of a lake the warm surface water has been found to form well-defined fronts distinct from the main body of open lake water. From an analysis of this phenomenon in Lake Ontario Simons (1978) concluded that downwelling fronts are internal surges associated with the transient response of the lake to wind.

Application of three-dimensional numerical modeling of upwelling in a lake by Hollan & Simons (1978) has permitted a quantitative evaluation of our present understanding of upwelling in lakes. A three-dimensional time-dependent numerical simulation of a storm event results in the best overall agreement with current and temperature observations along and across the axis of Lake Constance when compared with two-dimensional models of upwelling.

It is appropriate to conclude this section with discussion of the question of the specification of wind drag over water. As lake models have become progressively more refined, the drag coefficients employed by modelers to bring their simulations of current and water level into line with observation continue to exceed by a factor of two or more those directly measured in the boundary layer over water. For example, Graf & Prost (1979) found that, on the average, the boundary-layer data indicated an aerodynamic drag coefficient, c_{10} of 1.15×10^{-3} over Lake Geneva, whereas comparison of simulated currents and observation indicated a c_{10} of 4×10^{-3} (Bauer & Graf 1979).

9. SUMMARY

This review has shown that a great many mechanisms may be identified as actively contributing to the state of a lake. Because of the enormous diversity of length and time scales encountered in these mechanisms, universal averaging techniques (closure schemes) are not very suitable for most applications with present computing capacities and it is felt that phenomenological modeling has a greater chance of success, provided it is aimed at a specific task. The prediction of the vertical temperature distribution within a lake is an example. This methodology must now be applied to the modeling of momentum and of conservative tracers or biological constituents that are dependent for their distribution on the subtleties of mixing in the water column. Indeed, the task of interconnecting all the mechanisms reviewed above still lies ahead.

Certain major deficiencies in our understanding of lake dynamics have been identified:

1. Vertical mixing in the hypolimnion with a special emphasis on the role of the parameter W in determining the relative strength of boundary mixing, internal mixing, and mass transfer by double-diffusive processes. The transition between turbulence influenced by the stratification and turbulence unaffected by buoyancy must also be clarified.
2. The dynamics of the metalimnion under upwelling conditions is unknown.

3. Horizontal mixing in the epilimnion, metalimnion, and hypolimnion. Only empirical results appear to be available and the role of fronts, intrusions, and gravitational adjustments is at present only speculation.
4. The understanding of the influence of rotation on the withdrawal layer, on the inflowing stream with and without cabeling, and on intrusions is very rudimentary. The stability of these flows requires careful examination.
5. The influence of Langmuir circulation, variable wind stress, and periodic wind stress on the mixed-layer deepening process has not been investigated.
6. Excitation of the baroclinic-wave field by a variable-wind field in a lake of arbitrary shape requires much further work.
7. Interbasin flows driven by wind, buoyancy, or inertia (separation) are important and should be analyzed.
8. The interplay between physical processes and water quality and the feedback mechanisms must be better understood if water quality models are to be used as a predictive tool.
9. Research on the gravitational circulation in a side arm due to diurnal heating cycle, for the case where longitudinal transport becomes first order, must be pursued as it is the dominant mass-transport mechanism in sheltered embayments.
10. More accurate specification of wind-drag dependence in lakes upon fetch, water depth, sheltering, and atmospheric stability is required.

ACKNOWLEDGMENT

The authors would like to thank Dr. John Brubaker for discussions on the material treated in Section 4. One of us wishes to acknowledge the financial support of the National Water Research Institute, Burlington, and the Gledden Foundation of the University of Western Australia which made this collaboration possible.

Literature Cited

Allender, J. H. 1977. Comparison of model and observed currents in Lake Michigan. *J. Phys. Oceanogr.* 7:711–18

Armi, L. 1978. Some evidence for boundary mixing in the deep ocean. *J. Geophys. Res.* 83:1971–79

Armi, L., D'Asaro, E. 1980. Flow structures of the benthic ocean. *J. Geophys. Res.* 85:469–84 (1 insert)

Bauer, S. N., Graf, W. H. 1979. Wind-induced water circulation of Lake Geneva. In *Marine Forecasting*, ed. J. C. J. Nihoul, pp. 219–33 Amsterdam: Elsevier

Bennett, J. R. 1973. A theory of large-amplitude Kelvin waves. *J. Phys. Oceanogr.* 3:57–60

Bennett, J. R. 1975. Another explanation of the observed cyclonic circulation of large lakes. *Limnol. Oceanogr.* 20:108–10

Birchfield, G. E., Hickie, B. P. 1977. The time-dependent response of a circular basin of variable depth to a wind stress. *J. Phys. Oceanogr.* 7:691–701

Blanton, J. O. 1973. Vertical entrainment into the epilimnia of stratified lakes. *Limnol. Oceanogr.* 18(5):697–704

Bowman, M. J., Esaias, W. E., eds. 1978. Oceanic fronts in coastal processes. *Proc. Workshop Mar. Sci. Res. Cent.*, 1977. 114 pp.

Boyce, F. M. 1977. The response of the coastal boundary layer on the north shore of Lake Ontario to a fall storm. *J. Phys. Oceanogr.* 7:719–32

Brubaker, J. M. 1979. *Space-time scales of temperature variability in the seasonal thermocline of Lake Tahoe*. PhD thesis. Oregon State Univ., Oregon. 104 pp.

Buchan, S. J. 1975. *Possible dispersion mechanisms of a saline inflow into a temperature stratified reservoir*. Hons. thesis. Univ. Western Australia, Nedlands. 160 pp.

Caldwell, D. R., Chriss, T. M. 1979. The viscous sublayer at the sea floor. *Science* 205:1131–32

Caldwell, D. R., Dillon, T. M., Brubaker, J. M., Newberger, P. A., Paulson, C. A. 1980. The scaling of vertical temperature gradient spectra. *J. Geophys. Res.* 85:1917–24

Carmack, E. C. 1979. Combined influence of inflow and lake temperatures on spring circulation in a riverine lake. *J. Phys. Oceanogr.* 9:422–34

Carmack, E. C., Killworth, P. D. 1979. Observations on the dispersal of saline groundwater in the Beaver Creek diversion system, 1976–78. *Syncrud. Environ. Res. Monogr. 1979-2*. 83 pp.

Carmack, E. C., Gray, C. B. J., Pharo, C. H., Daley, R. J. 1979. Importance of lake-river interactions on the seasonal patterns in the general circulation of Kamloops Lake/British Columbia. *Limnol. Oceanogr.* 24(4):634–44

Chen, C. W., Orlob, G. T. 1978. Ecologic simulation for aquatic environments. In *Systems Analysis and Simulation in Ecology*, 3:475–588. New York: Academic

Csanady, G. T. 1975. Hydrodynamics of large lakes. *Ann. Rev. Fluid Mech.* 7:357–86

Csanady, G. T. 1977. Intermittent "full" upwelling in Lake Ontario. *J. Geophys. Res.* 82(3):397–419

Davis, R. E., de Szoeke, R., Halpern, D., Niiler, O. 1981. Variability and dynamics of the upper ocean during MILE. *J. Phys. Oceanogr.* In press

Delisi, D. P., Orlanski, I. 1975. On the role of density jumps in the reflexion and breaking of internal gravity waves. *J. Fluid Mech.* 69(3):445–64

Denton, R. A. 1978. *Entrainment by penetrative convection at low Péclet number*. PhD thesis. Univ. Canterbury, Christchurch. 202 pp.

Dillon, T. M., Caldwell, D. R. 1978. Catastrophic events in a surface mixed layer. *Nature.* 276(5688):601–2

Dillon, T. M., Caldwell, D. R. 1980. The Batchelor spectrum and dissipation in the upper ocean. *J. Geophys. Res.* 85:1910–16

Dillon, T. M., Powell, T. M. 1979. Observations of a surface mixed layer. *Deep Sea Res.* 26A:915–32.

Dillon, T. M., Richman, J. G., Hansen, C. G., Pearson, M. D. 1981. Near-surface turbulence measurements in a lake. *J. Phys. Oceanogr.* In press

Emery, K. O., Csanady, G. T. 1973. Surface circulation of lakes and nearly landlocked seas. *Proc. US Natl. Acad. Sci. 70*, pp. 93–97

Farmer, D. M. 1978. Observations of long nonlinear internal waves in a lake. *J. Phys. Oceanogr.* 8:63–73

Farmer, D. M. 1981. Mixed layer dynamics in a lake near the temperature of maximum density. *Proc. Int. Symp. Stratified Flows, 2nd, Trondheim, 1980* 2:998–1007

Fedorov, K. N. 1978. *The Thermohaline Finestructure of the Ocean*. Oxford: Pergamon. 170 pp.

Fischer, H. B., List, E. J., Koh, R. C. Y., Imberger, J., Brooks, N. H. 1979. *Mixing in Inland and Coastal Waters*. New York: Academic. 483 pp.

Ford, D. E., Johnson, M. C. 1980. Field observations of density currents in impoundments. *Proc. Symp. Surface Water Impoundments, Minneapolis.* In press

Garrett, C. 1979a. Comment on "Some evidence for boundary mixing in the deep ocean" by Laurence Armi. *J. Geophys. Res.* 84:5096–96

Garrett, C. 1979b. Mixing in the ocean interior. *Dynam. Atmos. and Oceans* 3:239–66

Garrett, C., Munk, W. 1979. Internal waves in the ocean. *Ann. Rev. Fluid Mech.* 11:339–70

Gartrell, G. Jr. 1979. Studies on the mixing in a density-stratified shear flow. *W. M. Keck Laboratory Rep. No. KH-R-39*. 447 pp.

Gibson, C. H. 1980. Fossil temperature, salinity, and vorticity turbulence in the ocean. In *Marine Turbulence*, ed. J. C. J. Nihoul, pp. 221–57. Amsterdam: Elsevier. 378 pp.

Graf, W. H., Prost, J. P. 1979. The aerodynamic drag; experiments on Lake Geneva. In *Hydrodynamics of Lakes*, ed. W. H. Graf, C. H. Mortimer, pp. 303–12. Amsterdam: Elsevier

Grant, W. D., Madsen, O. S. 1979. Combined wave and current interaction with a rough bottom. *J. Geophys. Res.* 84:1797–808

Gregg, M. C., Briscoe, M. G. 1979. Internal waves, finestructure, microstructure and mixing in the ocean. *Rev. Geophys. Space Phys.* 17(7):1524–48

Hamblin, P. F. 1976. Seiches, circulation and storm surges of an ice-free Lake Winnipeg. *J. Fish. Res. Bd., Canada.* 33(10):2377–91

Hamblin, P. F. 1977. Short-period internal waves in the vicinity of a river-induced shear zone in a fjord lake. *J. Geophys. Res.* 82(21):3167–74

Hamblin, P. F., Carmack, E. C. 1978. River-induced currents in a fjord lake. *J. Geophys. Res.* 83:885–99

Hamblin, P. F., Hollan, E. 1978. On the gravitational seiches of Lake Constance and their generation. *Swiss J. Hydrol.* 40:119–54

Hamblin, P. F., Muhleisen, R., Bosenberg, U. 1977. The astronomical tides of Lake Constance. *Deutsche Hydrograph. Z.* 30(4):105–16

Hebbert, B., Imberger, J., Loh, I., Patterson, J. 1979. Collie River underflow into the Wellington Reservoir *J. Hydraul. Div. ASCE* 105:533–45

Hollan, E., Simons, T. J. 1978. Wind-induced changes of temperature and currents in Lake Constance. *Arch. Meteorol. Geophys. u. Bioklimatol. Ser. A* 27(3,4):333–73

Holyer, J. Y., Huppert, H. E. 1980. Gravity currents entering a two-layer fluid. *J. Fluid Mech.* 100(4):739–67

Huppert, H. E., Simpson, J. E. 1980. The slumping of gravity currents. *J. Fluid Mech.* 99(4):785–99

Imberger, J. 1974. Natural convection in a shallow cavity with differentially heated end walls. Part 3. Experimental results. *J. Fluid Mech.* 65(2):247–60

Imberger, J. 1976. Dynamics of a longitudinally stratified estuary. *Proc. Coastal Engrg. Conf., 5th, Honolulu, 1976* 4:1308–23

Imberger, J. 1977. On the validity of water quality models for lakes and reservoirs. *Proc. Cong. IAHR, 17th, Baden-Baden, 1977* 6:293–302

Imberger, J. 1980. Selective withdrawal: A review. *Proc. Int. Symp. Stratified Flows, 2nd, Trondheim, 1980* 1:381–400

Imberger, J., Patterson, J. C. 1980. A dynamic reservoir simulation model — DYRESM:5. Presented at Int. Symp. Predictive Abilities of Surface Water Flow and Transport Models, Berkeley

Imberger, J., Thompson, R., Fandry, C. 1976. Selective withdrawal from a finite rectangular tank. *J. Fluid Mech.* 78(3):489–512

Imberger, J., Patterson, J., Hebbert, B.,

Loh, I. 1978. Dynamics of reservoir of medium size. *J. Hydraul. Div. ASCE* 104:725–43

Imboden, D. M., Emerson, S. 1977. Natural radon and phosphorus as limnologic tracers: Horizontal and vertical eddy diffusion in Greifensee. *Limnol. Oceanogr.* 23(1):77–90

Ivey, G. N. 1980. *Boundary mixing in density-stratified fluids.* PhD thesis. Univ. Calif., Berkeley. 123 pp.

Ivey, G., Imberger, J. 1978. Field investigation of selective withdrawal. *J. Hydraul. Div. ASCE.* 104:1225–37

Jassby, A., Powell, T. 1975. Vertical patterns of eddy diffusion during stratification in Castle Lake, California. *Limnol. Oceanogr.* 20(4):530–43

Jirka, G. H., Harleman, D. R. 1979. Cooling impoundment: Classification and analysis. *J. Energy Div. ASCE* 105:291–309

Jirka, G. H., Watanabe, M. 1980. Thermal structure of cooling ponds. *J. Hydraul. Div. ASCE.* 106:701–15

Johnson, C. L., Cox, C. S., Gallagher, B. 1978. The separation of wave-induced and intrusive oceanic finestructure. *J. Phys. Oceanogr.* 8(5):846–60

Johnson, N. M., Merritt, D. H. 1979. Convective and advective circulation of Lake Powell, Utah-Arizona, during 1972–75. *Water Resources Res.* 15(4):873–84

Joyce, T. M. 1977. A note on the lateral mixing of water masses. *J. Phys. Oceanogr.* 7:626–29

Kanari, S. 1975. Long-period internal waves in Lake Biwa. *Limnol. Oceanogr.* 20(4):544–59

Kao, T. W., Pao, H. P. 1979. Wake collapse in the thermocline and internal solitary waves. *J. Fluid Mech.* 97(1):115–27

Koh, R. C. Y. 1976. Buoyancy-driven gravitational spreading. *Proc. Coastal Engrg. Conf., 5th, Honolulu, l976* 4:2956–75

Kraus, E. B., ed. 1977. *Modelling and Prediction of the Upper Layers of the Ocean.* Oxford: Pergamon. 325 pp.

Kullenberg, G., Murthy, C. R., Westerberg, H. 1974. Vertical mixing characteristics in the thermocline and hypolimnion regions of Lake Ontario (IFYGL). *Proc. Conf. Great Lakes Res., 17th, Internat. Assoc. Great Lakes Res.* 1:425–34

Lazier, J., Sandstrom, H. 1978. Migrating thermal structure in a freshwater thermocline. *J. Phys. Oceanogr.* 8:1070–79

LeBlond, P. H., Mysak, L. A. 1978. *Waves in the Ocean.* Amsterdam: Elsevier. 602 pp.

Lee, A. C., Mysak, L. A. 1979. Transverse upwelling in a long narrow lake, with application to Babine Lake and Lake Michigan. *Atmos. Ocean* 17(3):200–18

Lick, W. 1976. Numerical modeling of lake currents. *Ann. Rev. Earth Planet. Sci.* 4:49–74

Liu, P. C. 1971. Normalized and equilibrium spectra of wind waves in Lake Michigan. *J. Phys. Oceanogr.* 1:249–57

Lyne, V. D. 1976. *Horizontal intrusions in a bounded stratified fluid.* Hons. thesis. Univ. Western Australia, Nedlands

Marmorino, G. O. 1978. Inertial currents in Lake Ontario, winter 1972–73. (IFYGL) *J. Phys. Oceanogr.* 8:1104–20

Maxworthy, T. 1980. On the formation of nonlinear internal waves from the gravitational collapse of mixed regions in two and three dimensions. *J. Fluid Mech.* 96(1):47–64

Mellor, G. L., Durbin, P. A. 1975. The structure and dynamics of the ocean surface mixed layer. *J. Phys. Oceanogr.* 5:718–28

Middleton, J. H., Foster, T. D. 1980. Fine structure measurements in a temperature-compensated halocline. *J. Geophys. Res.* 85:1107–22

Mortimer, C. H. 1971. Large-scale oscillatory motions and seasonal temperature changes in Lake Michigan and Lake Ontario. *Spec. Rep. No. 12.* Center for Great Lake Studies, Univ. Wis., Milwaukee

Mortimer, C. H. 1974. Lake hydrodynamics. *Mitt. Int. Ver. Limnol.* 20:124–97

Mortimer, C. H. 1977. One of Lake Michigan's responses to the sun, the wind and the spinning earth: the thermal bar. *Contrib. No. 116.* Center for Great Lake Studies, Univ. Wis., Milwaukee

Murthy, C. R. 1976. Horizontal diffusion characteristics in Lake Ontario. *J. Phys. Oceanogr.* 6:76–84

Newman, F. C. 1976. Temperature steps in Lake Kivu: a bottom heated saline lake. *J. Phys. Oceanogr.* 6:157–63

Niiler, P. P. 1975. Deepening of the wind-mixed layer. *J. Mar. Res.* 33:405–22

Orlob, G. T. 1977. *Mathematical Modeling of Surface Water Impoundments, Vol. I & II.* Resource Management Associates, Inc. 154 pp.

Osborn, T. R. 1980. Estimates of the local rate of vertical diffusion from dissipation measurements. *J. Phys. Oceanogr.* 10:83–89

Osmidov. R. V. 1965. On the turbulent exchange in a stably stratified ocean. *Atmos. Oceanic Phys.* 1:493–97

Ou, H. W., Bennett, J. R. 1979. A theory of the mean flow driven by long internal waves in a rotating basin with application to Lake Kinneret. *J. Phys. Oceanogr.* 9(6):1112–25

Patterson, J., Imberger, J. 1980. Unsteady natural convection in a rectangular cavity. *J. Fluid Mech.* 100(1):65–86

Phillips, O. M. 1970. On flows induced by diffusion in a stably stratified fluid. *Deep-Sea Res.* 17:435–43

Price, J. F. 1979. Observations of a rain-formed mixed layer. *J. Phys. Oceanogr.* 9(3):643–49

Price, J. F., Mooers, C. N. K., Van Leer, J. C. 1978. Observations and simulation of storm-induced mixed-layer deepening. *J. Phys. Oceanogr.* 8(4):582–99

Rayner, K. N. 1980. *Diurnal energetics of a reservoir surface layer. Rep. No. ED-80-005.* Univ. of Western Australia, Nedlands. 227 pp.

Robarts, R. D., Ward, P. R. B. 1978. Vertical diffusion and nutrients transport in a tropical lake (Lake McIlwaine, Rhodesia). *Hydrobiologia* 59(3):213–21

Ruddick, B. R., Turner, J. S. 1979. The vertical length scale of double-diffusive intrusions. *Deep-Sea Res.* 268A:903–14

Saylor, J. H., Huang, J. C. K., Reid, R. O. 1980. Vortex modes in Southern Lake Michigan *J. Phys. Oceanogr.* 10:1814–23

Schwab, D. J. 1978. Simulation and forecasting of Lake Erie storm surges. *Monthly Weather Rev.* 106(10):1476–87

Serruya, S. 1974. The mixing patterns of the Jordon River in Lake Kinneret. *Limnol. Oceanogr.* 19(2):175–81

Sherman, F. S., Imberger, J., Corcos, G. M. 1978. Turbulence and mixing in stably stratified waters. *Ann. Rev. Fluid Mech.* 10:267–88

Simons, T. J. 1978. Generational and propagation of downwelling front. *J. Phys. Oceanogr.* 8:571–81

Smith, J. D., Long, C. E. 1976. The effects of turning in the bottom boundary layer on continental shelf sediment transport. *Mem. Soc. R. Sci. Liège* 10:369–96

Spigel, R. H., Imberger, J. 1980. The classification of mixed-layer dynamics in lakes of small to medium size. *J. Phys. Oceanogr.* 10(7):1104–21

Steele, T. D., Stefan, H. 1979. Water quality. *Rev. Geophys. Space Phys.* 17(6):1306–36

Stefan, H., Ford, D. E. 1975. Mixed layer depth and temperate dynamics in temperature lakes. *Verh. Internat. Verein. Limnol.* 19:149–57

Stern, M. E. 1980. Geostrophic fronts, bores, breaking and blocking waves. *J. Fluid Mech.* 99(4):687–703

Stewart, K. M., Hollan, E. 1975. Meromixis in Ulmener Maar (Germany). *Verh. Internat. Verein. Limnol.* 19:1211–19

Svensson, U. 1978. *A mathematical model of the seasonal thermocline. Rep. No. 1002.* Dept. Water Resources Engrg., Univ. Lund. 187 pp.

Svensson, U. 1979. The structure of the turbulent Ekman layer. *Tellus* 31:340–50

Thompson, E. F. 1978. An evaluation of two Great Lakes waves models. *US Army Corps of Engineers. Tech. Rep. No. 78–1*

Thompson, R. O. R. Y., Imberger, J. 1980. Response of a numerical model of a stratified lake to wind stress. *Proc. Int. Symp. Stratified Flows, 2nd, Trondheim, 1980* 1:562–70

Thorpe, S. A. 1974. Near-resonant forcing in a shallow two-layer fluid: a model for the internal surge in Loch Ness? *J. Fluid Mech.* 63:509–27

Thorpe, S. A. 1977. Turbulence and mixing in a Scottish Loch. *Philos. Trans. R. Soc. London Ser. A* 286:125–81

Thorpe, S. A., Hall, A. J., Taylor, C., Allen, J. 1977. Billows in Loch Ness. *Deep-Sea Res.* 24:371–79

Toole, J. M. 1980. *Wintertime convection and frontal interleaving in the Southern Ocean.* PhD thesis. Mass. Inst. Technol./Woods Hole Oceanogr. Inst. *WHOI-80-25*. 325 pp.

Turner, J. S. 1973. *Buoyancy Effects in Fluids.* Cambridge Univ. Press. 367 pp.

Turner, J. S. 1978. Double-diffusion intrusions into a density gradient. *J. Geophys. Res.* 83:2887–2901

Turner, J. S., Chen, C. F. 1974. Two-dimensional effects in double-diffusion convection. *J. Fluid Mech.* 63:577–92

Tyler, P. A., Buckney, R. T. 1974. Stratification and biogenic meromixis in Tasmanian reservoirs. *Aust. J. Mar. Freshwat. Res.* 25:299–313

Ward, P. R. B. 1977. Diffusion in lake hypolimnia. *Proc. Cong. Int. Assoc. Hydraul. Res., 17th, Baden-Baden* 2:103–10

Weinstock, J. 1978. Vertical diffusion in a stably stratified fluid. *J. Atmos. Sci.* 35:1022–27

Weiss, W., Lehn, H., Munnich, K. O., Fischer, K. H. 1979. On the deep-water turnover of Lake Constance. *Arch. Hydrobiol* 86:405–22

Whitaker, R. E., Reid, R. O., Vastano, A. C. 1975. An analysis of drag coefficient at hurricane windspeeds from a numerical simulation of dynamic water level changes in Lake Okachobee, Florida. *US Army Corps of Engineers. Tech. Memo. No. 56.*

Wiegand, R. C., Carmack, E. C. 1980. A wintertime temperature inversion in Kootenay Lake, British Columbia. *Nat. Water Res. Inst. British Columbia, Canada*

Wunderlich, W. O., Fan, F. N. 1971. Turbulent transfer in stratified reservoir. Presented at Ann. Specialty Conf. ASCE Hydraul. Div., 19th, Iowa City

Wunsch, C. I. 1973. On the mean drift in large lakes. *Limnol. Oceanogr.* 18:793–95

TURBULENT JETS AND PLUMES

E. J. List

Environmental Engineering Science, California Institute of Technology, Pasadena, California 91125

INTRODUCTION

Turbulent jets are fluid flows produced by a pressure drop through an orifice. Their mechanics, although studied for over fifty years, has recently received research attention that has resulted in a much improved understanding of the process by which they entrain surrounding fluid. Turbulent plumes are fluid motions whose primary source of kinetic energy and momentum flux is body forces derived from density inhomogeneities. Plumes have not been studied in the same detail as jets but nevertheless there have been some recent gains in the understanding of their mechanics. In this article we will review this progress, especially in relation to how jets and plumes interact with environmental factors, such as density stratification or uniform motion of the ambient fluid. As will become evident, many problems remain and, in some circumstances, we simply cannot describe precisely what does occur. In such cases we will try to provide current references and suggest approaches for future research.

TURBULENT JETS

Near Jet Region

There is now overwhelming evidence that the initial growth of turbulent jets is a direct consequence of large-scale motions generated at the jet boundaries. These large-scale motions are primarily responsible for jet noise production and the initial entrainment of ambient fluid.

The basic sequence for axisymmetric jets seems to be as follows: In the immediate neighborhood of the orifice, the high-speed jet flow causes a laminar shear layer to be produced. The shear layer is unstable and grows very rapidly, forming ring vortices that carry turbulent jet fluid into the irrotational ambient fluid and irrotational ambient fluid into the jet, as shown clearly by Crow & Champagne (1971) and Becker & Massaro (1968; Figure 1).

The motion induced in the fluid by each vortex affects other vortices in such a way that adjacent vortices pair off, as discovered by Wehrmann & Wille (1957) and shown in the photograph by Freymuth (1966; Figure 2). The vortex motion develops a secondary circumferential instability that causes each vortex to break up, as discovered by Schneider (1980) and Yule (1978) and shown in Figure 3. The situation is even more complicated in some cases by the presence and growth of helical waves that develop into helical vortices. These have been observed by Brown (1935), Wehrmann & Wille (1957), Plaschko (1979), and Freeman & Tavlarides (1979).

For plane jets there are also two modes of vortex growth, on alternate sides as shown in Figure 4, or simultaneously on both sides, as observed

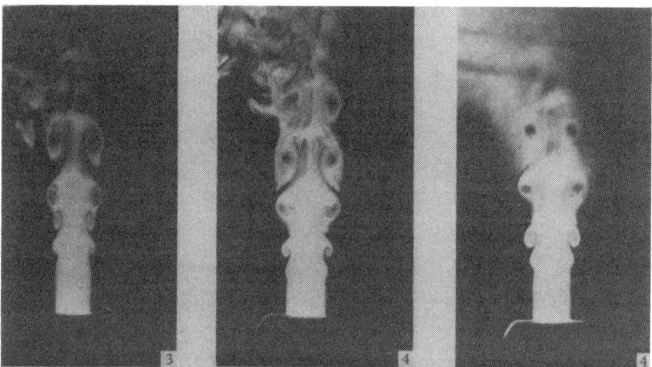

Figure 1 Illustration of ring vortex production and fusion in a turbulent jet. From Becker & Massaro (1968), courtesy of Cambridge University Press.

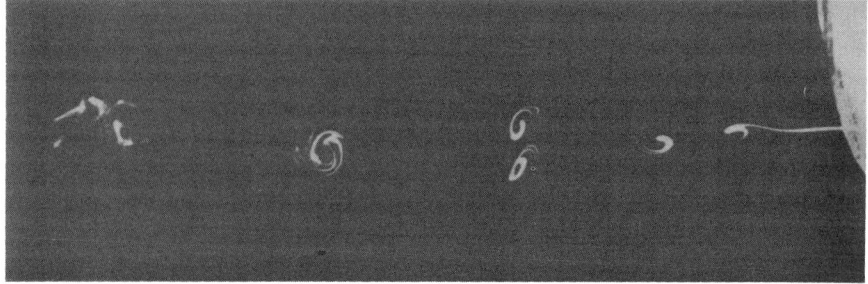

Figure 2 Vortex pairing in a turbulent jet shear layer. From Freymuth (1966), courtesy of Cambridge University Press.

by Beavers & Wilson (1970) and Rockwell & Niccolls (1972). These have been called the "flapping" and "puffing" modes of oscillation. The existence of a secondary instability of the vortices in two-dimensional jets has not yet been observed, but current work in shear layers would suggest that it is most likely present.

The investigation of these "gulping" modes of entrainment, where ambient fluid is carried into the jet and "digested" by the smaller-scale turbulence, has proceeded using hot wires, laser-Doppler systems, thermocouples, and pressure transducers. The results are interesting in that so many careful prior studies of jets, e.g. Corrsin & Uberoi (1950), Sami, Carmody & Rouse (1967), Wygnanski & Fiedler (1969), Gutmark & Wygnanski (1976), and Bradbury (1965) did not discuss the vortex nature of the initial entrainment.

The primary impetus to recent work in the near field has been a strong interest in noise production by jets. Early work by Davies et al. (1962), Bradshaw et al. (1964), and Mollø-Christensen (1957) indicated that more structured motions than had previously been believed were possibly present, while instability work by Batchelor & Gill (1962), and later by Grant (1974), Chan (1977), Liu (1974), Crighton & Gaster (1976), and

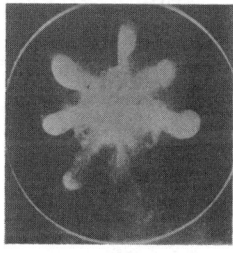

Figure 3 Instability of ring vortices as shown by Yule (1978). Courtesy of Cambridge University Press.

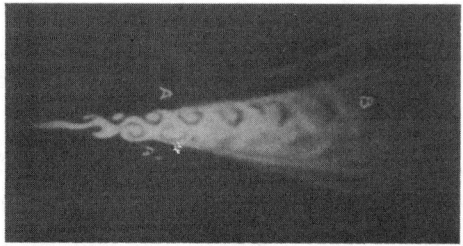

Figure 4 One mode of vortex growth in plane jets. From Brown (1935), courtesy of Cambridge University Press.

Plaschko (1979), showed analytically that such flows could be present. Recent experimental work [see, for example, Davis & Davies (1979), Bruun (1977, 1979), Maestrello & Fung (1979), Armstrong et al. (1977), Moore (1977), Sreenivasan et al. (1979), and Chevray & Tutu (1978)] confirms the basic nature of the coherency in the near flow field and shows that the vortex production frequency f is given by a Strouhal number fD/U in the range 0.3–0.5, where D is the jet orifice diameter and U the mean efflux velocity. Readers interested in the acoustic aspect of jets are referred to the comprehensive paper by Moore (1977); the subject will not be discussed further here.

It is clear that the production of vortices is the key element in initial jet dilution. Each vortex wraps ambient fluid about itself; then, as the vortices pair, the fusion process mixes the ambient and jet fluid. Circumferential instability, and possible interaction with helical modes, leads to the apparent eventual self destruction of the large-scale structures and this, in turn, generates the subsequent small-scale turbulent mixing. Downstream of the initial interacting vortex zone, the flow appears to be more in line with the general preconception of random and chaotic turbulent motion, although recent work by de Gortari & Goldschmidt (1981) shows very clear evidence of a flapping mode persisting far downstream in plane jets $(x/D \sim 100)$.

Experimental Measurements

Most experimental velocity measurements in turbulent jets have been made with hot-wire instrumentation. However, other studies have recently been performed using laser-Doppler velocimetry (LDV) and the results form a very interesting comparison that offers some insight on the mechanics of jets. Kotsovinos (1975) and Lau et al. (1979), both using LDV, measured probability density functions for velocity in plane and round jets respectively. They observed that in the outer regions of a jet there are a significant fraction of negative velocities (Figure 5). This result is important since it means that in measurements with a conventional hot wire, which cannot resolve directional ambiguity, there will be an underestimation of the velocity variance $\overline{u'^2}$ and an overestimation of the mean velocity U_m. The errors near the jet centerline should be minimal with the discrepancy increasing with distance from the jet axis. Thus, it is clear that hot-wire measurements of relative turbulent momentum and heat flux, $\overline{u'^2}/U_m^2$ and $\overline{u'\theta'}/U_m\theta_m$ will also be increasingly in error where the relative fraction of negative velocities increases. This is precisely what was found by Lau et al. (1981), although they ascribe the difference in hot-wire and laser measurements to counting errors in the laser and deformation of the hot wire. We believe the argument above is also plausible.

Antonia et al. (1980) have also recently pointed out that temperature measurements taken with cold wires used in conjunction with hot wires would be contaminated in reverse flows. This is likely to cause further errors in the measurement of temperature variances and heat fluxes.

Experimental Data

Experimental data for $(\overline{u'^2})^{1/2}/U_m$ and $(\overline{\theta'^2})^{1/2}/\theta_m$, obtained in axisymmetric jets at distances in excess of 40 jet diameters downstream, make it clear that at this distance from the orifice a turbulent jet is, to a good approximation, in a fully developed self-similar state. Wygnanski & Fiedler (1969) and Antonia et al. (1975) found values of $(\overline{u'^2})^{1/2}/U_m = 0.28$–$0.29$ on the jet axis, which agrees well with the value of 0.285 found by Birch et al. (1978) for the relative concentration fluctuations, $(\overline{\theta'^2})^{1/2}/\theta_m$. The relative concentration-fluctuation profiles measured by Birch et al. (1978), Becker et al. (1967), Shaughnessy & Morton (1977), Antonia et al. (1975), and Chevray & Tutu (1978) all show a well-defined peak in the profile at $r/r_{1/2} = 1.0$. The relative velocity-fluctuation profiles measured by Wygnanski & Fiedler (1969), Antonia et al. (1975), and Chevray & Tutu (1978) have no well-defined maximum, and in relative magnitude lie essentially everywhere below the concentration-fluctuation curves (see Figure 6). As explained above, although this could be primarily an artifact of the experimental technique, it could also be a fundamental difference in the production, advection, and dissipation of $\overline{\theta'^2}$ and $\overline{u'^2}$. Wygnanski & Fiedler (1969) and Antonia et al. (1975) have measured budgets of $\overline{u'^2}$ and $\overline{\theta'^2}$, respectively, and although these appear qualitatively similar, the

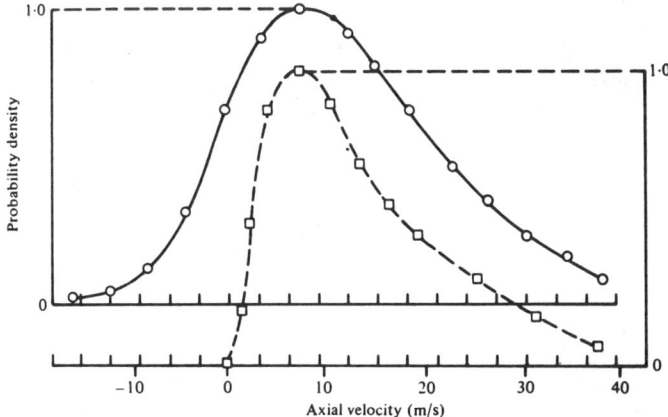

Figure 5 Probability density functions for velocity measured by Lau et al. (1979) at $r/r_o = 1.4$ in turbulent jet. Full line: laser-Doppler velocimeter; broken line: hot wire anemometer. Courtesy of Cambridge University Press.

errors are such that it is not possible to discern if the pressure terms are significant in the $\overline{u'^2}$ transport.

The distribution of mean velocity across a jet appears to be quite self-similar with a reasonable data fit given by

$$\overline{u} = U_m \exp[-(r/b_u)^2] . \qquad (1)$$

The evaluation of how U_m and b_u depend upon x, the distance from the jet origin, depends upon the evolution of the momentum flux specified by an integrated momentum equation

$$\frac{\partial}{\partial x} \int_0^{b(x)} \left[\overline{u}^2 + \overline{u'^2} + \frac{\overline{p} - p_\infty}{\rho_0} - \frac{\tau_{xx}}{\rho_0} \right] 2\pi r dr$$

$$= -\lim_{r \to b(x)} \left[2\pi r \left(\overline{u}\,\overline{v} + \overline{u'v'} - \frac{\tau_{rx}}{\rho_0} \right) \right] \qquad (2)$$

$$+ \frac{\partial b}{\partial x} 2\pi b(x) \left[\overline{u}^2 + \overline{u'^2} + \frac{\overline{p} - p(\infty)}{\rho_0} - \frac{\tau_{xx}}{\rho_0} \right]_{b(x)},$$

where $b(x)$ is some radial distance from the jet axis. Assuming viscous stresses negligible and no external flow, the equation reduces to

$$\int_0^{b(x)} [\overline{u}^2 + \overline{u'^2} + (\overline{p} - p(\infty))/\rho_0] 2\pi r dr = M . \qquad (3)$$

For a given self-similar distribution of mean velocity $\overline{u}$, stress $\overline{u'^2}$, and pressure, then

$$b_u^2 U_m^2 / M = \text{constant}. \qquad (4)$$

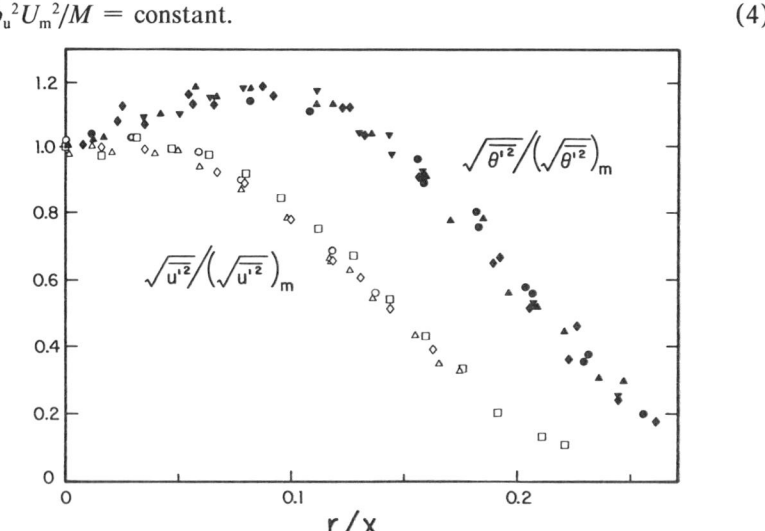

Figure 6 Profiles of relative turbulent velocity and concentration in round jets. Open symbols Wygnanski & Fiedler (1969), closed symbols Becker et al. (1967).

The constant can be determined from experimental measurements of b_u and U_m. Experimental results (Fischer et al. 1979) give $b_u/x = 0.107$, so the distribution of mean velocity [Equation (1)] enables us to write

$$U_m = kM^{1/2}x^{-1}. \tag{5}$$

The value of k computed ignoring the turbulent flux $\overline{u'^2}$ is about 7.5; including $\overline{u'^2}$, based on measured profiles of $\overline{u'^2}$, gives $k = 6.9$, which agrees well with the experimental results (Fischer et al. 1979).

The same arguments applied to the flux, Y, of a conserved tracer material lead to the experimental results (Fischer et al. 1979)

$$b_\theta \simeq 0.126x \tag{6}$$

and

$$\theta_m \simeq 5.8Y\, M^{-1/2}\, x^{-1} \tag{7}$$

as shown in Figure 7.

If μ is the volume flux within the jet, then the rate of entrainment

$$d\mu/dx = C_j M^{1/2} \tag{8}$$

and, as shown in Fischer et al. (1979), C_j has a value of about 0.25.

Equations (5) and (7), coupled with the results for $\overline{(u'^2)}^{1/2}/U_m$ and $\overline{(\theta'^2)}^{1/2}/\theta_m$, make it clear that the turbulent fluctuations also scale with distance from the orifice. Measurements of integral scales, based on velocity and concentration correlations by Wygnanski & Fiedler (1969) and

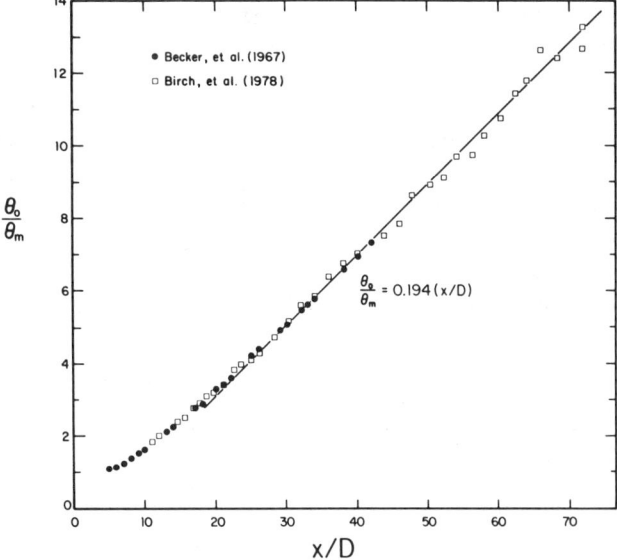

Figure 7 Mean concentration on the axis of a round turbulent jet.

Becker et al. (1967), show Δ_u and $\Delta_\theta \sim 0.042$–$0.045\ x$. Measurements of maximum turbulent shear stress intensities $\overline{u'v'}/U_m^2$ by Wygnanski & Fiedler (1969) and Antonia et al. (1975) differ by a factor of two. There is no ready explanation. Wygnanski & Fiedler's value of 0.017 is more consistent with values in the range 0.009–0.013 measured closer to the jet orifice by Boguslawski & Popiel (1979), Lau et al. (1979), and Chevray & Tutu (1978).

Turbulent heat-flux measurements by Antonia et al. (1975) and Chevray & Tutu (1978) both suggest, by direct computation, that about 7–8% of the total axial heat flux is carried by the turbulent flux. However, direct computations of the flux by the mean advection give a turbulent heat flux of about 14%, in agreement with the turbulent total momentum flux estimated above. Clearly, improved heat-flux measurements using instrumentation capable of resolving the directional ambiguity are a high priority.

Plane Jets

The basic difference with the mechanics of plane jets is the existence of two modes of large-scale vortex formation as shown by Brown (1935), Beavers & Wilson (1970), and Rockwell & Niccolls (1972). However, experimental attempts to prove the existence of "flapping" or "puffing" modes by spectral measurements have been equivocal. Goldschmidt & Bradshaw (1973), Everitt & Robins (1978), and de Gortari & Goldschmidt (1981) concluded there was flapping; Wygnanski & Gutmark (1971) and Moum et al. (1979) concluded there was not.

Measurements of relative turbulent intensities of velocity and temperature fluctuations in the self-similar regime for $x/D > 40$ are given in Table 1. The measured values are not very different from axisymmetric jets, although the relative amplitude of velocity fluctuations appears to be slightly lower than the 0.27 value for round jets.

The only known axial heat-flux measurements in plane jets are those by Kotsovinos (1975), who measured $\bar{u}$, $\bar{\theta}$, $\overline{u\theta}$ using a laser-Doppler and microbead thermistor, and derived the turbulent heat flux $\overline{u'\theta'}$ by computation. The distribution of $\overline{u\theta}$ was found to be self-similar for $x/D = 20.8$,

Table 1 Equilibrium relative velocity and temperature fluctuations in plane jets

Investigator	x/D	$(\overline{u'^2})^{1/2}/U_m$	$(\overline{T'^2})^{1/2}/T_m$
Bashir & Uberoi (1975)	40	0.24	0.26
Bradbury (1965)	40	0.25	—
Gutmark & Wygnanski (1976)	40	0.23	—
Heskestad (1965)	40	0.22	—
Kotsovinos (1975)	37	0.23	0.30

37, 58.75, and 93.75. The integrated area of the heat-flux profiles was within 5% of the known input heat flux. Kotsovinos's results suggest a total turbulent heat flux of about 6%. Measurements of the radial heat flux $\overline{v'\theta'}$ by Bashir & Uberoi (1975) were in poor agreement with computations from the mean transfer. The entrainment in a plane turbulent jet is determined by the specific momentum flux at the jet orifice to give a jet invariant

$$C_j = \mu^2/(Mx) \ . \tag{9}$$

Measurements by Kotsovinos (1975) show C_j has a value of 0.29. Equation (9) obviously implies that the rate of entrainment

$$d\mu/dx = 1/2 \ (C_j M/x)^{1/2} \ . \tag{10}$$

TURBULENT PLUMES

Plume Buoyancy

If a plume source is thermal, the weight force driving the plume is established by the volumetric expansion of the fluid. We define a specific buoyancy flux

$$B = g\sigma H_0 \ / \ (\rho c_p) \ , \tag{11}$$

where σ is the volumetric coefficient of thermal expansion, H_0 is the heat flux, c_p the specific heat of the fluid at constant pressure. If the plume arises from a source with mass flux, $\rho_0 Q$, then the specific buoyancy flux is

$$B = g(\rho_a - \rho_0)Q/\rho_0 \ , \tag{12}$$

where ρ_a is the ambient fluid density and ρ_0 the source fluid density. For a thermal source buoyancy conservation depends upon the temperature distribution in the environment and the temperature dependence of σ. For a mass source, the ambient fluid density must be constant. We assume that B is conserved in the discussion following.

Unless the buoyancy source is sustained an isolated "thermal" forms; the mushroom cloud of a nuclear explosion or a volcanic eruption is an example. Thermals are discussed in detail by Turner (1973). When the source of buoyancy is sustained, the initial starting cap advances with a displacement proportional to $B^{1/4}t^{3/4}$, as shown initially by Turner (1962). Recently, Delichatsios (1979) has extended this work to include variable-strength sources of buoyancy, such as would be generated by a fast-growing fire. The results show an advancing front moving with a maximum velocity of 0.42 of the maximum velocity behind the front; Turner found 0.38 experimentally. Delichatsios found also that the normalized temperature history behind the front is relatively independent of the rate at which the buoyancy flux increases.

For a discussion of starting plumes readers are referred to the reviews by Turner (1969, 1973) and the paper by Delichatsios (1979).

Steady Plumes

If the initial geometry of the plume is ignored, then the mean properties of the convection generated by the buoyancy flux must be a function of B, x, v, κ, where x measures the distance above the source, v/κ is the Prandtl number. Dimensional analysis then prescribes that the mean velocity U_m on the axis of a round plume is given by

$$U_m = (B/x)^{1/3} f(B^{1/3}x^{2/3}/v, v/\kappa), \qquad (13)$$

where $B^{1/3}x^{2/3}/v$ is a local Reynolds number. Self-similarity in this Reynolds number and independence from the Prandtl number implies

$$U_m = k_R (B/x)^{1/3}. \qquad (14)$$

Similar reasoning for plane plumes gives

$$U_m = k_P B^{1/3}, \qquad (15)$$

where, in Equation (15), B is specific buoyancy flux per unit length of plume (dimensions $L^3 T^{-3}$) and k_R and k_P are constants.

Few detailed experiments exist to test these results and, while it is agreed that the basic idea is confirmed, there is some debate over the value of the constants k_R and k_P. Relatively crude measurements by Rouse et al. (1952) suggest a value of 4.7 for k_R. Other measurements by George et al. (1977), Nakagome & Hirata (1976), and Beuther (1980) indicate values of k_R in the range 3.4–3.9. While for any given experiment the dependence of U_m on $x^{-1/3}$ may be quite evident, the value of the constant k_R depends upon knowledge of B. Rouse et al. found k_R by equating their integrated heat flux to the known heat input. George et al. (1977) evaluated B by integrating both the mean and turbulent fluxes measured with hot wires. However, as previously discussed, turbulent flux measurements with hot wires in jets are subject to significant error. In plumes the problem is exacerbated by the relatively larger proportion of negative velocities that occur in the velocity distribution. Velocity probability density function estimates by Kotsovinos (1975) in plane plumes show quite clearly that negative velocities exist. Underestimates of the heat flux would therefore underestimate B and suggest a larger value of k_R than is appropriate.

For plane plumes the only known velocity measurements on the axis are by Kotsovinos (1975), who used a laser-Doppler velocimeter (Figure 8). Kotsovinos's (1977) measurements of turbulence quantities in plane jets have been criticized by Chen & Rodi (1980) as being "unrealistic." However, as noted in the previous section, his results for both velocity and temperature fluctuations in plane jets were within the range of hot-wire measurements on the axis and his heat-flux measurements were consistent with heat conservation. The same is true for his measurements of the total heat flux in plane plumes. The difficulty apparently arises from a comparison of Kotsovinos's turbulent velocity measurements in plane plumes relative to those measured by George et al. (1977) and Nakagome & Hirata (1976) for round plumes. Figure 9 gives the distributions of $(\overline{u'^2})^{1/2}/U_m$ and $(\overline{T'^2})^{1/2}/T_m$ measured by these investigators. Also included are the respective heat-flux profiles. The temperature fluctuation intensities recorded by the three investigators are much the same, with a peak value of about 0.40. Kotsovinos's measurements of $(\overline{u'^2})^{1/2}/U_m$ also have a peak value of about 0.40 whereas the others found values of 0.27, as for jets. Although the resolution of this discrepancy will come from other independent measurements in plane and round plumes, it is clear that the hot wire will underestimate $\overline{u'^2}$ while the laser will overestimate it (Lau et al.

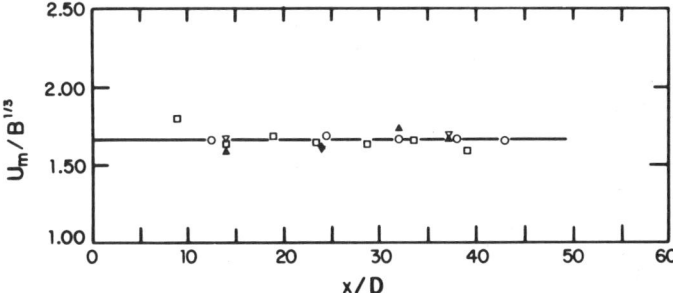

Figure 8 Velocity on the axis of a plane plume (Kotsovinos 1977).

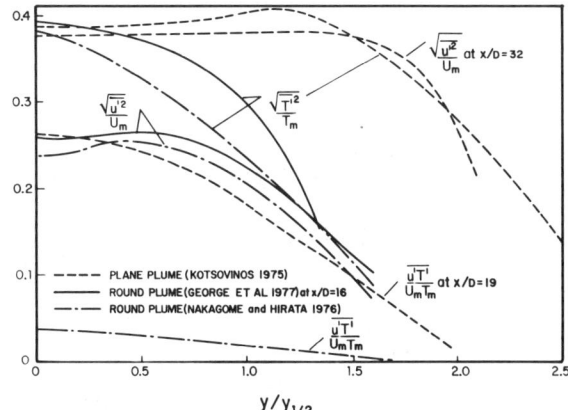

Figure 9 Distributions of $(\overline{u'^2})^{1/2}/U_m$ and $(\overline{T'^2})/T_m$ and heat flux for plumes. From Chen & Rodi (1980).

1981). The heat-transfer measurements by Nakagome & Hirata (1976) are clearly at variance with other hot-wire measurements, such as by Chevray & Tutu (1978) or Antonia et al. (1975).

In the above analyses it has been assumed that self-similarity in the local Reynolds number ($B^{1/3}x^{2/3}/v$ in round plumes) is valid. Pera & Gebhart (1971) and Mollendorf & Gebhart (1973) addressed this question by finding instability estimates for laminar plumes. Bill & Gebhart (1975) studied the laminar to turbulent transition process using the Grashof number

$$\mathrm{Gr} = gx^3\sigma(T_m - T_\infty)/v^2 . \tag{16}$$

For turbulent round plumes, dimensional reasoning gives

$$g\sigma(T_m - T_\infty) \sim B^{2/3} x^{-5/3}, \tag{17}$$

and for plane plumes

$$g\sigma(T_m - T_\infty) \sim B^{2/3} x^{-1}, \tag{18}$$

so that the Grashof number is equivalent to the local Reynolds number. Bill & Gebhart found that the transition in plane plumes was complete at a Grashof number of about 3×10^8.

For fully developed plumes it can be shown that the total specific momentum flux is given asymptotically by

$$m = k_m B^{2/3} x^{4/3} \tag{19}$$

and the volume flux by

$$\mu = k_\mu B^{1/3} x^{5/3}, \tag{20}$$

where k_m and k_μ are constants. These two results can be used to derive plume invariants

$$C_p = \mu/(m^{1/2} x) \tag{21}$$

and

$$R_p = \mu B^{1/2}/m^{5/4} . \tag{22}$$

Rouse et al.'s (1952) experiments suggest values of $C_p = 0.25$ and $R_p = 0.56$ (see List & Imberger 1973, 1975). The local Reynolds number, based on the local length scale $\mu/m^{1/2}$ and velocity scale m/μ, is $m^{1/2}/v$, which increases with m, in contrast to the jet Reynolds number which remains constant. For plane plumes the equivalent results are (Kotsovinos & List 1977)

$$C_p = \mu^2/(mx) = 0.29, \tag{23}$$

$$R_p = \mu^2 B^{2/3}/m^2 = 0.735 . \tag{24}$$

The above results can be used to deduce the rates of entrainment into plumes. For round plumes

$$d\mu/dx = (5/3)C_p m^{1/2} \tag{25}$$

and for plane plumes
$$d\mu/dx = (C_p m/x)^{1/2}.\qquad(26)$$
These are remarkable results since they specify that the rates of entrainment into jets and plumes are both defined by the local specific momentum flux and the distance from the source. The coefficients of proportionality are slightly different in each case: compare Equations (8) and (25) and Equations (10) and (26).

TURBULENT BUOYANT JETS

Vertical Motions

The effect of the weight deficiency induced by the buoyancy flux of a plume is to increase the momentum flux. Thus a turbulent buoyant jet with an initial specific momentum flux M will have this momentum flux continuously increased with distance from the source. The momentum flux will be specified asymptotically by Equation (19) so that when $M^{3/4}/B^{1/2} \ll x$ the initial momentum flux will have been overwhelmed by that produced by the buoyancy. The length scale $\ell_M = M^{3/4}/B^{1/2}$, defined initially by Morton (1959), therefore controls the degree of plumelike behavior at a given distance from the source. There is a further length scale, $\ell_Q = Q/M^{1/2}$, which effectively defines the initial jet geometry. The ratio of these two scales $R = \ell_Q/\ell_M = QB^{1/2}/M^{5/4}$ is the initial jet Richardson number. Obviously, if R is equal to R_p, defined by Equation (22), the turbulent buoyant jet will be initiated as a plume.

The original analyses of turbulent buoyant jets were made by Morton, Taylor & Turner (1956) and Priestley & Ball (1955). Their integral techniques are well known and the reader is referred to Chapter 9 of the recent book by Fischer et al. (1979) for a current summary.

It is worth noting here that the research literature is sadly deficient in the description of turbulence within buoyant jets in transition from jets to plumes. If we regard the ratio x/ℓ_M as the factor defining the degree of plumelike behavior of a buoyant jet ($\ell_M = M/B^{2/3}$ for plane jets) then apparently there is only one experimental study that covers the range from jet to fully developed plume. This work, by Kotsovinos (1975), detailed both the mean and turbulence quantities of a buoyant jet as it evolved into a plume. Kotsovinos found several results of importance. First, the asymptotic mean description of jets and plumes derived by dimensional analysis was confirmed. Second, the integral analysis as defined by Morton et al. (1956), and applied to plane buoyant jets by Koh & Brooks (1975), gives good results provided the entrainment is modified to include the Richardson-number effect defined by Priestley & Ball (1955). Third, significant negative velocities were found to occur in both jets and plumes. Fourth,

Gaussian statistics, as defined by skewness and kurtosis, were appropriate on jet and plume axes. Fifth, a fraction of the total heat flux borne by the turbulent fluctuations in a turbulent buoyant jet increased significantly with x/ℓ_M (see Figure 10). In other words, the velocity and temperature fluctuations in plumes are more highly correlated than in jets. Since a plume is a buoyancy-driven flow, this does not appear too surprising. Sixth, an intermittency analysis of the temperature fluctuations for both plumes and jets showed that the contribution to the temperature variance from entrainment of ambient fluid was approximately the same in both jets and plumes. The contribution from fluctuations within the "turbulent" fluid was significantly higher in plumes.

Although Kotsovinos did not perform a turbulent kinetic energy balance in the plume flow, it is clear that there must be a fundamental difference between jets and plumes. In jets there is turbulent kinetic energy production from the mean flow and loss by viscous dissipation. In plumes, the buoyancy has two effects: not only does it increase the mean momentum flux, and therefore the mean shear production, but it also leads to a direct transfer from potential energy to turbulent kinetic energy. Since the only dissipation is viscous, it is to be expected that the equilibrium turbulence intensities will be significantly higher in plumes than in jets. Direct experimental confirmation of this reasoning would be an interesting study, for there are few motions in which the buoyancy flux leads to an increase in both shear and buoyancy production of turbulent kinetic energy.

Angled Jets, Negative Buoyancy

Positively buoyant jets discharged other than vertically upward have a curved trajectory, since the body forces apply a vertical acceleration to the discharged fluid. If a jet with negative buoyancy is directed vertically upward, a terminal height is reached at which point the flow reverses and

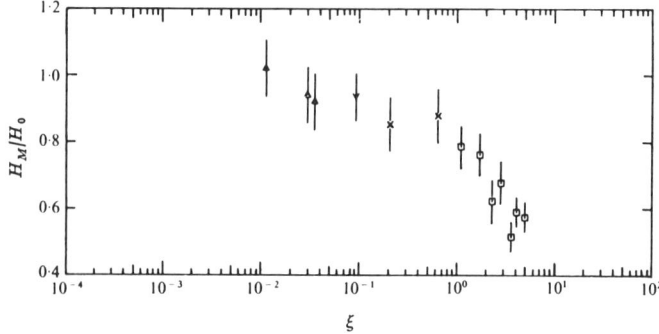

Figure 10 Advective heat flux, H_M, in turbulent buoyant jets in transition to turbulent plumes (Kotsovinos & List 1977).

the jet folds back on itself, just as in a water fountain. This particular problem has been considered by Abraham (1967), Anderson et al. (1973), Morton (1959), Turner (1966), Vergara & James (1979), and Zeitoun et al. (1970). The results all seem to indicate that the terminal height is proportional to the length scale $M^{3/4}/B^{1/2}$, although there is some disagreement on the coefficient of proportionality; values vary from 1.54 to 2.06.

The application of the integral analysis to the situation where there is a directly reversing flow does not seem appropriate since the entrainment involves flow previously discharged. However, when the negatively buoyant jet is angled to the vertical the integral method seems to work and the model by Hofer & Hutter (1981) appears to give reliable estimates of the fall-back zone.

Density-Stratified Environments

Positively buoyant turbulent jets invariably rise regardless of the discharge angle to the vertical. However, when the environment is density-stratified, jets reach a terminal height of rise. This is a direct result of the fact that dense fluid entrained in the lower regions of a buoyant jet, when carried to a less dense environment, will ultimately give rise to a negative buoyancy flux that destroys the upward momentum. The momentum of the jet may, in fact, carry dense fluid beyond a level of neutral stability so that a fall-back occurs.

Given a linear density stratification represented by the Brunt-Väisälä frequency

$$N^2 = -(g/\rho) \, \partial\rho/\partial z \, , \tag{27}$$

then the terminal height of rise of a round jet or plume can be shown to be proportional to $(M/N^2)^{1/4}$, or $(B/N^3)^{1/4}$, respectively. Equivalent results for plane jets and plumes are $(M/N^2)^{1/3}$ and $(B/N^3)^{1/3}$. Defining S by the ratio of these two length scales, where

$$S = (MN/B)^2 \tag{28}$$

gives the dimensionless parameter which, along with the Richardson number $R = QB^{1/2}/M^{5/4}$, will define the complete behavior of a turbulent buoyant jet in a linear density stratification. With normalized volume flux and elevation, as given in Table 2, then asymptotic results for $S \gg 1$ and $S \ll 1$ can be found quite easily. Table 3 presents the results of experimental studies for round jets by Abraham & Eysink (1969), Crawford & Leonard (1962), Fan (1967), Fox (1970), and Sneck & Brown (1974), and for plane jets by Wright & Wallace (1979) and Chen et al. (1980). Similar results can be deduced for the terminal height of rise of a thermal, although few experimental results are available. For a thermal release of E joule in the atmosphere the terminal height of rise is roughly given by 2.7 $[g\sigma E/(\rho c_p N^2)]^{1/4}$ (Fischer et al. 1979).

When the density stratification is not linear the only option is to integrate the entrainment equations numerically. Although there are no direct experimental studies of the rate of jet entrainment in stratified environments, other than that for sloping jets by Ellison & Turner (1959), the hypothesis that the entrainment is locally controlled by the specific momentum flux seems to give reasonable results. Clearly, when the jet reaches a level of neutral stability and spreads laterally within the density-stratified environment, almost all turbulent mixing will cease, as shown by studies of stably stratified shear layers (Gartrell 1979).

There are no known studies of the turbulence in jets discharged to a density-stratified environment. It is unlikely that self-similarity of the turbulence intensities will prevail since the buoyancy in this case ultimately acts to reduce turbulent kinetic energy levels.

Jets and Plumes in Crossflows

Round buoyant jets discharging vertically into a uniform density crossflow form a complex flow pattern whose basic features have only recently been described. The region near the jet orifice has been studied experimentally by Chassaing et al. (1974), Moussa et al. (1977), and Crabb et al. (1981). The two basic variables, so far as the gross flow features are concerned, are the jet momentum flux M and the mean crossflow velocity U. These specify a length scale $z_M = M^{1/2}/U$ that defines the relative influence of the jet and the crossflow. For axial distances z from the orifice such that

Table 2 Normalized volume flux and elevation used in Table 3

	Plane flow	Axisymmetric flow
Volume flux $\bar{\mu}$	$\mu B^{1/3} R_p^{-1/2} M^{-1}$	$\mu B^{1/2} R_p^{-1} M^{-5/4}$
Elevation ξ	$C_p x B^{2/3} (R_p M)^{-1}$	$C_p x B^{1/2} R_p^{-1} M^{-3/4}$

Table 3 Dimensionless terminal height of rise and dilution for round and plane jets and plumes in a linearly stratified environment.

Variable		Plane Source		Round Source	
		$S \ll 1$	$S \gg 1$	$S \ll 1$	$S \gg 1$
Terminal height for jet	h_M	—	$3.6(M/N^2)^{1/3}$	—	$3.8(M/N^2)^{1/4}$
Terminal height for plume	h_B	$2.5(B/N^3)^{1/3}$	—	$3.8(B/N^3)^{1/4}$	—
Normalized height of rise	ξ_T	$1.0S^{-1/2}$	$1.4S^{-1/3}$	$1.7S^{-3/8}$	$1.7S^{-1/4}$
Normalized dilution	$\bar{\mu}_T$	$1.0S^{-1/2}$	$0.82^a S^{-1/6}$	$1.5S^{-5/8}$	$1.2S^{-1/4}$

[a] Computed using Koh & Brooks (1975).

$z \ll z_M$ then the basic picture is vortex production by the shear layer at the jet orifice, as in the absence of a crossflow.

However, the crossflow appears to see the jet as an essentially solid object and a further vortex-shedding process occurs. As described by McMahon et al. (1971) this shedding occurs at a Strouhal number in agreement with those found for solid bodies of comparable size. This vortex shedding induces streamwise (with respect to the jet stream) vorticity which may become sufficiently intense to bifurcate the flow (Figure 11), as initially described by Scorer (1959) and Turner (1960) and more recently in great detail by Crabb et al. (1981). Buoyancy within the jet exacerbates the splitting (Abdelwahed & Chu 1978). If the vortices impact a boundary such as a density discontinuity in the ambient fluid, or a free water surface, then the splitting is dramatic (Hayashi 1972). The influence of jet-discharge angle relative to the free stream on the production of vortices is described by Krausche et al. (1978). The description of the mean trajectory of a buoyant jet in a crossflow appears to be unaffected by bifurcation, except that the role of buoyancy is unresolved (Abdelwahed & Chu 1978).

The fundamental parameter in the description of buoyant-jet trajectories in crossflows in the ratio of the length scale $z_M = M^{1/2}/U$ to the equivalent length scale for a plume in a crossflow, defined by $z_B = B/U^3$.

Two asymptotic cases occur, corresponding to $z_M > z_B$ and $z_M < z_B$, and within each case there are three regimes. For $z_M > z_B$ jet momentum is dominant and the regimes can be loosely described as "vertical" jet, "bent" jet, and "bent" plume. When $z_B > z_M$ jet buoyancy is important so that the regimes can be called vertical jet, vertical plume, and bent plume. In all cases, the buoyancy ultimately controls the trajectory.

The trajectory equations due initially to work by Priestley (1956), Scorer (1959), and Moore (1966), and confirmed in experimental work by Hoult et al. (1969), Chu & Goldberg (1974), and Wright (1977), can be deduced by an application of similarity theory to conservation equations on vertical and horizontal planes (see Fischer et al. 1979). The results are given in Table 4 where the constants C_1–C_4, as determined from experiments, are given in Table 5.

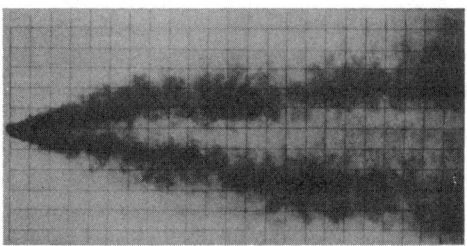

Figure 11 Bifurcation of a turbulent buoyant jet. Courtesy of Abdelwahed & Chu (1978).

It must be recognized that the results presented in Tables 3 and 4 are simplistic in that they account for only the grossest features of the flows. Nevertheless, they do indicate the importance of scaling in evaluating any particular flow configuration. Attempts have also been made to model the effect of crossflows on jets using versions of the integral analysis developed by Morton et al. (1956). Fan (1967) used a drag coefficient in the horizontal momentum equation. This gave satisfactory results to the region of maximum jet curvature. More comprehensive integral models such as by Schatzmann (1978, 1979) and Hofer & Hutter (1981) appear to be successful in reproducing most of the salient features of the flows, even in the presence of ambient density stratification.

Other models attempt to describe jet development at a more fundamental level, these including eddy viscosity, mixing length, and various second-order closure models. Tennankore & Steward (1978) conclude that the k-ϵ model shows the best promise. Chen & Nikitopoulos (1979) used a k-ϵ model for axisymmetric buoyant jets and claim reasonable agreement with the experimental hot-wire results of Nakagome & Hirata (1976). Teske, Lewellen & Segur (1978) used a second-order closure model applied to two-dimensional jets. The model appears to reproduce most of the basic features qualitatively, including the Brunt-Väisälä oscillations of jets and plumes into a flowing density-stratified environment. However, there is disaccord with measured temperature fluctuations in jets.

Table 4 Trajectory equations for turbulent buoyant jet: evaluation of $\bar{z}(x)$ for jet in a crossflow

	Cases for $z_M > z_B$		
RANGE	$D \ll \bar{z} \ll z_M$	$z_M \ll \bar{z} \ll \left(\dfrac{z_M}{z_B}\right)^{1/3} z_M$	$\ll \bar{z}$
$\bar{z}(x)/z_M$	$C_1(x/z_M)^{1/2}$	$C_2(x/z_M)^{1/3}$	$C_4(z_B/z_M)(x/z_B)^{2/3}$
	Cases for $z_B > z_M$		
RANGE	$D \ll \bar{z} \ll \ell_M$	$\ell_M \ll \bar{z} \ll z_B$	$z_B \ll \bar{z}$
$\bar{z}(x)/z_B$	$C_1(z_M/z_B)(x/z_M)^{1/2}$	$C_3(x/z_B)^{3/4}$	$C_4(x/z_B)^{2/3}$

Table 5 Constants used in asymptotic trajectory laws in Table 4

Investigator	C_1	C_2	C_3	C_4
Briggs[a] (1975)	—	1.8–2.1	—	0.85–1.3
Chu & Goldberg (1974)	—	1.44	—	1.14
Hoult et al. (1969)	1.8–2.5	—	—	—
Wright (1977)	1.8–2.3	1.6–2.1	1.4–1.8	$(0.85–1.4)(z_M/z_B)^2$

[a]Mean of 14 studies.

Jets and Plumes With Crossflows and Stratification

With ambient density stratification and crossflows, the problem gets complex and very little work has been done in either analysis or experiment. The problem is an essentially three-parameter one involving the Richardson number R, the stratification parameter S, and the crossflow-ratio number z_M/z_B. There is a horizontal length scale in the problem, $L = U/N$, which is the wavelength of Brunt-Väisälä waves supported by the stratification. The terminal height of rise of a buoyant jet can then be very roughly specified by replacing x by L in the trajectory equations of Table 4. This gives the results in Table 6, and there is some experimental support for the coefficients presented. E_1 and E_3 are shown by Briggs (1975) to have a value of about 4. From atmospheric studies Briggs (1975) also suggests a value of 3.8 for E_4. There appear to be no data to support a value for E_2. The integral model developed by Schatzmann (1978, 1979) should produce these solutions asymptotically but this has not been verified.

Manins (1979) has attempted to develop a model for the relative fraction of plume penetrating a strong pycnocline. His experiments clearly show the rapid reduction of mixing once the plume enters a region of strong stratification. Luti & Brzustowski (1977) have developed a model that describes the influence of a two-dimensional buoyant plume on a density-stratified crossflow, extending the work of Cederwall (1971) who described the role of two-dimensional plane jets in crossflows. The model by Luti & Brzustowski succeeds in describing most of the features observed in mass fires. The solution obtained indicates a series of standing internal

Table 6 Asymptotic heights of rise for a vertical turbulent buoyant jet discharging into a density-stratified crossflow (see text for values of E_i.)

CASE	$\dfrac{(U/N)^2}{M^{1/2}} \ll 1$	$1 \ll \dfrac{(U/N)^2}{M^{1/2}} \ll \dfrac{z_M}{z_B}$	$\dfrac{z_M}{z_B} \ll \dfrac{(U/N)^2}{M^{1/2}}$
$z_M > z_B$	$\dfrac{z_T}{h_M} = E_1$	$\dfrac{z_T}{z_M^{2/3} L^{1/3}} = E_2$	$\dfrac{z_T}{z_B^{1/3} L^{2/3}} = E_4$
CASE	$S^{-1/2} \ll 1$	$1 \ll S^{-1/2} \ll \left(\dfrac{z_B}{z_M}\right)^2$	$\left(\dfrac{z_B}{z_M}\right)^2 \ll S^{-1/2}$
$z_M < z_B$	$\dfrac{z_T}{h_M} = E_1$	$\dfrac{z_T}{h_B} = E_3$	$\dfrac{z_T}{z_B^{1/3} L^{2/3}} = E_4$

waves as found by List (1971) for vertical jets in density-stratified environments. The model developed is an interesting one in that it is concerned with the flow external to the plume but generated by the plume entrainment, a factor that seems to be a key ingredient in large-scale fires.

Little work appears to have been done on the influence of ambient shear flows and ambient turbulence on turbulent buoyant jet behavior. Slawson & Csanady (1967) and Akatnov (1977) have discussed the role of ambient turbulence but there is an absence of good experimental data.

CONCLUSION

If one can draw an overall conclusion from a review of the literature on jets and plumes, it is that some areas of the flow field have received attention quite out of proportion to other areas. Clearly, it will be a truly exceptional new research paper that contributes substantially to the understanding of jet flows in the region of flow within ten jet diameters of the discharge orifice. On the other hand, almost any additional piece of work describing the mechanics of turbulence in plumes will be essentially new work. There are many such gaps in the literature which deserve research attention. Some of these are:

1. A comparison of turbulence intensity and turbulent flux measurements in jets, where the data is obtained using both hot wires or films and laser-Doppler anemometry. This should help to establish the authenticity of existing data.
2. Measurements of heat and momentum fluxes in buoyancy-dominated flows to help establish the basic nature of such flows.
3. An assessment of the appropriateness of the self-similar hypothesis for jets and plumes in density-stratified and uniform flow environments.
4. An evaluation of entrainment and entrainment hypotheses for jets and plumes in density-stratified and uniform flow environments.
5. Establishment of mixing criteria at neutral density levels for jets and plumes in stratified environments, possibly by evaluation of turbulent kinetic energy sources.
6. Determination of the influence of ambient turbulence levels on dilution within buoyant jets.
7. Establishment of criteria for turbulent buoyant jet bifurcation.

Research contributions in all of these areas would provide much clearer guidelines and better insight to engineers and geophysicists concerned with buoyancy-driven and jetlike motions in the ocean and atmosphere.

ACKNOWLEDGMENTS

This article was written at the California Institute of Technology in Pasadena, California. Without the support of Caltech faculty colleagues, students, and staff it could not have appeared. Over a period of years many sponsors have supported the research program in the W. M. Keck Laboratories and this article has drawn substantially on work performed under those research grants. In particular, the support of the U.S. National Science Foundation, the Southern California Edison Company, and the Ford Energy Program at Caltech are gratefully acknowledged. The author is particularly appreciative of the assistance received from Joan Mathews and Melinda Hendrix-Werts in the preparation of this article.

Literature Cited

Abdelwahed, M. S. T., Chu, V. H. 1978. Bifurcation of buoyant jets in cross flow. *Tech. Rep. 78-1.* Dept. Civ. Eng., McGill Univ. 130 pp.

Abraham, G. 1967. Jet with negative buoyancy in homogeneous fluid. *J. Hydraul. Res.* 5(4):235–48

Abraham, G., Eysink, W. D. 1969. Jets issuing into fluid with a density gradient. *J. Hydraul. Res.* 7(2):145–75

Akatnov, N. J. 1977. Effect of external turbulence on the development of a turbulent jet. *Izv. Akad. Nauk SSSR Mech. Zhid i Gaza* 12(1):24–29

Anderson, J. L., Parker, F. L., Benedict, B. A. 1973. Negatively buoyant jets in a cross flow. *Rep. No. EPA-660/2-73-012.* 199 pp.

Antonia, R. A., Prabhu, A., Stephenson, S. E. 1975. Conditionally sampled measurements in a heated turbulent jet. *J. Fluid Mech.* 72:455–80

Antonia, R. A., Chambers, A. J., Hussain, A. K. M. F. 1980. Errors in simultaneous measurements of temperature and velocity in the outer part of a heated jet. *Phys. Fluids* 23(5):871–74

Armstrong, R. R., Michalke, A., Fuchs, H. V. 1977. Coherent structures in jet turbulence and noise. *AIAA J.* 15(7):1011–17

Bashir, J., Uberoi, M. S. 1975. Experiments on turbulent structure and heat transfer in a two-dimensional jet. *Phys. Fluids* 18(4):405–10

Batchelor, G. B., Gill, A. E. 1962. Analysis of the stability of axisymmetric jets. *J. Fluid Mech.* 14:529–51

Beavers, G. S., Wilson, T. A. 1970. Vortex growth in jets. *J. Fluid Mech.* 44:97–112

Becker, H. A., Massaro, T. A. 1968. Vortex evolution in a round jet. *J. Fluid Mech.* 31:435–48

Becker, H. A., Hottel, H. C., Williams, G. C. 1967. The nozzle-fluid concentration field of the round, turbulent, free jet. *J. Fluid Mech.* 30:285–303

Beuther, P. D. 1980. *Experimental investigation of the turbulent axisymmetric plume.* PhD thesis. SUNY at Buffalo

Bill, R. G., Gebhart, B. 1975. The transition of plane plumes. *Int. J. Heat Mass Transfer* 18:513–26

Birch, A. D., Brown, D. R., Dodson, M. G., Thomas, J. R. 1978. The turbulent concentration field of a methane jet. *J. Fluid Mech.* 88:431–49

Boguslawski, L., Popiel, C. O. 1979. Flow structure of the free round turbulent jet in the initial region. *J. Fluid Mech.* 90(3):531–39

Bradbury, L. J. S. 1965. The structure of a self-preserving plane jet. *J. Fluid Mech.* 23:31–64

Bradshaw, P., Ferriss, D. H., Johnston, R. F. 1964. Turbulence in the noise-producing region of a circular jet. *J. Fluid Mech.* 19:591–624

Briggs, G. A. 1975. Plume rise predictions. In *Lectures on Air Pollution and Environmental Impact Analysis,* Chap 3. Am. Meteorol. Soc.

Brown, G. B. 1935. On vortex motion in gaseous jets and the origin of their sensitivity to sound. *Proc. Phys. Soc.* 47:703–32

Bruun, H. H. 1977. A time-domain analysis of the large-scale flow structure in a circular jet. Part 1: Moderate Reynolds number. *J. Fluid Mech.* 83:641–71

Bruun, H. 1979. A time-domain evaluation of the large-scale flow structure in a turbulent jet. *Proc. Roy. Soc. Ser. A* 367:193–218

Cederwall, K. 1971. Buoyant slot jets into stagnant or flowing environments. *Tech. Rep. KH-R-25* W. M. Keck Lab. Hydraul. & Water Res., Calif. Inst. Tech., Pasadena, Calif. 95pp.

Chan, Y. Y. 1977. Wavelike eddies in a turbulent jet. *AIAA J.* 15(7):992–1001

Chassaing, P., George, J., Claria, A., Sananes, F. 1974. Physical characteristics of subsonic jets in a cross stream. *J. Fluid Mech.* 62:41–64

Chen, C. J., Nikitopoulos, C. P. 1979. On the near field characteristics of axisymmetric turbulent buoyant jets in a uniform environment. *Int. J. Heat Mass Transfer* 22:245–55

Chen, C. J., Rodi, W. 1980. *Vertical turbulent buoyant jets: A review of experimental data.* London/New York: Pergamon. 83 pp

Chen, J.-C., Papanicolaou, P. N., List, E. J. 1980. Discussion on: Two-dimensional buoyant jets in stratified fluid. *Proc. ASCE, J. Hydraul. Div.* 106(HY10):1720–22

Chevray, R., Tutu, N. K. 1978. Intermittency and preferential transport of heat in a round jet. *J. Fluid Mech.* 88:133–60

Chu, V. H., Goldberg, M. B. 1974. Buoyant forced-plumes in cross flow. *Proc. ASCE, J. Hydraul. Div.* 100(HY9):1203–14

Corrsin, S., Uberoi, M. S. 1950. Further experiments on the flow and heat transfer in a heated turbulent air jet. *NACA Rep. 998.* 17 pp.

Crabb, D., Durão, D. F. G., Whitelaw, J. H. 1981. A round jet normal to a crossflow. *Trans. ASME J. Fluids Engrg.* 103:142–53

Crawford, T. V., Leonard, A. S. 1962. Observations of buoyant plumes in calm stably stratified air. *J. App. Meteorol.* 1:251–56

Crighton, D. G., Gaster, M. 1976. Stability of slowly diverging jet flow. *J. Fluid Mech.* 77:397–413

Crow, S. C., Champagne, F. H. 1971. Orderly structure in jet turbulence. *J. Fluid Mech.* 48:547–91

Davies, P. O. A. L., Fisher, M. J., Barratt, M. J. 1962. The characteristics of the turbulence in the mixing region of a round jet. *J. Fluid Mech.* 15:337–67

Davis, M. R., Davies, P. O. A. L. 1979. Shear fluctuations in a turbulent jet shear layer. *J. Fluid Mech.* 93:281–303

de Gortari, J. C., Goldschmidt, V. W. 1981. The apparent flapping motion of plane jet—further experimental results. *Trans. ASME J. Fluids Engrg.* 103:119–26

Delichatsios, M. A. 1979. Time similarity analysis of unsteady buoyant plumes. *J. Fluid Mech.* 93:241–50

Ellison, T. H., Turner, J. S. 1959. Turbulent entrainment in stratified flows. *J. Fluid Mech.* 6:423–48

Everitt, K. W., Robins, A. G. 1978. The development and structure of turbulent plane jets. *J. Fluid Mech.* 88:563–83

Fan, L.-N. 1967. Turbulent buoyant jets into stratified or flowing ambient fluids. *Tech. Rep. No. KH-R-18* W. M. Keck Lab. Hydraul. & Water Res., Calif. Inst. Tech., Pasadena, Calif. 104 pp.

Fischer, H. B., List, E. J., Koh, R. C. Y., Imberger, J., Brooks, N. H. 1979. *Mixing in Inland and Coastal Waters.* New York: Academic

Fox, D. G. 1970. Forced plume in a stratified fluid. *J. Geophys. Res.* 75:6818–35

Freeman, R. W., Tavlarides, L. L. 1979. Observations of the instabilities of a round jet and the effect of concurrent flow. *Phys. Fluids* 22(4):782–83

Freymuth, P. 1966. On transition in a separated laminar boundary layer. *J. Fluid Mech.* 25:683–704

Gartrell, G. 1979. *Studies on the mixing in a density-stratified shear flow.* PhD thesis. Calif. Inst. Technol., Pasadena, Calif. 446 pp.

George, W. K., Alpert, R. L., Tamanini, F. 1977. Turbulence measurements in an axisymmetric buoyant plume. *Int. J. Heat Mass Transfer* 20:1145–54

Goldschmidt, V. W., Bradshaw, P. 1973. Flapping of a plane jet. *Phys. Fluids* 16(1):354–55

Grant, A. J. 1974. A numerical model of instability in axisymmetric jets. *J. Fluid Mech.* 66:707–24

Gutmark, E., Wygnanski, I. 1976. The planar turbulent jet. *J. Fluid Mech.* 73:465–95

Hayashi, T. 1972. Bifurcation of bent-over plumes in the ocean. *Coastal Engrg. in Japan* 15:153–65

Heskestad, G. 1965. Hot wire measurements in a plane turbulent jet. *J. Appl. Mech.* 32(4):721–34

Hofer, K., Hutter, K. 1981. Turbulent jet diffusion in stratified quiescent ambients. Part 1: Theory. *J. Non-Equilibr. Thermodyn.* 19:31–48

Hoult, D. P., Fay, J. A., Forney, L. J. 1969. A theory of plume rise compared with field observations. *J. Air Pollut. Control Assoc.* 19:585–90

Koh, R. C. Y., Brooks, N. H. 1975. Fluid mechanics of waste water disposal in the ocean. *Ann. Rev. Fluid Mech.* 7:187–211

Kotsovinos, N. E. 1975. *A study of the entrainment and turbulence in a plane buoyant jet.* PhD thesis. Calif. Inst. Technol., Pasadena, Calif. 306 pp.

Kotsovinos, N. E. 1977. Plane turbulent buoyant jets. Part 2: Turbulence structure. *J. Fluid Mech.* 81:45–62

Kotsovinos, N. E., List, E. J. 1977. Plane turbulent buoyant jets. Part I: Integral properties. *J. Fluid Mech.* 81:25–44

Krausche, D., Fearn, R. L., Weston, R. P. 1978. Round jet in a crossflow: Influence of injection angle on vortex properties. *AIAA J.* 16:636–37

Lau, J. C., Morris, P. J., Fisher, M. J. 1979. Measurements in subsonic and supersonic free jets using a laser velocimeter. *J. Fluid Mech.* 93:1–27

Lau, J. C., Whiffen, M. C., Fisher, M. J., Smith, D. M. 1981. A note on turbulence measurements with a laser velocimeter. *J. Fluid Mech.* 102:353–66

List, E. J. 1971. Laminar momentum jets in a stratified fluid. *J. Fluid Mech.* 45:561–74

List, E. J., Imberger, J. 1973. Turbulent entrainment in buoyant jets. *Proc. ASCE, J. Hydraul. Div.* 99:1461–74

List, E. J., Imberger, J. 1975. Closure of discussion to: Turbulent entrainment in buoyant jets and plumes. *Proc. ASCE, J. Hydraul. Div.* 101:617–20

Liu, J. T. C. 1974. Developing large-scale wavelike eddies and the near jet noise field. *J. Fluid Mech.* 62:437–64

Luti, F. M., Brzustowski, T. A. 1977. Flow due to a two-dimensional heat source with cross flow in the atmosphere. *Comb. Sci. Tech.* 16:71–87

Maestrello, L., Fung, Y.-T. 1979. Quasiperiodic structure of a turbulent jet. *J. Sound Vib.* 64(1):107–22

Manins, P. C. 1979. Partial penetration of an elevated inversion layer by chimney plumes. *Atmos. Environ.* 13:733–41

McMahon, H. M., Hester, D. D., Palfery, J. G. 1971. Vortex shedding from a turbulent jet in a cross wind. *J. Fluid Mech.* 48:73–80

Mollendorf, J. C., Gebhart, B. 1973. An experimental and numerical study of the viscous stability of a round laminar vertical jet with and without thermal buoyance for symmetric and asymmetric disturbances. *J. Fluid Mech.* 61:367–99

Mollø-Christensen, E. 1967. Jet noise and shear flow instability seen from an experimenter's viewpoint. *J. Appl. Mech. Trans. ASME* 34:1–7

Moore, C. J. 1977. The role of shear-layer instability waves in jet exhaust noise. *J. Fluid Mech.* 80:321–67

Moore, D. J. 1966. Physical aspects of plume models. *Air & Water Pollut.* 10:411–17

Morton, B. R. 1959. Forced plumes. *J. Fluid Mech.* 5:151–63

Morton, B. R., Taylor, G. I., Turner, J. S. 1956. Turbulent gravitational convection from maintained and instantaneous sources. *Proc. R. Soc. Ser. A.* 234:1–23

Moum, J. N., Kawall, J. G., Keffer, J. F. 1979. Structural features of the plane turbulent jet. *Phys. Fluids* 22(7):1240–44

Moussa, Z. M., Trischka, J. W., Eskinazi, S. 1977. The near field in the mixing of a round jet with a cross stream. *J. Fluid Mech.* 80:49–80

Nakagome, H., Hirata, M. 1976. The structure of turbulent diffusion in an axisymmetric thermal plume. *Proc. 1976 ICHMT Seminar on Turbulent Buoyant Convection,* pp. 361–72 Hemisphere Publ.

Pera, L., Gebhart, B. 1971. On the stability of laminar plumes: Some numerical solutions and experiments. *Int. J. Heat Mass Transfer* 14:975

Plaschko, P. 1979. Helical instabilities of slowly divergent jets. *J. Fluid Mech.* 92:209–15

Priestley, C. H. B. 1956. A working theory of the bent-over plume of hot gas. *Q. J. R. Meteorol. Soc.* 82:165–76

Priestley, C. H. B., Ball, F. K. 1955. Continuous convection from an isolated source of heat. *Q. J. R. Meteorol. Soc.* 81(348):144–57

Rockwell, D. O., Niccolls, W. O. 1972. Natural breakdown of planar jets. *Trans. ASME, J. Basic Engrg.* 94:720–30

Rouse, H., Yih, C.-S., Humphreys, H. W. 1952. Gravitational convection from a boundary source. *Tellus* 4:201–10

Sami, D., Carmody, T., Rouse, H. 1967. Jet diffusion in the region of flow establishment. *J. Fluid Mech.* 27:231–52

Schatzmann, M. 1978. The integral equations for round buoyant jets in stratified flows. *J. Appl. Math & Physics (ZAMP)* 29:608–20

Schatzmann, M. 1979. An integral model of plume rise. *Atmos. Environ.* 13:721–31

Schneider, P. E. M. 1980. Sekundärwirbelbildung bei Ringwirbeln und in Freistrahlen. *Z. Flugwiss. Weltraumforsch* 4(5):307–18

Scorer, R. S. 1959. The behaviour of chimney plumes. *Int. J. Air Pollut.* 1:198–220

Shaughnessy, E. J., Morton, J. B. 1977. Laser light-scattering measurements of particle concentration in a turbulent jet. *J. Fluid Mech.* 80:129–48

Slawson, P. R., Csanady, G. T. 1967. On the mean path of buoyant bentover chimney plumes. *J. Fluid Mech.* 28:311–22

Sneck, H. J., Brown, D. H. 1974. Plume rise from large thermal sources such as dry cooling towers. *Trans. ASME, J. Heat Transfer* 96:232–38

Sreenivasan, K. R., Antonia, R. A., Britz, D. 1979. Local isotropy and large structures in a heated turbulent jet. *J. Fluid Mech.* 94:745–75

Tennankore, K. N., Steward, F. R. 1978. Comparison of several turbulence models for predicting flow patterns within confined jets. *Can. J. Chem. Engrg.* 56:673–78

Teske, M. E., Lewellen, W. S., Segur, H. S. 1978. Turbulence modeling applied to buoyant plumes. *EPA Report EPA-600/4-78-050.* 47 pp.

Turner, J. S. 1960. A comparison between vortex rings and vortex pairs. *J. Fluid Mech.* 1:419–32

Turner, J. S. 1962. The 'starting plume' in neutral surroundings. *J. Fluid Mech.* 13:356–68

Turner, J. S. 1966. Jets and plumes with negative or reversing buoyancy. *J. Fluid Mech.* 26:779–92

Turner, J. S. 1969. Buoyant plumes and thermals. *Ann. Rev. Fluid Mech.* 1:29–44

Turner, J. S. 1973. *Buoyancy effects of fluids.* Cambridge Univ. Press. 367 pp.

Vergara, I., James, W. P. 1979. Dilution of a dense, vertical jet in stagnant homogeneous fluid. *Tech. Rep. TAMU-SG-79-202, COE Rep. No. 211.* Texas A&M Univ.

Wehrmann, O., Wille, R. 1957. Beitrag zur Phänomenologie des laminar-turbulenten Übergange im Freistrahl bei kleinen Reynoldszahlen. In *Boundary Layer Research, IUTAM Sym., Frieburg,* ed. H. Görtler, pp. 387–403

Wright, S. J. 1977. Mean behavior of buoyant jets in a crossflow. *Proc. ASCE, J. Hydraul. Div.* 103(HY5):499–513

Wright, S. J., Wallace, R. B. 1979. Two-dimensional buoyant jets in a stratified fluid. *Proc. ASCE, J. Hydraul. Div.* 105(HY11):1393–1406

Wygnanski, I., Fiedler, H. 1969. Some measurements in the self-preserving jet. *J. Fluid Mech.* 38:577–612

Wygnanski, I., Gutmark, E. 1971. Lateral motion of the two-dimensional jet boundaries. *Phys. Fluids* 14(7):1309–11

Yule, A. J. 1978. Large scale structure in the mixing layer of a round jet. *J. Fluid Mech.* 89:413–32

Zeitoun, M. A., McIlhenny, W. F., Reed, R. O. 1970. Conceptual designs of outfall systems for desalination plants. *R&D Rept. 550.* Office of Saline Water, US Dept. Interior. 139 pp.

GRAVITY CURRENTS IN THE LABORATORY, ATMOSPHERE, AND OCEAN

John E. Simpson

Department of Applied Mathematics and Theoretical Physics, University of Cambridge, Cambridge, England

INTRODUCTION

A gravity current, or "density current," is the flow of one fluid within another caused by the density difference between the fluids. The difference in specific weight that provides the driving force may be due to dissolved or suspended material or to temperature differences. Gravity currents are primarily horizontal, occurring as either top or bottom boundary currents, or as intrusions at some intermediate level. The fluids are usually miscible and the mixing that results can play an important part in the dynamics of the flow. Since gravity currents are formed in many different natural situations and may also be man-made, knowledge of their properties is of importance in many scientific disciplines.

In the atmosphere, thunderstorm outflows and sea-breeze fronts are gravity currents of relatively cold dense air. Atmospheric-suspension gravity currents include avalanches of airborne snow particles, also fiery avalanches and base surges formed from gases and solids issuing from volcanic eruptions.

Gravity currents have important applications in aircraft safety, atmospheric pollution, entomology and pest control, and especially in dense-gas technology. Industrial accidents in which gravity currents may be formed include the spread of a dense gas from an accidental release.

In the ocean, gravity currents are driven by salinity and temperature inhomogeneities, or as turbidity currents whose density derives from suspended mud or silt. Lines of foam and debris on the ocean surface may indicate the front of a gravity current, frequently brought about by tidal processes. The latter also affect the behavior of gravity currents such as river plumes at the surface and salt wedges on a river bed. The important problems related to oil spillage on the sea have been the subject of a review paper in this series (Hoult 1972).

214   SIMPSON

In all the gravity currents we consider here the effect of the earth's rotation is neglected.

THE HEAD OF THE CURRENT

At the leading edge of a gravity current there is a characteristic "head," deeper than the following flow. This is a zone of breaking waves and intense mixing and plays an important part in the behavior of the current. In a gravity current flowing horizontally this head remains quasi-steady, and is about twice as deep as the following flow, but in a current flowing down a slope the size of the head continually increases. A current flowing along a horizontal surface usually has a "nose," or foremost point, raised a short distance above the ground.

A "universal profile" of a gravity current head does not exist, as even for flows into calm surroundings the value of the excess head height above the following flow varies with the fraction of the total depth occupied by the current. The form of the head is strongly modified by opposing and following ambient flows, and other physical effects.

Some early experimental work on gravity currents was carried out by Schmidt (1911), who used a laboratory water channel to model the front of an advancing cold squall in the atmosphere. The silhouettes of some of his gravity-current heads are still worth viewing and are shown in Figure 1. In Schmidt's shadow-pictures the views (a) and (b) are of flows with low Reynolds number, Uh/ν less than 100. (U is the velocity of the front, h the depth of the current, and ν the kinematic viscosity.) As the temperature difference, and hence the Reynolds number of the flow, increases, the shape of the head of the gravity current alters, the foremost point of the nose approaches the ground, and more intense mixing occurs at the front and top of the head. The final view (f) shows a profile typical of

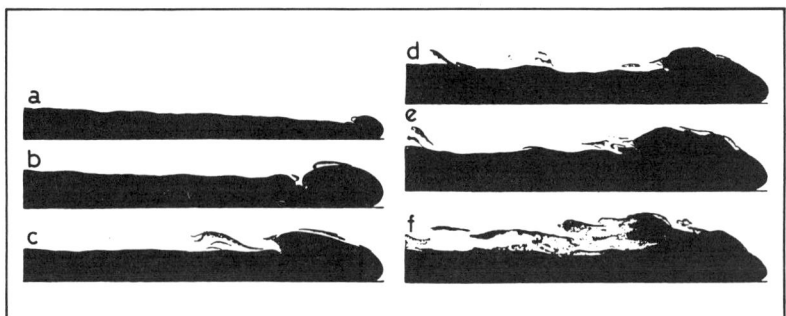

Figure 1   Profiles of the head of a gravity current in a laboratory tank (from Schmidt 1911). The temperature difference increases from a very small value in (a) to 35° in (f).

flows with Re greater than 1000 in which billows are seen streaming back from the upper surface of the gravity-current head.

Billows

In most cases the billows formed on the upper surface are broken up in a complex three-dimensional flow but slit-lighting reveals Kelvin-Helmholz instabilities (Simpson 1969). Winant & Bratkovich (1977) measured the fluxes in and out of the frontal zone of a gravity current from a carriage moving with the front. They found the mass flux of heavy liquid into the front to be of order 0.15 Q, where Q is the mass flux per unit width of the current. Simpson & Britter (1979a), holding a front stationary in a water channel, also found the ratio of the mean overtaking speed U_o to front speed U_f to be equal to 0.15.

Lobes and Clefts

At the head of a gravity current moving along a horizontal surface there is a complicated shifting pattern of lobes and clefts, believed to be caused by gravitational instability of the less dense fluid which is overrun by the nose of the current (Simpson 1972a). A rough estimate of the expected flux of light fluid into the head by entrainment of about 0.2 Q was made by Allen (1971), and measurements using dye lines ahead of the front give 0.13 Q as a probable upper limit (B. R. Morton and J. E. Simpson, unpublished).

THE DYNAMICS OF GRAVITY-CURRENT FRONTS

The interface between the two fluids at the head of a gravity current is a typical frontal zone, that is, a region in which, although there is intense motion and mixing, a high density gradient is maintained.

The motion of a gravity current has been described in terms only of the depth and density difference at the head itself by Keulegan (1957). He found from extensive measurements of the advance of saline currents along a horizontal floor into calm fresh water that the velocity $U = 1.05 (g'h)^{1/2}$, where $g' = g\Delta\rho/\rho$ and h is the depth of the following steady flow. These results were mainly obtained from flows occupying about 1/5 of the total depth H, but more recent work on gravity currents from a lock at the end of a water channel has shown that U is sensitive to changes in the value of h/H in the range 1/3 to 1/10 (Simpson & Britter 1979a).

Inviscid-Fluid Theory

Benjamin (1968) applied inviscid-fluid theory to study aspects of a steady gravity current; in particular, he showed the essential role of wave-breaking and the associated energy losses. He analyzed the front of a frictionless two-dimensional gravity current in terms of a "cavity flow" displacing a fluid beneath it (see Figure 2a.) In a closed channel, with axes moving with the front of the current, continuity and Bernoulli's equation applied along the interface give two equations involving the velocity and depth h_2 of the flowing layer. Being frictionless the "flow force" (total pressure force plus the momentum flux per unit span) is also conserved, resulting in a solution $h_2 = 1/2H$. Thus the only steady energy-conserving flow (implied by the use of Bernoulli's equation) is one in which the advancing layer fills half the channel. Flows in which $h > 1/2H$ are not possible, and if $h < 1/2H$, as is found in most practical situations, the loss of energy at the front exceeds that available by wave radiation so that "breaking" must occur.

Benjamin also showed the importance of the fractional depth $h/H = \phi$, say. The Froude number, $U/(g'h)^{1/2}$, based on the velocity U of the front, is $2^{-1/2}$ at $\phi = \frac{1}{2}$, equals about 1 at $\phi = 1/5$, increasing to $2^{1/2}$ as $\phi \to 0$. In Benjamin's treatment of the velocity of the head under "deep" water (i.e. $\phi \to 0$) if the pressure downstream in the wake is taken to be hydrostatic so the dynamic pressure $1/2\rho U^2$ at the stagnation point 0 equals the difference between the hydrostatic pressure at the boundary far upstream and downstream, it follows that

$$\tfrac{1}{2}\rho U^2 = g(\rho_1 - \rho_2)h$$

or $U = 2^{1/2}(g'h)^{1/2}$.

The same result had been obtained by von Kármán (1940) by applying the Bernoulli condition along the interface, but this is invalid in deep water since the interface must be dissipative. Prandtl (1952) deduced a front velocity half that of the layer behind, but his argument applies only to the initial transient phase of the motion.

Benjamin also calculated the approximate shape of the interface. The slope at the stagnation point on the ground is 60°, a result previously deduced by von Kármán.

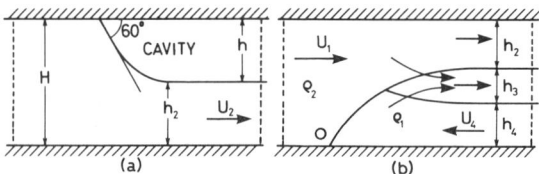

Figure 2 Theoretical treatment of a gravity-current front, with flow relative to the front. (*a*) Benjamin (1968); (*b*) Britter & Simpson (1978).

An Inviscid Current with Mixing

As a first step towards realistic gravity currents, in a semiempirical analysis by Britter & Simpson (1978), the flow relative to the gravity-current head is divided into three regions, as in Figure 2b. The bottom region is the flow into the gravity-current head of dense unmixed fluid, depth h_4. The top region contains only the less dense fluid, depth h_2. Between the two is a mixing region, whose depth h_3 is that of the wake of the collapsed billows. This region has nonuniform velocity and concentration profiles determined by experiment. Values of the Froude number $U/(g'h_4)^{1/2}$ were calculated using continuity and momentum equations and Bernoulli's equation applied along the floor to the foremost stagnation point O. This Froude number varied both with the fractional depth h_4/h_1 and also with $q = g'Q/U^3$, the nondimensional mixing rate. Values of q needed to close the equations could either be measured experimentally or else deduced from billow properties.

The mixing rate and billow properties were studied using a water channel with a moving floor. A front with no friction at the floor was modeled by holding it stationary on a fixed floor just downstream of the end of the conveyor-belt section. No fluid was overrun at the nose, hence the lobe and cleft structure was absent and the mixing billows above the head were nearly two-dimensional. The difference between the flow observed and the familiar arrested saline wedge (Riddell 1970) appears to be in the imposed velocity profile which has a critical effect on the mixing (see Figure 3). The nondimensional breakdown size of the billows, $R_L = g'h_3/(\Delta U)^2$, where ΔU is the velocity difference across the front, was found to be 0.33, close to that obtained by Thorpe (1973) for billows forming at an interface with very small initial Richardson number. The Froude number was about 1 for fractional depth h_4/h_1 about one fifth of the total depth. For smaller fractional depths, down to about 0.05, the value of the Froude number rose to approximately 2.

These results appear to model closely the behavior of a gravity current of fresh water moving along the free surface above saline surroundings,

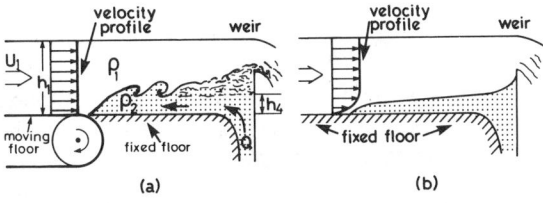

Figure 3 (a) An inviscid gravity current on a fixed floor downstream of a moving flow and floor. (b) An "arrested saline wedge" on a fixed floor, brought to rest by an opposing flow.

a situation of interest at much larger scales. Modeling such flows in the laboratory by small-scale currents along a free surface is difficult because it is hard to protect the free surface from contaminants that can modify the flow.

Front Advancing over a Horizontal Surface

In the next step, including friction on the boundary, the main difference from the inviscid model can be related to the height above the ground of the mean position of the foremost point of the head. This is considered as a stagnation point, and for large values of Re, both laboratory and atmospheric observations suggest a constant value for the nose height of about 1/8 of the total height of the gravity-current head (Simpson & Britter 1979a). An alternative approach, used by Kranenberg (1978), is to include an empirical head-loss coefficient for the stagnation streamline.

Head- and tail-flow effects on the front of a gravity current moving along a horizontal surface have been examined in a water channel with a moving floor by Simpson & Britter (1980). Head flows, or "head winds," are simulated by bringing the front to rest on the moving floor with an opposing flow at a greater speed than that of the floor. Tail winds are obtained by moving the floor faster than the opposing flow. With a head wind the head profile is longer and the nose height lower than in the familiar calm case. With a tail wind the head is shorter and the nose height is larger. A head or tail wind is found to change the speed along the ground by 3/5 of the applied wind.

Values of the overtaking speed relative to the front, U_4, were compared with U_1, the flow relative to the front and the values of U_4/U_1 were all found to be close to 0.15, independent of the values of the head or tail flows.

The value of $R_L = g'h_3/(\Delta U)^2$ was deduced to be 0.5, in all head and tail wind cases. This is greater than the value measured in the two-dimensional (inviscid) case, but close to that already found in the calm case along a horizontal floor, so it seems that the shape of the head and mixed region adjusts to maintain this approximately constant layer Richardson number.

Gravity Currents on Slopes

Georgeson (1942) made direct measurements of the advance of a quantity of methane along the roof of a passage in a mine. From these and later experiments, e.g. Wood (1965), Middleton (1966), Hopfinger & Tochon-Danguy (1977), it is clear that the motion of a gravity current down an incline is appreciably different from that on a horizontal surface. On slopes from a few degrees to 90°, for a given buoyancy flux, there is a nearly constant front velocity that is only about 60% of the mean velocity of the following flow. The head volume increases, both by direct entrainment and by addition from the following flow. There is little variation in front speed with slope, because although the gravitational force increases, so does the entrainment, both into the head and into the flow behind it. Direct entrainment increases with slope, leading to 1/10 of the head-growth at 10° and about 2/3 at 90°. With a steady buoyancy flux Q, Britter & Linden (1980) found a constant head velocity for slopes greater than 1/2°, and from 5° to 90° the front velocity was given by $U/(g'Q)^{1/3} = 1.5 \pm 0.2$.

The release of a finite quantity of dense fluid down a slope has practical applications and the resulting "inclined thermals" have been investigated by Beghin, Hopfinger & Britter (1981).

THE SPREAD OF NEGATIVELY BUOYANT FLUID

Release of a Fixed Quantity of Fluid

When a fixed volume of fluid is released into nonturbulent surroundings it spreads with a very little mixing except near the leading edge which advances at a velocity dependent on local parameters. However, a relationship between the current depth and the frontal speed is only part of the solution of the important practical problem of finding the rate of advance of a gravity current.

Fay (1969) determined the rate of advance in terms of the balance between the horizontal buoyancy forces and the inertia forces in the current. For the axisymmetric case, he found

$$R \sim (g'q)^{1/4} t^{1/2}$$

where R is the radial coordinate of the front of the current and q is the volume. With a fixed volume the current eventually becomes thin enough for viscous forces to take over from inertia forces to give the balance

$$R \sim (g'q^2 \nu^{-1/2})^{1/6} t^{1/4},$$

where ν is the kinematic viscosity. These two equations give a transition time $t_* = (q/\nu g')^{1/3}$. Fay identified a third stage of the spreading for oil-water flows when there is a balance between viscous forces and surface tension.

On the grounds that the length of the current greatly exceeds its vertical thickness, Hoult (1972) based his analysis on the depth-averaged shallow-water equations. Retaining only the buoyancy and inertia terms, he solved the equations and evaluated the coefficients from experiments on the spreading of oil over water. Hoult obtained

$$R = 1.3 \, (g'q)^{1/4} t^{1/2} \text{ (axisymmetric)},$$

$$L = 1.6 \, (g'q)^{1/3} t^{2/3} \text{ (two-dimensional)}.$$

The spreading laws when viscosity dominates inertia were obtained by Hoult as

$R = 0.94\ (g'q^4/\nu)^{1/12}t^{1/4}$ (axisymmetric),

$L = 1.5\ (g'^2q^4/\nu)^{1/8}t^{3/8}$ (two-dimensional).

Lock-Exchange Experiments

Since the pioneering experiments of O'Brien & Cherno (1934) many gravity currents have been released from behind lock-gates. However, most of these flows start with a constant-speed regime, showing very little agreement with the above power laws. One difference from Hoult's experiments is that in all these experiments the dense fluid released was originally the same depth as the fluid in the rest of the channel. When the partition was removed the initial flow pattern was identical to that of "Mutual Intrusion" experiments in which the partition is at the center of the tank. In such flows Yih (1980) showed from energy considerations that $U = \frac{1}{2}(g'H)^{1/2}$. Using much larger tanks, Barr (1967) found the constant speeds of the fronts were $0.47\ (g'H)^{1/2}$ along the ground and $0.58\ (g'H)^{1/2}$ at the free surface, until a viscous regime was reached.

Huppert & Simpson (1980) showed that if steady-flow results relating the head Froude number and the fractional depth of the gravity current were assumed to apply in a simple box model, then an almost constant velocity regime was to be expected in the early stages of a finite volume release. They called this the "slumping" regime and showed that most of Keulegan's 1957 experiments moved directly from this regime to a viscous one. The constant speed came to an end when the fractional depth had fallen to about one tenth, in their two-dimensional and axisymmetric experiments.

REFLECTION EFFECTS IN LOCK-EXCHANGE GRAVITY CURRENTS As the result of further experimental work an explanation can be given of the end of the constant-speed regime. When dense fluid is released into a channel from a lock of finite length the reverse upper flow is reflected from the rear wall of the lock as a depression behind the underflow, which then becomes a slug of fluid with only a thin layer of dense fluid following it (Keulegan 1957, Barr 1967). The head of this slug moves at nearly constant speed but its length decreases as it is eroded from the leading edge (see Figure 4). After the front has traveled about ten lock-lengths the reflection has almost reached the head, the slug form disappears, and the head is followed by a layer about one tenth of the total depth of the fluid. It now advances with distance L proportional to $t^{2/3}$, as described by inertia-buoyancy theory. This regime, which may be short or even nonexistent, is followed by a viscous-buoyancy regime in which L varies with $t^{3/8}$.

Replotting Barr's data, suitably nondimensionalised, for flows influenced by reflection confirms ten lock-lengths as the extent of the constant-velocity regime in currents along the ground. In his experiments on free-surface gravity currents the reflections appear to influence the head after traveling about twenty lock-lengths.

Axisymmetric gravity currents from finite locks also have a short constant-velocity regime, which ends when the reflection approaches the front after about four lock-lengths. In the collapse from a very short lock the reflection causes a concentration of mass in a narrow head annulus when this stage is reached.

Constant Buoyancy Flux

Consider a gravity current formed by a steady buoyancy flux either at a free surface or along the ground, first in a parallel-sided channel. An

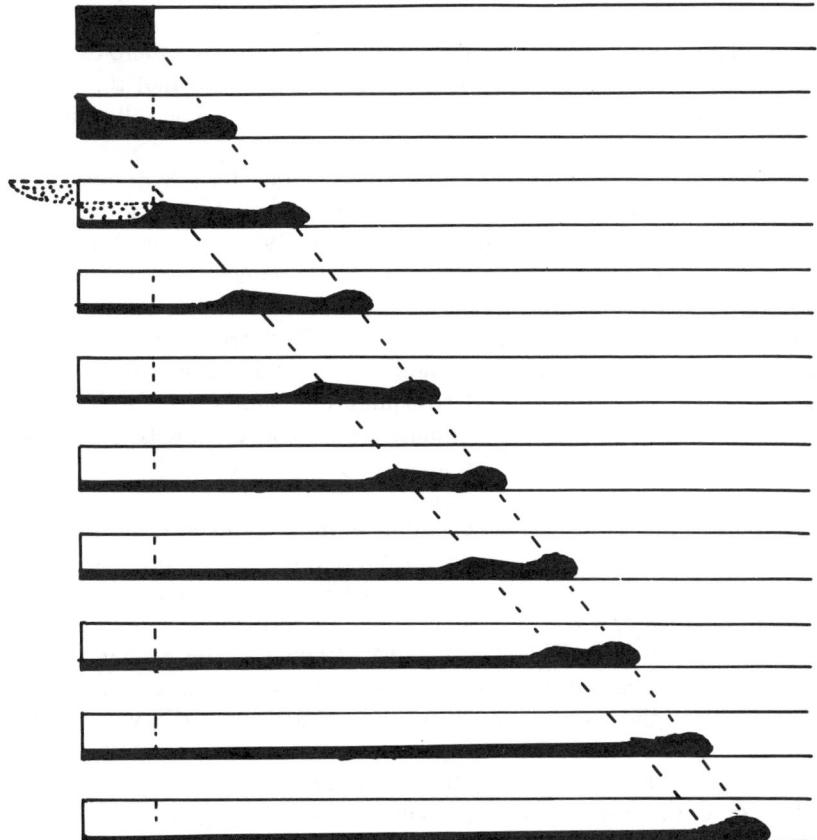

Figure 4 Development of the reflected upper current in a lock-exchange flow.

example is the warm outflow from a power station, where both the advance of the initial leading edge and the mixing during later stages are of interest.

These "starting plumes" are different from the lock-exchange flows we have considered, in which a barrier is removed in a tank containing fluids of different density on either side. In lock-exchange flows there is no total flow across any cross section of the tank, so a gravity current of large fractional depth will have a faster return current above it than will a starting plume of the same depth.

Wilkinson & Wood (1972) showed that for inviscid, non-mixing starting flows the Froude number $= [(2 - \phi)/(1 - \phi^2)]^{1/2}$ (ϕ is the fractional depth h/H of the current). This expression varies much less with ϕ than the corresponding one for lock-exchange flows $[(2 - \phi)(1 - \phi)/(1 + \phi)]^{1/2}$.

It is possible to set up a flow rate in which the layer stays uniform for some distance along the flume (Barr 1959), but for increasing distances the velocity decreases and the layer thickens. Lean & Whillock (1965) concluded that for an inlet Froude number greater than 1 there exists an entraining hydraulic jump at the interface, and it seems that the rate of entrainment here and the conditions downstream are controlled by the gravity-current head.

The analysis of Wood (1967) was similar to that for a hydraulic jump in which the amount of mixing depends on the turbulence and hence on the inflow Froude number. Wilkinson & Wood (1971) examined the characteristics of the hydraulic jump associated with some form of downstream control in a two-layer system and showed that this "density jump" is able to entrain the varying amounts of ambient fluid necessary to satisfy a range of downstream controls. However, along a horizontal floor with a gravity-current head as control no equilibrium state exists (Koh 1971), and the velocity of the flow continues to fall and eventually causes the density jump to flood.

AXISYMMETRIC FLOWS Analogous results to the two-dimensional flow have been obtained for axisymmetric flows, of interest in describing a spreading plume of sewage, or of cooling water from a power station.

The analysis of Chen & List (1976) described the spread of such a plume in separate regimes, first dominated by an inertia-buoyancy force balance and then by a friction-buoyancy balance. The rate of spreading in the first regime was shown to be $R(t) = k_1 (Qg')^{1/4} t^{3/4}$, where $k_1 = 0.75$ and Q is the mass flow per unit width. The changeover from inertia-buoyancy balance will occur on the viscous time scale $t_* = (Q/\nu g')^{1/2}$ and the form of the viscous-buoyancy solution is $R = k_2 (Q^3 g'/\nu)^{1/3} t^{1/2}$. Recent experiments by Britter (1979) gave a value for $k_1 = 0.84$ and for $k_2 = 0.67$.

The spread of a gravity current from a continuous source into a stratified cross-flow has been investigated by Jirka (1980). He determined the shape and velocity distribution enclosing the source, including the length of the upstream intrusion. His laboratory observations agree with the analysis for small enough values of the near-field momentum, in which the radially expanding jet is not entraining.

Flow at Very Low Reynolds Number

At the onset of the "viscous-buoyancy" regime discussed above, the Reynolds number based on the total head height, $Re = Uh/v$, is about 500. The raised head disappears when $Re \simeq 10$. At lower Re, in flows along a rigid boundary, cells of overrun fluid may not rise through the dense fluid, but can remain in a "rice-grain"–like pattern on the floor. In such flows through an unstable layer, interaction between the gravity-current front and thermals appears to be possible (Simpson 1972b).

Gravity currents at low Re occur naturally in some geological events. For example, H. E. Huppert (unpublished work) has had some success in the use of laboratory gravity-current models to explain observations of the flow of lava in the crater at the top of Soufriere, St. Vincent.

The Spread and Dilution of Heavy Gas Cloud

Notable safety hazards can arise from the accidental release of a large quantity of negatively buoyant gas that is combustible or toxic. After an almost instantaneous release of vapor a drifting cloud is formed, with diameter greater than its thickness, and it is important to know both the rate of advance of the cloud and its dilution.

A recent paper by Fay (1980) compares various models proposed for treating the dispersion of dense gas clouds. These are mostly based on the box model which considers the cloud to occupy initially a cylindrical box of radius R and height H. R increases with time, with radial speed proportional to excess hydrostatic height and $dR/dt = a \, (g'H)^{1/2}$. Some field tests and water-tank and wind-tunnel measurements are compared with model predictions.

Nearly all recent models of dense-cloud behavior have included major gravity-current effects and many different criteria have been proposed for separating the early stage when gravity currents dominate the mixing process from the later stages when atmospheric turbulence dominates (McQuaid 1979).

AMBIENT TURBULENCE Most of the mixing in a gravity current occurs in the region of the head and a stable layer is laid down above the following current. When the ambient level of turbulence is small it can have only a small effect on the mixing of the dense fluid; however, if we consider the spread of a finite amount of dense fluid into a turbulent environment, although at the start gravity current forces dominate, as the cloud thins and the velocity of the front is much reduced, the effect of external turbulence on the mixing can become appreciable. In the simple model of a gas cloud by van Ulden (1974) this transition was taken to occur when U_f, the speed of the front, became less than u_*, the friction velocity. Van Ulden argued that at this point the cloud, whose height had been decreasing, began to increase again.

TURBULENCE EXPERIMENTS   N. H. Thomas and J. E. Simpson (unpublished) have used an oscillating grid to maintain and enhance the shear turbulence in a streaming flow over a stationary gravity current. In this way they were able to impose turbulent intensities u'/U_f at any level from zero (the "standard" gravity current) up to about ten or more (as may be found in tidal flows).

At these large intensities the frontal zone appears as a wedge, with a fluctuating sharp, well-defined interface separating the undiluted current flow and the turbulent surroundings. The current, maintained by a continuous flow in at the rear, is eroded but not diffused. This picture is different from van Ulden's, in which the transition at $u'/U_f \geqslant 1$ is to upwards diffusion mixing of the current with its surroundings.

A conclusion from this work is that in the limit $u'/U_f \gg 1$, the mixing rate is independent of U_f, at least when the fractional depth is small, and that the entrainment speed (mixing flux per unit area) is scaled in accordance with Turner's (1968) result for the static two-layer system.

MATHEMATICAL MODELS   A number of mathematical models have been developed using solutions of the equations of fluid motion to elucidate the behavior of a gravity-current head or the collapse of a cloud of dense gas. The models of Daly & Pracht (1968) and Vreugdenhil (1976) both employ the marker and cell method, but the latter included the hydrostatic approximation. Others, eddy viscosity models, have been offered by England & Teuscher (1978), Thorpe et al. (1980), Vasiliev et al. (1973), and Kao, Park & Pao (1977, 1978) but a detailed discussion of these models is beyond the scope of this review.

IN THE ATMOSPHERE

Thunderstorm Outflows

The downdraft section of a thunderstorm plays an important part in the structure and regeneration of severe storms (Thorpe et al. 1980). Good agreement between observations of the frontal properties of the cold dense outflow resulting from the downdraft suggests that laboratory gravity currents are reasonable analogue systems (Idso et al. 1972, Simpson & Britter 1980). For example, referring to the head- and tail-wind results described above, the laboratory results give the relationship $U_0/U_\Delta = 0.91 + 0.62 \, U_2/U_\Delta$, where U_Δ is the densimetric velocity $(g'h_4)^{1/2}$, and U_2 is the external wind measured in the direction of advance of the front. Eighteen atmospheric observations from Miller & Betts (1977) and from Goff (1976), normalized in this way, gave a similar result.

Understanding the details of air flow near the front of this strong gravity current is important for safety of flight in large airplanes. Several accidents have been associated with flights through these small-scale fronts (Fujita & Byers 1977). Danger may occur through overstressing the aircraft by flying through vertical wind-shear, or through sudden large horizontal wind changes during takeoff or landing. For example, a case has been reported in which a wind discontinuity of 30 m s^{-1} lay across the runway (Fujita & Caracena 1977). The use of an acoustic sounder (Hall et al. 1976) has shown vertical wind-shears after the passage of a gravity current which were large enough to be hazardous to aircraft. Laboratory models give an estimate of similar value in the mixing zone after the breakdown of billows a few head-heights back from the leading edge.

A system has been installed in the neighborhood of Dulles Airport, Washington, D.C., to obtain warning of the arrival of such fronts, which often are invisible. A network of sensors is used to follow their progress by measuring the associated rapid change in pressure on the ground (Bedard et al. 1977).

Sea-Breeze Fronts

Sea breezes brought about by diurnal changes in air temperature over the land and sea are found at many of the world's coasts. The inland boundary of the sea breeze may often form a small cold front and detailed measurements from pilot balloons and instrumented aircraft in southern England show that in the late afternoon the behavior is similar to that of any other cold outflow or gravity current (Simpson, Mansfield & Milford 1977).

In Australia an extensive study of deeply penetrating sea-breeze fronts has been made by Clarke (1961, 1973). A mean speed of 3.5 m s^{-1} was measured, but fronts with onshore prevailing wind moved with speeds up to 7 m s^{-1}, traveling over 200 km from the coast. Many investigations of sea-breeze fronts have been motivated by pollution studies, and the sea-breeze circulation in some cases has been shown to recirculate pollutants emitted near a shoreline (Lyons & Keen 1976).

Airborne insects may follow similar flight paths near the coast, and Rainey (1963) has shown the importance in the life history of locusts of the convergent flow at small-scale fronts. Other pests, such as the spruce budworm moth in Canada, have been concentrated at sea-breeze fronts enabling airborne radar to clarify details of the sea-breeze frontal circulation (Schaefer 1979).

IN THE OCEAN

Conspicuous lines of foam or surface debris are often formed by surface convergence and such bands along the sea surface across which the density changes abruptly show a subsurface structure similar to that at atmospheric fronts. Such fronts in the ocean, known for many years to Japanese fishermen in their search for high concentrations of fish, were investigated scientifically by Uda (1938). Many fronts, especially those created by tidal processes, appear to be governed by simple gravity-current frontal dynamics.

Shelf-Sea Fronts

Shelf-sea fronts are induced at the transition between well-mixed and stratified water. Tidal currents in shallow seas cause high levels of turbulent dissipation which mix downward the seasonal buoyancy input and prevent stratification. The boundary between the mixed and stratified regions is often a well-defined front with a sharp change in surface temperature.

Such fronts have been studied on the European continental shelf. There are several regions where such fronts are regularly detected (J. H. Simpson et al. 1978, Pingree et al. 1974), and infrared satellite images have been tied up with observations from ships. These fronts have been shown to be nongeostrophic with flows of 2 to 16 cm s^{-1} in the mean direction of the fronts.

River Plumes and Salt Wedges

When the river-discharge currents closely match the tidal currents in an estuary, a tidal intrusion front may form, together with the associated salt wedge (Officer 1976). Well-documented measurements of fresh-water plumes and plume fronts have been made by Garvine (1974) and Garvine & Monk (1977) in the Connecticut River. Their measurements at the front are consistent with the gravity-current model described above. For example, in one set of measurements the plume was about 100 cm deep at the front, with a mean density difference of 1.2% and moved at $U = 50$ cm s^{-1}, giving a Froude number $U/(g'h)^{1/2} = 1.5$.

Fronts in Fjords

Deep-water renewal in fjords depends on several processes, one of which is the gravity-current mechanism. With a shallow sill entrance to a fjord, inflows may be intermittent and at spring tide a salty flow may flow down the slope past the sill. Currents flowing down a slope of 6° in a Scottish fjord had a front speed about 50% of the speed of the following flow, as in laboratory measurements (Edwards & Edelsten 1977).

SUSPENSION FLOWS

Nearly all the treatment of saline gravity currents can be applied to suspension flows, if we bear in mind that at any particular time a vertical gradient of density is likely to exist (Middleton 1966).

Turbidity Currents

Under-water muddy slumps can change into turbidity currents with density difference due to suspended particles. It is believed that they transport materials from shallow to deep water, causing erosion of underwater canyons, the building of submarine fans along the foot of the continental rise, and sedimentation on abyssal plain. They are responsible for damage to submarine cables and for the formation of tsunamis. A critical history of investigations of turbidity currents has been given by Longinov (1971), and many aspects of their behavior have been studied by Komar (1972, 1977).

Branching channels seen on the surface of Mars resemble some of those on the ocean bed of Earth, carved out by turbidity currents, and it has been suggested that some of the Martian channels could have been formed by atmospheric-suspension gravity currents (Komar 1979).

Allen (1971) related the lobe and cleft structure observed at the head of saline flows via the transverse variation of bed-shear stresses to observed sedimentary structures. The longitudinal branching current ridges seen beneath sandstones attributed to turbidity-current deposition (Dzulyinski 1965) are examples.

One difference from the gravity currents previously examined is in the manner of starting and maintaining the necessary suspension of particles. Reviews of the processes of transformation of the gravitational movements of sediments into a sediment-carrying current were given by Pykhov & Longinov (1972) and Pykhov (1973).

Avalanches

An avalanche consisting of airborne snow particles is a spectacular example of a gravity current running down a slope. Voellmy (1955) described avalanche speeds between 50 and 120 m s^{-1}, with a density close to the ground between two and twenty times that of air. Observations conform with results from laboratory gravity-current studies, but caution needs to be taken in scaling flows with a large density ratio in which the Boussinesq approximation is hardly realistic. Conditions for the occurrence of avalanches have been reviewed by Shen & Roper (1970) and the problems of avalanche research summarized by de Quervain (1966).

Nuées Ardentes

Surges called *nuées ardentes* can originate from explosive volcanic eruptions as extremely hot, opaque, billowing clouds rushing down the side of the mountain. Examples have been filmed from airplanes (Moore & Melson 1969, Stith et al. 1977) showing speeds approaching 50 m s^{-1} over distances of several kilometers. Fluidization caused by gases released from the particles or entrained by lobes and clefts may play an important part in generating these high speeds (Wilson 1980). Reimers & Komar (1979) have pointed out that some of the morphological features associated with volcanoes on the planet Mars are similar to those on Earth formed by explosive volcanic gravity currents, and could have a similar origin.

Base Surges

Base surges are also suspension flows resulting from volcanic activity. These result from explosive volcanic eruptions into a watery environment, and unlike the nuées described above, are cool and moist and usually annular in form (Moore 1967).

AMBIENT STRATIFICATION

Gravity currents flowing through stratified surroundings may generate internal waves, and some interesting current-wave interactions have been observed. These are strongly dependent on the relative depth of the gravity current h and the depth H through which the ambient stratification extends (see Figure 5).

Sharp Interfaces

INTERMEDIATE CURRENTS After a section of two-layer fluid has been thoroughly mixed and released, a symmetric current flows along the interface. When the interface is sharp, breaking billows appear at the head, provided the Reynolds number is large enough (Simpson & Britter 1979b). The structure and dynamics of the resulting front are similar to an inviscid gravity current reflected about a horizontal plane.

BOUNDARY CURRENTS Holyer & Huppert (1980) dealt with steady flows in which waves are only swept downstream from the head and cannot go upstream. Simpson (1980) released a finite amount of dense fluid along the lower boundary and showed how the behavior of the current varied in relation to the speed of long waves on the interface between the two fluids.

A range existed in which a sudden change appeared at the gravity-current head as the flow changed from super- to subcritical. This was associated with an interfacial wave propagating upstream.

Interface Lower Than the Gravity-Current Depth

INTRUSIONS Suppose the interface between the fluids is allowed to diffuse. After a time the density distribution approaches that assumed in the theoretical work of Benjamin (1967) and Davis & Acrivos (1967) on waves in deep layers, later examined experimentally by Hurdis & Pao (1975) and Ono (1975). The collapse of mixed fluid can result in large-amplitude waves traveling in the thick interface for long distances without change of form. The waves have closed streamlines and consist mainly of mixed fluid from the collapsed region. Their formation and propagation has been described in detail recently by Kao & Pao (1980) and Maxworthy (1980). The Red Spot of Jupiter has been shown to have features in common with this type of solitary wave (Maxworthy et al. 1978).

BOUNDARY CURRENTS Maxworthy also described boundary gravity currents flowing in two-dimensional channels and radially with a thin dense miscible layer on the floor. Simpson (1980) has described similar two-dimensional experiments in which a gravity current was run at super- and subcritical speed compared with the solitary-wave speed. At supercritical speed the gravity current appeared almost unaffected by the stratification, but at subcritical speed a train of solitary waves in the thin layer moved ahead of the front. At an intermediate speed, when the density of the fluid in the current was roughly equal to that of the surroundings at the floor, a solitary wave with closed streamlines built up.

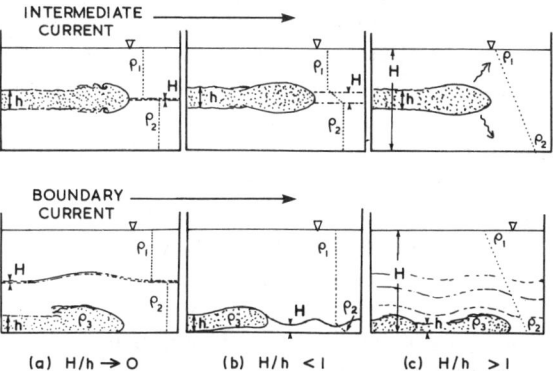

Figure 5 Gravity currents in stratified media and the formation of waves. The density of the intermediate currents is $(\rho_1 + \rho_2)/2$.

In the atmosphere, radar used in insect detection has outlined the presence of multiple frontal lines (Schaefer 1976). These are believed to have been formed by the advance of a gravity current through the strongly stratified nighttime boundary-layer inversion.

The propagation of sea-breeze fronts can be affected by waves in the developing nocturnal inversion. Extensive measurements of the progress of sea-breeze fronts in Australia by Clarke (1965) led him to suspect the formation late in the day of a vortex, cut off from the sea breeze, rolling on ahead of the main flow. The "cut-off vortex" described by Simpson, Mansfield & Milford (1977) seems consistent with laboratory observations of mixed fluid trapped within the leading solitary wave. Solitary waves measured by Christie et al. (1978), using pressure variations on the ground, were associated with sea-breeze fronts, and the "Morning Glory" phenomenon seen in Northern Australia seems to have features of both gravity currents and solitary waves (Clarke 1972, Neal et al. 1977). On the planet Mars a line of cloud resembling the Morning Glory has been seen at sunrise crossing the plain near Olympus Mons (Briggs et al. 1977, A. Pickersgill, unpublished).

Interface Thicker Than the Gravity-Current Depth

INTRUSIONS A continuous discharge of fluid into a linearly stratified fluid was studied by Masuda & Nagata (1974) and by Manins (1976). The intrusion velocities were very slow, and the heads of the intrusions were long and pointed. Forward-propagating disturbances and the ends of the tank played an important part in the flow.

Flows from a finite-volume mixed region into a linear stratification were described by Wu (1969). He found a strong internal-wave field generated by the collapse. Kao (1976) pointed out that the "principal stage" of the collapse described by Wu is a quasi-steady process—a gravity current in a stratified environment. The initial uniform-velocity regime observed in a gravity current released from a finite lock was negligible in Wu's experiments. However, Amen & Maxworthy (1980) observed an appreciable uniform regime in a flow into linear stratification.

A model for the evolution of such finite-volume mixed regions, or "turbulent spots," has been proposed by Barenblatt (1978). The waves radiated by these spots in their initial stages should interact between themselves and can also generate unstable shear flows which lead to the appearance of new turbulent spots of smaller scale. These, again in the initial stages of their collapse, radiate internal waves of small scale and so on in a cascade process. Barenblatt also obtained an expression for the radial spread of a spot in the long-lasting viscous regime, in which the distance r varies with $t^{1/8}$. Experimental confirmation of this work has been given by Zatsepin et al. (1978).

BOUNDARY CURRENTS   Boundary currents, both above and beneath a uniformly stratified fluid, were examined experimentally by Simpson (1980). He showed how, for flows with an overall Froude number U_f/NH less than $1/\pi$ (where N is the buoyancy frequency), first-mode waves were formed which interacted with the head of the advancing gravity current to destroy the original front and to cause a succession of gravity-current heads (Figure 5c).

Consecutive fronts forming at the leading edge of a gravity current moving above stratified fluid suggest a possible explanation for a series of frontal lines seen occasionally as foam lines in fjords (McClimans 1978). Regular discontinuities in surface-temperature structure at power-station cooling-water flows (Adams & Carr 1975) and at a river flowing out to sea (Scarpace & Green 1973, Gross 1972) also may have this explanation.

UNIVERSALITY OF GRAVITY CURRENTS

Gravity currents have wide-ranging applications from avalanches and atmospheric pollution at one end of the alphabet to vulcanology and zoology at the other. Their effects have been detected on at least two other planets as well as on Earth. Research undertaken on one scientific discipline in this topic is being applied to previously unrelated disciplines. For example, work done on saline flows under lock-gates in canals has explained both the behavior of locust swarms and of plankton in the ocean. This trend will continue as further natural events are identified involving gravity-current processes.

ACKNOWLEDGMENTS

I wish to thank numerous colleagues for helpful discussions and especially R. E. Britter, H. E. Huppert, P. F. Linden, J. W. Rottman, and N. H. Thomas. I am indebted to both the Central Electricity Generating Board and the Science Research Council for financial support.

Literature Cited

Adams, B. A., Carr, J. F. 1975. Report of an aerial infra-red survey of cooling water plume movements at Pembroke Power Station. *C.E.G.B., S.W. Region Job No. ZB 1406*

Allen, J. R. L. 1971. Mixing at turbidity current heads, and its geological implications. *J. Sediment. Petrol.* 41:97–113

Amen, R., Maxworthy, T. 1980. The gravitational collapse of a mixed region into a linearly stratified fluid. *J. Fluid Mech.* 96:65–80

Barenblatt, G. I. 1978. Dynamics of turbulent spots and intrusions in a stably stratified fluid. *Izv. Atmos. Oceanic Phys.* 14:139–45

Barr, D. I. H. 1959. Some observations of small-scale thermal density currents. *Proc. 8th Congress IAHR, Montreal, Paper 6-C.* 22 pp.

Barr, D. I. H. 1967. Densimetric exchange flow in rectangular channels. III. Large scale experiments. *Houille Blanche* 6/1967:619–31

Bedard, A. J., Hooke, W. H., Beran, D. W. 1977. The Dulles Airport pressure jump detector array for gust front detection. *Bull. Am. Meteorol. Soc.* 58:920–26

Beghin, P., Hopfinger, E. J., Britter, R. E. 1981. Gravitational convection from instantaneous sources on inclined boundaries. *J. Fluid Mech.* 107:407–22

Benjamin, T. B. 1967. Internal waves of permanent form in fluids of great depth. *J. Fluid Mech.* 29:559–92

Benjamin, T. B. 1968. Gravity currents and related phenomena. *J. Fluid Mech.* 31:209–48

Briggs, G. A., Klaasen, K., Thorpe, T., Wellman, J., Baum, W. 1977. Martian dynamical phenomena during June-November 1976: Viking Orbiter results. *J. Geophys. Res.* 82:4121–50

Britter, R. E. 1979. The spread of a negatively buoyant plume in a calm environment. *Atmos. Environ.* 13:1241–47

Britter, R. E., Linden, P. F. 1980. The motion of the front of a gravity current traveling down an incline. *J. Fluid Mech.* 99:531–43

Britter, R. E., Simpson, J. E. 1978. Experiments on the dynamics of a gravity current head. *J. Fluid Mech.* 88:223–40

Chen, J-C., List, E. J. 1976. Spreading of buoyant discharges. *Proc. 1st CHMIT Seminar on Turbulent Buoyant Convection, Dubrovnik, 1976*, pp.171–82

Christie, D. R., Muirhead, K. J., Hales, A. L. 1978. On solitary waves in the atmosphere. *J. Atmos. Sci.* 35:805–25

Clarke, R. H. 1961. Mesostructure of dry cold fronts over featureless terrain. *J. Meteorol.* 18:715–35

Clarke, R. H. 1965. Horizontal mesoscale vortices in the atmosphere. *Aust. Meteorol. Mag.* No. 50:1–25

Clarke, R. H. 1972. The Morning Glory: an atmospheric hydraulic jump. *J. Appl. Meteorol.* 11:304–11

Clarke, R. H. 1973. A numerical model of the sea-breeze. *1st Aust. Conf. Heat and Mass Transfer, Monash*, pp. 41–48

Daly, B. J., Pracht, W. E. 1968. Numerical study of density current surges. *Phys. Fluids* 11:15–30

Davis, R. E., Acrivos, A. 1967. Solitary internal waves in deep water. *J. Fluid Mech.* 29:593–607

de Quervain, M. R. 1966. Problems of avalanche research. *1st Int. Symp. Sci. Aspects of Snow and Ice Avalanches, Davos, 1965*, pp. 15–22

Dzulynski, S. 1965. New data on experimental production of sedimentary structures. *J. Sediment. Petrol.* 35:196–212

Edwards, A., Edelsten, D. J. 1977. Deep water renewal of Loch Etive; a three basin Scottish Fjord. *Estuarine and Coastal Mar. Sci.* 5:575–95

England, W. G., Teuscher, L. H., Hauser, L. E., Freeman, B. E. 1978. Atmospheric dispersion of liquefied natural gas vapor clouds, using SIGMET. *Proc. 1978 Heat Transfer and Fluid Mech. Inst.* Stanford Univ. Press

Fay, J. A. 1969. The spread of oil slicks on a calm sea. In *Oil on the Sea*, ed. D. P. Hoult, pp. 53–63. New York: Plenum

Fay, J. A. 1980. Gravitational spread and dilution of heavy vapor clouds. *2nd IAHR Symp., Trondheim, June 1980*, pp. 471–94

Fujita, T. T., Byers, H. R. 1977. Spearhead echo and downburst in the crash of an airliner. *Mon. Weather Rev.* 105:129–46

Fujita, T. T., Caracena, F. 1977. An analysis of three weather-related accidents. *Bull. Am. Meteorol. Soc.* 58:1164–81

Garvine, R. W. 1974. Dynamics of small-scale oceanic fronts. *J. Phys. Oceanogr.* 4:557–69

Garvine, R. W., Monk, J. D. 1977. Frontal structure of a river plume. *J. Geophys. Res.* 79:2251–59

Georgeson, E. M. H. 1942. The free streaming of gases in sloping galleries. *Proc. R. Soc. London Ser. A* 180:484–93

Goff, R. C. 1976. Thunderstorm-outflow kinematics and dynamics. *NOAA Tech. Memo ERL NSSL-75.* 63 pp.

Gross, M. G. 1972. *Oceanography: A view of the Earth.* Englewood Cliffs, NJ: Prentice-Hall

Hall, F. F., Neff, W. D., Frazier, T. V. 1976. Wind shear observations in thunderstorm density currents. *Nature* 264:408–11

Holyer, J. Y., Huppert, H. E. 1980. Gravity currents entering a two-layer fluid. *J. Fluid Mech.* 100:739–67

Hopfinger, E. J., Tochon-Danguy, J. C. 1977. A model study of powder-snow avalanches. *Glaciology* 19:343–56

Hoult, D. P. 1972. Oil spreading on the sea. *Ann. Rev. Fluid Mech.* 4:341–68

Huppert, H. E., Simpson, J. E. 1980. The slumping of gravity currents. *J. Fluid Mech.* 99:785–99

Hurdis, D. A., Pao, H-P. 1975. Experimental observation of internal solitary waves in a stratified fluid. *Phys. Fluids* 18:385–86

Idso, S. B., Ingram, R. S., Pritchard, J. M. 1972. An American haboob. *Bull. Am. Meteorol. Soc.* 53:930–35

Jirka, G. H. 1980. Two-dimensional density current from continuous source in stratified crossflow *2nd Int. Symp. Stratified Flows, Trondheim, June 1980*

Kao, T. W. 1976. Principal stage of wake collapse in a stratified fluid: two-dimensional theory. *Phys. Fluids* 9:1071–74

Kao, T. W., Pao, H-P. 1980. Wake collapse in the thermocline and internal solitary waves. *J. Fluid Mech.* 97:115–27

Kao, T. W., Park, C., Pao, H-P. 1977. Buoyant surface discharge and small-scale oceanic fronts: a numerical study. *J. Geophys. Res.* 82:1747–52

Kao, T. W., Park, C., Pao, H-P. 1978. Inflows, density currents and fronts. *Phys. Fluids* 21:1912–22

Keulegan, G. H. 1957. An experimental study of the motion of saline water from locks into fresh water channels. *US Natl. Bur. Stand., Rep. 5168*

Koh, R. C. Y. 1971. Two-dimensional surface warm jets. *J. Hydraul. Div. ASCE* 97(HY6):819–36

Komar, P. D. 1972. Relative significance of head and body spill from a channelized turbidity current. *Geol. Soc. Am. Bull.* 83:1151–56

Komar, P. D. 1977. Computer simulation of turbidity current flow and the study of deep-sea channels and fan sedimentation. In *The Sea, Vol. 6 Marine Modeling*, pp. 603–21. Chichester, England: Wiley

Komar, P. D. 1979. Comparisons of the hydraulics of water flows in Martian outflow channels with flows of similar scale on Earth. *Icarus* 37:156–81

Kranenberg, C. 1978. Internal fronts in two-layer flow. *J. Hydraul. Div. ASCE* 104(HY10):1449–53

Lean, G. H., Whillock, A. Z. 1965. The behaviour of a warm water layer flowing over still water. *Proc. 11th Cong. IAHR, Leningrad, Vol. 2.* 12 pp.

Longinov, V. V. 1971. The problem of turbidity currents in the lithodynamics of the ocean. *Oceanology* 11:303–10

Lyons, W. A., Keen, C. S. 1976. Computed 24-hour trajectories for aerosols and gases in a lake/land breeze circulation cell on the western shore of Lake Michigan. *6th Conf. Weather Forecasting, Albany, NY,* pp. 78–83

Manins, P. C. 1976. Intrusion into a stratified fluid. *J. Fluid Mech.* 74:547–60

Masuda, A., Nagata, Y. 1974. Water wedge advancing along the interface between two homogeneous layers. *J. Oceanogr. Soc. Jpn.* 30:289–97

Maxworthy, T. 1980. On the formation of nonlinear internal waves from the gravitational collapse of mixed regions in two and three dimensions. *J. Fluid Mech.* 96:47–64

Maxworthy, T., Redekopp, L. G., Weidman, P. 1978. On the production of planetary solitary waves: applications to the Jovian atmosphere. *Icarus* 33:338–409

McClimans, T. A. 1978. Fronts in fjords. *Geophys. Astrophys. Fluid Dyn.* 11:23–34

McQuaid, J. 1979. Dispersion of heavier-than-air gases in the atmosphere. *Health and Safety Executive Tech Paper No. 8.* 16 pp.

Middleton, G. V. 1966. Experiments on density and turbidity currents. I, Motion of the head. *Can. J. Earth Sci.* 3:523–46

Miller, M. J., Betts, A. K. 1977. Travelling convective storms over Venezuela. *Mon. Weather Rev.* 105:833–48

Moore, J. G. 1967. Base surge in recent volcanic eruptions. *Bull. Volcanol.* 30:337–63

Moore, J. G., Melson, W. G. 1969. Nuées ardentes of the 1968 eruption of Mayon volcano, Philippines. *Bull. Volcanol.* 33:600–20

Neal, A. B., Butterworth, I. J., Murphy, K. M. 1977. The Morning Glory. *Weather* 32:176–83

O'Brien, M. P., Cherno, J. 1934. Model law for motion of salt water through fresh. *Trans. Am. Soc. Civil Engrs.* 99:576–94

Officer, C. B. 1976. *Physical Oceanography of Estuaries and Associated Coastal Waters.* NY: Wiley

Ono, H. 1975. Algebraic solitary waves in stratified fluids. *J. Phys. Soc. Jpn.* 39:1082–91

Pingree, R. D., Forster, G. R., Morrison, G. K. 1974. Turbulent convergent tidal fronts. *J. Mar. Biol. Assoc. U.K.* 54:469–79

Prandtl, L. 1952. *Fluid Dynamics.* London: Blackie

Pykhov, N. V. 1973. The movements of sediments over an inclined sea floor after disturbance of their stability. *Oceanology* 13:893–96

Pykhov, N. V., Longinov, V. V. 1972. On the methods of computing the parameters of turbidity currents. *Oceanology* 12:761–71

Rainey, R. C. 1963. Meteorology and the migration of desert locusts. *WMO No. 138 TP. 64. Tech. Notes No. 54.* 115 pp.

Reimers, C. E., Komar, P. D. 1979. Evidence for explosive volcanic density currents on certain Martian volcanoes. *Icarus* 39:88–110

Riddell, J. C. 1970. Arrested saline wedge. *Houille Blanche* 4/1970:317–30

Scarpace, F. L., Green, T. 1973. Dynamic surface temperature structure of thermal plumes. *Water Resources Res.* 9:138–53

Schaefer, G. W. 1976. Radar observations of insect flight. In *Insect Flight: Symp. R. Entomol. Soc. London No. 7*, pp. 157–97

Schaefer, G. W. 1979. An airborne radar technique for the investigation and control of migrating pest insects. *Philos. Trans. R. Soc. London Ser. B* 287:459–65

Schmidt, W. 1911. Zur Mechanik der Böen. *Z. Meteorol.* 28:355–62

Shen, H. W., Roper, A. T. 1970. Dynamics of snow avalanche. *Bull. Int. Assoc. Sci. Hydrology* 15:7–26

Simpson, J. E. 1969. A comparison between laboratory and atmospheric density currents. *Q. J. R. Meteorol. Soc.* 95:758–65

Simpson, J. E. 1972a. Effects of the lower boundary on the head of a gravity current. *J. Fluid Mech.* 53:759–68

Simpson, J. E. 1972b. Instability patterns at the head of a cold outflow. *Proc. 1st Int. Symp. Tech. & Sci. Motorless Flight, MIT, Cambridge, Mass.*, pp. 8–18

Simpson, J. E. 1980. Experiments on gravity currents in stratified fluids. *Geophys. J. R. Astron. Soc.* 61:225 (Abstract only)

Simpson, J. E., Britter, R. E. 1979a. The dynamics of the head of a gravity current advancing over a horizontal surface. *J. Fluid Mech.* 94:477–95

Simpson, J. E., Britter, R. E. 1979b. The form of the head of an intrusive gravity current. *Geophys. J. R. Astron. Soc.* 57:289 (Abstract only)

Simpson, J. E., Britter, R. E. 1980. A laboratory model of an atmospheric mesofront. *Q. J. R. Meteorol. Soc.* 106:485–500

Simpson, J. E. Mansfield, D. A., Milford, J. R. 1977. Inland penetration of seabreeze fronts. *Q. J. R. Meteorol. Soc.* 103:46–76

Simpson, J. H., Allen, C. M., Morris, N. C. G. 1978. Fronts on the continental shelf. *J. Geophys. Res.* 83:4607–14

Stith, J. L., Hobbs, P. V., Radke, L. F. 1977. Observations of a nuée ardente from the St. Augustine volcano. *Geophys. Res. Lett.* 4:259–62

Thorpe, A. J., Miller, M. J., Moncrieff, M. W. 1980. Dynamical models of two-dimensional downdraughts. *Q. J. R. Meteorol. Soc.* 106:463–84

Thorpe, S. A. 1973. Experiments on instability and turbulence in a stratified shear flow. *J. Fluid Mech.* 61:731–51

Turner, J. S. 1968. The influence of molecular diffusivity on turbulent entrainment across a density interface. *J. Fluid Mech.* 33:639–56

Uda, M. 1938. Researches on 'Siome' or current rip in the seas and oceans. *Geophys. Mag.* 11:307–72

van Ulden, A. P. 1974. On the spreading of a heavy gas released near the ground. *1st Int. Loss Prevention Symp., The Hague, May 1974*, pp. 221–26

Vasiliev, O. F., Kvon, V. I., Chernyshova, R. T. 1973. Mathematical modelling of the thermal pollution of a water body. *IAHR Cong. , Istanbul, 1973, Pap. B17.*

Voellmy, A. 1955. Uber die Zerstörungskraft von Lawinen. *Schweiz. Bauzeitung* 73:159–65, 212–17

von Kármán, T. 1940. The engineer grapples with non-linear problems. *Bull. Am. Math. Soc.* 46:615–83

Vreugdenhil, C. B. 1976. Mathematical investigation of stratified flow. In *Rijkswaterstaat Communications No. 26, DHL, The Hague*, pp. 87–114

Wilkinson, D. L., Wood, I. R. 1971. A rapidly varying flow phenomenon in a two-layer flow. *J. Fluid Mech.* 47:241–56

Wilkinson, D. L., Wood, I. R. 1972. Some observations on the motion of the head of a density current. *J. Hydraul. Res.* 10:305–24

Wilson, C. J. N. 1980. The role of fluidization in the emplacement of pyroclastic flow: An experimental approach. *J. Volcanol. and Geothermal Res.* 8:231–49

Winant, C. D., Bratkovich, A. 1977. Structure and mixing within the frontal region of a density current. *6th Aust. Hydraul. & Fluid Mech. Conf., Adelaide, Australia, Dec. 1977*, pp. 9–12

Wood, I. R. 1965. Studies in unsteady self preserving turbulent flows. *Univ. N. S. Wales, Manly Vale; Australia, Rep. No. 81.* 152 pp.

Wood, I. R. 1967. Horizontal two-dimensional density current. *J. Hydraul. Div. ASCE* 93(HY2):35–42

Wu, J. 1969. Mixed region collapse with internal wave generation in a density-stratified medium. *J. Fluid Mech.* 35:531–44

Yih, C-S. 1980. *Stratified flows*, pp. 204–7. New York: Academic

Zatsepin, A. G., Fedorov, K. N., Voropayev, S. I., Pavlov, A. M. 1978. Experimental study of the spreading of a mixed region in a stratified fluid. *Izv. Atmos. and Oceanic Phys.* 14:170–3

THE FLUID DYNAMICS OF HEART VALVES: EXPERIMENTAL, THEORETICAL, AND COMPUTATIONAL METHODS

Charles S. Peskin

Courant Institute of Mathematical Sciences, New York, New York 10012

INTRODUCTION

The Role of Heart Valves in the Cardiac Cycle

The function of the valves on the left side of the heart is reviewed in Figure 1, which shows schematically the pressures, flows, and valve positions of the mitral (inflow) and aortic (outflow) valves of the left ventricle. Note that the cardiac cycle can be divided into two main parts according to whether the ventricular muscle is contracting (systole) or relaxing (diastole). During systole, the mitral valve is closed, the aortic valve is open, and the left ventricle forms a common chamber with the aorta. During diastole, the valve configurations are reversed, and the left ventricle forms a common chamber with the left atrium. Note that the pressure difference across the open valve is very small, and that there is no backflow except for a spurt of brief duration associated with the closure movement of the valve.

The valves themselves are thin membranous leaflets. To appreciate their remarkable anatomy, see the superb drawings of Netter (1969). Better still, dissect a fresh beef heart (available from wholesale meat markets and some retail stores).

Artificial Heart Valves

The principal motivation, direct or indirect, for most of the work that we describe in this review is the design of prosthetic heart valves. Nevertheless, we concentrate on fluid-dynamic principles and methods, so we make no attempt here to review the history or current status of these artificial valves. Instead, we refer the reader to the following books and papers which may serve as an introduction to the excitement, variety, and controversy of this rapidly developing field: Brewer (1969), Kalmanson (1976), Ionescu (1979), Yoganathan et al. (1979c), Roberts (1978), Bonchek (1980), Fowler & van der Bel-Kahn (1980).

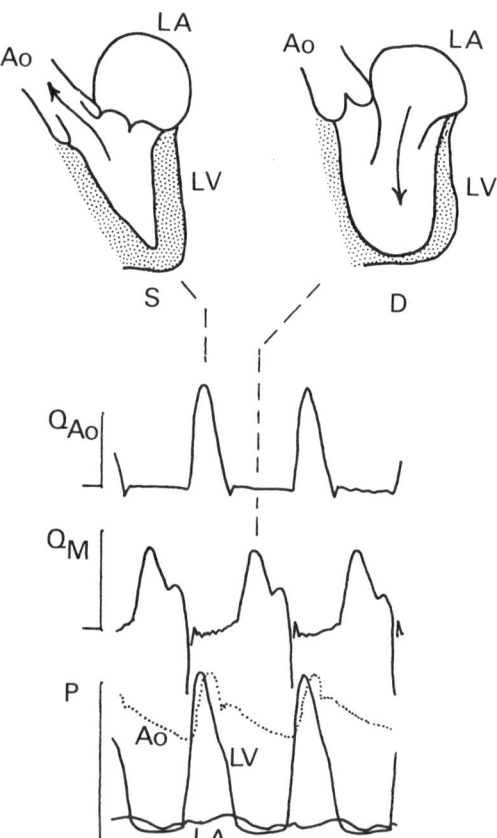

Figure 1 Left heart-valve configurations with corresponding waveforms of flow and pressure. Ao = Aorta, LA = Left Atrium, LV = Left Ventricle, S = Systole, and D = Diastole. Q_{Ao} = Aortic Flow, Q_M = Mitral Flow, and P = Pressure. Solid waveforms are traced from the data in Yellin et al. (1981), but the dotted waveform of aortic pressure is schematic only.

A BRIEF SURVEY OF EXPERIMENTAL METHODS

Animal Experiments

E. L. Yellin and R. W. M. Frater have developed a collection of experimental and surgical techniques that can be used to study the canine mitral valve (Frater et al. 1969, Laniado et al. 1973, 1975, Yellin et al. 1979, 1981). In a typical experiment they record instantaneous left atrial and ventricular pressures as well as instantaneous mitral flow (volume/time). Valve motion is simultaneously recorded on echocardiogram or cineangiogram. In the latter case, radiopaque sutures are used to opacify the mitral leaflets. Natural and prosthetic valves can be studied by the same techniques. In fact, several different prosthetic valves can be studied sequentially in the same animal (Frater et al. 1969).

These methods have been used by Yellin and his collaborators to establish the following description of the dynamics of the mitral valve:

(a) Forward mitral flow begins with a small component related to the unloading of the elastic valve apparatus as ventricular pressure falls at the end of systole. The valve is closed during this phase, and flow is produced by displacement as the closed valve moves towards the ventricle.

(b) Valve opening and rapid filling begin at the moment when the falling ventricular pressure crosses the atrial pressure at the onset of diastole.

(c) Valve opening is rapid: the valve reaches its maximally open position and begins to close while the flow is still accelerating.

(d) Because of fluid inertia, mitral flow lags behind the atrioventricular pressure difference. Thus peak flow occurs after peak forward-pressure difference both during rapid filling and again during atrial systole, and the pressure difference crosses zero at the onset of ventricular systole before the flow becomes zero.

(e) Valve closure is virtually synchronous with the moment of zero flow. A small spurt of backflow is seen by the flowmeter as fluid is displaced towards the atrium by the closed valve leaflets as pressure builds up in the ventricle. Oscillations in the flow trace following this spurt of backflow correspond to the vibrations of the first heart sound (Laniado et al. 1973).

Further information on the dynamics of the mitral valve in vivo has been obtained by Tsakiris et al. (1971, 1975, 1977) who used radiopaque markers to follow the motion of selected points on the valve leaflets and heart walls. As a result of this work, we have detailed records of cardiac chamber dimensions and valve positions as functions of time. A particularly interesting observation made by Tsakiris is that the mitral annulus does not have constant diameter. Instead, it begins to contract during atrial systole and continues to contract during early ventricular systole. The significance of this for the dynamics of mitral valve closure deserves further study (R. W. M. Frater, personal communication).

The most difficult information to obtain in vivo is a measurement of the flow pattern of blood in the heart. This problem was first attacked by Taylor & Wade (1973), who used thin carefully injected streams of contrast material to trace the flow pattern of blood around the atrioventricular valves on cineangiogram. These investigators also confirmed their findings using cine-endoscopic techniques with fine bubbles or calcium carbonate as fluid markers.

The principal finding of Taylor & Wade was that the flow pattern of blood in the heart forms a stable system with no evidence of large-scale mixing, disorder, or turbulence. This is an experimental result of great importance for any theoretical treatment of the fluid dynamics of heart valves.

According to Taylor & Wade, contrast material injected into the atrium during systole moves into the ventricle at the onset of diastole along a broad front with little evidence of lateral motion or early vortex formation. The contrast material eventually strikes the apex, moves up the ventricular walls towards the base of the heart, and turns back towards the ventricle again, washing the ventricular aspect of the atrioventricular valve.

On general principles, we may argue that such a large-scale circular motion developed within a single heartbeat must have a small, rapidly spinning vortex core at its center. The reasoning is as follows: circulation requires vorticity, vorticity can only be created at boundaries and then shed into the interior; the initial diameter of the vortex core (the region containing vorticity) is on the order of the boundary-layer thickness; and there is not sufficient time in a heartbeat for the diameter of the vortex core to exceed about 2 mm at most. Indeed, Tsakiris et al. (1975) report cineangiographic evidence of early formation of a small vortex core near the tips of the leaflets before the advancing jet of opacified blood has had time to strike the apex and turn up the ventricular wall.

Another approach to the difficult experimental task of determining the flow pattern around heart valves in vivo has recently been pioneered by Brun et al. (1980). Here, ultrasonic Doppler methods are used to obtain the local fluid velocity. The technical problems here are severe, but the method holds great promise as a noninvasive means of determining the flow pattern. Perhaps a mathematical study of the propagation of sound through the moving blood and its interaction with moving boundaries and scattering centers would be helpful for unraveling the information available in ultrasonic Doppler records.

Experiments in Physical Models

We may distinguish here between physical models designed to test artificial heart valves and physical models designed to elucidate the fluid mechanics of natural heart valves. In the first case, the emphasis is on the comparison between different valves in identical conditions. The valves are compared with respect to their flow patterns and pressure drops in steady or unsteady flow. Measurement of the flow pattern was pioneered by Wieting (1969) who used suspended particles and slit lighting to photograph a cross section of the flow (see also Dellsperger & Wieting 1978). A three-dimensional version of this technique was introduced by Peskin (1972b) who used an optical system that made it possible for a single cine camera to record two perpendicular views of the flow field.

Another approach to measuring the flow pattern is laser-Doppler anemometry (Yoganathan et al. 1978, 1979a,d, 1980) in which laser light is scattered from microscopic suspended particles and the resulting Doppler shift is used to measure the fluid velocity. By focusing the beam on a particular location, it is possible to record the local velocity as a function of time.

Using this technique, Yoganathan and his colleagues have demonstrated the existence of a relatively stagnant region in the flow pattern of a flat tilting disc-aortic valve; they have correlated this finding with pathological results on the site of thrombus formation; and they have demonstrated that curvature of the valve can alleviate the problem. These results not only demonstrate the effectiveness of laser-Doppler methods in measuring the flow pattern, but they also illustrate the importance of the flow pattern in prosthetic-valve design.

Measurement of the pressure drop across artificial heart valves during forward flow is also of great importance. Comprehensive studies of this kind have been carried out by Wright (Wright & Temple 1971, 1972, Wright 1979) and by Gabbay et al. (1978, 1979). Wright's test apparatus includes a transparent model ventricle with rigid walls. The ventricle is driven by a curved piston at the apex and valves can be tested in either the mitral or the aortic positions.

The apparatus used by Gabbay and his colleagues is simpler, but their study is remarkable for the methods that were used to analyze the data. These methods will be reviewed below, once the appropriate pressure-flow relation has been derived. The principal conclusions are that steady pressure drops can be used to predict energy losses during pulsatile flow and that significant differences between valves show up only at flow rates appropriate to exercise. The latter observation is consistent with clinical experience that prosthetic heart valves may be well tolerated at rest but may limit the patient's ability to exercise (Björk & Henze 1979, Frater et al. 1979).

We now turn to a consideration of physical models that have been used to study the dynamics of natural heart valves. To study the aortic valve, Bellhouse & Talbot (1969) constructed a flexible trileaflet valve and mounted it in a rigid chamber with a model aortic sinus behind each leaflet. Subsequently, Bellhouse (1972) constructed a transparent elastic model ventricle that could be used in physical model studies of the mitral valve. This kind of ventricle was also used by Lee & Talbot (1979), who modeled the mitral valve as a rigid, bileaflet, hinged structure.

In work along different lines, van Steenhoven & van Dongen (1979) constructed an essentially two-dimensional physical model of the aortic valve using a water-tunnel approach. The two-dimensional character of their system makes flow visualization easier. It also helps answer the question of whether three-dimensional fluid-dynamic mechanisms are fundamental to the function of heart valves.

These physical models of natural heart valves have all resulted in physical theories of heart-valve dynamics. These theories are reviewed in the next section.

PHYSICAL THEORIES OF HEART-VALVE DYNAMICS

The theories in this section are *physical* in the sense that they rest upon some physical assumptions that simplify the fluid dynamics. Such theories are useful for the physical insight they provide and also because they are helpful in the analysis of experimental data.

Pressure-Flow Relations for an Open Heart Valve

It is often observed that the pressure drop Δp across an open valve (or simply a fixed stenosis) is related to the volume rate of flow Q through the valve by an equation of the form

$$a \frac{dQ}{dt} + bQ^2 = \Delta p. \tag{1}$$

We shall give a derivation of (1) for the situation depicted in Figure 2. Our assumptions are as follows:

(a) Energy is conserved in the inflow region, $x_1 \leq x \leq x_v$.
(b) Momentum is conserved in the outflow region, $x_v \leq x \leq x_2$.
(c) There is a stagnant region behind the valve leaflets where the pressure p_v is equal to the pressure in the fluid between the leaflets.

(d) The velocity profile is flat over the flow path.
(e) The geometry of the flow path is independent of time. (This is a quasi-steady assumption which is clearly violated in highly transient situations.)

Note that we do not assume energy conservation downstream of the valve. In fact, energy is lost through dissipative processes, such as vortex formation, which do not appear explicitly in the formulation of the problem. As in the case of a collision between two bodies that stick together, momentum conservation can be used to figure out how much energy is lost.

Let $Q(t)$ be the volume rate of flow and let $A(x)$ be the area of the flow path. Then $u(x, t) = Q(t)/A(x)$ is the fluid velocity. Let $p(x, t)$ be the pressure and let ρ be the (constant) density of the fluid. Then assumptions (a) and (b) yield the equations

$$\frac{d}{dt}\int_{x_1}^{x_v}\frac{1}{2}\rho u^2 A\, dx = \left(p_1 + \frac{1}{2}\rho u_1^2 - p_v - \frac{1}{2}\rho u_v^2\right)Q, \quad (2)$$

$$\frac{d}{dt}\int_{x_v}^{x_2}\rho u A\, dx = A_2(p_v - p_2) + (\rho u_v - \rho u_2)Q. \quad (3)$$

After some manipulation, these equations can be combined to obtain

$$\frac{L\rho}{\langle A\rangle}\frac{dQ}{dt} + \frac{1}{2}\rho Q^2\left(\frac{1}{A_v} - \frac{1}{A_2}\right)^2 = \Delta p^0, \quad (4)$$

where

$$L = x_2 - x_1, \quad (5)$$

$$\frac{L}{\langle A\rangle} = \left(\int_{x_1}^{x_v}\frac{dx}{A}\right) + \frac{x_2 - x_v}{A_2}, \quad (6)$$

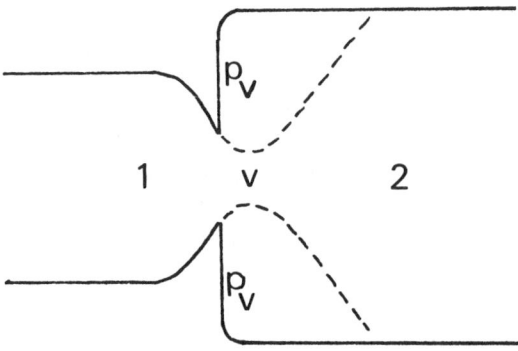

Figure 2 Pressure-flow relation for an open heart valve. The valve follows the separating streamline (dotted).

$$\Delta p^0 = \left(p_1 + \frac{1}{2}\rho u_1^2\right) - \left(p_2 + \frac{1}{2}\rho u_2^2\right),$$

$$= (p_1 - p_2) + \frac{1}{2}\rho Q^2 \left(\frac{1}{A_1^2} - \frac{1}{A_2^2}\right). \tag{7}$$

Equation (4) has the same form as (1) except that the right-hand side involves the drop in total head instead of the pressure drop. This distinction disappears when $A_1 = A_2$ and also when A_1 and A_2 are both large compared to A_v.

A formula for the rate of dissipation of energy δ can be derived as follows. Energy balance for the fluid between x_1 and x_2 may be written

$$\frac{d}{dt}\int_{x_1}^{x_2} \frac{1}{2}\rho u^2 A\, dx = \Delta p^0 Q - \delta. \tag{8}$$

Expressing u in terms of Q and using (4) we get

$$\frac{\delta}{Q} = \frac{1}{2}\rho Q^2 \left(\frac{1}{A_v} - \frac{1}{A_2}\right)^2 - \rho\frac{dQ}{dt}\int_{x_v}^{x_2}\left(\frac{1}{A} - \frac{1}{A_2}\right)dx. \tag{9}$$

The steady term here can be also written $\tfrac{1}{2}\rho\,(u_v - u_2)^2$, so it involves the relative velocity of the fluid at the valve compared with the fluid downstream.

The unsteady term has zero average over any cycle, but it asserts that the instantaneous dissipation rate is less for accelerating than for decelerating flows at the same flow rate. For large positive dQ/dt, the rate of dissipation can even be negative. It seems reasonable to interpret this as a breakdown in the assumption that the geometry of the flow path is independent of time.

We now consider the use of Equation (1) in the analysis of experimental data. The equation has been applied to experiments on flow through a fixed stenosis by Yellin & Peskin (1975) and to flow through prosthetic valves by Gabbay et al. (1978, 1979). The latter investigators demonstrated the important methodological point that steady-flow experiments on prosthetic heart valves can be used to predict the dissipative pressure drop in pulsatile flow through these valves (see also Yoganathan et al. 1979b). This was accomplished using Equation (1) in the following way.

In steady flow, the parameter b may be measured simply as the ratio $b = \Delta p/Q^2$. In pulsatile flow, if Equation (1) is integrated between two times t_1 and t_2 such that $Q(t_1) = Q(t_2)$, then the unsteady term drops out and we get $b = \langle\Delta p\rangle/\langle Q^2\rangle$ where

$$\langle\Delta p\rangle = \frac{1}{t_2 - t_1}\int_{t_1}^{t_2} \Delta p(t)\, dt, \tag{10}$$

$$<Q^2> = \frac{1}{t_2-t_1} \int_{t_1}^{t_2} Q^2(t)\, dt . \tag{11}$$

In the experiment of Gabbay et al., the values of b measured in these two ways for a given valve were statistically indistinguishable over a considerable range of flow rates and frequencies. This was no longer true when $<Q^2>$ was replaced by $<Q>^2$ in the analysis of the pulsatile data. A systematic error is expected in the latter case, since $<Q^2> > <Q>^2$. We may conclude that

(a) Steady flow measurements are adequate to predict the dissipative part of the pressure drop across prosthetic heart valves during pulsatile flow.
(b) Pulsatile flow should be analyzed in terms of the mean of the square of the flow (not the square of the mean), the mean being taken over a time interval such that the flow has the same value (e.g. zero) at both ends. The mean pressure difference should be evaluated over the same interval of time as the flow.

Applying these principles in vivo may be difficult if flow is not directly measured but inferred from cardiac output. In that case, it seems reasonable to propose that the times of zero flow should be estimated from the heart sounds and that a correction (based on an assumed flow waveform) should be used to convert $<Q>^2$ to $<Q^2>$.

Motion of Valves with a Single Degree of Freedom

The work that we review in this section includes that of Bellhouse & Talbot (1969), Bellhouse & Bellhouse (1971), van Steenhoven & van Dongen (1979), and Lee & Talbot (1979). This body of work encompasses both the aortic and the mitral valve. In each case the theory is based on physical-model experiments.

The common threads running through these theories are that the valve can be treated as a structure with a single degree of freedom and that the fluid between the leaflets can be described by the one-dimensional form of the incompressible Euler equations. (The latter assumption is stated explicitly only by van Steenhoven & van Dongen, but in every case it leads to the same results as the other authors derive from conservation arguments.)

The theories differ primarily in the assumptions about the fluid behind the valve leaflets. Bellhouse & Talbot and Bellhouse & Bellhouse model the flow behind the leaflets by a Hill spherical vortex, and they match this vortex to the main stream in different ways for the aortic and mitral valve.

Van Steenhoven & van Dongen and Lee & Talbot use simpler assumptions. For the aortic valve they both assume that the pressure in the aortic sinus is uniform and equal to the pressure at the cusp margins. For the mitral valve, Lee & Talbot assume a uniform distribution of kinetic energy in the ventricle.

We describe here the simplest of these closely related theories. This is the aortic-valve case with a uniform pressure in the aortic sinus. The equations are as follows:

$$\rho\left(\frac{\partial u}{\partial t} + u\frac{\partial u}{\partial x}\right) + \frac{\partial p}{\partial x} = 0, \qquad (12)$$

$$\frac{\partial A}{\partial t} + \frac{\partial}{\partial x}(Au) = 0, \qquad (13)$$

$$A(x,t) = A_0\left\{1 - [1 - \lambda(t)]\frac{x}{L}\right\}^2, \qquad (14)$$

$$-\int_0^L p(x,t)\frac{\partial A}{\partial x}dx = [A_0 - A(L,t)]\,p(L,t). \qquad (15)$$

In these equations u is the axial velocity, p is the pressure, A is the cross-sectional area of the valve, L is the axial length of the valve, and $\lambda(t)$ is the single degree of freedom. The meaning of λ is given by $\lambda^2(t) = A(L,t)/A_0$.

Note that Equations (12–13) are the one-dimensional form of the incompressible Euler equations. Equation (14) states that the valve has the geometry of a truncated cone. The valve is assumed to have zero mass, so the forces acting on it must balance. Equation (15) expresses this for the x-component of the force on the valve under the assumption that the pressure in the fluid behind the valve is uniform and equal to the pressure at the cusp margins $p(L,t)$.

These equations can be combined to obtain an ordinary differential equation for $\lambda(t)$ in terms of the inflow velocity $u_0(t) = u(0,t)$. The derivation of this equation is carried out exactly by Lee & Talbot and approximately by van Steenhoven & van Dongen. The latter authors restrict consideration to times when $\lambda \simeq 1$. This can be expressed by setting $\lambda = 1 + \epsilon\lambda_1$ and then retaining only terms up to first order in ϵ. We shall follow this approach.

To first order in ϵ, Equations (14–15) become

$$A(x,t) = A_0\left[1 + 2\epsilon\lambda_1(t)\frac{x}{L}\right], \qquad (16)$$

$$\frac{1}{L}\int_0^L p(x,t)\,dx = p(L,t). \qquad (17)$$

Now substitute (16) in (13) and integrate once in x to get an expression for $u(x,t)$ in terms of $\lambda_1(t)$. Then substitute this expression into (12) and integrate once in x to get $p(x,t)$ in terms of $\lambda_1(t)$. Finally, substitute this into (17) to get an ordinary differential equation for λ_1. Throughout these calculations, retain only terms up to first order in ϵ. The result, expressed in terms of $\lambda(t)$ is

$$\frac{2u_0'}{L} = -(1-\lambda)\left(\frac{8}{3}\frac{u_0'}{L} + \frac{4u_0^2}{L^2}\right) + \frac{16}{3}\lambda'\frac{u_0}{L} + \lambda'' \tag{18}$$

as given by van Steenhoven & van Dongen.

When u_0 is constant the valve has a stable equilibrium configuration given by $\lambda = 1$. When $u_0' < 0$, the valve is driven towards closure even though u_0 is still positive. This shows the importance of deceleration in closing the valve.

Convection Waves and the Motion of Flexible Leaflets

In the foregoing, the most questionable assumption is that the valve has a single degree of freedom. Lee & Talbot actually constructed a rigid, hinged valve, but the other experimental valves were flexible as are the natural valves of the heart. These considerations led the present author to the following physical theory, which is outlined here because it is such a natural extension of the work described above.

As before we use the one-dimensional Euler equations (12–13) and we retain the assumption that the fluid behind the cusps has a uniform pressure. Here, however, we assume that the leaflets are flexible (and massless, as before) so that the pressure difference across them is zero at every point, not just on the average. It follows from this and the assumption of uniform pressure behind the cusps that $\partial p/\partial x = 0$, and we have simply

$$\frac{\partial u}{\partial t} + u\frac{\partial u}{\partial x} = 0, \tag{19}$$

$$\frac{\partial A}{\partial t} + \frac{\partial}{\partial x}(Au) = 0. \tag{20}$$

Note that these are also the equations of one-dimensional flow in a jet with a free boundary on which $p = 0$.

We shall solve these equations subject to the boundary conditions $u(0,t) = u_0(t)$, $A(0,t) = A_0$. It may be checked by direct substitution that the solution is given implicitly by

$$u\left[x, t_0 + \frac{x}{u_0(t_0)}\right] = u_0(t_0), \tag{21}$$

$$A\left[x, t_0 + \frac{x}{u_0(t_0)}\right] = A_0 / \left[1 - x\frac{u_0'(t_0)}{u_0^2(t_0)}\right]. \tag{22}$$

Here t_0 is a parameter that may be interpreted as the time at which a signal leaves the valve ring in order to arrive at position x at time t given by

$$t = t_0 + \frac{x}{u_0(t_0)}. \tag{23}$$

Thus, we see that disturbances in fluid velocity and valve configuration propagate as a wave whose speed is the fluid velocity itself. It seems reasonable to call these "convection waves."

Here, as in the previous section, we see that $u_0 =$ constant implies $A = A_0$ and that deceleration tends to close the valve.

To study the mode of closure in more detail, consider the case where

$$u_0(t) = \begin{cases} c_0, & t \leq 0, \\ c_0\left(1 - \dfrac{t}{\tau}\right), & t \geq 0. \end{cases} \tag{24}$$

In this case (23) becomes a quadratic equation which we can solve explicitly for $t_0(x,t)$. This gives the following formula for the motion of the valve

$$A(x,t) = \begin{cases} A_0, & t < x/c_0, \\ A_0\left(\dfrac{1+\alpha}{2+\alpha}\right), & t > x/c_0, \end{cases} \tag{25}$$

where

$$\alpha = \frac{2}{\theta}\left(1 \pm (1 + \theta)^{1/2}\right), \tag{26}$$

$$\theta = 4\frac{x}{c_0\tau}\frac{1}{\left(1 - \dfrac{t}{\tau}\right)^2}. \tag{27}$$

In (26), the $+$ sign holds for $t < \tau$ and the $-$ sign holds for $t > \tau$.

Note that the valve suffers a kink that propagates along $x = c_0\tau$. In practice, the length of the valve is short compared with $c_0 t$, so this kink has small amplitude and it rapidly propagates off the end of the valve. After this, the contours of constant A in the (x,t) plane are the parabolas $\theta =$ const.

All of these parabolas meet at $x=0$, $t=\tau$, where the solution is highly singular. To study the behavior near this point, we first consider $x>0$, $t=\tau$. Then $\theta = \infty$, $\alpha=0$, and $A=A_0/2$ for all x. Thus the valve is uniformly half-closed at $t=\tau$. Next we evaluate

$$\lim_{x \to 0} A(x, t) = \begin{cases} A_0, & t < \tau, \\ 0, & t > \tau. \end{cases} \tag{28}$$

We conclude that the valve snaps shut near $x=0$ at the moment when the flow at the valve ring crosses zero.

Vortex Dynamics of the Aortic Sinus

The first theory of the aortic sinus vortex was that of Bellhouse & Talbot (1969) who assumed that the flow in each sinus is described by half of a Hill spherical vortex. A similar theory was applied by Bellhouse & Bellhouse (1971) to the mitral valve. This work was reviewed above, although we emphasized the more recent forms of the theory which, ironically, omit any explicit reference to the vortex.

Here we describe an alternative theory (Peskin & Wolfe 1978) based on the dynamics of a point vortex in an otherwise irrotational flow. This approach is restricted to two dimensions, but it has the advantage that the motion of the vortex can be studied and the sense in which the aortic sinus "traps" a vortex can be defined in terms of the stability of the equilibrium position of the vortex.

Another advantage is that the approach can be generalized so that a large population of point vortices is considered. This leads to a numerical method (Chorin 1973) which is also applied to the aortic sinus problem in Peskin & Wolfe (1978).

The point vortex theory of the aortic sinus is based on a conformal mapping from the domain occupied by fluid in the z-plane to the upper half ζ-plane. The fluid domain is constructed in the z-plane as the union of the upper half-plane and an overlapping circular disc whose boundary forms an arc from $z = -1$ to $z = +1$. In this simple model, the upper half-plane represents the aorta and the bulge formed by the disc corresponds to an aortic sinus. The sinus depth is controlled by a parameter α such that π/α gives the interior angle at $z = \pm 1$ where the boundary of the sinus meets the wall of the aorta. Thus $\alpha = 2/3$ generates a semicircular sinus, and with $\alpha = 1$ the sinus is obliterated.

The conformal mapping that takes this domain onto the upper half-plane with ± 1 and ∞ as fixed points is

$$\zeta = f(z) = B\{A[B(z)]\}, \tag{29}$$

where

$$B(z) = \frac{z + 1}{z - 1}, \tag{30}$$

$$A(z) = z^\alpha. \tag{31}$$

Using this conformal mapping we can write down the complex velocity potential corresponding to a point vortex of strength K at z_0 in an otherwise irrotational flow of strength U on our fluid domain. The formula is

$$\Phi(z) = U\zeta + \frac{K}{2\pi i} \log(\zeta - \zeta_0) - \frac{K}{2\pi i} (\zeta - \bar{\zeta}_0), \tag{32}$$

where $\zeta = f(z)$ and $\zeta_0 = f(z_0)$. Note that Φ can be written

$$\Phi(z) = \Phi_0(z) + \frac{K}{2\pi i} \log(z - z_0), \tag{33}$$

where

$$\Phi_0(z) = U\zeta + \frac{K}{2\pi i} \log \frac{\zeta - \zeta_0}{z - z_0} - \frac{K}{2\pi i} \log(\zeta - \bar{\zeta}_0). \tag{34}$$

The decomposition (33) is useful because Φ_0 is analytic at z_0 and $\Phi - \Phi_0$ corresponds to a symmetrical flow around $z = z_0$. Thus the motion of the vortex is generated by Φ_0 alone according to

$$\frac{dz_0}{dt} = \overline{\frac{d\Phi_0}{dz}(z_0)}. \tag{35}$$

The results of this analysis are as follows. First, the vortex has an equilibrium position along the midline of the aortic sinus, $\text{Re}(z) = 0$. The equilibrium position is uniquely determined if we impose the condition that there is a streamline that separates from the upstream border of the sinus ($z = -1$) and reattaches at the downstream border ($z = +1$). (This constraint plays a role analogous to that of the Joukowski condition in the theory of airfoils.) In this situation, the equilibrium vortex strength is proportional to the free-stream velocity, but the equilibrium position is independent of the free-stream velocity.

Next, consider the vortex trajectories, which must be carefully distinguished from the streamlines of the fluid. In the neighborhood of the equilibrium point, the vortex trajectories form closed orbits. This is a stability result which may be described by saying that the vortex is "trapped" in the sinus. It is the strongest kind of stability that we could hope for in an inviscid problem where the equations are invariant under a change of sign of the time. The size of the region filled by closed-vortex trajectories is an increasing function of the sinus depth.

Finally, consider the changes that occur if the free stream is suddenly shut off ($U=0$). In that case, the closed fluid streamlines surrounding the vortex are no longer confined to the sinus, and they fill the entire fluid domain. In particular, they cross the position of an open aortic leaflet in such a way that the leaflet is driven towards its closed position. (Recall that streamlines cross moving immersed boundaries. In fact, such a boundary cannot move except tangentially unless streamlines cross it.)

Moreover, with the potential flow removed, the balance of the vortex equilibrium is upset, and the vortex starts to move upstream and out of the sinus, chasing the aortic leaflet towards closure! This upstream shift of the vortex has, in fact, been observed in the physical-model experiments of Bellhouse & Talbot (1969) and van Steenhoven & van Dongen (1979).

Remarks on the Mechanism of Heart-Valve Closure

How do heart valves initiate closure during forward flow so that they are nearly closed by the moment of flow reversal? This is a very old question: it goes back to Leonardo da Vinci (see O'Malley & Saunders 1952) and to Henderson & Johnson (1912). Briefly, Leonardo believed that the valve was closed during forward flow by the "revolving impetus" of a vortex that forms behind the valve leaflets. Henderson & Johnson developed an alternative theory. Their idea was that the jet of forward flow is broken during flow deceleration and that fluid rushing in from the sides to fill the gap in the jet closes the valve.

The theories that we reviewed above do not resolve this question, since some of them are founded on vortex dynamics and others do not appear to contain a vortex at all. Indeed, the history of the subject shows that vortex theories can appear in their next incarnation lacking a vortex but making very similar predictions. How can this be?

The answer is that there is a vortex hidden in all of the theories of valve closure where it is not explicit. To see this, recall that these theories assume a stagnant, constant-pressure region behind the valve leaflets. Now consider the contour shown in Figure 3. The circulation around this contour is clearly nonzero, so there is vorticity concentrated along the valve leaflets themselves.

Moreover, this vorticity is essential for deceleration to achieve heart-valve closure. To see this consider the alternative, which is a potential flow. Under potential flow, the valve would keep opening as long as the flow is turned on. In fact, for the potential flow to close the valve it would have to reverse its motion and undo what was accomplished during forward flow. The net stroke volume in this hypothetical situation would be zero. This follows from the reversible character of potential flow.

Fundamentally, a valve is an irreversible machine. As such, viscosity must somewhere play a fundamental role in its operation. The generation of vorticity at boundaries by the viscous forces is the irreversible process that is needed for efficient valve closure.

We now turn to a related aspect of heart-valve closure: the question of what role (if any) the chordae tendineae play in the closure of the mitral valve. The obvious role of the chordae is to support the mitral leaflets in their closed position during ventricular systole, but the idea that they are also under some tension during diastole and that they play a role in the normal closure of the mitral valve goes back to Rushmer et al. (1956). This concept was also used by Frater et al. (1965) in the design of a tissue valve constructed at surgery.

By now, there is impressive evidence from the physical models and theories we have reviewed that heart valves can close efficiently without anything analogous to chordal restraint. Despite this, the numerical experiments of McQueen et al. (1981) reviewed below show definite abnormalities leading to an ineffective closure movement of the mitral leaflets when diastolic tension is omitted from the model. Briefly, the leaflets open more widely than usual during rapid filling and again during atrial systole, the vortex flow pattern is not reinforced during atrial systole, and the leaflets are wide open at the onset of ventricular systole.

These results have sometimes been regarded as an artifact of the model, but they have recently received striking experimental support from a fortuitous experiment of Yellin et al. (1981) in which a papillary muscle that had been sutured to the ventricular wall suddenly broke loose while mitral flow and valve motion were being monitored. The principal finding was increased excursion of the anterior leaflet as predicted by the computer model for the situation where diastolic tension on the chordae has been deleted.

This experiment suggests that the chordae of the mitral valve are under tension during diastole. How important that tension is to the normal function of the mitral valve remains to be seen.

A COMPUTATIONAL METHOD FOR THE HEART-VALVE PROBLEM

The work described in this section has been carried out by the author and his colleagues with the idea of developing a practical method that can be used in a variety of computational experiments on natural and artificial

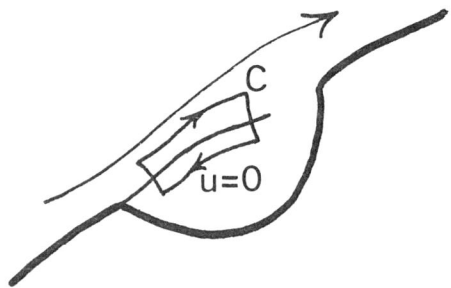

Figure 3 The hidden vortex in deceleration theories of heart-valve closure. Although the fluid in the aortic sinus is regarded as stagnant, the circulation around the contour C is certainly nonzero.

valves. The key papers that describe the current state of the method are Peskin (1977), Peskin & McQueen (1980), and McQueen et al. (1981). Further background can be found in Peskin (1972a,c, 1973, 1975, 1980, 1981).

Equations of Motion

In this section we state the equations of motion of a viscous incompressible fluid that contains a massless elastic boundary in its interior. The fluid is blood, and the boundary is a heart-valve leaflet. With minor modification, the same equations can be used to describe an artificial valve and also the muscular heart wall.

The fluid is described in Eulerian form by the velocity $\mathbf{u}(\mathbf{x},t)$ and pressure $p(\mathbf{x},t)$. The boundary is described in Lagrangian form by giving its configuration $\mathbf{X}(s,t)$. Here s stands for a pair of parameters (s_1,s_2), and fixed s marks a material point.

The coupled equations of the fluid and the immersed boundary may be written

$$\rho\left(\frac{\partial \mathbf{u}}{\partial t} + \mathbf{u}\cdot\nabla\mathbf{u}\right) = -\nabla p + \mu\Delta\mathbf{u} + \mathbf{F}, \qquad (36)$$

$$\nabla\cdot\mathbf{u} = 0, \qquad (37)$$

$$\mathbf{F}(\mathbf{x},t) = \int_B \mathbf{f}(s,t)\,\delta[\mathbf{x}-\mathbf{X}(s,t)]\,ds, \qquad (38)$$

$$\frac{\partial \mathbf{X}}{\partial t}(s,t) = \mathbf{u}[\mathbf{X}(s,t),t] = \int_\Omega \mathbf{u}(\mathbf{x},t)\,\delta[\mathbf{x}-\mathbf{X}(s,t)]\,d\mathbf{x}, \qquad (39)$$

$$\mathbf{f}(\ ,t) = S[\mathbf{X}(\ ,t)]. \qquad (40)$$

Equations (36–40) are the Navier-Stokes equations of a viscous incompressible fluid with external force per unit volume $\mathbf{F}(\mathbf{x},t)$. In our case, $\mathbf{F}$ is a distribution (δ-function layer) that represents the force applied by the immersed boundary to the fluid. The specific form of $\mathbf{F}$ is given by Equation (38) in which $\mathbf{f}(s,t)$ is the density of the boundary force with respect to the measure $ds = ds_1 ds_2$, $\delta(\mathbf{x}) = \delta(x)\delta(y)\delta(z)$ is the three-dimensional δ-function, and the integral extends over the immersed boundary B.

Equation (39) states that the immersed boundary moves at the local fluid velocity. The motion of the boundary is not given in advance, so this is an equation of motion for the boundary rather than a constraint on the motion of the fluid. Since the fluid is viscous, the velocity is continuous across the immersed boundary, and there is no ambiguity in the definition of $\mathbf{u}[\mathbf{X}(s,t),t]$. The second equality in (39) is just the definition of the three-dimensional δ-function. The integral extends over the entire domain Ω occupied by the fluid and the immersed boundary, and $d\mathbf{x}$ stands for the volume element $dx\,dy\,dz$.

This integral formula for the local fluid velocity is written out to emphasize the symmetry with (38) in which the δ-function also appears as the kernel of a local interaction. Note that (38) is not merely the definition of the δ-function, however. Since the integral in (38) extends over a two-dimensional domain, it does not completely remove the singularity in the three-dimensional δ-function, and $F(x,t)$ has the character of a δ function layer.

Finally (40) states that the boundary force at time t is determined by the boundary configuration at time t. This follows from the assumption that the boundary is elastic and massless. The details depend, of course, on the elastic properties of the boundary.

As remarked above, these equations can also be used for the rigid occluders of artificial valves and for the muscular heart wall. The rigid occluder is treated as a stiff elastic structure, and its shape is maintained by appropriate restoring forces. If the occluder is thin, it is treated as infinitely thin and massless. If it is thick, it is modeled as a fluid-filled elastic shell so that the model occluder is neutrally buoyant.

In the case of the muscular heart wall, Equation (40) is modified to take into account the time-dependent properties of heart muscle. The thickness of the wall can also be included in the formulation by generalizing the idea of an immersed boundary so that it includes thick, neutrally buoyant, incompressible regions where extra forces are applied to the fluid.

Numerical Method

The numerical solution of Equations (36–40) is accomplished as follows (Peskin 1977, Peskin & McQueen 1980). Up to now the method has only been implemented in two dimensions, and we describe it accordingly.

The fluid velocity and pressure are defined on a fixed, square computational mesh, and the configuration of the immersed, elastic boundary is specified in terms of the coordinates of a collection of moving points. Links joining specified pairs of boundary points are used for the computation of the boundary forces. These links are elastic in the heart-valve leaflets, but they have active contractile properties in the muscular heart walls. For artificial valves, three-point forces (Peskin & McQueen 1980) are also used to provide bending rigidity and to enforce the constraints on the motion of the valve.

Each time step of the numerical method proceeds as follows. First, the boundary forces are computed from the boundary configuration. The method is only stable when the forces are computed implicitly, by solving a nonlinear system of equations for an estimate of the forces at the end of the time step.

Next, the boundary forces are applied to the nearby mesh points of the fluid. This is accomplished by means of a discretization of Equation (38) in which a carefully constructed approximation to the δ-function is used.

With a boundary force-field defined on the fluid mesh, the fluid velocity is updated for one time step under the influence of boundary and fluid forces. The subroutine used for this purpose is the finite-difference method of Chorin (1968, 1969). Note that this subroutine does not see the immersed boundary at all except in terms of a force-field defined on the fluid mesh.

Finally, the new fluid velocity is interpolated to the boundary points, which are moved at the local fluid velocity to complete the time step. The interpolation is based on the integral form of (39) and it uses the same approximation to the δ-function that was used in applying the boundary forces to the fluid mesh.

Applications

For a detailed account of the work summarized in this section, see McQueen et al. (1981). Some representative results are shown in Figures 4–5.

Within the framework of the method outlined above, McQueen and Peskin have developed a computer model of the left heart for computational experiments on the mitral valve. The model has contractile walls with the physiological properties of heart muscle, and its parameters have been adjusted so that the model produces physiological pressures and velocities as judged by direct comparison with Yellin's experimental data on the canine mitral valve.

At this writing, the model has been used in computer experiments on the diastolic role of the chordae tendineae, the etiology of mitral valve prolapse syndrome, and the fluid dynamics of straight and curved pivoting disc valves in the mitral position.

The computer study of pivoting disc valves is described here in some detail because it illustrates the practical application of this work to prosthetic valve design. This study was aimed at answering a series of questions: Can fluid-mechanical forces limit the angle of opening of pivoting disc valves? If so, how does the angle of opening depend on the position of the pivot point? Is it changed by curvature of the valve? Is there an optimal pivot point for flat pivoting disc valves and an optimal combination of curvature and pivot point for curved valves?

These questions were attacked by constructing a computer model of a pivoting disc valve with no built-in mechanical constraint on its angle of opening. Computer experiments on this valve revealed that the angle of opening is indeed limited by fluid-mechanical forces, that this angle is always less than 90°, and that the fluid-mechanical limitation on the opening angle gets more severe as the pivot point is moved closer to the center of the valve. These results are consistent with the torque measurements of Köhler (1975).

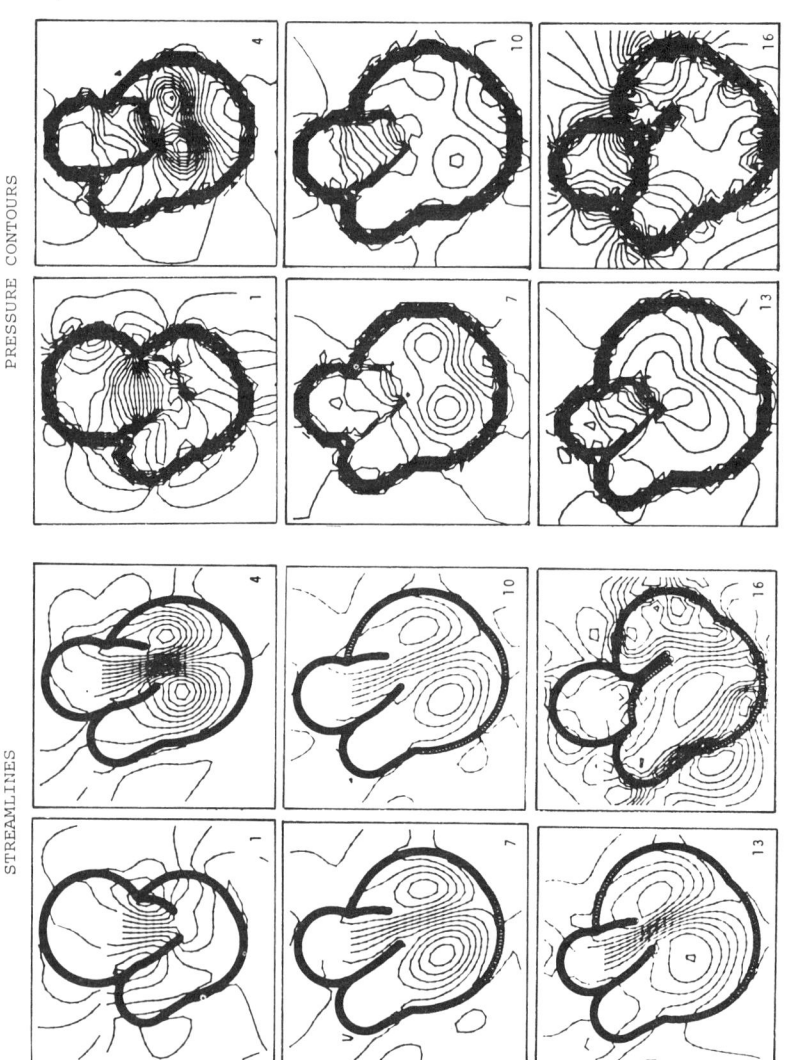

Figure 4 Computed flow patterns and pressure contours of the natural mitral valve during ventricular diastole and early ventricular systole. Pressure difference between adjacent contours is 0.1 mm Hg. The heart walls and valve leaflets are not drawn in the pressure contour plots; when they appear it is because of the steep pressure gradients across these internal boundaries (McQueen et al. 1981).

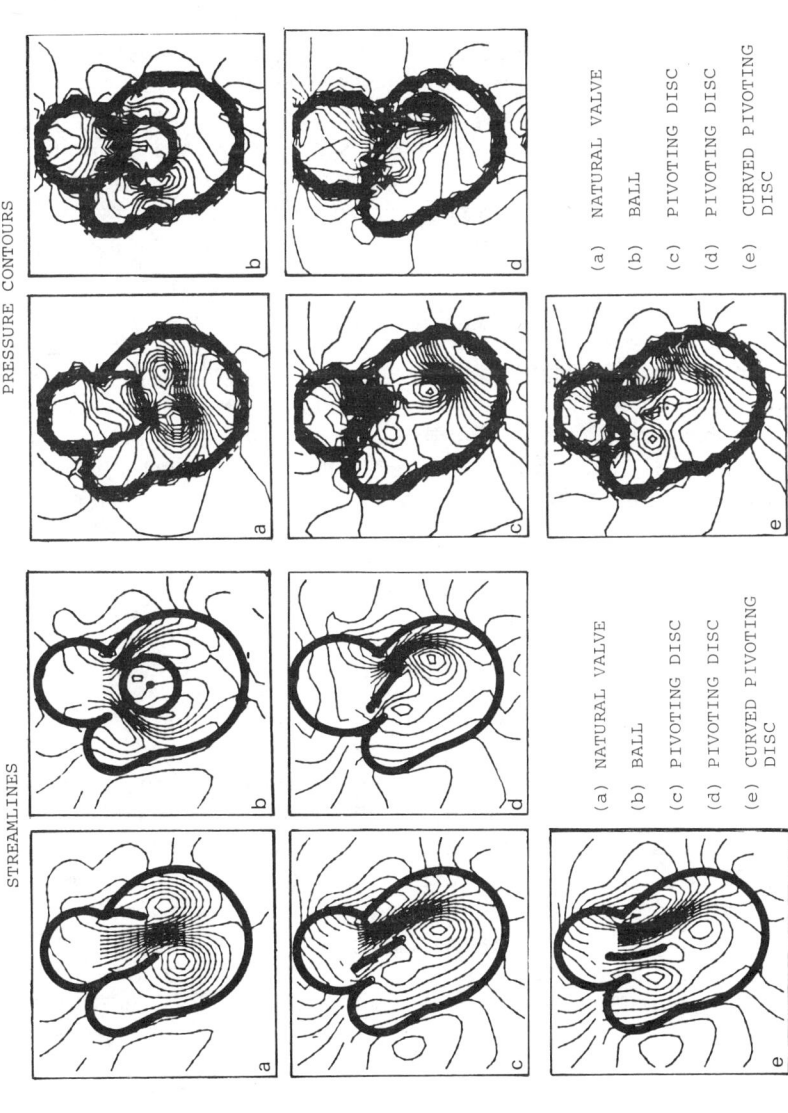

Figure 5 Computed flow patterns and pressure contours near the time of peak flow for the natural mitral valve and several prosthetic valves tested under identical conditions. The valve in (*d*) is pivoted closer to the center of the disc than in (*c*), and it opens less. The curved disc (*e*) is pivoted at the same point as (*c*), but it opens fully to 90° because of the curvature (McQueen et al. 1981).

Next, a computer model of a curved pivoting disc was tested. Small amounts of curvature turned out to have a substantial effect on valve motion, and, at a given pivot point, it was possible to adjust the curvature to achieve full opening to an angle of 90°.

These studies are continuing, and current work involves the optimal choice of curvature and pivot point, where the design criterion is based on the fluid velocity in the smaller opening of the valve. This criterion was motivated by the observations of Yoganathan et al. (1978) on stagnation and thrombus formation on the side of the disc facing the smaller opening.

This study illustrates the use of the computational method in the choice of optimal parameters for a prosthetic valve design. Another major use of the method should be to explore the sensitivity of valve performance to changes in the physiological state of the heart such as exercise.

Limitations

The current limitations on the work that we have described are that the model is two-dimensional and that the computations are performed at a Reynolds number which is 1/25 of the Reynolds number appropriate for the canine heart (where heart-valve experiments are often done.) Neither of these limitations is inherent in the method; they both result from practical limits of available computer time.

How serious these limitations are should be judged from the results of the computational experiments. In this respect, the agreement reported in McQueen et al. (1981) between the computational and animal experiments is very encouraging.

Moreover, the computational results contain internal evidence that the Reynolds number is not so low that viscous forces dominate the computation. If, for example, we follow a streamline from the atrium to the ventricle in the computational results, we find that the pressure falls as the fluid accelerates entering the valve and that it rises again as the fluid decelerates in the ventricle. Pressure recovery, a purely inertial effect, is about 75% complete. There are also strong pressure minima produced by centrifugal force, again an inertial effect, at the centers of the vortices that are shed from the tips of the leaflets.

Despite these arguments it would certainly be desirable to perform heart-valve computations at physiological Reynolds numbers and also in three dimensions. Progress in raising the Reynolds number is described in Mendez (1977), Peskin & Wolfe (1978), and McCracken & Peskin (1980). These papers are based on the vortex method (Chorin 1973). Three-dimensional heart-valve computations may have to await the development of large-scale parallel computers. Such computers are not far away (Schwartz 1980, Gottlieb & Schwartz 1981).

FUTURE DIRECTIONS FOR HEART-VALVE RESEARCH

We conclude this review with three major research challenges:
(a) Develop a quantitative experimental method for determining the flow pattern of blood in the heart, in vivo.
(b) Develop a theory of the fluid dynamics of blood clotting that can be used to predict the thrombogenic potential of an artificial heart valve.
(c) Develop a practical computational method for the three-dimensional heart-valve problem at physiological Reynolds numbers.

ACKNOWLEDGMENTS

The author's work on heart valves is supported by the National Institutes of Health under research grant HL17859. Marion Peskin was extremely helpful with library research related to this review.

Literature Cited

Bellhouse, B. J. 1972. Fluid mechanics of a model mitral valve and left ventricle. *Cardiovasc. Res.* 6:199–210

Bellhouse, B. J., Bellhouse, F. H. 1971. The fluid mechanics of a model mitral valve and left ventricle. *Eng. Sci. Rep. 1031/71.* Univ. Oxford

Bellhouse, B. J., Talbot, L. 1969. The fluid mechanics of the aortic valve. *J. Fluid Mech.* 35:721–35

Björk, V. O., Henze, A. 1979. Prosthetic heart valve replacement. Nine years' experience with the Björk-Shiley tilting disc valve. See Ionescu 1979, pp. 1–28

Bonchek, L. I. 1980. Indications for surgery of the mitral valve. *Am. J. Cardiol.* 46:155–58

Brewer, L. A. III, ed. 1969. *Prosthetic Heart Valves.* Springfield, Ill: Charles C. Thomas. 909 pp.

Brun, P., Oddou, C., Dantan, P., Laporte, J. P., Laurent, F., Perrot, P. 1980. Blood flow dynamics during the human left ventricular filling phase. In *Cardiac Dynamics,* ed. J. Baan, A. C. Arntzenius, E. L. Yellin, 3.3:169–81. The Hague/Boston/London: Martinus Nijhoff. 549 pp.

Chorin, A. J. 1968. Numerical solution of the Navier-Stokes equations. *Math. Comp.* 22:745–62

Chorin, A. J. 1969. On the convergence of discrete approximations to the Navier-Stokes equations. *Math. Comp.* 23:341–53

Chorin, A. J. 1973. Numerical study of slightly viscous flow. *J. Fluid Mech.* 57:785–96

Dellsperger, K. C., Wieting, D. W. 1978. An in vitro fluid dynamic comparison of the new St. Jude Medical prosthetic mitral valve with Starr-Edwards, Björk-Shiley, and Lillehei-Kaster prostheses. In *Advances in Bioeng. Symp. ASME,* pp. 31–33

Fowler, N. O., van der Bel-Kahn, J. M. 1980. Operations on the mitral valve: A time for weighing the issues. *Am. J. Cardiol.* 46:159–62

Frater, R. W. M., Berghuis, J., Brown, A. L. Jr., Ellis, F. H. Jr. 1965. The experimental and clinical use of autogenous pericardium for the replacement and extension of mitral and tricuspid valve cusps and chordae. *J. Cardiovasc. Surg.* 6:214–28

Frater, R. W. M., Wexler, H. R., Yellin, E. 1969. The in vivo comparison of hemodynamic function of ball, disc, and eccentric monocusp artificial mitral valves. See Brewer 1969, pp. 262–77

Frater, R. W. M., Gabbay, S., McQueen, D. M., Becker, R. M., Borg, M., Strom, J., Lin, T. Y., Oka, Y., Yellin, E. L. 1979. Bioprostheses: Hemodynamic performance, thromboembolic incidence, and symptomatic benefits. See Yoganathan et al. 1979c, pp. 341–55

Gabbay, S., McQueen, D. M., Yellin, E. L., Becker, R. M., Frater, R. W. M. 1978. In vitro hydrodynamic comparison of mitral valve prostheses at high flow rates. *J. Thorac. Cardiovasc. Surg.* 76:771–87

Gabbay, S., McQueen, D. M., Yellin, E. L., Frater, R. W. M. 1979. In vitro hydrodynamic comparison of mitral valve bioprostheses. *Circulation* 60: Suppl. 1, pp. I62-I70

Gottlieb, A., Schwartz, J. T. 1981. Networks and algorithms for very large scale parallel computation. *Comp. Sci. Tech. Rep. No. 030.* New York Univ.

Henderson, Y., Johnson, F. E. 1912. Two modes of closure of the heart valves. *Heart* 4:69–82

Ionescu, M. I., ed. 1979. *Tissue Heart Valves.* Boston, Mass: Butterworths. 373 pp.

Kalmanson, D., ed. 1976. *The Mitral Valve.* Acton, Mass: Publishing Sciences Group.

Köhler, J. 1975. Opening angle and torque of the Björk-Shiley and Lillehei-Kaster heart valve prostheses. *Proc. Eur. Soc. Art. Organs,* Vol. II, pp. 33–35. W. Berlin: Westkreuz

Laniado, S., Yellin, E. L., Miller, H., Frater, R. W. M. 1973. Temporal relation of the first heart sound to closure of the mitral valve. *Circulation* 47:1006–14

Laniado, S., Yellin, E., Kotler, M., Levy, L., Stadler, J., Terdiman, R. 1975. A study of the dynamic relations between the mitral valve echogram and phasic mitral flow. *Circulation* 51:104–13

Lee, C. S. F., Talbot, L. 1979. A fluid-mechanical study of the closure of heart valves. *J. Fluid Mech.* 91:41–63

McCracken, M. F., Peskin, C. S. 1980. A vortex method for blood flow through heart valves. *J. Comput. Phys.* 35:183–205

McQueen, D. M., Peskin, C. S., Yellin, E. L. 1981. Fluid dynamics of the mitral valve: Physiological aspects of a mathematical model. *Am. J. Physiol.* In press

Mendez, R. H. 1977. *Numerical study of incompressible flow in a region bounded by elastic walls.* PhD thesis. Univ. Calif., Berkeley (Univ. Microfilms #7812684). 121 pp.

Netter, F. H. 1969. *CIBA Collection of Medical Illustrations. 5. Heart* pp. 8–12. Summit, NJ: Ciba. 295 pp.

O'Malley, C. D., Saunders, J. B. de C. M. 1952. *Leonardo da Vinci on the Human Body,* pp. 258–75. New York: Henry Schuman. 506 pp.

Peskin, C. S. 1972a. *Flow patterns around heart valves: A digital computer method for solving the equations of motion.* PhD thesis. Albert Einstein College Med., Yeshiva Univ., Bronx, NY (Univ. Microfilms #72-30, 378). 211 pp.

Peskin, C. S. 1972b. Three-dimensional generalization of Wieting's apparatus. See Peskin 1972a, pp. 25–27

Peskin, C. S. 1972c. Flow patterns around heart valves: A numerical method. *J. Comput. Phys.* 10:252–71

Peskin, C. S. 1973. Flow patterns around heart valves. *Lecture Notes in Physics* 19:214–21

Peskin, C. S. 1975. *Mathematical Aspects of Heart Physiology.* New York: Courant Institute. 278 pp.

Peskin, C. S. 1977. Numerical analysis of blood flow in the heart. *J. Comput. Phys.* 25:220–52

Peskin, C. S. 1980. Fluid dynamics of the heart and the ear. In *Computing Methods in Applied Sciences and Engineering,* ed. R. Glowinski, J. L. Lions, pp. 587–613. Amsterdam/New York/Oxford: North Holland. 724 pp.

Peskin, C. S. 1981. Lectures on mathematical aspects of physiology. In *Mathematical Aspects of Physiology, Lectures in Applied Mathematics,* Vol. 19, pp. 1–107, ed. F. Hoppensteadt. Providence, R.I.: Am. Math. Soc. 394 pp.

Peskin, C. S., McQueen, D. M. 1980. Modeling prosthetic heart valves for numerical analysis of blood flow in the heart. *J. Comput. Phys.* 37:113–32

Peskin, C. S., Wolfe, A. W. 1978. The aortic sinus vortex. *Fed. Proc.* 37:2784–92

Roberts, W. C. 1978. Prosthetic heart valves: Which ones and why. *Hosp. Prac.* 13(1):63–69

Rushmer, R. F., Finlayson, B. L., Nash, A. A. 1956. Movements of the mitral valve. *Circ. Res.* 4:337–42

Schwartz, J. T. 1980. Ultracomputers. *ACM Trans. Prog. Lang. Syst.* 2:484–521

Taylor, D. E. M., Wade, J. D. 1973. Pattern of blood flow within the heart: A stable system. *Cardiovasc. Res.* 7:14–21

Tsakiris, A. G., von Bernuth, G., Rastelli, G. C., Bourgeois, M. J., Titus, J. L., Wood, E. H. 1971. Size and motion of the mitral valve annulus in anesthetized intact dogs. *J. Appl. Physiol.* 30:611–18

Tsakiris, A. G., Gordon, D. A., Mathieu, Y., Lipton, I. 1975. Motion of both mitral valve leaflets: A cineroentgenographic study in intact dogs. *J. Appl. Physiol.* 39:359–66

Tsakiris, A. G., Padiyar, R., Gordon, D. A., Lipton, I. 1977. Left atrial size and geometry in the intact dog. *Am. J. Physiol.* 232:H167–72

van Steenhoven, A. A., van Dongen, M. E. H. 1979. Model studies of the closing behaviour of the aortic valve. *J. Fluid Mech.* 90:21–32

Wieting, D. W. 1969. *Dynamic flow characteristics of heart valves.* PhD thesis. Univ. Texas, Austin (Univ. Microfilms #69–21, 904). 308 pp.

Wright, J. T. M. 1979. Hydrodynamic evaluation of tissue valves. See Ionescu 1979, pp. 29–87

Wright, J. T. M., Temple, L. J. 1971. An improved method for determining the flow characteristics of prosthetic mitral heart valves. *Thorax* 26:81–88

Wright, J. T. M., Temple, L. J. 1972. Relationship between the physical size, incompetence, and stenosis of prosthetic mitral valves. *Thorax* 27:287–303

Yellin, E. L., Peskin, C. S. 1975. Large amplitude pulsatile water flow across an orifice. *Trans. ASME J. Dyn. Syst. Meas. Contr.* 97G:92–95

Yellin, E. L., Yoran, C., Sonnenblick, E. H., Gabbay, S., Frater, R. W. M. 1979. Dynamic changes in the canine mitral regurgitant orifice area during ventricular ejection. *Circ. Res.* 45:677–83

Yellin, E. L., McQueen, D., Gabbay, S., Strom, J. A., Becker, R. M., Frater, R. W. M. 1980. Pressure-flow relations and energy losses across prosthetic mitral valves: In vivo and in vitro studies. In *Cardiac Dynamics,* ed. J. Baan, A. C. Arntzenius, E. L. Yellin, 6.6:509–19. The Hague/Boston/London: Martinus Nijhoff. 549 pp.

Yellin, E. L., Peskin, C., Yoran, C., Koenigsberg, M., Matsumoto, M., Laniado, S., McQueen, D., Shore, D., Frater, R. W. M. 1981. Mechanisms of mitral valve motion during diastole. *Am. J. Physiol.* In press

Yoganathan, A. P., Corcoran, W. H., Harrison, E. C., Carl, J. R. 1978. The Björk-Shiley aortic prosthesis: Flow characteristics, thrombus formation, and tissue overgrowth. *Circulation* 58:70–76

Yoganathan, A. P., Corcoran, W. H., Harrison, E. C. 1979a. In vitro velocity measurements in the vicinity of aortic prostheses. *J. Biomech.* 12:135–52

Yoganathan, A. P., Corcoran, W. H., Harrison, E. C. 1979b. Pressure drops across prosthetic aortic heart valves under steady and pulsatile flow—in vitro measurements. *J. Biomech.* 12:153–64

Yoganathan, A. P., Harrison, E. C., Corcoran, W. H., eds. 1979c. *Prosthetic Heart Valves: Proc. Symp. AAMI, 14th, Las Vegas, Nev.* 460 pp.

Yoganathan, A. P., Reamer, H. H., Harrison, E. C., Corcoran, W. H. 1979d. Laser-Doppler anemometer to study velocity fields in the vicinity of prosthetic heart valves. *Med. Biol. Eng. Comput.* 17:38–44

Yoganathan, A. P., Reamer, H. H., Corcoran, W. H., Harrison, E. C. 1980. The Björk-Shiley aortic prosthesis: Flow characteristics of the present model vs. the convexo-concave model. *Scand. J. Thor. Cardiovasc. Surg.* 14:1–5

THE COMPUTATION OF TRANSONIC POTENTIAL FLOWS

David A. Caughey

Sibley School of Mechanical and Aerospace Engineering, Cornell University, Ithaca, New York 14853

1. INTRODUCTION

In recent years, the ability to predict transonic flow fields past complete, aircraft-like configurations and components has played an increasingly important role in the design of aircraft. The widespread use of computer programs designed to solve discrete approximations to the equations of mixed subsonic/supersonic flows has been a result primarily of the development of efficient and geometrically general numerical algorithms to solve the equations of potential flow. The widespread use of these methods by the aircraft industry is a testament to the usefulness of the potential-flow model in many design problems. In this article, I review some of the ideas leading to these methods, including the formulation of finite-difference approximations to the potential equation, the treatment of the problem of geometrical complexity, and the development of efficient iterative schemes for solving the difference equations. Finally, I review some of the experience obtained using these methods, and describe some approaches to their use in design problems. I concentrate on methods based upon finite-difference approximations to the potential equation, since these have been more successful for transonic flow problems than the related finite-element methods; a discussion of the latter has been presented in this series by Shen (1977).

The scope of this article is limited to methods appropriate for aerodynamic problems that are steady in time, and for which solutions of the potential equation provide good approximations to the flow in most of the domain of interest. Viscous stresses will be considered to be important only in thin boundary layers, and it will be assumed that their effects can be incorporated within the framework of relatively classical boundary-layer corrections. Unsteady transonic flows have been the subject of a recent article in this series by Tijdeman & Seebass (1980), and an important class of problems relating to the inclusion of viscous effects—that of shock-wave/boundary-layer interactions—has been discussed by Adamson & Messiter (1980). This article might be considered as a report on the status of methods that were in their early stages of development when Nieuwland & Spee (1973) described the status of transonic airfoil theory and experiment nearly ten years ago.

2. FORMULATION OF POTENTIAL PROBLEMS

The geometrical complexity of problems solvable with a reasonable expenditure of computer resources varies inversely with the degree of fluid-mechanical complexity required to model the important features of the flow field. Thus, while it is practical at one extreme to determine the incompressible potential flow past a reasonably complete aircraft configuration, solutions of the complete (even time-averaged) Navier-Stokes equations for the high Reynolds numbers of aerodynamic interest have been presented, with few exceptions, only for rather simple geometries, such as the interaction of a straight, oblique shock with the boundary layer on a flat plate. This is a reflection of both the increased requirements on computer storage imposed by the necessity to store several independent variables at each point in a grid of sufficient density to resolve the finest features of the flow field, and the greater computing time required per mesh point to achieve the solution of the more complicated equations.

The ultimate goal of computational aerodynamicists is to solve the Navier-Stokes equations for a complete aircraft configuration, consisting of wing, fuselage, tail, nacelles, etc. Although those equations describe turbulent fluctuations of all length scales required for practical aerodynamic calculations, solutions in such detail are beyond the capabilities of present (and probably next-generation) computers, and it is necessary to model at least some of the physically important turbulent fluctuations. Even so, the solution of the resulting (time-averaged) Navier-Stokes equations presents a formidable problem, especially when the Reynolds number is high so that the length scales relevant to the parabolic (viscous) terms and to the convective terms become widely disparate. Physically, this stiffness of the equations can be identified with a tendency for the flow to behave almost inviscidly throughout the flow field, with the exception of thin boundary layers within which the viscous stresses are important.

This inviscid flow is described by the Euler equations, obtained by neglecting the viscous terms altogether. If the flow can be approximated further as being isentropic, the Euler equations can be simplified further to a single equation in a single scalar unknown. For the case of inviscid compressible flows of constant energy, this requirement is met if no strong shocks appear in the flow field. Under these conditions the flow is irrotational and a velocity potential ϕ can be introduced such that the velocity vector **q**

$$\mathbf{q} = \nabla \phi. \tag{2.1}$$

If x, y are chosen to be Cartesian coordinates, the velocity potential satisfies the equation

$$(\rho\phi_x)_x + (\rho\phi_y)_y = 0, \tag{2.2}$$

where

$$\rho = [1 + \frac{k-1}{2} M_\infty^2 (1 - q^2)]^{\frac{1}{k-1}} \tag{2.3}$$

is a function only of the magnitude of the gradient of ϕ. In the last equation, k is the ratio of specific heats of the fluid, and M_∞ is the Mach number of the free stream. Note that here, and throughout the paper, equations are generally written for two-dimensional flows for the sake of brevity and clarity. Extensions to three-dimensional flows are straightforward unless otherwise noted.

Equation (2.2) is in divergence, or conservation, form. It is important to use this form of the equation when shock waves are present, since this ensures that solutions of the difference equations will approximate the proper weak solutions of the equations. These will contain discontinuities satisfying jump relations that are consistent with the conservation laws from which the equations were derived. Discontinuities appearing in weak solutions of Equation (2.2) are mass-conserving and isentropic, and do not correspond exactly to the Rankine-Hugoniot shocks which the weak form of the Euler Equations allows. These isentropic discontinuities closely approximate their Rankine-Hugoniot counterparts, however, if the shocks are sufficiently weak (Steger & Baldwin 1972).

3. ANALYSIS OF TRANSONIC FLOW FIELDS

If the dependence of the density upon ϕ is explicitly included using Equation (2.3), the potential equation can be written in the quasi-linear form

$$(a^2 - u^2)\phi_{xx} - 2uv\phi_{xy} + (a^2 - v^2)\phi_{yy} = 0. \tag{3.1}$$

If we further introduce a local Cartesian coordinate system (s, n) with the s-direction aligned with the local flow velocity, Equation (3.1) can be written

$$(a^2 - q^2)\phi_{ss} + a^2\phi_{nn} = 0. \tag{3.2}$$

This form of the equation more clearly illustrates how the type of the equation must change from elliptic to hyperbolic as the local Mach number $M = q/a$ changes from subsonic to supersonic.

A small-disturbance approximation to Equation (3.1) which maintains the essential nonlinearity required to describe transonic flows can be derived (von Kármán 1947). The earliest relaxation solutions for transonic potential flows were obtained using this small-disturbance equation and its associated linearized boundary conditions by Murman & Cole (1971), and developed for three-dimensional wings and wing-fuselage combinations by a number of investigators (see, for example, Ballhaus & Bailey 1972, Ballhaus 1978, and Schmidt 1978). The linearization of the body surface boundary condition allows treatment of rather complex wing-fuselage geometries using relatively simple, nearly Cartesian coordinate systems, but also introduces spurious singularities at blunt or rounded leading edges. If these were accurately resolved by the numerical scheme, they would result in inaccurate solutions (Keyfitz et al. 1979). Fortunately, for the meshes frequently used in practice, the truncation error nearly cancels the effect of the singularity, resulting in quite good agreement with experiment for airfoil pressure distributions at transonic Mach numbers. Agreement with experiment for swept wings apparently suffers when the component Mach number is not near unity, which is the case for many practical designs (Hinson & Burdges 1980). For these reasons, I concentrate on solutions of the complete potential equation, making no assumptions about the magnitude of the perturbations from the free-stream velocity.

Geometrical Considerations

The solution of the full potential equation is complicated by the need to treat the boundary conditions on curved body surfaces more exactly than when using the small-disturbance theory. This requires either that the computation be done in a boundary-conforming (or body-fitted) coordinate system, or that special interpolation formulas be used in the difference equations near boundaries. For several reasons, the use of a boundary-conforming coordinate system is probably preferable. Such coordinate systems have the advantages of (a) eliminating the need for cumbersome, and possibly destabilizing or inaccurate, interpolation formulas at the body surface—i.e., precisely where one is usually interested in an accurate solution and where flow gradients are likely to be the largest, and (b) allowing relatively easy and efficient clustering of mesh points in the vicinity of the body surface. That the difficulties associated with (a) can be overcome has been demonstrated by Carlson (1975) and Reyhner (1977), but the efficient clustering of mesh points remains a critical problem in these Cartesian mesh methods.

The generation of a suitable boundary-conforming coordinate system is often a major obstacle to solving problems for geometries of practical interest. Three approaches have received relatively wide use: 1. conformal mapping of the flow field to a simple domain, 2. numerical generation of transformations by solving systems of elliptic equations, and 3. a sequential approach in which complicated transformations are built up using relatively simple conformal and shearing transformations as building blocks. Of these three approaches, only the latter two are generally applicable to very complex geometries because of the difficulty of constructing conformal transformations for general shapes. A fourth approach, developed by Eiseman (1979), is attractive, but has not yet been developed to treat highly complex geometries. It is an algebraic technique, so it is relatively economical in terms of computer resources, and allows the continuation of coordinate systems defined in boundary surfaces into the interior of the domain with specified orthogonality properties, at least in the neighborhood of the boundaries.

For those geometries for which they can be found, conformal transformations to simple domains result in grids having many desirable properties. These include orthogonality and constant cell aspect ratio. The idea of performing a compressible airfoil calculation in a conformally mapped "circle-plane" was used first by Sells (1968) and subsequently by Garabedian & Korn (1972) and by Jameson (1971, 1974). (See also Bauer et al. 1972, 1975). Conformal transformations have also been used to determine the transonic flow past nacelles (Arlinger 1975), multi-element airfoils (Grossman & Melnik 1976), and two-dimensional cascades of airfoils (Ives & Luitermoza 1977).

The second method listed above has been developed largely by Thompson et al. (1974) and has been used to generate grids for transonic wing-body computations by Yu (1980b). The method is, in principle, very general. A set of Poisson equations for the computational variables, say ξ and η, is solved:

$$\xi_{xx} + \xi_{yy} = f(\xi, \eta), \qquad (3.3)$$

$$\eta_{xx} + \eta_{yy} = g(\xi, \eta),$$

where the source functions f and g are introduced to allow clustering of the mesh in desired regions. This formulation in the physical plane is used to ensure that the computational variables have no local extrema within the domain, i.e. to prevent self-intersections of the mesh. Since the mesh-generating equations must ultimately be solved numerically in the computational domain, however, this formulation has the disadvantage that the equations actually solved for the mesh coordinates are

$$Ax_{\xi\xi} - 2Bx_{\xi\eta} + Cx_{\eta\eta} = \tilde{f}(\xi, \eta), \qquad (3.4)$$

$$Ay_{\xi\xi} - 2By_{\xi\eta} + Cy_{\eta\eta} = \tilde{g}(\xi, \eta),$$

where

$$A = x_\eta^2 + y_\eta^2,$$
$$B = x_\xi x_\eta + y_\xi y_\eta, \qquad (3.5)$$
$$C = x_\xi^2 + y_\xi^2.$$

These are no longer Poisson equations and, in fact, are now nonlinear and coupled. The solution of these equations to determine the mesh distribution can pose a problem of the same order of difficulty as the fluid-flow problem to be solved. Holst (1979) used this method to generate a mesh for two-dimensional airfoil calculations and indicated that, in spite of his use of an efficient approximate-factorization method to solve Equations (3.4), the mesh-generation step required the same order of computation time as the transonic-flow calculation. Camarero & Younis (1980) have applied the multi-level adaptive grid technique (Brandt 1973) to accelerate the iterative convergence of the solution of Equations (3.4), but if this technique is also used to solve the flow problem, the fraction of time spent on the grid-generation step should remain approximately the same. Finally, there is no guarantee that the mesh will be particularly well behaved. A change in the manner in which the computational points are distributed along the boundaries can result in highly nonorthogonal meshes. On the other hand, a careful choice of the boundary conditions for Equations (3.4) can result in meshes that are nearly conformal when $f = g = 0$. In this case, one need not worry about self-intersections of the mesh, and it makes sense to solve the system

$$x_{\xi\xi} + x_{\eta\eta} = 0,$$
$$y_{\xi\xi} + y_{\eta\eta} = 0. \qquad (3.6)$$

This approach was used by Chen & Caughey (1979) to generate grids for inlet nacelles with centerbodies, and is applicable to other classes of geometries as well.

Although it is difficult to find conformal transformations that map very complicated geometries to simple domains, it is often relatively easy to find a conformal transformation that very nearly maps the geometry of interest to a simple domain (Caughey 1978). The resulting domain can then be regularized by the use of simple shearing transformations which are everywhere weak if the conformal mapping has been carefully chosen. It follows that the resulting transformation is nearly conformal. This idea forms the basis of the third technique, which has been used for the calculation of transonic flows past yawed wings (Jameson 1974), airfoils in wind tunnels (Caughey & Jameson 1977a), swept wings (Jameson & Caughey 1977a), and inlet nacelles (Caughey & Jameson 1977b). The method has also been used to generate grids for use with the finite-volume method (to be described shortly) for the calculation of transonic potential flows past inlets (Chen & Caughey 1980), wing-cylinder (Jameson & Caughey 1977b and Caughey & Jameson 1979a) and wing-fuselage combinations (Caughey & Jameson 1979a,b), and cascades and propellers (Dulikravich 1979 and Dulikravich & Caughey 1980).

As an example of this technique, we consider the transformation sequence used to generate a boundary-conforming coordinate grid for an arbitrary wing-fuselage combination by Caughey & Jameson (1979a,b). The method is capable of treating a swept, tapered wing of arbitrary planform, dihedral and section shape mounted upon a finite fuselage of varying cross-sectional area and shape. The grid is generated using nearly conformal transformations in each of a series of quasi-cylindrical coordinate surfaces wrapping around the fuselage and cutting through the wing at a number of spanwise stations. A perspective view of the wing and fuselage grids produced in this manner for a high-wing, fighter-type aircraft is shown in Figure 1.

Finite-Volume Formulation

The earliest solutions to the full potential equation (e.g. Garabedian & Korn 1972, Jameson 1971, 1974, and Jameson & Caughey 1977a) were

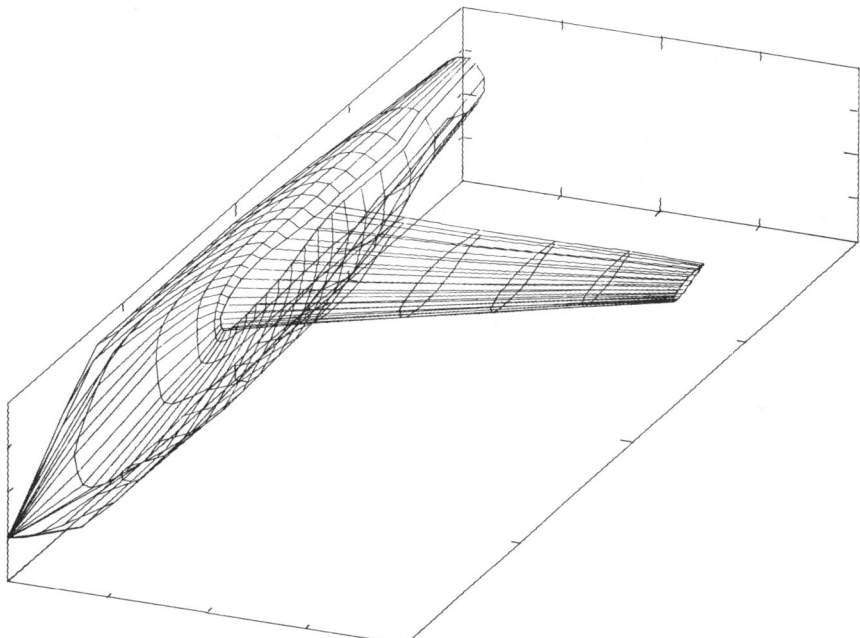

Figure 1 Finite-difference grids in wing and fuselage surfaces for high-wing, fighter-type aircraft.

obtained by transforming the equation analytically, and solving it in the transformed computational plane. This procedure is attractive because no accuracy is lost in the transformation process, but if the transformation becomes complicated, the computational labor of evaluating the coefficients of the transformed equation becomes great, and the resulting code is not particularly efficient. In addition, the algebraic labor of deriving the transformed equations becomes tedious, and the chance for error grows; this problem is compounded by the fact that the transformations must be rederived for each new class of geometries to be analyzed. For these reasons, it is attractive to have a transonic potential algorithm that is independent of the details of the coordinate transformation used to generate the finite-difference mesh. This is the concept behind the finite-volume techniques (Jameson & Caughey 1977b and Caughey & Jameson 1979a), which evaluate numerically the transformation metric in each mesh cell using the Cartesian coordinates of its corner points. The method is also appropriate when the mesh has been generated numerically, and the transformation derivatives cannot be evaluated analytically.

Consider a transformation to a new set of coordinates X, Y. Let the Jacobian matrix of the transformation be defined by

$$H = \begin{Bmatrix} x_X & x_Y \\ y_X & y_Y \end{Bmatrix}, \tag{3.7}$$

and let h denote the determinant of H. The metric tensor of the new coordinate system is given by the matrix $G = H^T H$, and the contravariant components of the velocity vector U, V are given by

$$\begin{Bmatrix} U \\ V \end{Bmatrix} = H^{-1} \begin{Bmatrix} u \\ v \end{Bmatrix} = G^{-1} \begin{Bmatrix} \phi_X \\ \phi_Y \end{Bmatrix}. \tag{3.8}$$

The potential equation, upon multiplication by h, can then be written

$$(\rho h U)_X + (\rho h V)_Y = 0. \tag{3.9}$$

The fully-conservative finite-volume approximation to Equation (3.9) is constructed by assuming separate bilinear variations of the independent and dependent variables within each mesh cell. If we introduce the averaging and differencing operators

$$\mu_X f_{i,j} = 1/2(f_{i+1/2, j} + f_{i-1/2, j}), \tag{3.10}$$

$$\delta_X f_{i,j} = (f_{i+1/2, j} - f_{i-1/2, j}),$$

where i, j are the mesh indices in the X, Y directions, then the transformation derivatives, evaluated at the cell centers, can be expressed by formulas such as

$$x_X = \mu_Y \delta_X x \tag{3.11}$$

$$x_Y = \mu_X \delta_Y x$$

with similar expressions for the derivatives of y and the potential. Such formulas can be used to determine ρ, h, U, and V at the center of each cell using Equations (2.3), (3.7), and (3.8). Equation (3.9) is represented by conserving fluxes across the boundaries of auxiliary cells whose faces are chosen to be midway between the faces of the primary mesh cells. This can be represented as

$$\mu_Y \delta_X(\rho h U) + \mu_X \delta_Y(\rho h V) = 0 . \tag{3.12}$$

This formula can also be obtained by applying the Bateman variational principle that the integral of the pressure

$$I = \int p \, dx \, dy \tag{3.13}$$

is stationary, and then approximating I by a simple one-point integration scheme in which the pressure at the center of each grid cell is multiplied by the cell area. In this way, the finite-volume method can equally well be regarded as a finite-element method with isoparametric bi-linear elements for subsonic flow. The use of the one-point integration scheme leading to Equation (3.12) has the advantage of requiring only one density evaluation per mesh point, but also has the undesirable effect of tending to decouple the solution at odd- and even-numbered points of the grid, and suitable recoupling terms can be added to improve the stability of the solution.

The scheme is stabilized in supersonic regions by the explicit addition of an artificial viscosity, chosen to emulate the directional bias introduced by the rotated difference scheme of Jameson (1974). In that scheme, the equation is considered to be written in the quasi-linear form of Equation (3.2), and the differences contributing to ϕ_{ss} are evaluated using upwind, or retarded, formulas. This can be considered equivalent to evaluating the equation entirely with central differences and adding a term approximately equal to $-\Delta s(a^2 - q^2)\phi_{sss}$ at points where the local flow velocity is supersonic. The upwinding of the differences in the ϕ_{ss} term is thus equivalent to adding an artificial longitudinal viscosity to the equation being solved. Since the added term is of the order of the mesh spacing, it is small wherever the solution is smooth, yet provides enough dissipation to allow shocks to be captured and expansion shocks eliminated.

The artificial viscosity is introduced by defining (for $U, V > 0$)

$$P = \rho h \sigma / a^2 (U^2 \delta_{XX} + UV \delta_{XY}) \phi , \tag{3.14}$$

$$Q = \rho h \sigma / a^2 (UV \delta_{XY} + V^2 \delta_{YY}) \phi ,$$

where the switching function

$$\sigma = \max[0 . , 1 - (M_c/M)^2] \tag{3.15}$$

is nonzero only for values of the local Mach number M greater than some critical Mach number M_c (usually taken near unity). Then an upwind form of $\delta_X P + \delta_Y Q$ is added to Equation (3.12).

It is of interest to note that this artificial viscosity can equally well be introduced by defining a fictitious density.
Since

$$\partial \rho / \partial q^2 = -\rho/2a^2 ,\qquad(3.16)$$

and from Equations (3.8) we have

$$q^2 = (U\ V)\ G \begin{bmatrix} U \\ V \end{bmatrix},$$

it follows that

$$\rho_X = -\rho/a^2\ (U\phi_{XX} + V\phi_{XY})\ .$$

Thus the leading terms of $-\partial/\partial X(\sigma h U \delta_X \rho)$ are equivalent to $\delta_X P$. The analogous formula for ρ_Y completes the verification that the introduction of the modified density

$$\tilde{\rho} = \rho - \partial/\partial X(\sigma h U \delta_X \rho) - \partial/\partial Y(\sigma h V \delta_Y \rho) \qquad(3.17)$$

into Equation (3.12) results in the same leading terms for the viscosity.

A related method was applied by Harten (1977) to a single hyperbolic conservation law and proved to be effective in capturing shocks while not smearing contact discontinuities as badly as other methods. The idea has been used in transonic computations by Hafez et al. (1979), who experimented with a variety of standard elliptic schemes to solve the resulting difference equations, and by Holst & Ballhaus (1979), who used an approximate factorization method.

Solution of the Difference Equations

The difference equations resulting from Equation (3.12) can be solved by carefully constructed successive-line-over-relaxation (SLOR) schemes (Jameson & Caughey 1977b, Caughey & Jameson 1979a). These schemes have the advantages of being quite stable, and of rapidly eliminating any large local errors in the initial estimates for the potential field. Their rates of convergence decrease as the local errors become smaller, however, with the result that convergence to very small residuals is excruciatingly slow, especially when the mesh spacing is small.

Approximate factorization (A-F) techniques have been used by Holst & Ballhaus (1979), Holst (1979), and Hafez et al. (1979). In these schemes the correction C to the potential at each point is calculated according to

$$(\alpha\tilde{\rho} - \delta_x\tilde{\rho}\delta_x)(\alpha\tilde{\rho} - \delta_y\tilde{\rho}\delta_y)C = \alpha\sigma\tilde{\rho}[(\tilde{\rho}\phi_x)_x + (\tilde{\rho}\phi_y)_y]\ ,\qquad(3.18)$$

where α, σ are acceleration parameters, and $\tilde{\rho}$ is taken to be its value from the preceding iteration. In order to maintain stability when the supersonic zone is large and/or shocks are strong it is generally necessary to add terms to the time-dependent equation describing the iterative process that are proportional to ϕ_{xt} and ϕ_{yt}. This can be done easily (Hafez et al. 1979) by dividing Equation (3.18) through by $\tilde{\rho}^2$ and writing

$$(\alpha + \beta_1 \delta_x - \delta_{xx})(\alpha + \beta_2 \delta_y - \delta_{yy})C = \alpha\sigma/\tilde{\rho}[(\tilde{\rho}\phi_x)_x + (\tilde{\rho}\phi_y)_y] \,. \tag{3.19}$$

The β_1 and β_2 parameters are chosen so that a term proportional to ϕ_{st} is added in supersonic zones, and the $\beta_1 \delta_x$, $\beta_2 \delta_y$ terms are upwind differenced.

Either Equation (3.18) or Equation (3.19) can be solved efficiently by sequentially inverting the two factors, each of which requires only a tridiagonal inversion along each line. At first glance this might seem to be twice the work per iteration of SLOR methods, but because the amount of work required to evaluate the residuals is usually many times larger than that of the actual inversion, a single A-F sweep actually requires little more work than an SLOR sweep.

Even using these approximate factorization methods, several hundreds of iterations are sometimes necessary to achieve convergence in difficult cases. A still more efficient alternative has been demonstrated by Jameson (1979), based upon the multi-level adaptive-grid technique first proposed by Fedorenko (1962) and developed and popularized by Brandt (1973). The concept behind the multi-grid method is to eliminate each band of wavenumbers in the error spectrum on a finite-difference grid which is, in a sense, optimal for that component. Thus, low-wavenumber errors are eliminated on coarse grids, while high-wavenumber errors are eliminated on fine grids. Alternatively, the use of coarse grids to eliminate the low-wavenumber component of the error can be thought of as allowing a very high signal speed for this component of the error to be transmitted across the grid during the relaxation process. The multi-grid method was first applied to the transonic small-disturbance problem by South & Brandt (1977) and has been used in a Cartesian mesh, full potential equation solution for the flow past inlet nacelles by McCarthy & Reyhner (1980). The latter work represents a formidable programming achievement, and points up another disadvantage of the Cartesian-mesh/interpolated-boundary-condition approach, since the boundary locations are not independent of grid level in the multi-level scheme.

The method is applied within the finite-volume framework as follows. Let
$$L^h \phi = 0 \tag{3.20}$$
represent Equation (3.12) on a grid with a spacing proportional to h. Given an initial estimate $\phi^{(n)}$ of the solution to Equation (3.20), the basis of the multiple-grid method is to calculate an improved estimate $\phi^{(n+1)}$ on a coarser grid according to
$$L^{2h} \phi^{(n+1)} = f, \tag{3.21}$$

where

$$f = L^{2h}\phi^{(n)} - I_h^{2h}L^h\phi^{(n)}, \tag{3.22}$$

and I_h^{2h} is a collection operator which averages the residuals over the fine mesh points in the neighborhood of each coarse mesh point. The fine grid solution is then improved according to

$$\phi^{(n+1)} = \phi^{(n)} + I_{2h}^h(\phi^{(n+1)} - \phi^{(n)}), \tag{3.23}$$

where I_{2h}^h is an interpolation operator.

The success of the multiple-grid method generally depends upon the use of a relaxation method to eliminate high-frequency errors on any given grid. Since point and line relaxation schemes do not necessarily provide efficient smoothing of all high-wavenumber errors on nonuniform grids, a generalized alternating-direction-implicit (ADI) scheme was used by Jameson. He introduced the difference operator

$$S = \alpha_0 + \alpha_1 \delta_X^- + \alpha_2 \delta_Y^-, \tag{3.24}$$

where δ_X^-, δ_Y^- represent one-sided difference operators, and solved

$$(S - A_X \delta_X^2)(S - A_Y \delta_Y^2)(\phi^{(n+1)} - \phi^{(n)}) = SL^h\phi^{(n)}, \tag{3.25}$$

where A_X and A_Y are the coefficients of ϕ_{XX} and ϕ_{YY} in the expanded form of Equation (3.12). This scheme can be considered a discrete approximation to the time-dependent equation

$$\beta_0 \phi_t + \beta_1 \phi_{Xt} + \beta_2 \phi_{Yt} = A_X \phi_{XX} + A_Y \phi_{YY}, \tag{3.26}$$

where the coefficients β_0, β_1, and β_2 are related to the parameters α_0, α_1, and α_2. The advantage of using the operator defined in Equation (3.24) in place of the simple constant of conventional ADI schemes is that Equation (3.26) remains hyperbolic even when the signs of A_X or A_Y change. To ensure consistency, the δ_X^2 in Equation (3.25) is replaced by an upwind operator when $A_X < 0$, and the δ_X^- in S is chosen to be upwind. It is interesting to note that Hafez et al. (1979) claim the upwinding of the δ_X^2 term to be unnecessary.

The relative efficiencies of the A-F and multi-grid methods can be described approximately as follows. To reduce all wavenumbers of error by a given amount, the A-F method uses a sequence of 6 to 8 acceleration parameters α, each chosen to work effectively on a particular range of wavenumbers. The multi-grid method uses a sequence of grids to achieve a similar result, but, on each grid, the work performed is 1/4 of that on the preceding grid (for two-dimensional problems!). If one ADI sweep on the fine grid is defined to be one unit of work, a multi-grid method consisting of two sweeps on each coarser grid per fine grid sweep uses less than $1 + 2/3$ work units to reduce all wavenumbers of the error, while 6 to 8 work units are required by the A-F method alone.

One additional factor that should be mentioned in this regard is the developing need to consider machine characteristics when comparing algorithms, which the increasing availability of parallel- and/or vector-processing computers makes necessary. In particular, the A-F methods are very attractive on vector machines, while for very large three-dimensional problems they are problematical because of the difficulty in accessing the data in the third dimension (as when, for example, large arrays must be stored on a disk). Some considerations on the use of vector-processing machines for transonic potential-flow problems are discussed by Redhed et al. (1979) and by South et al. (1980).

4. RESULTS AND DESIGN USE

In this section I present selected results which illustrate the accuracy of the methods we have been describing, the effects of boundary-layer corrections, and the use of these methods in design problems.

Although the emphasis of the previous section has been on the more recently developed finite-volume methods, I also describe here results of earlier methods that solve the analytically-transformed equation. In particular, much practical experience has been obtained by a number of investigators with the computer program called FLO-22 (Jameson & Caughey 1977a). That program solves the quasi-linear form (Equation 3.1) of the three-dimensional potential equation for swept wings of arbitrary planform and section shape, mounted on a wall (or symmetry plane). The potential equation is transformed analytically to a boundary-conforming coordinate system which is nearly conformal in planes parallel to the symmetry plane.

This analysis is also important because of the close similarity of its treatment of the vortex sheet to that used in the finite-volume codes. In each of these methods, convection and roll-up of the vortex sheet trailing behind lifting wings are neglected. A linearized model is used that assumes the vortex filaments trail downstream from the wing in the free-stream direction in a surface whose shape is determined *a priori*. While this is clearly only an approximation to the true behavior of the vortex sheet, it is useful because it greatly simplifies the task of grid generation, and because it provides results that are in good agreement with both experiment and other analyses. In particular, a recent lifting-line analysis of three-dimensional wings by Cheng & Meng (1980), which properly accounts for the far-field behavior of the trailing vorticity in a consistent manner using matched asymptotic expansions, provides results that are in excellent agreement with results from FLO-22 calculations, except in the immediate vicinity of the wing leading edge, where the asymptotic theory breaks down.

For the high Reynolds numbers of practical interest to aircraft designers, the results of the potential methods I have been describing agree very well with experimental data, especially when corrections are applied for the displacement effect of the boundary layer. Excellent agreement has been illustrated for supercritical airfoils at both design and off-design conditions by Bauer & Korn (1975) and Bauer et al. (1975) with classical boundary-layer techniques, particularly when a non-conservative formulation of the potential equation was used. Melnik et al. (1977, 1981; see also Melnik

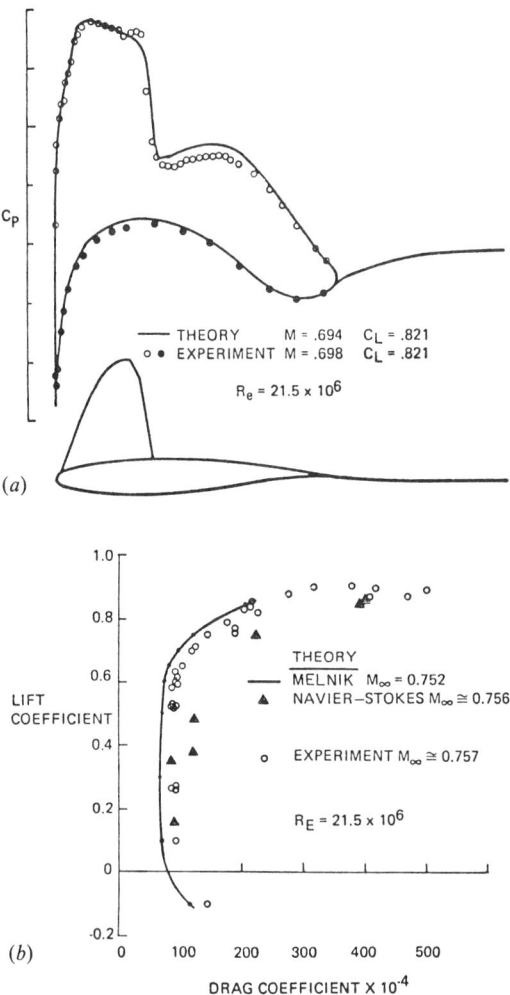

Figure 2 (*a*) Calculated and experimentally measured surface pressure distributions on Korn airfoil. (*b*) Calculated and experimentally determined drag polars for Korn airfoil at constant Mach number.

1980) found that it was necessary to include wake curvature effects and trailing-edge interactions to achieve a consistent formulation of the Kutta condition, and their results, using the fully conservative potential code of Jameson (1975) compare well with a variety of experimental data. The trailing-edge corrections are particularly important for aft-loaded airfoils. Figure 2a and 2b show the results of their calculations (taken from Melnik et al. 1981) for the original Korn airfoil (Bauer et al. 1972) compared with the experimental results of Kacprzynski (1972). The experiment was performed in the Canadian NAE high-Reynolds-number, transonic wind tunnel, which is known to require large Mach-number and angle-of-attack corrections due to tunnel-wall interference. In the absence of calibration data for the conditions of these tests, the calculations were performed at the angle of attack giving the same lift coefficient as the experiment, and a Mach-number correction of the order of -0.005 was applied to provide agreement in shock location. Figure 2a shows a comparison of the calculated surface-pressure distribution with that determined experimentally at a free-stream Mach number of 0.698 (calculation performed at 0.694); Figure 2b shows a comparison of the calculated and experimentally determined drag polars at a fixed Mach number of 0.757 (calculations performed at 0.752). Also shown on the drag polar are the results of calculations performed using the complete, time-averaged Navier-Stokes equations at the NASA Ames Research Center (G. S. Deiwert, unpublished results, 1977). The lack of agreement between these results and both the experiment and Melnik's calculations is attributed by Melnik to a lack of resolution in the Navier-Stokes calculations, since only about 50 mesh points were used on the airfoil surface.

Comparisons of several finite-difference codes with wind-tunnel data for three supercritical wings of differing aspect ratio have been made by Hinson & Burdges (1980). They also made a careful study of the effects of wind-tunnel wall interference upon their results using a combination of far-field data taken from static pressure probes near the tunnel walls in their experiment, and calculations that included the effect of the tunnel walls. They found that a small constant Mach-number correction could be used to account for blockage effects in their tunnel, and determined an angle-of-attack correction by matching the calculated surface pressures near the wing leading edge with those experimentally determined. Samples of comparisons for the intermediate aspect-ratio wing of their tests at a free-stream Mach number of 0.90 and a nominal lift coefficient of 0.50 are shown in Figure 3a–c. Figure 3a shows comparisons of results of both the non-conservative full-potential code FLO-22 (Jameson & Caughey 1977a) and the Bailey-Ballhaus extended small-disturbance code (Ballhaus et al. 1976) with experimental data; Figure 3b shows comparisons of results of the fully conservative full-potential code FLO-27 (Jameson

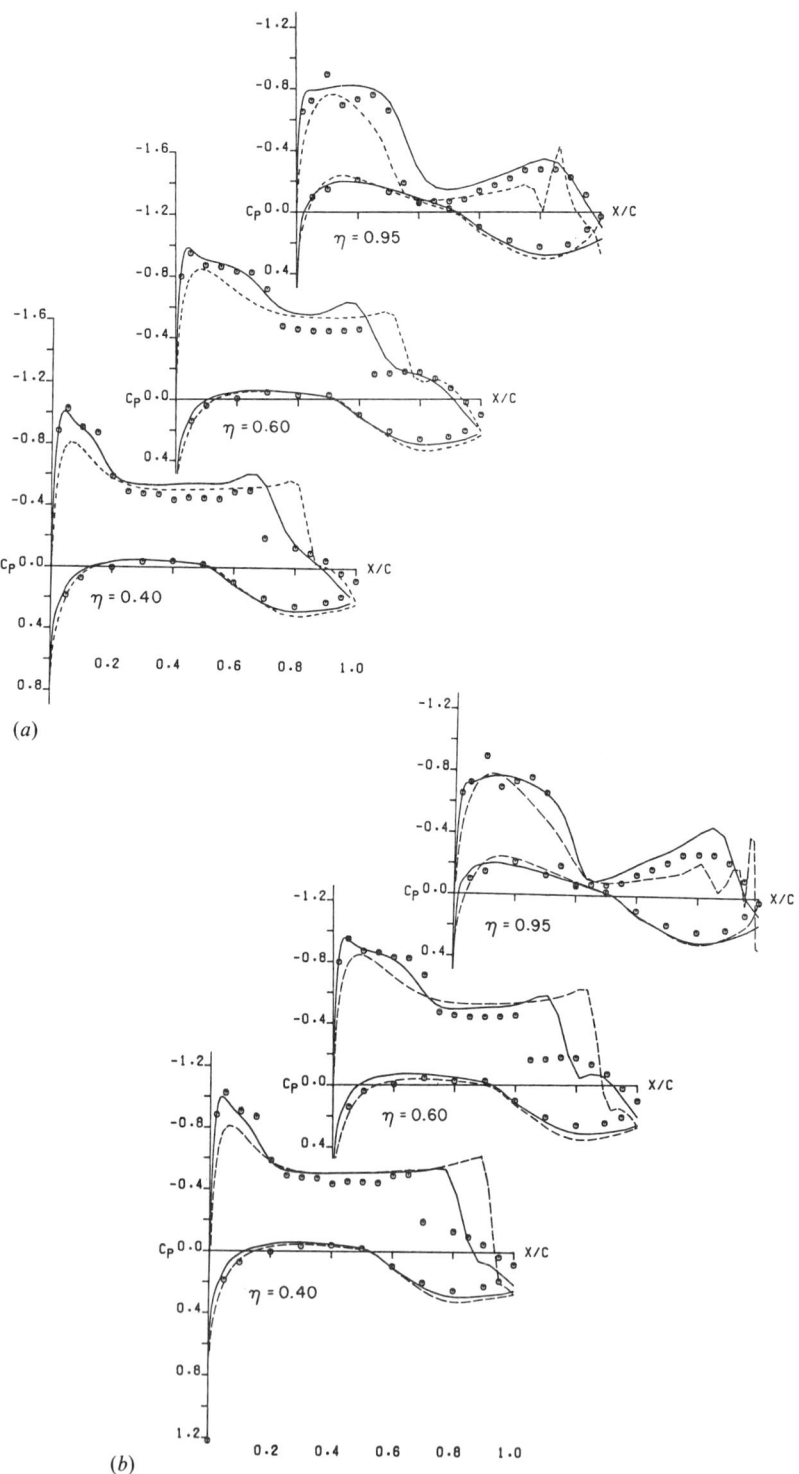

(a)

(b)

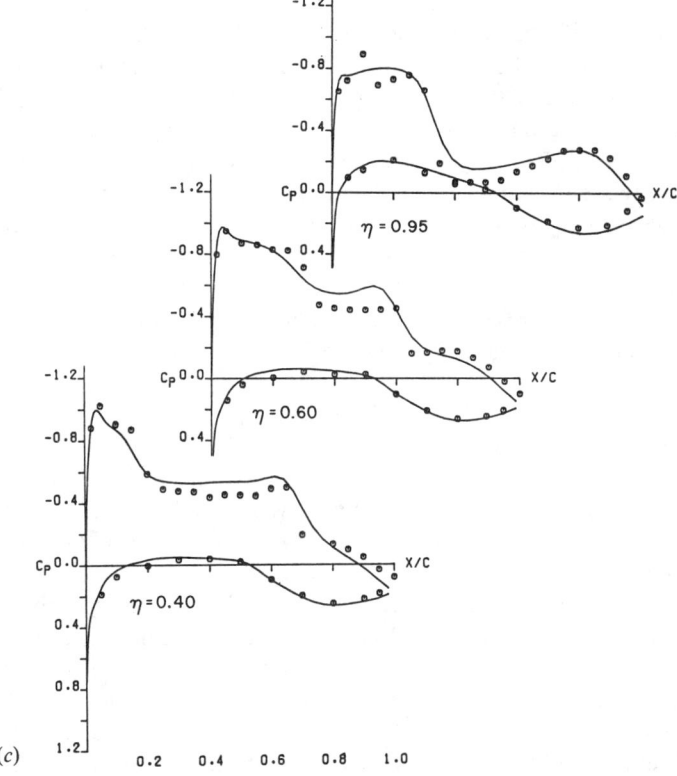

Figure 3 Measured and calculated chord-wise distributions of wing surface pressure at 40, 60, and 98 percent semi-span stations. Symbols represent experimental data. (*a*) Results of non-conservative calculations; solid lines are FLO-22 results; broken lines represent results of non-conservative Bailey-Ballhaus code. (*b*) Results of fully conservative calculations; solid lines represent FLO-27 results; broken lines represent results of fully conservative Bailey-Ballhaus code. (*c*) Results of calculations with boundary-layer effect; solid lines represent results of FLO-22 coupled with two-dimensional strip boundary-layer model.

& Caughey 1977b) and a fully-conservative version of the Bailey-Ballhaus code with experiment; Figure 3c shows comparisons with FLO-22 results when viscous effects are modeled in the computation by adding a displacement thickness calculated using a two-dimensional strip boundary-layer method. Hinson & Burdges found that, of the codes studied, the non-conservative FLO-22, when coupled with the boundary-layer method, provided the most reliable agreement with experimental data. The full conservative codes did not agree as well with experiment, but no attempt was made to include the trailing-edge viscous effects shown important in two-dimensional calculations by Melnik et al. (1977, 1981).

Both small-disturbance and full potential transonic codes have been widely used for analysis and design of new wings. Mann (1979) has designed a low-aspect-ratio fighter wing using both FLO-22 and FLO-27, and Henne (1980) has made wide use of FLO-22 in the analysis of new wings for commercial transport aircraft. A sample of the development of isobars in the wing planform during a series of systematic design alterations to the wing geometry to reduce its transonic drag is shown in Figure 4 (after Lynch 1978). The figure shows the isobars of the original wing (A), the effect of a planform modification designed to eliminate the strong shock appearing near the trailing-edge break of the original wing (B), the effect of an airfoil section modification (C), and, finally, the effect of addition of a leading-edge glove to eliminate the unsweeping of the isobars near the wing root (D). This last modification also eliminated a shock appearing near mid-chord between the 70 and 85 percent semi-span stations of the wing.

Optimization procedures have also been coupled with FLO-22 by Hicks & Henne (1978) and by Haney et al. (1980). These authors showed that the results of the finite-difference calculations were sufficiently sensitive to small changes in wing geometry that it was possible to perform a multidimensional, gradient-type search to reduce wave drag and/or improve low-speed characteristics of three-dimensional wings.

An ingenious way to use flow-analysis methods to design shock-free airfoils and wings has been suggested by Sobieczky (Sobieczky et al. 1979). The method is based upon determining, for a given wing geometry, a solution of Equation (2.2) for a fluid that obeys a fictitious gas law in supersonic zones. The fictitious behavior is chosen such that the equation remains elliptic everywhere in the domain, with the result that its solution

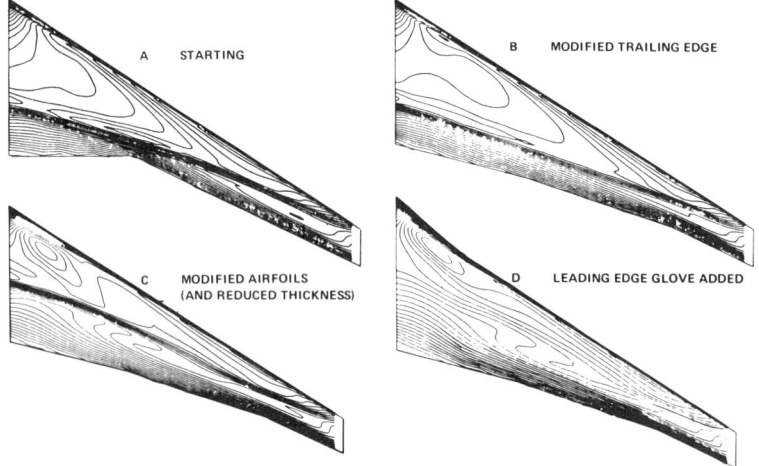

Figure 4 Predicted upper-surface isobars for several configuration modifications (after Lynch 1978).

is smooth. An initial-value problem is then solved in the supersonic zone for the correct gas law, marching from the sonic line to the wing surface. This procedure determines the shape of the wing in the supersonic zone that produces a flow with a smooth recompression through the sonic velocity. The initial-value problem is ill-posed for three-dimensional flows, but the method has, nevertheless, been applied with success for the design of wings by Yu (1980a), who used a modification of FLO-27 as the analysis tool for the first step of the design procedure.

Henne (1980) has applied the iterative procedure developed by Tranen (1974) for airfoil design to three-dimensional wings using another modification of FLO-22 and demonstrated its ability to improve the drag characteristics of supercritical wings. The method alternates between direct solutions for a given wing geometry using Neumann and Dirichlet boundary conditions for the potential on the wing surface. The Dirichlet condition is used to specify a desired pressure distribution over (at least part of) the wing surface, and the resulting solution gives the transpiration velocity distribution over the wing surface required for that solution. This is interpreted as an effective change in wing geometry (in the same manner that a change in boundary-layer displacement thickness can be interpreted as a blowing rate). A solution using the Neumann boundary condition and the modified wing geometry is then performed to verify that the desired pressure distribution has been achieved. The iteration between the two types of solutions continues until the required transpiration velocities vanish and the desired pressure distribution has been achieved. It is known that it is impossible in incompressible flow to design a closed profile having a given pressure distribution over the surface unless the corresponding velocity distribution satisfies certain integral constraints first described by Lighthill (1945). (See, for example, Thwaites 1960.) Presumably, related conditions must be satisifed for the compressible problem but, as in the case of Sobieczky's method, this problem of ill-posedness does not seem to affect the usefulness of the method, perhaps because the solutions obtained are not exact.

The finite-volume methods have been applied to geometries as complex as the A-7 attack fighter wing-fuselage, which was illustrated in Figure 1. Predicted wing pressure distributions agree quite well with experimental data for this configuration even in the absence of viscous corrections. Figure 5 (after Caughey & Jameson 1979b) shows the results of two calculations, one in which the presence of the fuselage was accounted for using a grid similar to that in Figure 1, but with many more mesh cells (FLO-30), and another in which the fuselage was neglected and the wing was assumed to be mounted on a plane wall (FLO-28). The results are compared with data from wind-tunnel tests performed at the NASA Ames Research Center (R. M. Hicks, private communication, 1977). For these

280 CAUGHEY

comparisons, the wind-tunnel data were apparently nearly interference-free, and the calculations were performed at the same Mach number and angle of attack as the experiment. The results clearly indicate the importance of modeling the fuselage presence to accurately predict the wing surface pressures, at least for this configuration.

5. CONCLUSION

I have reviewed some of the ideas that have led to our ability to predict, with a reasonable expenditure of computer resources, the transonic potential flow past configurations of interest to aircraft designers. The development of efficient and generally applicable finite-difference methods for solving the potential equation has provided the designer with a useful tool for a number of practical problems, but many important phenomena remain to be included in the class of problems we can usefully attack by computation. It is unlikely that computational aerodynamics will ever be developed to the point where we can treat computer models as a numerical wind tunnel from which we can extract data without a clear understanding of the limitations of the models and of the fundamental physical processes occurring in the flows we are trying to predict. Nor is that a desirable goal. Experience with transonic potential computation has shown, however, that when the model is an appropriate one, computational methods can

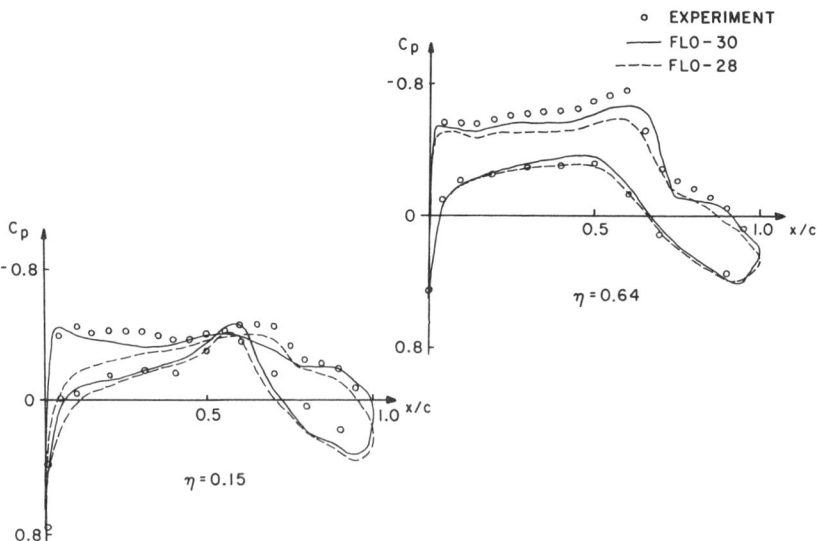

Figure 5 Chord-wise distributions of wing surface pressure at 15 and 64 percent semi-span stations. Symbols are experimental data; solid lines are results of wing-body calculations (FLO-30); broken lines are results of wing-alone calculations (FLO-28).

provide the aerodynamic designer with the flexibility to investigate a far greater number of candidate configurations than was previously possible, and ensure that no catastrophic surprises will be encountered when the final design is tested in the wind tunnel.

ACKNOWLEDGMENTS

Most of the author's work presented in this review has been supported by the Office of Naval Research under Contract N00014-77-C-0033. It is also a pleasure to acknowledge the debt that my education in computational methods, and the development of computational methods for transonic problems in general, owe to Professor Antony Jameson of Princeton University.

Literature Cited

Adamson, T. C. Jr., Messiter, A. F. 1980. Analysis of two-dimensional interactions between shock waves and boundary layers. *Ann. Rev. Fluid Mech.* 12:103–38

Arlinger, B. G. 1975. Calculation of transonic flow around axisymmetric inlets. *AIAA Journal* 13:1614–21

Ballhaus, W. F. 1978. Some recent progress in transonic flow computations. *Numerical Methods in Fluid Dynamics*, ed. H. J. Wirz, J. J. Smolderen pp. 155–236. Washington, DC: Hemisphere Publ. Corp.

Ballhaus, W. F., Bailey, F. R. 1972. Numerical calculation of transonic flow about swept wings. *AIAA Pap. 72-677, Fluid and Plasma Dynamics Conf., 5th, Boston*

Ballhaus, W. F., Bailey, F. R., Frick, J. 1976. Improved computational treatment of transonic flow about swept wings. *NASA CP-2001*. Moffett Field, Ca: NASA Ames Research Center

Bauer, F., Garabedian, P. R., Korn, D. G. 1972. *Supercritical Wing Sections*. New York: Springer

Bauer, F., Garabedian, P. R., Korn, D., Jameson, A. 1975. *Supercritical Wing Sections II*. New York: Springer

Bauer, F., Korn, D. 1975. Computer simulation of transonic flow past airfoils with boundary layer correction. *Proc. AIAA Computational Fluid Dynamics Conf., 2nd, Hartford*, pp. 184–189

Brandt, A. 1973. Multi-level adaptive grid technique (MLAT) for fast numerical solution to boundary value problems. *Proc. Int. Conf. Numer. Methods in Fluid Mechanics, 3rd*, pp. 82–89. New York: Springer

Camarero, R., Younis, M. 1980. Efficient generation of body-fitted coordinates for cascades using multigrid. *AIAA J.* 18:487–88

Carlson, L. A. 1975. Transonic airfoil analysis and design using Cartesian coordinates. *Proc. AIAA Computational Fluid Dynamics Conf. 2nd, Hartford*, pp. 175–83

Caughey, D. A. 1978. A systematic procedure for generating useful conformal mappings. *Int. J. Numer. Methods in Engineering* 12: 1651–57

Caughey, D. A, Jameson, A. 1977a. Calculation of transonic potential flow fields about complex, three-dimensional configurations. *Transonic Flow Problems in Turbomachinery*, ed. T. C. Adamson, M. Platzer, pp. 274–91. Washington: Hemisphere Publ. Corp.

Caughey, D. A., Jameson, A. 1977b. Accelerated iterative calculation of transonic nacelle flowfields. *AIAA J.* 15:1474–80

Caughey, D. A., Jameson, A. 1979a. Numerical calculation of transonic potential flow about wing-body combinations. *AIAA J.* 17:175–81

Caughey, D. A., Jameson, A. 1979b. Progress in finite-volume calculations for wing-fuselage combinations. *AIAA J.* 18:1281–88

Chen, L. T., Caughey, D. A. 1979. Transonic inlet flow calculations using a general grid generation scheme. *Proc. Conf. on Flow in Primary, Non-rotating Passages in Turbomachinery*, pp. 125–32. ASME Winter Annual Meeting, New York

Chen, L. T., Caughey, D. A. 1980. Calculation of transonic inlet flowfields using generalized coordinates. *J. Aircraft* 17:167–74

Cheng, H. K., Meng, S. Y. 1980. The oblique wing as a lifting-line problem in transonic flow. *J. Fluid Mech.* 97:531–56

Dulikravich, D. J. 1979. *Numerical calculation of inviscid, potential, transonic flows through rotors and fans*. PhD thesis. Cornell Univ., Ithaca, New York

Dulikravich, D. J., Caughey, D. A. 1980. Finite-volume calculation of transonic potential flow through rotors and fans. *Rep. FDA-80-03.* Sibley School of Mechanical and Aerospace Engineering, Cornell University, Ithaca, NY

Eiseman, P. R. 1979. A multi-surface method of coordinate generation. *J. Computational Phys.* 33:118–50

Fedorenko, R. P. 1962. A relaxation method for solving elliptic difference equations. *USSR Comput. Math. & Math. Phys.* 1:1092–96

Garabedian, P. R., Korn, D. G. 1972. Analysis of transonic airfoils. *Comm. Pure & Appl. Math.* 24:841–51

Grossman, B., Melnik, R. E. 1976. The numerical computation of the transonic flow over two-element airfoil systems. *Proc. Int. Conf. Numerical Methods in Fluid Dynamics, 5th, Lecture Notes in Physics* 59:220–27. Berlin: Springer

Hafez, M., South, J., Murman, E. 1979. Artificial compressibility methods for numerical solution of transonic full potential equation. *AIAA J.* 17:838–44

Haney, W. P., Johnson, R. R., Hicks, R. M. 1980. Computational optimization and wind-tunnel test of transonic wing designs. *J. Aircraft* 17:457–63

Harten, A. 1977. The artificial compressibility method for computation of shocks and contact discontinuities. *ICASE Rep. No. 77-2.* Institute for Computer Appl. in Science and Engineering, Hampton, VA

Henne, P. A. 1980. An inverse transonic wing design method. *AIAA Pap. 80–0330, Aerospace Sciences Meeting, 18th, Pasadena*

Hicks, R. M., Henne, P. A. 1978. Wing design by numerical optimization. *J. Aircraft* 15:407–12

Hinson, B. L., Burdges, K. P. 1980. An evaluation of three-dimensional transonic codes using new correlation-tailored test data. *AIAA Pap. 80–0003, Aerospace Sciences Meeting, 18th, Pasadena*

Holst, T. L. 1979. Implicit algorithm for conservative transonic full potential equation using an arbitrary mesh. *AIAA J.* 17:1038–45

Holst, T. L., Ballhaus, W. F. 1979. Conservative implicit schemes for the full potential equation applied to transonic flows. *AIAA J.* 17:145–52

Ives, D. C., Luitermoza, J. F. 1977. Analysis of transonic cascade flow using conformal mapping and relaxation techniques. *AIAA J.* 15:647–57

Jameson, A. 1971. Transonic flow calculations for airfoils and bodies of revolution. *Aerodynamics Rep. 391-71-1.* Grumman Aerospace Corp, Bethpage, NY

Jameson, A. 1974. Iterative solution of transonic flows over airfoils and wings including flows at Mach 1. *Comm. Pure & Appl. Math.* 27:283–309

Jameson, A. 1975. Transonic potential flow calculations in conservation form. *Proc. AIAA Computational Fluid Dynamics Conf. 2nd, Hartford,* pp.148–61

Jameson, A. 1979. A multi-grid scheme for transonic potential calculations on arbitrary grids. *Proc. AIAA Computational Fluid Dynamics Conf. 4th, Williamsburg,* pp. 122–46

Jameson, A., Caughey, D. A. 1977a. Numerical calculation of the transonic flow past a swept wing. *ERDA Rep. C00-3077-140.* New York Univ., New York

Jameson, A., Caughey, D. A. 1977b. A finite-volume method for transonic potential flow calculations. *Proc. AIAA Computational Fluid Dynamics Conf. 3rd, Albuquerque,* pp 35–54

Kacprzynski, J. J. 1972. A second series of wind-tunnel tests on shockless, lifting airfoil no. 1. *NAE Project Rep. 5×5/0062.* National Aircraft Establishment, Ottawa, Canada

Keyfitz, B. L., Melnik, R. E., Grossman, B. 1979. Leading-edge singularity in transonic small-disturbance theory: Numerical resolution. *AIAA J.* 17:296–97

Levy, L. L. 1978. Experimental and computational steady and unsteady transonic flows about a thick airfoil. *AIAA J.* 16:564–72

Lighthill, M. J. 1945. A new method of two-dimensional aerodynamic design. *Rep. Mem. Aero. Res. Council London 2112*

Lynch, F. T. 1978. Recent applications of advanced computational methods in the aerodynamic design of transport aircraft configurations. *Proc. Cong. ICAS, 11th, Lisbon,* pp. 270–84

Mann, M. J. 1979. The design of supercritical wings by the use of three-dimensional transonic theory. *NASA Tech. Pap. 1400.* NASA Langley Research Center, Hampton, VA

McCarthy, D. R., Reyhner, T. A. 1980. A multi-grid code for three-dimensional transonic potential flow about axisymmetric inlets at angle of attack. *AIAA Pap. 80–1365, Fluid and Plasma Dynamics Conference, 13th, Snowmass, CO*

Melnik, R. E. 1980. Turbulent interactions on airfoils at transonic speeds—recent developments. *Pap. 10, AGARD CPP-291, Conf. on Computation of Viscous-Inviscid Interactions, Colorado Springs*

Melnik, R. E., Chow, R. R., Mead, H. R. 1977. Theory of viscous transonic flow over airfoils at high Reynolds numbers. *AIAA Pap. 77–680, Fluid and Plasma Dynamics Conference, 10th, Albuquerque*

Melnik, R. E., Chow, R. R., Mead, H. R., Jameson, A. 1981. An improved viscid/inviscid interaction procedure for transonic flow over airfoils. *NASA CR,* to be released. NASA Langley Research Center, Hampton, VA

Murman, E. M., Cole, J. D. 1971. Calculation of plane, steady transonic flows. *AIAA J.* 9:114–21

Nieuwland, G. Y., Spee, B. M. 1973. Transonic airfoils: Recent developments in theory, experiment, and design. *Ann. Rev. Fluid Mech.* 5:119–50

Redhed, D. D., Chen, A. W., Hotovy, S. G. 1979. New approach to the 3-D transonic flow analysis using the STAR-100 computer. *AIAA J.* 17:98–99

Reyhner, T. A. 1977. Cartesian mesh solution for axisymmetric transonic potential flow around inlets. *AIAA J.* 15:624–31

Schmidt, W. 1978. Progress in transonic flow computations: Analysis and design methods for three-dimensional flows. In *Numerical Methods in Fluid Dynamics* ed. H. J. Wirz, J. J. Smolderen, pp. 299–338. Washington: Hemisphere Publ. Corp.

Sells, C. C. L. 1968. Plane subcritical flow past a lifting airfoil. *Proc. R. Soc. London Ser. A* 308:377–401

Shen, S. F. 1977. Finite element methods in fluid mechanics. *Ann. Rev. Fluid Mech.* 9:421–45

Sobieczky, H., Yu, N. J., Fung, K. Y., Seebass, A. R. 1979. New method for designing shock-free transonic configurations. *AIAA J.* 17:722–29

South, J. C. Jr., Brandt, A. 1977. Application of a multi-level grid method to transonic flow calculations. *Transonic Flow Problems in Turbomachinery,* ed T. C. Adamson, M. Platzer, pp. 180–297. Washington, DC: Hemisphere Publ. Corp.

South, J. C. Jr., Keller, J. D., Hafez, M. M. 1980. Vector processor algorithms for transonic flow calculations. *AIAA J.* 18: 786–92

Steger, J. L., Baldwin, B. S. 1972. Shock waves and drag in the numerical calculation of isentropic, transonic flow. *NASA TN D–6977.* NASA Ames Research Center, Moffett Field, CA

Thompson, J. F., Thames, F. C., Mastin, C. W. 1974. Automatic numerical generation of body-fitted curvilinear coordinate system for field containing any number of arbitrary two-dimensional bodies. *J. Computational Phys.* 15: 299–319

Thwaites, B. 1960. *Incompressible Aerodynamics,* pp. 143–45. Oxford: Clarendon

Tijdeman, H., Seebass, R. 1980. Transonic flow past oscillating airfoils. *Ann. Rev. Fluid Mech.* 12:181–222

Tranen, T. L. 1974. A rapid computer-aided transonic airfoil design method. *AIAA Pap. 74–501, Fluid and Plasma Dynamics Conf., 7th, Palo Alto*

von Kármán, T. 1947. The similarity law of transonic flow. *J. Math & Phys.* 26:182–90

Yu, N. J. 1980a. Efficient transonic shock-free wing redesign procedure using a fictitious gas method. *AIAA J.* 18:143–48

Yu, N. J. 1980b. Grid generation and transonic flow calculations for three-dimensional configurations. *AIAA Pap. 80–1391, Fluid and Plasma Dynamics Conf. 13th, Snowmass, CO*

UNSTEADY AIRFOILS[1]

W. J. McCroskey

U.S. Army Aeromechanics Laboratory and National Aeronautics and Space Administration, Moffett Field, California 94035

INTRODUCTION

Flow over streamlined lifting surfaces has been a major topic of fluid mechanics during most of the twentieth century. For simplicity, unsteady effects have often been ignored. However, most modern fluid-dynamic devices encounter or generate unsteady flow, whether intentional or not, for at least part of their operating conditions. In this review, we focus on the role of unsteady effects in an important class of flow problems, namely, two-dimensional oscillating airfoils, and on the advances that have been made within the past decade toward understanding these special and challenging flows.

Studies of unsteady-airfoil flows have been motivated mostly by efforts to avoid or reduce such undesirable effects as flutter, vibrations, buffeting, gust response, and dynamic stall. This requires predicting the magnitude and phase (or time lag) of the unsteady fluid-dynamic loads on thin lifting surfaces. Some attention has also been given to potentially beneficial effects of unsteadiness, such as the propulsive efficiency of flapping motion, controlled periodic vortex generation, and stall delay, and to improving the performance of turbomachinery, helicopter rotors, and wind turbines by controlling the unsteady forces in some optimum way. Most of these studies and applications concern either periodic motion of an airfoil in a uniform stream or periodic fluctuations in the approaching flow. We follow this pattern in this review, to the exclusion of impulsive or other nonperiodic motions. The scope is further limited to two-dimensional flows past streamlined shapes, notwithstanding the practical importance of three-dimensional effects and bluff-body configurations.

Our plan is first to explain and illustrate some basic concepts and unsteady phenomena by means of thin-airfoil theory for an inviscid incom-

---

[1] The US Government has the right to retain a nonexclusive, royalty-free license in and to any copyright covering this paper.

pressible fluid. This background is a valuable framework against which we may then view recent and current research on the various complicating effects of compressibility and viscosity, with particular emphasis on unsteady transonic flow and dynamic stall. We conclude with a discussion of probable developments in the future.

A MODEL PROBLEM—THE OSCILLATING THIN AIRFOIL

Many important features of unsteady-airfoil behavior can be described by linearized thin-airfoil theory. The fluid-dynamic pressure forces acting on a thin lifting surface inclined at a small angle relative to the approaching flow are proportional to the effective angle of attack and to the square of the speed of the flow. If either the body or the flow fluctuates, so do the circulation and the pressure distribution; and each change in circulation around spanwise sections of the body is accompanied by the shedding of free vorticity from the trailing-edge region into the wake. This time-dependent vortical wake is an important distinguishing feature of unsteady airfoils.

Among the best known and most enlightening analyses of this class of problems are those by Theodorsen (1935) and von Kármán & Sears (1938), who considered a thin flat plate and a trailing flat wake of vorticity in an incompressible fluid. For periodic oscillations of the airfoil or the flow, the fluid-dynamic phenomena are characterized by a nondimensional frequency parameter $k = \omega c/2U_\infty$, and the solution can be expressed in terms of combinations of standard Bessel functions whose argument is k. Here ω is the circular frequency of the oscillation, c is the chord of the airfoil, and U_∞ is the mean free-stream velocity. We examine several illustrative results for inviscid flow in the following paragraphs.

Sinusoidal Oscillations in Pitch

Our reference problem for this review is a flat-plate airfoil oscillating sinusoidally in pitch about an axis located at $X = A$, where the airfoil is defined by $0 \leq X \leq 1$ and the angle of attack is $\alpha = \alpha_1 \mathcal{R}e\, e^{i\omega t}$. Theodorsen (1935) divided the flow into two components. The first is a "noncirculatory" component of sources and sinks, which satisfies the boundary conditions on the oscillating plate and which therefore includes the apparent-mass effects. The second is a "circulatory" component, which includes bound vortices and wake vortices and which is matched to the noncirculatory component at the trailing edge in such a manner as to enforce the Kutta condition of nonsingular flow there. We can distill his results for the nondimensional pressure distribution, lift, and pitching moment about the axis of rotation into the following form:

$$C_P = -2\alpha \sqrt{\frac{1-X}{X}} [f_1 + ig_1], \qquad (1)$$

$$C_L = 2\pi\alpha [f_2 + ig_2], \qquad (2)$$

$$C_M = 2\pi\alpha \left(A - \frac{1}{4}\right) [f_3 + ig_3]. \qquad (3)$$

The terms outside the square brackets are immediately recognized as the steady solution for a flat plate at instantaneous incidence α. All of the unsteady effects are included in the functions $f_n + ig_n$, which contain the Bessel functions and which depend only on k, A, and in the case of the pressure, X. Details of the unsteady solution can be found in Theodorsen (1935), Postel & Leppert (1948), Fung (1969), and McCroskey (1973). The real parts of Equations (1)–(3), represented by f_n, designate the "in-phase" components of C_P, C_L, and C_M that follow directly the motion of the airfoil; and the imaginary parts, g_n, are the "quadrature" components that lag or lead the motion by $\omega t = \pi/2$. Alternatively, the airloads are sometimes expressed in terms of the modulus $(f_n^2 + g_n^2)^{1/2}$ and phase angle $\phi = \tan^{-1}(g_n/f_n)$, as shown in Figure 1. In the quasi-steady limit of $k \to 0$, the solutions for f_n approach unity as $(1 - \pi k/2)$, and g_n vanishes as $k \ln k$. This logarithmic behavior of g_n is unique to the two-dimensional case, however.

The important point here is that relatively simple, known functions modify the classical steady-state solution for a thin airfoil when the incidence changes at a finite rate. Equations (1)–(3) give explicitly the relative amplitude and phase of the airloads with respect to the airfoil motion. For example, the solid line in Figure 1 shows the locus of the solution for C_L/α at various values of reduced frequency for the important case of rotation about the static aerodynamic center, $A = 1/4$. For this case, the real part of the lift function f_2 decreases monotonically with increasing k, whereas the imaginary part g_2 changes sign at $k \simeq 0.15$ and increases thereafter. The behavior of both f_2 and g_2 is dominated by noncirculatory terms at high frequency (see Fung 1969).

The pitching moment for $A = 1/4$ is given by $C_M/2\pi\alpha = (3/32)k^2 - i(k/4)$, and the fact that the imaginary part is always negative is significant. This means that C_M always lags the airfoil motion and that the aerodynamic damping in pitch, $-\oint C_M d\alpha$, tends to suppress torsional oscillations at all frequencies. In subsequent sections, we examine cases in which $\mathscr{I}m(C_M)$ can become positive, leading to the possibility of flutter in pure pitching motion. Normally such behavior is most likely to occur when shock waves or separation introduce strong nonlinearities in the flow that violate the assumptions of classical thin-airfoil theory.

Other Types of Motion

Analogous to the preceding example, linear theory has been applied to other types of motion that occur in practical applications. For incompressible fluids, well-known solutions exist for translational oscillations normal to the flight path (plunging motion), longitudinal oscillations, vertical and horizontal gusts, step changes in incidence, oscillating control surfaces, and multiple airfoils (cascades). Lists of references on these important topics can be found in Carta (1978), Ashley (1977), and McCroskey

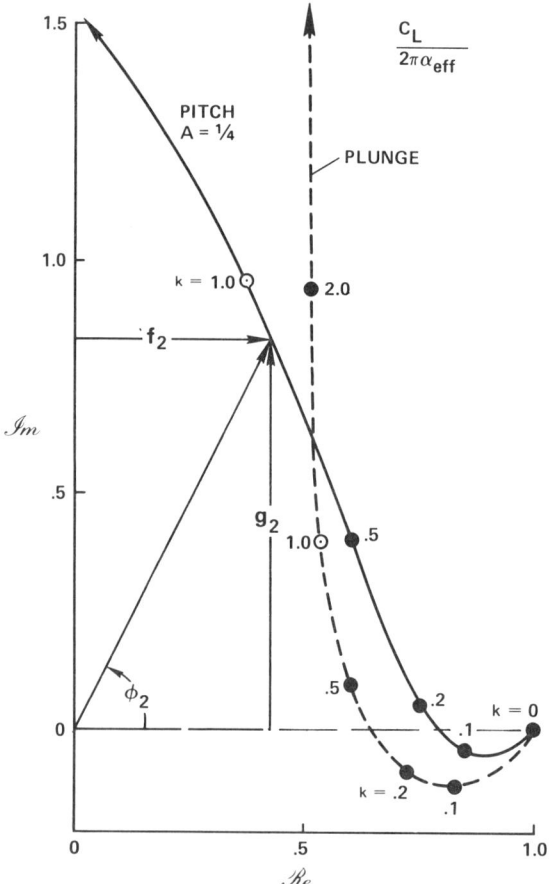

Figure 1 Loci of the real and imaginary components of unsteady lift on an oscillating flat plate as a function of reduced frequency.

(1977). Excluding the case of oscillating control surfaces for the moment, we can express the pressure distribution, lift, and moment for any of these motions in the general form of Equations (1)–(3). That is, the unsteady effects can always be lumped into unsteady factors $f_n + ig_n$ that multiply the classical steady-state solutions.

VERTICAL AIRFOIL OSCILLATIONS   In this case the periodic airfoil displacement $h = h_1 \mathcal{R}e\, e^{i\omega t}$ creates an effective incidence $\alpha_{\text{eff}} = -\dot{h}/U_\infty$, and this quantity is used to normalize C_L in Figure 1. The unsteady lift behavior with respect to α_{eff} generally resembles the pitching case, but quantitative differences can be seen as k increases.

The lift always lags the plunging *displacement*; therefore, as in the previous case of pure pitching motion, single-degree-of-freedom flutter does not occur. However, *combined plunging and pitching* can produce interactions and phasings between the two degrees of freedom that extract energy from the air stream. This occurs in practice as classical bending-torsion flutter of flexible aircraft wings, which is one of the principal topics of aeroelasticity (see Garrick & Reed 1981).

OSCILLATING CONTROL SURFACES   The sinusoidal deflection $\delta = \delta_1 \mathcal{R}e\, e^{i\omega t}$ of a plain flap on a flat plate produces a pressure distribution that consists of the sum of three parts. The first has the same basic form as Equation (1), containing unsteady functions $f_4 + ig_4$ and an inverse square-root singularity with respect to X at the leading edge. The second is algebraically rather complicated (see Postel & Leppert 1948), and contains a logarithmic singularity at the hinge point X_c, where the dominant behavior is given by

$$C_P(X_c) \to \frac{2\delta}{\pi} \ln(X - X_c)^2 [f_5 + ig_5]. \tag{4}$$

The quasi-steady behavior is again given by $f_4 = f_5 \to 1$ and $g_4 = g_5 \to 0$. The third part, which vanishes as $k \to 0$, is nonsingular and it generally tends to reduce the amplitude of the oscillating pressure caused by the flap motion as k increases. This is illustrated in Figure 2, adapted from Tijdeman & Seebass (1980), which shows representative instantaneous pressure distributions and their decomposition into amplitude and phase by means of harmonic analysis. Unsteady effects also produce a chord-wise distribution of phase angle relative to the flap deflection that is almost linear.

The ultimate interest in this problem lies in the forces and moments that the oscillatory pressure distribution creates. The moment on the flap lags the flap motion, and once again, single-degree-of-freedom flutter does not occur within the framework of linear theory. However, flap motion *combined* with pitching or plunging motion can easily lead to negative aerodynamic damping and catastrophic flutter. Also, as we shall see in subsequent sections, moving shock waves in transonic flow can drastically alter the phase of the hinge moment, leading to the single-degree-of-freedom phenomenon known as "aileron buzz."

Compressibility Effects

Oscillating airfoils at subsonic Mach numbers are not fundamentally different from their incompressible counterparts described in the previous sections, but the solutions are much more difficult to express analytically. Numerical solutions of the linearized potential equation abound, and a series of new analyses in the mid-1970s essentially put the finishing touches on the theoretical understanding of this problem (see Amiet 1975 and Kemp & Homicz 1976, for example). Experiments have provided adequate verification of the two-dimensional theory for small-amplitude oscillations at low and moderate reduced frequencies. However, it should be mentioned that three-dimensional effects on wings become increasingly important as M_∞ increases.

The general form of Equations (1)–(4) is preserved for subcritical Mach numbers, that is, when the flow remains subsonic everywhere. The most important quantitative details are that the pressure, lift, and moment scale inversely with the Prandtl-Glauert factor $\beta = (1 - M_\infty^2)^{1/2}$, as expected

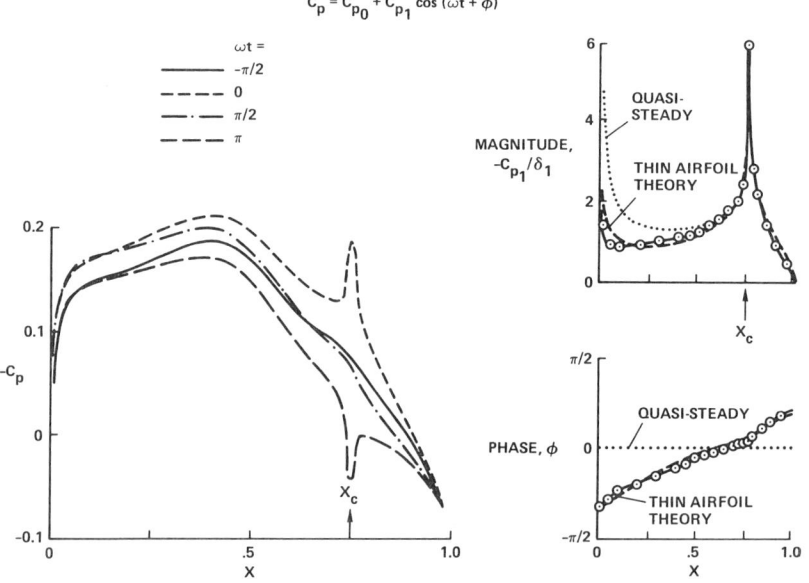

Figure 2 Pressure distributions on an airfoil with an oscillating flap: $M_\infty = 0.5$, $k = 0.39$. —o— experiment.

from classical steady theory, and the relevant reduced-frequency parameter becomes $k^* = k/\beta^2$. Then the functions $f_n(k^*)$ are relatively insensitive to M_∞ at small values of the quantity k^*M_∞, but the g_n's are generally less so. The curves in Figure 3 show some representative examples, kindly supplied by S. R. Bland in a private communication.

The pressure distribution in *supersonic* flow is, of course, altogether different, and this significantly alters the lift curve slope $dC_L/d\alpha$ and the location of the aerodynamic center. The frequency parameter $\hat{k} = kM_\infty^2/\beta^2$ plays a role similar to k^* in defining unsteady functions $f_n + ig_n$ that multiply the corresponding steady-state solutions. Generally speaking, f_n and g_n vary less with reduced frequency and Mach number than their subsonic counterparts. In both cases, if M_∞ is close to one, linear theory predicts significant reductions in aerodynamic damping, and single-degree-of-freedom flutter can occur for pitch axes near the leading edge (see Lambourne 1968). These cases border on the transonic regime, however, where nonlinear equations have to be used.

In the hypersonic limit, Newtonian flow theory yields approximate results (see Hui & Tobak 1981, for example). Important effects of leading-edge bluntness and surface curvature arise as $M_\infty \to \infty$, including negative

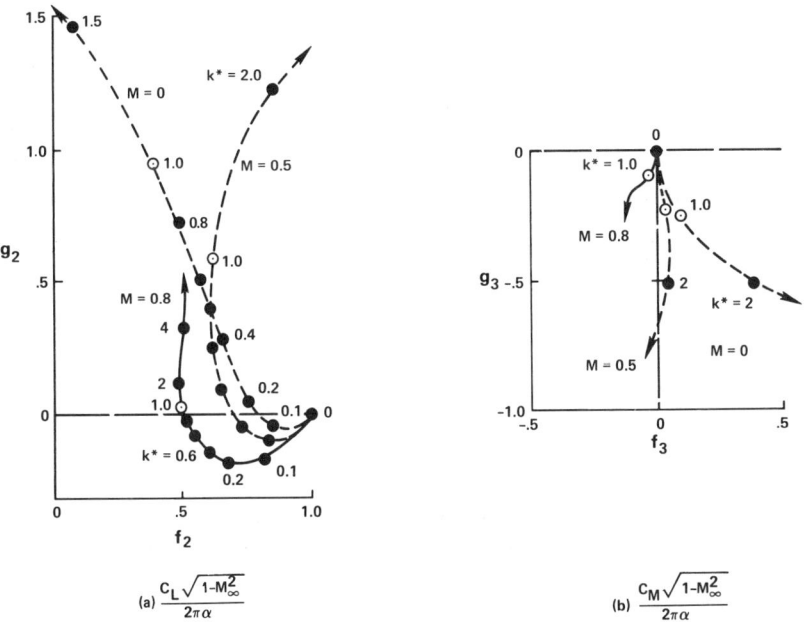

Figure 3 Real and imaginary components of unsteady lift and moment for pitch oscillations: $A = 1/4$.

aerodynamic pitch damping on thin airfoils with sharp leading edges and large concave curvature.

REAL AIRFOILS IN REAL FLUIDS

Modern developments in unsteady-airfoil research can be viewed essentially as attempts to relax the various assumptions that underlie thin-airfoil theory. We first consider effects that are relatively straightforward perturbations of the linear solutions described in the previous section; then we examine highly nonlinear phenomena that drastically alter the nature of the flow.

Second-Order Effects

Numerous authors have analyzed various complicating factors that can exist on real airfoils, such as airfoil thickness and camber, finite mean angle of attack, large-amplitude motion, boundary-layer displacement thickness effects, and wind-tunnel wall corrections. A few representative examples include the early theoretical analyses of thickness effects by Küssner (1960) and van de Vooren & van de Vel (1964); the finite-difference programs of Giesing (1968) and Desopper (1981) for arbitrary geometry and airfoil motion, the latter including boundary-layer effects; the superposition methods of McCroskey (1973) and Garner (1975); the second-order perturbation analysis of the interaction between a thick cambered airfoil and a gust field by Hamad & Atassi (1980); and the analysis of unsteady wind-tunnel interference by Fromme & Golberg (1980). One general trend that emerges from these and many other studies is that the individual surface pressures are usually affected much more than the *difference* in pressure across the airfoil, which produces the fluctuating lift and pitching moment. It may be mentioned, moreover, that the various approaches generally agree in their predictions of the *trends* of C_P, but there are discrepancies in the predictions of the relatively smaller deviations of C_L and C_M from thin-airfoil theory.

In the linear theory, boundary conditions on the airfoil and in the wake are applied either along the x-axis or along the chordline of the airfoil and a mean streamline in the wake. Provided this mean surface approximation is not violated too strenuously, the various effects classified herein as secondary can be described rather well by superposition of individual corrections to the thin-airfoil solutions discussed previously. This is illustrated in Figure 4, where the superposition method of McCroskey (1973) is applied to the experimental results of McCroskey & Pucci (1981) for a 17%-thick airfoil with moderate camber at $M_\infty = 0.3$. Here the motion is given by $\alpha = \alpha_0 + \alpha_1 \cos \omega t$, where $\alpha_0 = \alpha_1 = 5°$, and the pressure coefficient can be decomposed into mean, first, and second harmonic components as follows:

$$C_P = C_{P_0} + C_{P_{1c}} \cos \omega t + C_{P_{1s}} \sin \omega t + C_{P_2} \cos(2\omega t + \phi_2). \quad (5)$$

Also shown in Figure 4 is the flat plate result, Equation (1), which is clearly unsatisfactory for C_P; nevertheless, Δp across the airfoil and consequently C_L (not shown) are given reasonably well by Equations (1) and (2), respectively. For this example the best results were obtained by superposing the individual inviscid contributions (thickness, camber, mean angle, and amplitude) to the surface velocity, correcting for boundary-layer displacement thickness effects, and then using the nonlinear Bernoulli equation to calculate the theoretical unsteady pressure coefficient. Linear theory was used to correct the data for wind-tunnel interference.

What can be said about the results of these refinements? The answer depends on what information is required. Obviously, the details of the individual pressure distributions in Figure 4 are improved, but in this and many other practical cases, thin-airfoil theory gives adequate results for the overall forces and moments. This is somewhat analogous to the steady

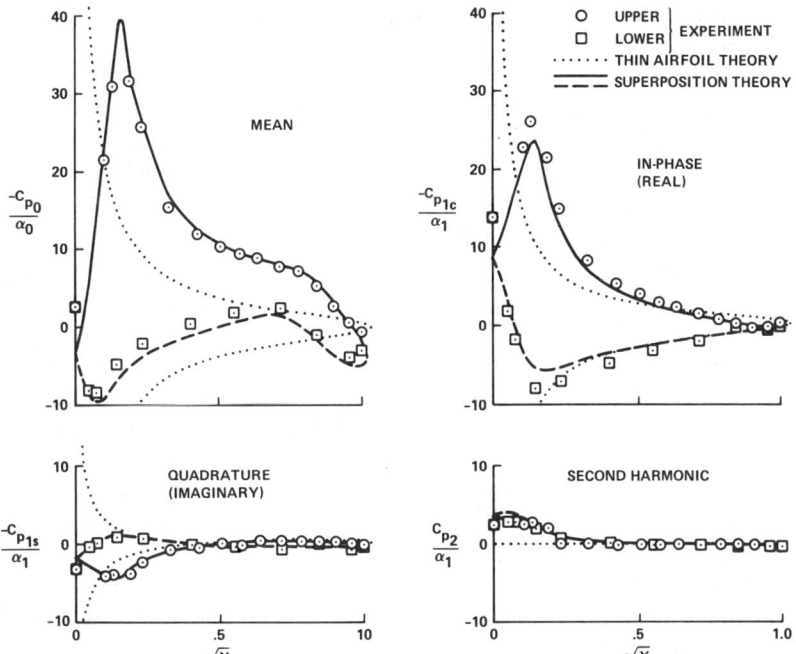

Figure 4 Pressure distributions on an oscillating airfoil: NLR-7301 airfoil, $\alpha = 5° + 5° \cos \omega t$.

case, in which thickness effects tend to be approximately cancelled by viscous effects in the absence of separation or strong shock waves. However, if boundary-layer calculations are to be done, this requires a realistic pressure distribution. Therefore, meaningful refinements to linearized thin-airfoil theory require that the actual airfoil geometry *and* viscous effects be considered together. The near-term trends are probably best exemplified by Desopper (1981); an unsteady integral boundary-layer method utilizing a quasi-steady turbulence model is coupled to a potential-flow numerical code, at little additional computational expense. This approach is probably adequate so long as no boundary-layer separation occurs.

Trailing-Edge and Wake Effects

For uniqueness, inviscid airfoil theory invokes the so-called Kutta condition of regular flow in the neighborhood of a sharp trailing edge. In the steady case, this condition is variously interpreted as fixing the rear stagnation point at the trailing edge, avoiding infinite velocities there, or establishing pressure or velocity continuity (or both) across the trailing edge. Refinements that correct for boundary-layer interactions in the trailing-edge region and for wake thickness and curvature effects have been reviewed recently by Melnik (1980), but comparable extensions to the unsteady case remain to be done.

In the more complex unsteady case, the trailing-edge flow situation is poorly understood, physically and mathematically. The usual assumption in inviscid analyses is that no pressure difference can be sustained across the trailing edge and the wake, as illustrated in the upper half of Figure 5. In this model, attributed to E. C. Maskell by Basu & Hancock (1978), the streamlines adjacent to the airfoil leave tangentially to either the upper or lower surface, according to the sign of $d\Gamma/dt$. Furthermore, the wake is usually assumed to remain thin and straight, or at least to follow a steady-flow streamline.

Numerous flow visualizations on oscillating airfoils indicate that the wake has a strong tendency to organize itself into a series of vortices, as indicated in the lower half of Figure 5 and discussed by McCroskey (1977). Few unsteady-airfoil theories include this phenomenon, but more disturbing from the standpoint of theoretical modeling are the experimental indications that finite pressure loading and abrupt streamline curvature can exist in the trailing-edge region under some conditions. Also, the phase lag of the absolute pressure near the trailing edge can deviate significantly from linear theory (see Commerford & Carta 1974 and Lorber & Covert 1981, for example).

Notwithstanding the recent analyses by Daniels (1978) and Yates (1978), a theoretical understanding of this problem is lacking, the experimental evidence on the range of validity of the Kutta condition is mixed, and the dominant parameters have not been clearly identified. For example, the breakdown of the Kutta-Joukowski condition generally seems to occur as the reduced frequency increases, more so for blunt trailing edges than for sharp (see McCroskey 1977 and Brooks & Hodgson 1979). However, Satyanarayana & Davis (1978) reported the breakdown for $k \simeq 1$ for a thin airfoil oscillating in pitch, whereas Fleeter (1980) concluded that the zero-loading condition prevailed to k in excess of 7 on a flat plate but not for a highly cambered blade in a cascade. Other possibly significant parameters, such as amplitude and type of model motion or flow oscillation, have not been studied systematically.

Further detailed studies are necessary to clarify the nature of the trailing-edge conditions, with or without moderate amounts of trailing-edge separation. The correct theoretical modeling of this region is important in determining the acoustic radiation from wings and rotating blades, in predicting hydrofoil flutter, in analyzing the flow in turbomachinery, in predicting trailing-edge stall on oscillating airfoils, and in understanding certain types of bird and insect flight.

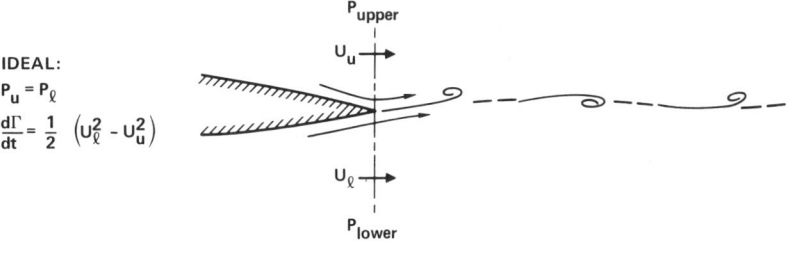

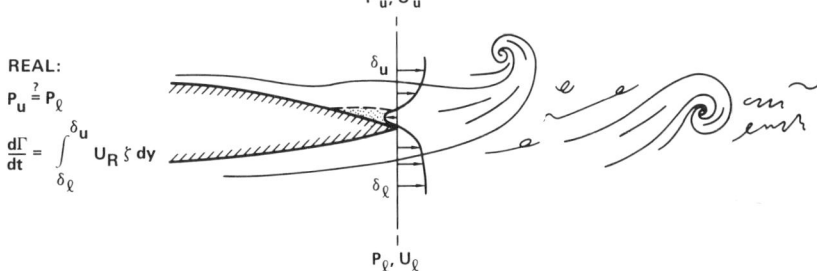

Figure 5 Sketches of some possible trailing-edge conditions in unsteady flow; ζ is the vorticity convected into the wake at relative velocity U_R.

UNSTEADY TRANSONIC FLOWS

The highly nonlinear flows in this category contain supersonic zones embedded in subsonic fields. The supersonic regions on the airfoil usually terminate with shock waves, whose position and strength are sensitive to small changes in the flow conditions or airfoil parameters. The airloads are much more complicated, and flutter in either single or multiple degrees of freedom is much more likely to occur in the transonic regime than for flows at M_∞ significantly different from unity.

A very large number of papers on unsteady transonic flow has appeared in the recent literature, including a lucid and comprehensive review by Tijdeman & Seebass (1980). Mehta & Lomax (1981) and Nixon (1981) have reviewed computational aspects, Yates & Olsen (1980) have discussed recent experiments, and Borland (1979) has prepared a bibliography that lists more than 200 publications between 1970 and 1979. Since then, over 70 more papers have appeared. With this much recent information readily available, we restrict this section to a brief discussion of the distinguishing features of unsteady transonic flows and to the general trends of current research.

Although it is very difficult to generalize about transonic flows, some idea of the differences from the linear regimes discussed earlier can be obtained from Figure 6. The flow-field development does not pass uniformly from the purely subsonic to completely supersonic regimes; instead, a wide range of patterns can occur in the intermediate stages. Figure 6 merely indicates some representative ones.

As M_∞ is increased above the critical value, shock waves normally form near the middle of the airfoil (Figure 6b) and move fore and aft as the airfoil oscillates. In the simplest case, the shock-wave motion is approximately sinusoidal, although it lags the airfoil motion appreciably (see Tijdeman & Seebass 1980, for example). Points on the airfoil that are traversed by the shock wave alternately experience the low pressure ahead of it and the higher pressure behind it. It is important to realize that at any given time, the instantaneous pressure distribution qualitatively resembles the plot of C_{P_0}; but the harmonic components of $C_P(t)$ reflect the large fluctuations due to the moving shock wave, as illustrated by the peaks in the curves of C_{P_1} in Figure 6b. This relatively concentrated apparent pressure disturbance, sometimes called a shock doublet, is centered about the mean shock-wave position, and it is a distinguishing feature of periodic transonic flows.

As $M_\infty \rightarrow 1$, the shock waves tend to move to the trailing edge, as in Figure 6c. A fishtail pattern of shock waves develops, and the shock doublet tends to disappear. At supersonic speeds (Figure 6d) the pressure coefficients, especially the unsteady components, are more nearly constant over the airfoil.

Other and more complex flow patterns are possible in the transonic regime, depending on the airfoil parameters, such as mean angle, geometry, and flap configuration; the parameters of the motion, such as amplitude, reduced frequency, and pitch axis; and an interrelated aspect, the extent and severity of shock-wave and boundary-layer interaction. In the following section, we consider conditions for which the shock-wave motion and mean position are crucial, but unsteady viscous effects are secondary. This is followed by a brief review of unsteady shock-induced separation, which, of course, is much more difficult to analyze and predict.

Nearly Inviscid Flows

In the absence of boundary-layer separation, the motion of the shock wave is the essential feature that makes the unsteady transonic problem nonlinear (see Tijdeman & Seebass 1980, Ashley 1980, and Nixon 1981). The collective experience from theoretical analyses, experiments, and numerical computations indicates that the shock-wave motion is greatest at low to moderate reduced frequencies and in the low transonic regime (Figure 6b, for example). The nonlinearity of this regime poses a potentially severe complication when the fluid-dynamic calculations must be coupled to the

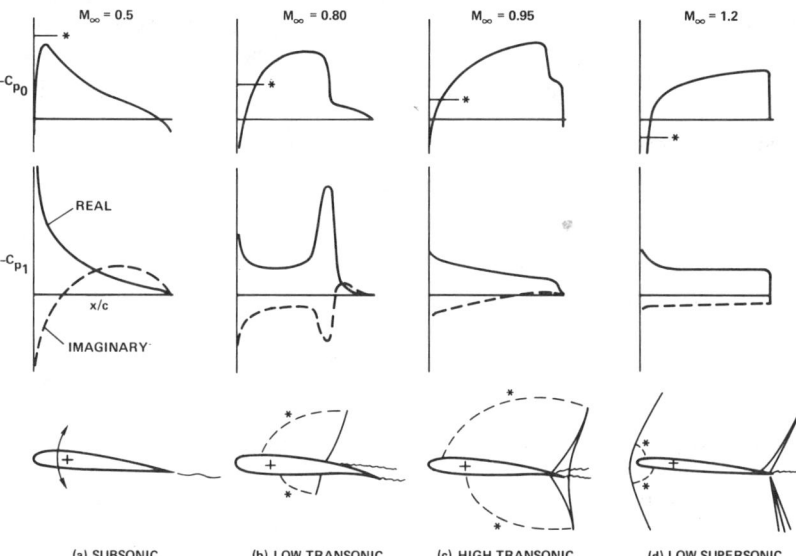

Figure 6 Representative flow fields and pressure distributions at various Mach numbers; * denotes sonic conditions.

motion of a flexible dynamic structure and many cases analyzed, such as in flutter predictions.

In many cases, however, the most difficult part of the problem is determining the correct *mean* position and strength of the shock wave. The integrated *fluctuating* lift and, to a somewhat lesser extent, the pitching moment, often tend to remain more nearly linear with respect to the motion than do the individual pressures (see Tijdeman & Seebass 1980 and Davis & Malcolm 1980a,b, for example). This is somewhat analogous to the lesser importance of secondary effects with respect to C_L and C_M than to C_P in the purely subsonic domain, discussed previously. Consequently, the superposition principles that are invaluable to aeroelastic analyses tend to remain valid to a surprising extent. When this is true, these analyses can continue to use either harmonic coefficients or indicial functions, even though classical subsonic and supersonic linear theory breaks down and other aerodynamic methods are needed. The requirements for this fortunate situation appear to be that the shock waves not induce boundary-layer separation and that the amplitude of the unsteady motion remain small. However, additional research is needed to establish this more definitively in terms of the airfoil conditions and motion parameters.

If classical linear theory is inadequate, what alternatives are available? Table 1 lists the hierarchy of possible unsteady equations; the equations themselves and detailed discussions of various solution techniques are given by McCroskey (1977), Tijdeman & Seebass (1980), Ashley & Boyd (1980), Nixon (1981), and Mehta & Lomax (1981). Most of the recent and current research activity concerns the small-disturbance equations,

Table 1 Hierarchy of equations for unsteady transonic flow

Equations	Remarks
Navier-Stokes with turbulence modeling	Strong shock waves and separation
Thin-layer Navier-Stokes with turbulence modeling	Strong shock waves, moderate separation
Euler[a]	Inviscid, rotational, strong shock waves
Full potential[a]	Weak shock waves, exact airfoil boundary conditions
Small disturbance	Simplified grids and boundary conditions
Nonlinear[a]	Moderate amplitude
Time linearized[b]	Small amplitude shock motion, harmonic or indicial
Local linearizations[b]	Small amplitudes, harmonic or indicial
Transonic linear theory	High-frequency, stationary shock waves

[a] With or without coupled viscous layer equations.
[b] With or without simplified viscous corrections.

which may be expressed in many forms (see Ashley & Boyd 1980, for example). For many practical problems, this general level represents a reasonable compromise between acceptable accuracy and the computational efficiency that is required for aeroelastic calculations.

Dowell et al. (1981) recently commented on the suitability of the various equations for engineering applications, based on their study of the boundaries between linear and nonlinear regimes, that is, where superposition is and is not valid for C_L and C_M. Noting that unsteady nonlinear effects diminish with increasing frequency, they suggested seeking the simplest method that would suffice in a given reduced-frequency domain. Figure 7 illustrates this concept for a representative small-amplitude oscillation of an airfoil in a flow with weak shock waves.

In particular, the need for nonlinear unsteady equations might well be limited to the low-frequency region for many applications. For a given airfoil, mean angle, and Mach number, the extent of the nonlinear domain would be expected to diminish with decreasing amplitude of oscillation, and vice versa. Furthermore, Chyu & Schiff (1981) described a technique that could perhaps further reduce the number of nonlinear calculations required. In any case, Dowell et al. (1981) correctly conclude that no method can be useful unless the mean steady flow that it either uses or predicts is correct.

Meanwhile, three other facets of contemporary research are noteworthy. The first is that numerical algorithms and grid-generation techniques for the more accurate full-potential, Euler, and Navier-Stokes equations are

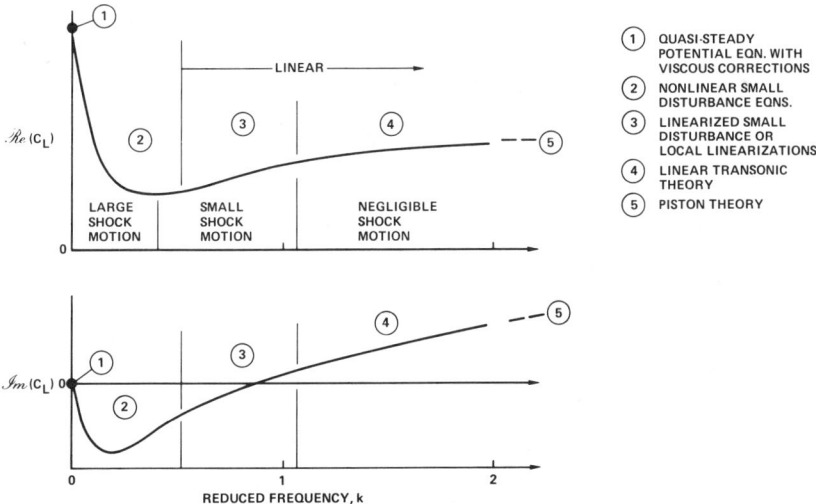

Figure 7 Unsteady flow regimes for a transonic airfoil with weak shock waves.

being vigorously developed for future applications. The second concerns the use of approximations to the far-field boundary conditions that permit the size of the computational domain to be reduced significantly. For example, Kwak (1981) and Fung (1981) have incorporated nonreflecting boundary conditions and linearized far-field solutions, respectively, into small-disturbance codes to accomplish this.

The third important trend is the coupling of viscous effects to various inviscid formulations. Viscous effects can have a major effect on the mean shock-wave position (see Melnik 1980 and LeBalleur 1980 for steady flows), and, in the light of the previous remarks, this should certainly be included. Beyond that, unsteady viscous phenomena are just beginning to be incorporated, but preliminary indications are that they are more important in the transonic than in the subsonic regime. The preliminary studies by Desopper (1981), Couston et al. (1980), and Rizzetta & Yoshihara (1980) presage new results that can be anticipated in the future for relatively mild interactions.

Strong Viscous Effects

Certain ranges of flow conditions and airfoil-motion parameters produce unsteady shock-induced boundary-layer separation, trailing-edge separation, or various combinations of strong viscous-inviscid interactions that are well beyond the scope of the techniques of the previous section. Some of the more common examples are separation-induced transonic flutter, buffet, and aileron buzz.

Aerodynamic data relevant to flutter problems are normally generated on airfoils undergoing forced oscillations. The experiments of Davis & Malcolm (1980a,b) provide striking examples of the effects of shock-induced separation on the motion of the shock wave and on the pressure loading downstream of the shock. The lift and moment coefficients are also affected dramatically. Accurate predictions of these flows will probably require numerical solutions of the Navier-Stokes equations, but to date this has not been accomplished.

Self-induced flow oscillations, alternating in an approximately regular and periodic manner between trailing-edge and shock-induced separation, have been studied on stationary, biconvex airfoils of 14% and 18% thickness and at zero incidence (see Seegmiller et al. 1978 and Levy 1981). For these cases, numerical solutions of the Reynolds-averaged Navier-Stokes equations were obtained, using an algebraic eddy-viscosity model of the turbulent stresses. These matched the qualitative features of the measured flows and were consistent with the experimental reduced frequencies and Mach-number boundaries for the self-induced oscillations. The review by Mehta & Lomax (1981) may be consulted for additional comments on these investigations.

Although these biconvex airfoil examples are relevant to flutter, they are actually special cases of the important problem of transonic buffet. The term "buffeting" is used to describe the structural response of aircraft or missile components to the aerodynamic excitation, or "buffet," that is produced by flow separation. In contrast to flutter, the overall body motion is approximately uniform, and the structure normally flexes less under the influence of the buffet airloads.

Buffeting predictions require a knowledge of the fluctuating pressures, which are characterized primarily by the rms pressure coefficient, the dimensionless frequency spectrum, and a frequency parameter $n = \pi k L/c$, where L is a measure of the size of the separated zone and k is the reduced frequency defined previously. Although many experimental results indicate that the spectra may span $0.01 < n < 100$, typically only a small fraction of the total buffet energy is contained above $n \simeq 1$ (see Mabey 1973). This value corresponds very approximately to k of the order of 10 for many typical transonic applications and to frequencies of the same order but somewhat less than the mean frequencies of turbulent eddies quoted by Chapman (1979), assuming L to be of the order of 10 times the viscous layer thickness.

The conditions for the onset of buffet and the spectra of the pressure fluctuations are normally obtained experimentally. However, Mehta & Lomax (1981) have reported on thin-layer, Reynolds-averaged Navier-Stokes calculations by H. E. Bailey and L. L. Levy, Jr., in which buffet-onset boundaries were estimated from the unsteady lift behavior. In accord with the general recommendation of Chapman (1979), it would seem that this and similar Navier-Stokes calculations should be continued and extended to examine the extent to which the broad-band frequency *spectra* of the fluctuating pressures can be predicted, using current quasi-steady turbulence models in the unsteady Reynolds-averaged equations, and whether relatively straightforward extensions can produce useful quantitative results.

The final topic of this section concerns aileron buzz. Aerodynamically excited oscillations of control surfaces can occur when the shock-wave motion and the associated shock doublet introduce negative aerodynamic damping into the hinge moment (see Lambourne 1968 for an excellent discussion of this problem). Although strong viscous effects are not necessarily essential to initiate buzz, in most cases they appear to be crucial in order to sustain its limit-cycle behavior. The most successful analysis of the phenomenon to date is the series of thin-layer Navier-Stokes calculations by Steger & Bailey (1980). Figure 8 illustrates their attempts to predict the measured buzz boundary on a P-80 wing. The numerical simulation of this long-standing problem was not perfect, but the results are unique and quite encouraging.

It is interesting to note that the extent of the separated regions in Steger & Bailey's results was rather limited. This raises the question of whether comparable results might be obtained from simpler sets of equations that also couple the nonlinear inviscid flow with a thin viscous layer. Based on Lambourne's observations (1968), there are probably a number of buzz conditions for which other zonal modeling, as well as the full Navier-Stokes approach, should be explored.

DYNAMIC STALL AND STALL FLUTTER

A certain degree of unsteadiness always accompanies the flow over an airfoil or other streamlined lifting surface at high angle of attack, but the stall of a thin body undergoing unsteady motion is even more complex than static stall. If the angle of attack oscillates around a mean value α_0 that is of the order of the static-stall angle, large hystereses develop in the fluid-dynamic forces and moments with respect to the instantaneous angle $\alpha(t)$. The maximum values of the lift, drag, and pitching-moment coefficients can greatly exceed their static counterparts, and not even the qualitative behavior of C_L, C_D, and C_M can be reproduced by neglecting the unsteady motion of the airfoil.

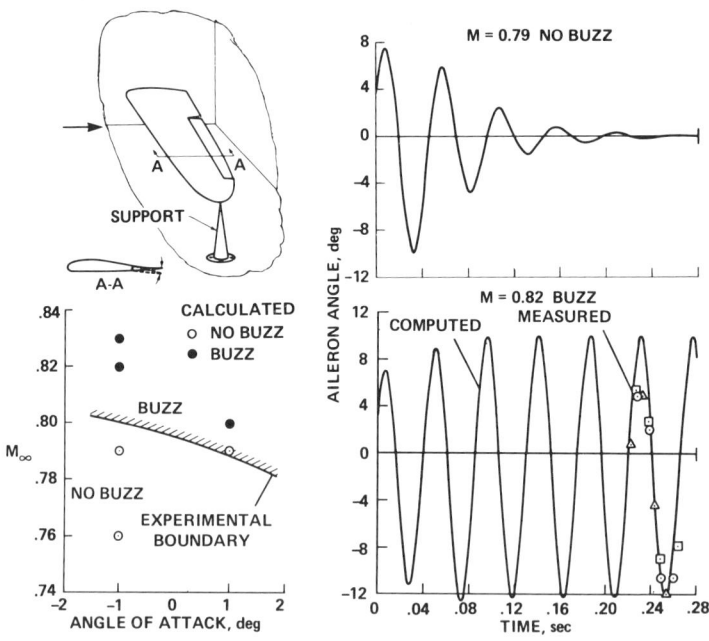

Figure 8 Comparison of measurements and calculations for transonic aileron buzz.

In addition, the instantaneous aerodynamic damping can become negative during a portion of the cycle. This is illustrated by the dotted shading in Figure 9, adapted from McCroskey & Pucci (1981). If the net damping over the cycle $\zeta = -\oint C_M d\alpha$ is negative, the airfoil extracts energy from the flow, and the pitch oscillations will tend to increase in amplitude, unless restrained. This, of course, is the condition for flutter, and unlike the linear domain described in the first part of this review, the hysteresis in the unsteady separation and reattachment permits flutter to occur in a single degree of freedom of oscillatory body motion.

In practical applications, a distinction is sometimes made between stall flutter and dynamic stall. Stall flutter refers to oscillations of an elastic body that are caused by separated flow which would be nominally steady in the absence of any motion of the body, but which is made unsteady by the flow-induced body oscillations. The term dynamic stall, on the other hand, usually refers to unsteady separation and stall phenomena on airfoils that are forced to execute time-dependent motion, oscillatory or otherwise, or to cases where flow-field perturbations induce transitory stall. Stall flutter and dynamic stall share many common features; the primary fluid-dynamic difference is that the amplitude of the motion is often smaller in stall flutter.

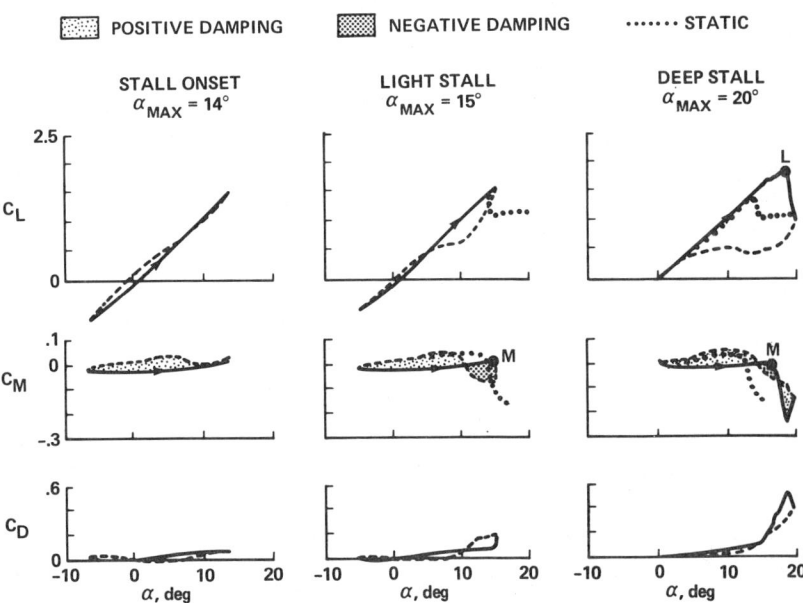

Figure 9 Unsteady forces and moments in three dynamic-stall regimes: $M_\infty = 0.3$, $\alpha = \alpha_0 + 10° \cos \omega t$, $k = 0.10$.

Stall Regimes

One of the reasons that dynamic stall is more difficult to analyze and predict than static stall is its dependence on a much larger number of parameters. Table 2 lists the more important ones and gives some idea of their significance in delineating the distinctions between static and dynamic stall. Experiments in the last few years have shown that the flow field around an oscillating airfoil in subsonic flow can be characterized by the degree or extent of flow separation and that for a given airfoil, Mach number, and Reynolds number, the primary parameter that determines the degree of separation is the maximum angle of attack ($\alpha_{max} = \alpha_0 + \alpha_1$ for sinusoidal oscillations). This contrasts with the hierarchy of viscous effects on oscillating airfoils at transonic speeds and low angles of attack, where the scale of the interaction is governed primarily by the strength and motion of the shock wave. So far, prediction methods for this class of problems have not been successful for the low-speed, high-angle problems, nor have prediction methods for low-speed, high-angle problems been successful for transonic flows.

The importance of α_{max} is illustrated in Figure 9, which portrays three important regimes of viscous-inviscid interaction for oscillating airfoils. In this particular instance, very little separation occurred when α_{max} was 13° or less, although some viscous effects were noted. When α_{max} was increased to 14°, a limited amount of separation occurred during a small fraction of the cycle and distorted the hysteresis loops of the unsteady pressures and airloads (Figure 9a). From a practical standpoint, the effect on C_M and C_D is particularly important. This stall-onset condition represents the limiting case of the maximum unsteady lift that can be obtained with no significant penalty in pitching moment or drag. Further increases in α_{max} produce the dynamic-stall conditions described below, which are charac-

Table 2 Importance of the dynamic-stall parameters

Stall parameter	Effect
Airfoil shape	Large in some cases
Mach number	Small below $M_\infty \sim 0.2$; large above $M_\infty \sim 0.2$
Reynolds number	Small (?) at low Mach number; unknown at high Mach number
Reduced frequency	Large
Mean angle, amplitude	Large
Type of motion	Virtually unknown
3-D effects	Virtually unknown
Tunnel effects	Virtually unknown

terized by large phase lags and hystereses in the separation and reattachment of the viscous flow.

LIGHT STALL   Figure 9b illustrates the next level of viscous-inviscid interaction, which was obtained at a slightly greater value of α_{max} with all other conditions remaining the same. This category of dynamic stall shares some of the general features of classical static stall, such as loss of lift and significant increases in drag and nose-down pitching moment compared with the theoretical inviscid values, when α exceeds a certain critical value. However, the *unsteady* stall behavior is characterized by growing hysteresis in the airloads. Also, the tendency toward negative aerodynamic damping, as discussed above, is strongest in this regime.

Another distinguishing feature of light dynamic stall is the scale of the interaction. The vertical extent of the viscous zone tends to remain of the order of the airfoil thickness, illustrated in Figure 10a, and this is generally less than for static stall. Consequently, this class of oscillating-airfoil problems should be more within the scope of zonal methods or thin-layer Navier-Stokes calculations with relatively straightforward turbulence modeling than either static stall or deep dynamic stall.

The qualitative behavior of light stall is known to be especially sensitive to airfoil geometry, reduced frequency, maximum incidence, and Mach number; also, three-dimensional effects and the type of motion are probably important. The quantitative behavior is closely related to the boundary-layer separation characteristics, for example, leading-edge versus trailing-edge separation, and to the changes in this separation behavior with α_{max}, k, and M_∞. The effects of these parameters are described in more detail by McCroskey (1981) and McCroskey et al. (1981).

DEEP STALL   Figure 9c shows the effects of increasing the maximum incidence to values well in excess of the static-stall angle. The initial breakdown of the flow in the deep-stall regime begins with the formation of a strong vortex-like disturbance in the leading-edge region. This vortex is shed from the boundary layer and moves downstream over the upper surface of the airfoil, producing values of C_L, C_M, and C_D that are far in excess of their static counterparts when α is increasing; large amounts of hysteresis occur during the rest of the cycle. The scale of the viscous interaction zone is also large; the thickness of the viscous layer is of the order of the airfoil chord during the vortex-shedding process, as illustrated in Figure 10b.

Figures 9 and 10 illustrate some of the qualitative and quantitative differences in light and deep dynamic stall. Moment stall, denoted by M in Figure 9, occurs rather abruptly in both cases, but the deep-stall drop in lift after $C_{L_{max}}$, denoted by L, is not evident in the light-stall case. The large negative values of C_M in deep stall are due to the vortex.

A combination of large amplitudes of the oscillation (or large pitch rates in the case of nonperiodic motion) *and* large maximum angles is essential to the development of deep stall. If these conditions are fulfilled, and provided that M_∞ is low enough that leading-edge shock waves do not develop, then the qualitative features are relatively insensitive to the details of the airfoil motion, airfoil geometry, Reynolds number, and Mach number. However, Mach number becomes an important parameter if the flow becomes supersonic in the leading-edge region. The quantitative airloads depend primarily on the time history of the angle of attack for the portion of the cycle when α exceeds the static stall angle, α_{ss}. Again, this feature and other details of the flow behavior are discussed at greater length by McCroskey (1977, 1981) and McCroskey et al. (1981).

Methods of Calculation

Theoretical progress remains rather slow on this difficult problem, although a number of prediction methods have been developed and are still

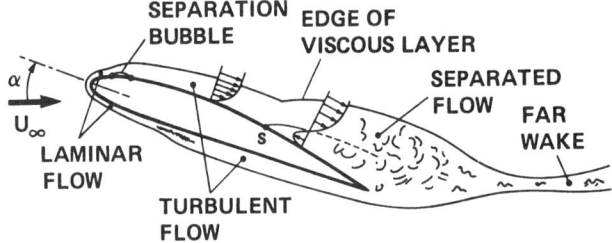

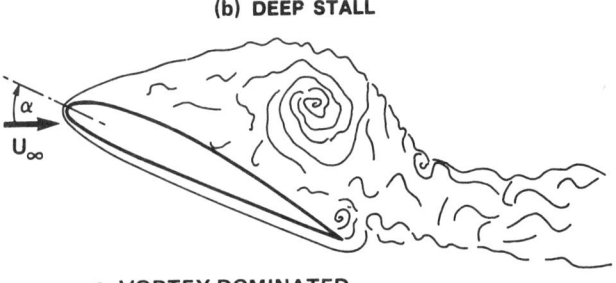

Figure 10 Sketches of flow fields during dynamic stall.

being refined. The techniques that exist all invoke restrictive assumptions and approximations, and they are tailored to the specific features of some particular stall regime or to a relatively narrow range of parameters. The brief descriptions below are condensed from the recent review by McCroskey (1981) which also cites the original references for the various approaches.

The organized vortex-shedding phenomenon of the deep-stall regime has motivated several attempts to model it by discrete potential vortices, analogous to the vortex methods for bluff-body flows. The crux of this general approach lies in choosing the strength and location of the vortex emissions, and in relating their properties to the boundary-layer separation characteristics on the body. Also, computational expense limits the time step size and the total number of vortices in the flow field. The qualitative features of deep stall have been simulated well by this approach, but further refinements are required for suitable engineering accuracy at reasonable computational efficiency.

Zonal methods of viscous-inviscid interaction, similar to those mentioned in the previous section, are being developed and refined in several laboratories, but no entirely satisfactory dynamic-stall predictions are available thus far. The same can be said of numerical solutions to the Reynolds-averaged Navier-Stokes equations for high-Reynolds-number flows, although the recent calculations of Tassa & Sankar (1981) are encouraging and several laminar calculations have been published (see McCroskey 1977). An increasing output of both zonal methods and Navier-Stokes calculations can be expected in the next few years.

The helicopter industry has developed several engineering prediction techniques based on empirical correlations of wind-tunnel data for estimating the unsteady airloads on oscillating airfoils. These methods seek to correlate force and moment data obtained from relatively simple wind-tunnel tests as functions of the numerous relevant parameters, such as airfoil shape, Mach number, amplitude and frequency of oscillation, mean angle, and type of motion.

Common to all the available literature relevant to dynamic stall is the observation that unsteady effects increase with increasing pitch rate, that is, with rate of change of airfoil incidence. It is also evident that the dynamic-stall events require finite times to develop. Therefore, some form of the nondimensional parameters $\dot{\alpha}c/U_\infty$ and $U_\infty \Delta t/c$ appears in all of the empirical methods. Another common aspect is that the empirical correlations are used as corrections to steady-airfoil data, so that the geometrical, Reynolds-number, and Mach-number effects are included only insofar as they determine the static section characteristics.

The most recent method to come to the author's attention is that of Tran & Petot (1980). In contrast to the earlier methods that essentially curve-fit experimental data with various algebraic or transcendental functions, this method utilizes a system of ordinary differential equations. Evaluating the empirical coefficients requires a large quantity of small-amplitude data over a wide range of reduced frequencies, but a systematic way of generating these data very efficiently has been developed. The results obtained so far have been encouraging, but the method fails to treat large-amplitude deep-stall cases adequately (e.g. Figure 9c). More satisfactory results from this and the other empirical-correlation techniques await a better theoretical understanding of the basic separation and reattachment phenomena on oscillating airfoils.

CONCLUDING REMARKS

As we have seen, the major thrusts in unsteady-airfoil research are directed at understanding and predicting nonlinear phenomena that are beyond the scope of classical thin-airfoil theory. The two main classes of unsteady-flow separation indicated in Figure 11 represent the greatest challenges for future research and development.

Considerable progress has been made since the author's 1977 review in developing numerical-solution techniques for the compressible Navier-Stokes equations, which are required for complex transonic flows with extensive separation. Further refinements are needed, of course, but with this new capability it would seem that the problem of buffet airloads should be addressed next.

On the other hand, a wide range of approximate techniques and efficient numerical algorithms have become available for the simpler cases that can

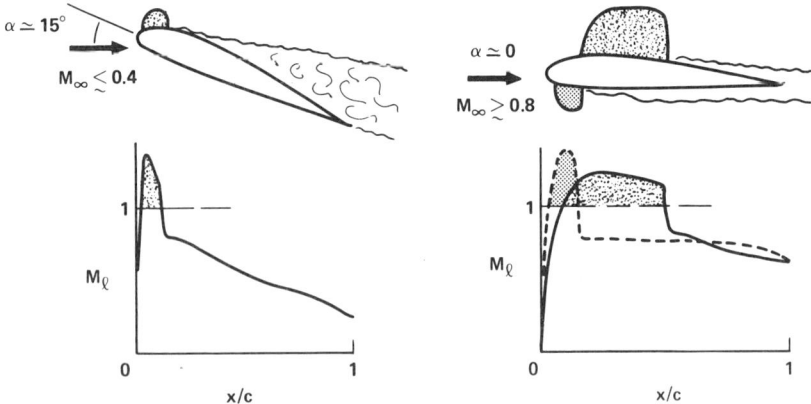

Figure 11 The twin challenges of oscillating-airfoil research.

be treated with the inviscid, transonic small-disturbance equations. For more complex flows with limited separation, vigorous development of zonal methods that couple thin viscous-layer equations with the full-potential or Euler equations seems essential and, in fact, inevitable.

Progress has lagged for flows at high angles of attack, with or without shock waves. Thus far, experiments have provided most of the fundamental knowledge and quantitative information, and the viscous behavior in the presence of large leading-edge pressure gradients has not been analyzed satisfactorily. Additional complicating factors that are important in this case are the transition and early stages of turbulent-flow development in the leading-edge region.

The qualitative features of dynamic stall are incorporated in the current empirical prediction methods that have been developed for helicopter applications, but all of them need further improvements. The recent identification of the distinct features of light and deep dynamic stall provides clearer targets for the different analytical and numerical approaches that are currently under development. Therefore, we can expect more rapid progress on this problem in the next few years.

Finally, we must appreciate the limitations of the two-dimensional viewpoint taken in this review. Useful concepts and theoretical models of unsteady phenomena can be obtained from such an approximation, and many engineering prediction methods are based on judicious applications of "strip theory" and nominally two-dimensional data. However, almost all real problems are three-dimensional to some extent, and there are often significant quantitative and occasionally major qualitative differences between two- and three-dimensional unsteady effects. Therefore, the main thrusts of unsteady fluid dynamics must ultimately be directed to three dimensions.

Literature Cited

Amiet, R. K. 1975. Effects of compressibility in unsteady airfoil lift theories. In *Unsteady Aerodynamics*, ed. R. B. Kinney, pp. 631–54. Tucson: Univ. Ariz.

Ashley, H. 1977. Unsteady subsonic and supersonic inviscid flow. *Pap. No. 1, AGARD Conf. Proc. 227, Neuilly-sur-Seine, France*

Ashley, H. 1980. Role of shocks in the "subtransonic" flutter phenomenon. *J. Aircraft* 17:187–97

Ashley, H., Boyd, W. N. 1980. On choosing the best approximations for unsteady potential theory. *Proc. Colloq. on Unsteady Aerodynamics, Deutsche Forschungs-und Versuchsanstalt für Luft- und Raumfahrt, Göttingen, Germany*

Basu, B. C., Hancock, G. J. 1978. The unsteady motion of a two-dimensional airfoil in incompressible inviscid flow. *J. Fluid Mech.* 87:159–78

Borland, C. J. 1979. A bibliography of recent developments in unsteady transonic flow. *USAF AFFDL-TR-78-189*, Vol. 1. Flight Dynam. Lab., Wright-Patterson AFB, Ohio

Brooks, T. F., Hodgson, T. H. 1979. An experimental investigation of trailing-edge noise. *NASA TM-80134*

Carta, F. O. 1978. Aeroelasticity and Unsteady Aerodynamics. In *Aerothermodynamics of Aircraft Gas Turbine Engines*, ed. G. C. Oates, Ch. 22. *USAF AFAPL TR-78-52*. Applied Prop. Lab., Wright-Patterson AFB, Ohio

Chapman, D. R. 1979. Computational aerodynamics development and outlook. *AIAA J.* 17.1293-1313

Chyu, W. J., Schiff, L. B. 1981. Nonlinear aerodynamic modeling of flap oscillations in transonic flow—a numerical validation. *AIAA Pap. 81-0073*

Commerford, G. L., Carta, F. O. 1974. Unsteady aerodynamic response of a two-dimensional airfoil at high reduced frequency. *AIAA J.* 12:43–48

Couston, M., Angelini, J. J., Le Balleur, J. C., Girodoux-Lavigne, Ph. 1980. Prise en compte d'éffets de couche limite instationnaire dans un calcul bidimensional transsonique. *Pap. No. 6, AGARD Conf. Proc. 296, Neuilly-sur-Seine, France*

Daniels, P. G. 1978. On the unsteady Kutta condition. *J. Mech. Appl. Math* 31:49–75

Davis, S. S., Malcolm, G. N. 1980a. Unsteady aerodynamics of conventional and supercritical airfoils. *AIAA Pap. 80-0734CP*

Davis, S. S., Malcolm, G. N. 1980b. Transonic shock-wave/boundary-layer interactions on an oscillating airfoil. *AIAA J.* 18:1306–12

Desopper, A. 1981. Influence of the laminar and turbulent boundary layers in unsteady two-dimensional viscous-inviscid coupled calculations. *Proc. IUTAM Colloq. Unsteady Turbulent Shear Flows, Toulouse, France*

Dowell, E. H., Bland, S. R., Williams, M. H. 1981. Linear/nonlinear behavior in unsteady transonic aerodynamics. *AIAA Pap. 81-0643CP*

Fleeter, S. 1980. Trailing edge conditions for unsteady flows at high reduced frequency. *AIAA J.* 18:497–503

Fromme, J. A., Golberg, M. A. 1980. Aerodynamic interference effects on oscillating airfoils with controls in ventilated wind tunnels. *AIAA J.* 18:417–26

Fung, K. Y. 1981. Far field boundary conditions for unsteady transonic flows. *AIAA J.* 19:180–83

Fung, Y. C. 1969. *Theory of Aeroelasticity*, pp. 401–17. New York: Dover

Garner, H. C. 1975. A practical approach to the prediction of oscillatory pressure distributions on wings in supercritical flow. *RAE TR-74181*, Farnborough, England

Garrick, I. E., Reed, W. H. III. 1981. Historical development of flutter. *AIAA Pap. 81-0591CP*

Giesing, J. P. 1968. Nonlinear two-dimensional unsteady potential flow with lift. *J. Aircraft* 5:135–43

Hamad, G., Atassi, H. 1980. Aerodynamic response of an airfoil with thickness to a longitudinal and transverse periodic gust. *AIAA Pap. 80-0151*

Hui, W. H., Tobak, M. 1981. Unsteady Newton-Busemann flow theory, Part I: Airfoils. *AIAA J.* 19:311–18

Kemp, N. H., Homicz, G. 1976. Approximate unsteady thin airfoil theory for subsonic flow. *AIAA J.* 14:1083–89

Küssner, H. G. 1960. Nonstationary theory of airfoils of finite thickness in incompressible flow. AGARD *Manual on Aeroelasticity*, Part II, Ch. 8, Neuilly-sur-Seine, France

Kwak, D. 1981. Nonreflecting far-field boundary conditions for unsteady transonic computation. *AIAA J.* In press

Lambourne, N. C. 1968. Flutter in one degree of freedom. AGARD *Manual on Aeroelasticity*, Part V, Ch. 5, Neuilly-sur-Seine, France

Le Balleur, J. C. 1980. Calcul des écoulements à forte interaction visqueuse au moyen de méthodes de couplage. *Pap. No. 1, AGARD Conf. Proc. 291, Neuilly-sur-Seine, France*

Levy, L. L., Jr. 1981. Predicted and experimental steady and unsteady transonic flows about a biconvex airfoil. *NASA TM-81262*

Lorber, P. F., Covert, E. E. 1981. Unsteady airfoil pressures produced by periodic aerodynamic interference. *AIAA J.* In press

Mabey, D. G. 1973. Beyond the buffet boundary. *Aeronaut. J.* 77:201–15

McCroskey, W. J. 1973. Inviscid flowfield of an unsteady airfoil. *AIAA J.* 11:1130–37

McCroskey, W. J. 1977. Some current research in unsteady fluid dynamics. *J. Fluids Engrg.* 99:8–38

McCroskey, W. J. 1981. The phenomenon of dynamic stall. *NASA TM-81264*

McCroskey, W. J., McAlister, K. W., Carr, L. W., Pucci, S. L., Lambert, O., Indergand, R. F. 1981. Dynamic stall on advanced airfoil sections. *J. AHS* 26(3):40–50

McCroskey, W. J., Pucci, S. L. 1981. Viscous-inviscid interaction on oscillating airfoils in subsonic flow. *AIAA Pap. 81-0051*

Mehta, U., Lomax, H. 1981. Reynolds-averaged Navier-Stokes computations of transonic flows—the state of the art. *Proc. Symp. Transonic Perspective*, ed. D. Nixon, Moffett Field, Calif.

Melnik, R. E. 1980. Turbulent interactions on airfoils at transonic speeds—recent developments. *Pap. No. 10, AGARD Conf. Proc. 291, Neuilly-sur-Seine, France*

Nixon, D. 1981. Prediction of aeroelasticity and unsteady aerodynamic phenomena in transonic flow. *Nielson Engineering and Research, Pap. 134*, Mountain View, Calif.

Postel, E. E., Leppert, E. L., Jr. 1948. Theoretical pressure distributions for a thin airfoil oscillating in incompressible flow. *J. Aeronaut. Sci.* 15:486–92

Rizzetta, D. P., Yoshihara, H. 1980. Computations of the pitching oscillation of an NACA 64A-010 airfoil in the small disturbance limit. *AIAA Pap. 80-0128*

Satyanarayana, B., Davis, S. S. 1978. Experimental studies of unsteady trailing edge conditions. *AIAA J.* 16:125–29

Seegmiller, H. L., Marvin, J. G., Levy, L. L., Jr. 1978. Steady and unsteady transonic flow. *AIAA J.* 16:1262–70

Steger, J. L., Bailey, H. E. 1980. Calculation of transonic aileron buzz. *AIAA J.* 18:249–55

Tassa, Y., Sankar, N. L. 1981. Dynamic stall of NACA 0012 airfoil in turbulent flow—numerical study. *Proc. IUTAM Symp. Unsteady Turbulent Shear Flows, Toulouse, France, also AIAA Pap. 81-1289*

Tijdeman, H., Seebass, R. 1980. Transonic flow past oscillating airfoils. *Ann. Rev. Fluid Mech.* 12:181–222

Theodorsen, T. 1935. General theory of aerodynamic instability and the mechanism of flutter. *NACA Rep. 496*

Tran, C. T., Petot, D. 1980. Semi-empirical model for the dynamic stall of airfoils in view of the application of responses of a helicopter blade in forward flight. *Pap. No. 48, Proc. 6th European Rotorcraft and Powered-Lift Aircraft Forum, Bristol, England*

van de Vooren, A. I., van de Vel, H. 1964. Unsteady profile theory in incompressible flow. *Arch. Mech. Stosowanej.* 16:709–35

von Kármán, T., Sears, W. R. 1938. Airfoil theory for nonuniform motion. *J. Aeronaut. Sci.* 5:379–90

Yates, J. E. 1978. Viscous thin airfoil theory and the Kutta condition. *AIAA Pap. 78-152*

Yates, E. C., Jr., Olsen, J. J. 1980. Aerodynamic experiments with oscillating lifting surfaces—review and preview. *AIAA Pap. 80-0450*

LOW-GRAVITY FLUID FLOWS

Simon Ostrach

Department of Mechanical and Aerospace Engineering, Case Western Reserve University, Cleveland, Ohio 44106

INTRODUCTION

As with many topics in fluid mechanics, the study of flows under low-gravity conditions was strongly motivated by practical considerations. With the advent of space flight, serious attention began to be given to the problems related to the behavior of fluids in low-gravity environments. In particular, capillary forces can be significant under low-gravity conditions. Thus, at an early stage it was recognized that the effects of capillary forces on fluids had to be understood to deal with the problems associated with the handling and control of fluids in liquid-propulsion systems and satellites.

In liquid-rocket-powered space vehicles in orbit or in transit to the moon or other planets, residual propellant can migrate away from the fuel tank drains (engine feed lines) which would preclude the restarting of engines after a period of time at low gravity. Another concern centered around the reorientation of liquid fuels after an engine is started and the liquid is subjected to accelerations. The sloshing of liquid in a partially filled tank in low gravity constituted another problem. Also, even in space the cryogenic liquids used as propellants are cooler than their surroundings and can gain heat which can lead to evaporation. Under normal gravity conditions the liquid and gas phases separate into two well-defined volumes, but in a low-gravity environment it is possible that the vapor produced at the tank wall will stay there. To design proper tank pressurization and venting systems it is necessary to know where the liquid and the vapor are distributed within the tank.

Although work on capillary-dominated fluid motions started early in the 19th century, serious attention to problems of liquid settling and interface dynamics as related to fuel storage and management began in the early 1960s. Many of the problems were, however, resolved by "practical" means. At first, for example, small rockets providing sufficient acceleration to reorient the fuel in the tank were utilized to solve the fuel-positioning problem. Later, passive retention systems consisting of baffles, which take advantage of capillary forces, were used. Mechanical stirrers were employed to overcome the phase-separation problem.

The work done in conjunction with those problems can be considered a separate topic and will not be summarized herein. Extensive listings of reports and papers dealing with the problems mentioned above, including summaries of some of the work, are presented by Stark et al. (1974a, b).

Materials science provided the second major thrust for low-gravity fluid-mechanics research. More specifically, with the increasing need for better single crystals, attention began to be focused on the transport phenomena that occur during the growth of crystals. In almost all of the commonly used techniques the parent phase is a liquid or a gas and the growth is essentially a mass-transfer process, driven by temperature and concentration gradients. In the presence of a gravitational field these gradients can lead to bulk flow of the fluid (natural convection) which in turn affects the crystal morphology and growth rate.

For the most part attempts at elucidating the fluid motions were based on ad hoc models or analogies to similar (but not identical) flow problems. In practice it was observed that uncontrolled convection often results in poor crystal morphology. Thus, many crystal deficiencies are attributed to natural convection, where that term is used in a generic sense to refer to unstable, unsteady, untameable, and undesirable flows (Carruthers 1973). Critical evaluations of the work on transport phenomena in crystal growth from a fluid-mechanics viewpoint have recently been made by Solan & Ostrach (1979) and Pimputkar & Ostrach (1981).

With the opportunity to grow crystals aboard a spacecraft in the highly reduced (*but not zero*) gravitational environment it was believed that the detrimental effects of convection would be eliminated. Several crystal-growth experiments were, therefore, conducted aboard Skylab but the results were much less conclusive than expected. It thus became evident that the complex fluid motions that occur in crystal growth warrant more systematic and intensive analysis than had previously been made. This is especially true when evaluating the potential use of low-gravity environments for crystal growth. Before proceeding with such expensive ventures, the role of convection and the parameters that control it should be fully understood on earth. Such an understanding would not only improve earth-based crystal growth but would also indicate its limitations and delineate the types of growth processes that would benefit most from the space environment.

Some work has been started to deal properly with the transport phenomena in crystal growth on earth but it still is in its initial stages. Nevertheless, the imminence of spaceflight opportunities with a seemingly simpler environment served as a stimulus for studying the various types of fluid flows that could occur under low-gravity conditions. In the present article emphasis is given to that work. Some attention is also given to flows other than those associated with materials processing that would be unique to the space environment. A number of such topics were identified by a committee (see Dodge 1975).

In order to identify and define clearly the meaningful low-gravity fluids problems it is necessary first to review the effects of gravity and its consequences (natural convection) on fluids. Then, a discussion of the importance of convection in materials processing helps indicate the basis for the topics that are receiving most attention. The environment in a spacecraft is next described to indicate the actual conditions to which the fluids will be subject. After those preliminaries, the various low-gravity fluids topics are discussed and the existing work summarized.

Gravity Effects on Fluids

The National Research Council Report, "Materials Processing in Space" (cited in the reference list), is used as background for this and the following subsections. More extensive discussion of the topics as they relate to materials processing can be found therein.

The macroscopic effects of gravity on fluids are, generally, well known. In fluids at rest hydrostatic pressure results from gravity. The buoyancy force on matter submerged in a fluid is proportional to both the gravitational force and the difference in density between the matter and the fluid. Thus, suspended particles such as gas bubbles, liquid drops, and solids move upward if they are less dense and downward if they are more dense than the surrounding fluid. Similarly, non-homogenized miscible but different liquids become stratified with the density decreasing upwards. Immiscible liquids become aligned in similar layers of phases. Such liquids are stably stratified.

If heat or a solute is added to a fluid column, density gradients are established in the fluid and the ultimate result is that in the presence of the gravitational force (or, more generally, a body force) fluid motions are generated that are called natural, buoyancy-driven, or density-gradient-driven convection. The flow can influence transport phenomena, such as heat transfer and solute redistribution. It is, in general, difficult to determine natural-convection flows in detail.

Thus, in summary, the macroscopic consequences of gravity on fluids are

1. the need for a container or levitation,
2. hydrostatic pressure in fluids,
3. the sedimentation of freely suspended particles, and
4. buoyancy-driven convection.

Each of these, which is directly proportional to the magnitude of gravity, can obviously be reduced in proportion to the reduction of gravity. However, the question now arises as to whether they can be controlled or eliminated on earth. To this end it is necessary to consider the other forces with which buoyancy competes in fluid systems.

Forces that Compete with Buoyancy in Fluids

Even when buoyancy alone is the driving force, pressure, viscous, and inertial forces also influence the fluid motion. Surface or interfacial tension and electric and magnetic forces also affect fluid flows and alter fluid interfaces. Mechanical pumping or surface shears are other ways of generating flows (forced convection). The importance of buoyancy depends not so much on the magnitude of gravity as on the ratio of the buoyancy force to each of the other forces involved. Dimensionless parameters, such as the Grashof number and the Bond number, are indicators of the ratios of such forces.

It is important to note that when the gravitational force is reduced, as in spaceflight experiments, the buoyancy force will be reduced and, thereby, buoyancy-driven convection. However, this may result in other forces, ordinarily of secondary importance, becoming significant so that different types of flows may ensue.

To alter the effect of buoyancy under normal gravitational conditions the magnitude of one of the other existing forces can be increased. For example, the viscous force can be increased by increasing the fluid viscosity or decreasing the size of the container. In relatively simple cases in which only a few forces act and the configuration is symmetric this can be an effective way of simulating low-gravity conditions. However, even then a desired prototype situation may not be properly modeled, i.e. the ratios of the forces to the viscous forces are altered thereby. The ability to simulate in this way diminishes as the number of forces acting are increased, as occurs in many materials-processing and separation techniques and in combustion. Another way to reduce buoyancy effects is to impose a stably stratified density gradient, but this, too, has its limitations.

Convection in Materials Processing

Heating, cooling, latent-heat and reaction-heat release, solute exclusion, and compositional gradients are commonplace in the processing of materials, i.e. in any process involving phase transformations. The associated density gradients in a gravitational field will induce flows (natural convection) that can be steady or time-dependent, weak or strong, laminar or turbulent. Although, as has been mentioned above, convection is usually considered to be detrimental it actually influences the processes in two different ways, one of which is beneficial and the other of which is not.

Convection increases the overall transport and, hence, the growth rate, which is desirable. On the other hand, it seems to affect the morphology of the solid adversely. The nature of the solid is largely determined by what occurs in the vicinity of the fluid-solid interface. In that thin, somewhat curved zone physical and/or chemical transformations occur that greatly change the composition across the interface. The heat release, density change, and other processes that take place in the vicinity of a transformation front result in nonuniformities along the front that cause its shape to change, steadily under some circumstances, periodically in others, and chaotically in still others. The result, in any case, is density gradients in the fluid that vary in direction and generate buoyancy-driven convection that can alter the transport of heat, constituent chemicals, and charge.

Materials processing can also involve particles suspended in a fluid, as in melt solidification and in cell-separation processes like electrophoresis, where distribution can be affected by natural convection. In some melt processes there are gas-liquid or liquid-liquid interfaces so that surface-tension effects can be present.

There is interest in isolating materials during their processing to eliminate contamination and other interferences due to container walls. To accomplish this it is necessary to offset the weight by another (levitating) force. Several techniques for levitating small melts in terrestrial conditions have been developed to some degree. Of these, electromagnetic levitation has been most successfully utilized for small melts of electrically conducting materials. The disadvantage of that method is that the levitating force is inherently coupled with heating so that independent control is not possible. Other means for levitating liquids in the presence of gravity include acoustic standing waves and gas streams. All of these are limited as to the size of the material that can be levitated. Fluids in low gravity are self contained by surface tension but are still subjected to small forces so that some force may still be needed for positioning. Thus, for successful containerless processing in space it is necessary to understand flows generated by surface-tension gradients due to temperature or concentration gradients. Also, the effects of levitating or positioning forces on the material must be assessed. Surprisingly very little work has been done on such problems.

It is, therefore, clear that complex physiochemical fluid flows are encountered in the processing of materials and that they are not well understood or, as yet, describable even under normal gravitational conditions. Nevertheless, from the discussion presented it would appear that among the potential advantages of a low-gravity environment are

1. The reduction of buoyancy-driven convection for substantial periods of time,

2. The reduction of the settling of particles for substantial periods of time,
3. The levitation and isolation of larger samples for containerless processing.

Offsetting these is the possibility that other forces may become significant and generate flows.

It should also be said that, in some situations, it may be desirable to have fluid flows in space, e.g. to stir fluids for cooling or for mixing to achieve uniform composition and eliminate bubbles in glasses, or to prevent radial segregation of a dopant in semiconductor crystal growth. In any event, it is important to know the extent and nature of fluid flows at low gravity and the factors on which they depend in order to minimize their effects or to utilize them to advantage.

The Near-Earth Gravity Environment

Of the various environmental parameters associated with a spacecraft in near-earth orbit gravity will have the greatest influence on the static and dynamic behavior of fluids. The force of gravity can be offset by the linear acceleration of free fall. The distance available for free fall near the earth's surface is limited and, therefore, the reduced-gravity time is short. For example, in a 400-foot drop tower free-fall time is about five seconds. For an aircraft flying a parabolic trajectory the low-gravity time is about 25 seconds because of practical limitations on speed and altitude. In sounding rockets free fall for several minutes can be obtained after the launch thrust is terminated. All of these methods have been used for reduced-gravity experiments.

On the other hand, an orbiting spacecraft is unique because it can provide a condition of essentially continuous free fall. The force of the earth's gravity on a mass in a spacecraft in near-earth orbit is only slightly less than at the earth's surface. The mass in orbit, however, experiences a force due to centripetal acceleration that virtually balances the earth's gravitational force. Both gravity and centripetal acceleration occur from conservative fields and depend on radial distance in almost exactly opposed ways. In orbital flight their effects almost completely cancel.

However, in near-earth orbit a spacecraft experiences other forces that alter the nearly weightless condition. One such force is drag that is caused by the outer fringes of the earth's atmosphere. An estimate has been made by NASA of the acceleration due to drag and perturbing forces for a typical Shuttle flight, the vehicle of primary current interest. Tables summarizing low-g accelerations for other missions (like Skylab) and for the Apollo flights are presented by Grodzka & Bannister (1974) and Ostrach (1977a). For minimum atmospheric drag the Space Shuttle will be operated in an attitude that is stabilized by the earth's gravity gradient, with the plane of the spacecraft wings in the orbital plane and the nose pointing outward from the center of the earth. In this condition the acceleration caused by drag will range from $-3 \times 10^{-6} g_0$ at 170 km altitude to $-2 \times 10^{-8} g_0$ at 560 km, where g_0 is the terrestrial gravity.

The oblateness of the earth and irregularities in its mass distribution produce small perturbations in the Shuttle's orbit. These perturbations will not influence the effective gravity at the Shuttle's center of mass because it will be in free fall. However, the forces of constraint that make everything in the Shuttle follow the same trajectory as the center of mass will produce accelerations of the order of $8 \times 10^{-7} g_0$ per meter of distance from the orbital-path center of mass and $6 \times 10^{-6} g_0$ per meter of separation from the center of the mass. These accelerations will be periodic, with the same period as the orbital motion.

Without regard to maneuvers for orbital adjustments, the only thruster firings will be those for the attitude-control system. Those will produce rotation of the spacecraft that will generally be of short duration, unless a uniform rotation is needed for thermal control of the vehicle. It is believed that firing of the attitude-control thrusters could be delayed for periods of time up to a day. The accelerations associated with attitude-control maneuvers are expected to range between 3.6×10^{-4} and $4.4 \times 10^{-5} g_0$.

Time-varying perturbations to the acceleration environment aboard a spacecraft, which are not as easily inhibited as maneuvers, can be caused by rotating and reciprocating machinery and crew motions (see Conway 1972). The net effect of crew motions, as estimated from Skylab data and corrected for mass differences between that spacecraft and the Shuttle, is expected to be a random "acceleration noise" distributed over a frequency range roughly from 0.1 to 10 Hz with a peak amplitude near 1 Hz. The amplitude of this background noise can probably be limited to approximately $3 \times 10^{-5} g_0$ when the crew is asleep. The amplitude is likely to be about ten times as large when the crew is active even when precautions are taken to maintain a "quiet ship."

Gas venting, fluid dumps, and flash-evaporator operation (to reject heat) also can produce accelerations of the order of $10^{-5} g_0$. Again, all of these can, more or less, be scheduled so that they can be delayed for periods up to a day.

From the above it is clear that the acceleration (or gravity) environment aboard a spacecraft is not uniform. To conduct meaningful scientific experiments in such a vehicle it is, therefore, necessary to have more accurate and detailed information on the background accelerations (both magnitude and direction) than is currently available. The nominal steady-state acceleration level in the Shuttle is currently given as between 10^{-3} and $10^{-4} g_0$.

Some of the gravity perturbations in the Shuttle could be avoided by placing experiments outside the vehicle in free flight, but this presents its own engineering difficulties at least equal to those aboard the spacecraft.

LOW-GRAVITY FLOWS

It has been established that the static consequences of a low-gravity environment is that the containerless processing of larger samples may be possible. The dynamic effect is that the levels of natural convection may be small enough so that flow-initiating mechanisms, which are usually negligible on earth except under special circumstances, may become dominant. It is, therefore, erroneous to think of near-earth orbits as providing "zero-gravity" conditions that result in no fluid flows. It has also been indicated that the acceleration environment aboard a spacecraft is not quiescent. A number of flows have been identified as being possible under low-gravity conditions and these, as well as others, will now be described. Some of these may appear simultaneously, but since practically no consideration has been given to such cases, the coupling of different flows cannot be discussed in much detail.

Buoyancy-Driven Flows

Natural convection even in a normal environment is one of the more complex fluid phenomena and in order to assess properly its effects in a reduced-gravity environment it is essential to be aware of what is known about it at present. To this end the key aspects are briefly reviewed here. For more details see the comprehensive reviews of that subject made by Ostrach (1964, 1972).

Buoyancy-driven flows can result from two basic configurations. The first, referred to as a stable configuration, is one in which a density gradient (due to thermal, concentration, or other effects) is normal to the body-force (gravitational) vector. In such a case flow (conventional convection) results immediately and transport may or may not be affected. The second configuration is an unstable one in which the density gradient is parallel but in an opposite sense to the body force. In that situation the fluid remains in a state of unstable equilibrium (due to heavier fluid being above lighter fluid) until a critical density gradient is exceeded. A spontaneous flow then ensues, which relatively quickly becomes steady. This motion usually takes the form of cells or vortex rolls and, hence, causes more mixing than laminar conventional convection. It must be emphasized that this instability is different from the one that leads to turbulence in laminar flows. The steady flow obtained after the critical condition is exceeded is laminar; this flow as well as all others due to buoyancy can also become turbulent under the proper conditions. The onset of unstable motion is highly dependent on the boundary conditions and on the confining boundaries, i.e. the motion can be delayed markedly by proper design of the configuration (see Ostrach & Pnueli 1963 and Sherman & Ostrach 1967). The stable and unstable modes of convection can also interact if the density gradient and the body-force vector alignment differs from those described above (Hart 1971, 1973).

If the density gradient is parallel to and has the same direction as the body-force vector the fluid is stably stratified. Again, this configuration can be superposed on the one leading to conventional convection (Ostrach & Raghavan 1979, Fu & Ostrach 1981).

Most of the configurations related to materials processing are ones in which the fluids are confined by rigid boundaries. Internal natural-convection flows are considerably more complex than external ones because the fluid boundary layer and core are intimately coupled (Ostrach 1972). This constitutes the main source of difficulty in predicting the resulting flow and transfer. More than one flow pattern seems possible for a given set of conditions and there is, at present, no way of determining a priori what it will be. Furthermore, the entire flow is sensitive to the configuration geometry and imposed boundary conditions.

Further complexities arise when materials-processing applications are considered because the density gradients can be due to both temperature and concentration gradients. The latter can have various orientations relative to themselves and to the gravity vector so that they can augment or oppose each other in coupled ways. Relatively little work exists on natural convection with coupled driving gradients, particularly for geometric configurations that relate to materials processing (Ostrach 1980).

As has already been said, the space environment complicates matters still more because of the possibility of additional driving forces. One important class of low-gravity flow problems, therefore, is that in which the buoyancy-induced flow acts as a perturbation on a process established by some other means.

In view of all the above, it is perhaps not surprising that the nature and effects of natural convection under reduced-gravity conditions were either neglected or misinterpreted by people not intimately familiar with that subject. It should now be clear that because different modes of convection are possible and are extremely sensitive to the configuration and boundary conditions the following information must be known specifically and explicitly in order to assess convection effects at reduced gravity: (a) both the magnitude and direction of any accelerations, (b) the geometric configurations, (c) the imposed boundary conditions, and (d) the material properties. Unfortunately, most existing literature on this subject is deficient in one or more of these. A delineation of the above considerations for materials scientists was made by Ostrach (1977a).

For the reasons just given, it is usually necessary to study each problem by itself to obtain detailed information on the transport processes due to buoyancy. However, considerable insight into the qualitative nature of the problem can be obtained from dimensional analysis. The first step in the understanding of complex phenomena, therefore, is to obtain the relevant dimensionless parameters. These are best determined from the basic equations and boundary conditions (Ostrach 1966).

PARAMETERS Although the dimensionless parameters for buoyancy-driven flows are seemingly well understood, until recently (Ostrach 1980) no general derivation of them existed when the density gradients are due to both temperature and concentration effects. The parameters can be used to obtain estimates of the velocities, but the existing literature is not completely clear on how this should be done properly. Therefore, since it is important to determine the magnitude of flows for a large variety of fluids under reduced gravity conditions the parameters are rederived here and methods of obtaining flow estimates are clearly delineated.

The basic (dimensional) equations expressing the conservation of mass, momentum, energy, and species are, respectively,

$$\frac{\partial U_j}{\partial X_j} = 0, \tag{1}$$

$$\rho U_j \frac{\partial U_i}{\partial X_j} = \mu \frac{\partial^2 U_i}{\partial X_j \partial X_j} + (\rho_\infty - \rho)g_i - \frac{\partial P_D}{\partial X_i}, \tag{2}$$

$$\rho c_p U_j \frac{\partial T}{\partial X_j} = k \frac{\partial^2 T}{\partial X_j \partial X_j}, \tag{3}$$

$$U_j \frac{\partial C}{\partial X_j} = D \frac{\partial^2 C}{\partial X_j \partial X_j}, \tag{4}$$

where U_j are the velocity components in the X_j directions, ρ and ρ_∞ are the fluid density in the flow field and at a reference location, respectively, μ is the absolute viscosity, g_i is the negative of the ith body force component, P_D is the pressure defined as $P - P_\infty$, c_p is the specific heat at constant pressure, T the temperature, D the diffusion coefficient, and C the concentration. The equations contain a number of assumptions that are made for convenience but are reasonable for present purposes. Density variations are considered only in the buoyancy term (the Boussinesq approximation). Compressive work, viscous dissipation, and Soret (thermal diffusion) and Dufour (diffusion thermal) effects are neglected.

A series expansion of $(\rho_\infty - \rho)$ in terms of P, T, and C indicates that pressure effects can be neglected and that only linear terms in the other variables need be retained if $\beta(T - T_\infty) \ll 1$ and $\bar{\beta}(C - C_\infty) \ll 1$, where

β is the volumetric-expansion coefficient and $\bar{\beta}$ is the concentration-densification coefficient. Introduction of the series into the buoyancy term in Equation (2) yields

$$U_j \frac{\partial U_i}{\partial X_j} = v \frac{\partial^2 U_i}{\partial X_j \partial X_j} + \beta g_i (T - T_\infty) + \bar{\beta} g_i (C - C_\infty) - \frac{1}{\rho} \frac{\partial P_D}{\partial X_i}. \quad (2a)$$

The variables are normalized by

$$u_i = U_i/U_R, \quad x_i = X_i/L, \quad \theta = (T - T_\infty)/(T_w - T_\infty),$$

$$\phi = (C - C_\infty)/(C_w - C_\infty), \quad p = P/\rho U_R^2,$$

where U_R denotes a (unknown) reference velocity, L a characteristic length, and the subscripts w and ∞ indicate two different reference values. The nondimensional equations that contain parameters are

$$u_j \frac{\partial u_i}{\partial x_j} = \frac{v}{U_R L} \frac{\partial^2 u_i}{\partial x_j \partial x_j} + \frac{g_i L \beta \Delta T}{U_R^2} (\theta + N\phi) - \frac{\partial p_D}{\partial x_i}, \quad (5)$$

$$u_j \frac{\partial \theta}{\partial x_j} = \frac{\alpha}{U_R L} \frac{\partial^2 \theta}{\partial x_j \partial x_j}, \quad (6)$$

$$u_j \frac{\partial \theta}{\partial x_j} = \frac{D}{U_R L} \frac{\partial^2 \phi}{\partial x_j \partial x_j}, \quad (7)$$

where $\Delta T = T_w - T_\infty$, $\Delta C = C_w - C_\infty$, α is the thermal diffusivity, and the ratio of concentration to thermal buoyancy is

$$N \equiv \bar{\beta} \Delta C / \beta \Delta T. \quad (8)$$

The parameter N can vary over a large range of values and for positive values the combined driving forces augment each other whereas for negative values they oppose each other. The parameters that appear as factors of the first term on the right of Equations (5) to (7) can be interpreted, respectively, as reciprocals of the Reynolds number, the product of the Prandtl and Reynolds numbers or the Péclet number, and the product of the Schmidt and Reynolds numbers. To determine the reference velocity, U_R, it is necessary to know the dominant forces involved and those, in turn, depend on the parameters. For the present, the flow will be considered to be due primarily to thermal effects, i.e. $N \ll 1$. Table 1 lists all possible combinations of the remaining parameters. The symbol S, for small, means less than unity, U denotes unit order of magnitude, and L, for large, means greater than unity. The fourth column of Table 1 indicates dominant forces and other qualitative features of the flow. Note that since the buoyancy force is generating the flow it must always be one of the dominant forces. For cases 1 to 3 a balance of the coefficients of the buoyancy and viscous forces in Equation (5) yields

$$U_R = \beta g \Delta T L^2 / v = \text{Gr}(v/L), \quad (9)$$

where $Gr = \beta g \Delta T L^3/\nu^2$ is the Grashof number. Whenever a thermal boundary layer exists (cases 4, 7, and 9 to 11) its thickness is the fundamental length because it is over that length that the buoyancy force acts. Therefore, a coordinate stretching is required to make both terms in Equation (6) of the same order. After this is done it is found that [see Ostrach (1964), pp. 578–80] for small Pr

$$U_R = (\beta g \Delta T L)^{1/2} = (Gr)^{1/2}\nu/L \tag{10}$$

and for large Pr

$$U_R = (\beta g \Delta T L/\text{Pr})^{1/2} = (Gr/\text{Pr})^{1/2}\nu/L . \tag{11}$$

When there is only a velocity boundary layer (case 8) buoyancy and inertia forces must balance and the reference velocity is just that given by Equation (10). In situations where all forces are of the same order (cases 5 and 6) either Equation (9) or (10) can be used.

If the appropriate velocity, as given by Equations (9) to (10), is substituted into the Reynolds and Péclet numbers, criteria, in terms of buoyancy-driven flow parameters, for the use of those velocities are obtained. Thus for

$Gr \leq 1$ and $Ra \leq 1$ use Equation (9),

$\sqrt{Gr} > 1$ and $Pr < 1$ use Equation (10),

$\sqrt{Gr} > 1$ and $Pr > 1$ use Equation (11);

here Ra denotes the Rayleigh number, PrGr. For situations in which the flows are primarily due to concentration differences, $N \gg 1$, $\bar{\beta}\Delta C$ should replace $\beta \Delta T$ in the definition of the Grashof number. When both thermal and concentration effects are of the same order, $N \approx 1$, $\triangle \rho/\rho$ should replace $\beta \Delta T$. Note the Grashof number can also be defined in terms of the temperature gradient as $\beta g (\partial T/\partial X) d^4/\nu^2$ and, correspondingly, for concentration gradients.

Table 1 Qualitative nature of the flow for different ranges of parametric values.

Case	Re	Pr	Pe	Comments
1	S	S	S	Buoyancy and viscous forces
2	S	U	S	Buoyancy and viscous forces
3	S	L	U	Buoyancy and viscous forces
4	S	L	L	Thermal boundary layer
5	U	S	S	All forces of same order
6	U	U	U	All forces of same order
7	U	L	L	Thermal boundary layer
8	L	S	S	Velocity boundary layer
9	L	S	L	Velocity and thermal boundary layers
10	L	U	L	Velocity and thermal boundary layers
11	L	L	L	Velocity and thermal boundary layers.

STABLE-MODE CONVECTION The procedure to estimate fluid velocitites can now be outlined. From the direction of the acceleration and the configuration the mode of convection to be expected must first be determined. The Grashof, Prandtl, and Rayleigh numbers are then computed. For conventional convection (stable configurations) the appropriate velocity equation is used according to the conditions given above.

From Equations (9) to (11) it appears that fluid velocities vary either directly with gravity or as the square root of it. However, this is not entirely true since for a given set of conditions different equations may apply as g is reduced. In any event, reducing gravity by factors of 10^{-3} to 10^{-5} does not give zero velocities. For example, under relatively mild conditions, viz. temperature differences of $10°K$ at a level of $20°C$ and a characteristic length of 10 cm, the Grashof number (for thermal convection) at one g_0 for a gas (like air) is of the order of 10^6, for a liquid (water) is like 10^7, and for a liquid metal (mercury) is about 10^9. Thus, at $10^{-5}g_0$ (about the lowest value anticipated for the Shuttle) velocities of the order of tenths of a millimeter per second would occur in the liquid and liquid metal under the given conditions and about four times as large in the gas. Clearly, increases in the temperature (or density) difference or the length would lead to higher velocities. Similar considerations apply for solutal (concentration-driven) convection. Concentration gradients in materials processing can be considerably greater than temperature gradients so that for a given g_0 level, solutal convection would be more significant than thermal convection. Whether the transport will be affected by the flows is, of course, determined from the appropriate parameters in the transport equations.

The considerations presented above are limited to situations in which there is heat (or mass) flow in only one direction (normal to gravity). Most existing work on buoyancy-induced flows has dealt only with that type of situation. Since in many practical cases heat fluxes are actually imposed simultaneously in more than a single direction and since there is interest in reducing natural convection by imposing stabilizing thermal gradients, the latter problem has recently received some attention (Ostrach & Raghavan 1979 and Fu & Ostrach 1981). It was found that stabilizing thermal gradients do retard natural convection flows appreciably, e.g. up to 62% for the ratio of vertical to horizontal temperature differences of 6. The implication for the present discussion is that velocities estimated by the method described herein would be higher than actual. Similarly, if a destabilizing thermal gradient were imposed on a conventional convection flow, velocities larger than those estimated would most likely result. Some indication of this is presented by Ostrach (1954). The interacting fluxes could be due to combinations of concentration and temperature gradients as mentioned previously. The limited state of knowledge for such problems is summarized by Ostrach (1980).

Because the configurations in which crystals are grown by the closed-tube vapor deposition and Bridgman melt methods have low aspect (height to width) ratios a study was made of the natural convection in such enclosures over a five-decade range of Grashof number and an order-of-magnitude range of aspect ratio (Ostrach et al. 1980). The flow patterns were found to vary from parallel unicellular ones, to ones that are skewed, and to ones in which there are secondary cells at the ends. The Rayleigh numbers were generally larger than those in actual processes so that the experiments did not simulate the real situations and cannot be interpreted as representative of conditions in a low-gravity environment. Nevertheless, the results do show that large changes in the Grashof number, as brought about by reduced gravity, can significantly alter the flow pattern. Reductions of the Grashof number due to low-gravity could also result in flows being laminar that are turbulent under terrestrial conditions.

There is, unfortunately, no worthwhile experimental data on reduced-gravity convection. Indirect evidence is, however, available from studies made for the cryogenic fuel-storage tanks for the Apollo 14 and subsequent flights. Although these ground-based and spaceflight studies of low-gravity convection are limited specifically to an unusual fluid (supercritical oxygen) and rather complex configurations (the Apollo tanks) the results are interesting. For example, it was found that the low-gravity convection in the Apollo tanks was sufficient to obviate the need for forced convection by mixing fans to resolve the fuel-stratification problems (Rice 1971). Data from the Apollo 14 flight (Fineblum et al. 1971) indicate increased convection as the vehicle rotation rate increased to 3 rpm.

UNSTABLE-MODE CONVECTION   If the acceleration direction and configuration are such that the unstable mode of convection is anticipated, the proper critical Rayleigh number for the configuration must be determined to see if such a flow will actually occur. In addition to the reference previously cited, there are numerous papers in which critical Rayleigh numbers are determined for various geometric configurations subject to different conditions. Space limitations preclude a detailed discussion of them. A summary of some of that work is given by Ostrach (1977a). The main point to be made, for present considerations, is that, depending on conditions, the critical Rayleigh number can be considerably less (Saville & Ostrach 1978) than 1708 (the value for a doubly infinite horizontal fluid layer bounded above and below by rigid conducting planes) and have values as high as 10^6 (due to highly confining boundaries). In any case, at low gravity the critical density gradients for the onset of flow will be much larger than on earth so that this type of convection is less likely to occur. However, as will be shown below, care must be taken not to compromise this advantage by improper orientation of the acceleration and the configuration; this may not always be possible on a spacecraft.

COMBINED-MODE CONVECTION   When a rectangular cavity with one wall hotter than the other is rotated with respect to the gravity vector (or vice versa) each of the two natural convection modes (stable and unstable) is obtained as limits and, for angles in between, both modes interact. This problem was studied by Hart (1971, 1973) and the results are summarized in Figure 1. The Rayleigh number is based on the distance and temperature difference between hot and cold walls. When the hot surface is below, $\delta = 90°$, there is no flow until the classical Bénard cells appear at Ra = 1708. At $\delta = -90°$ the hot surface is on top and the fluid layer is stably stratified so that, again, there is no flow. When the slot is nearly vertical, or inclined with the hot wall above, there is a steady circulating two-dimensional unicellular flow in which fluid rises along the hot surface and descends near the cold surface. The other types of flows that can occur when the two modes interact are delineated on the figure.

Other aspects of the combined modes for shallow boxes are treated by Unny (1972), Hollands & Konicek (1973), and Clever (1973). The combined effects of aspect ratio and inclination on the circulation and heat transfer in a finite rectangular region were theoretically investigated by Ozoe et al. (1974a, b) and experimentally by Ozoe et al. (1975). It was found that in the unstable configuration (zero inclination with the hot side below) the flow pattern was a series of roll cells with axes parallel to each other and perpendicular to the long axis of the channel. When the inclination angle was increased slightly in steps, a series of roll cells persisted with their axes in the upslope, but the average Nusselt number (heat

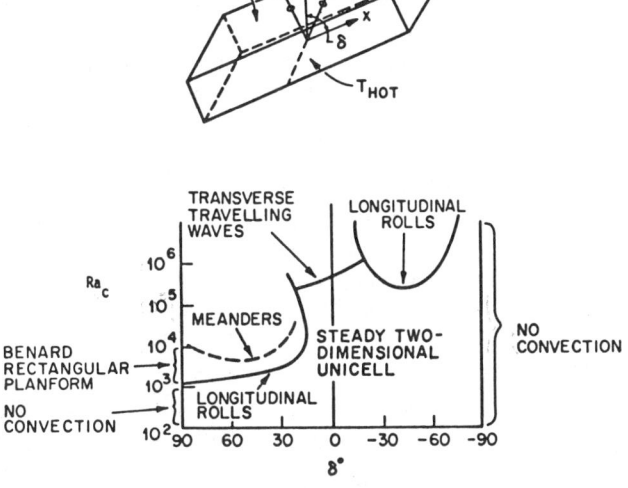

Figure 1   Combined-mode convection. From Hart (1971).

transfer) decreased. As the inclination was further increased a minimum Nusselt number was attained. Beyond that angle the flow pattern changed to a single roll cell with its axis in the long dimension of the channel and the associated Nusselt number increased. This flow pattern persisted as the angle was increased until the heated surface was horizontal but above the cold one. The Nusselt number passed through a maximum and then decreased to unity for the stratified configuration. This behavior was qualitatively the same for channels with aspect ratios from unity to 15, but the inclination angles corresponding to the Nusselt number extrema depend strongly on aspect ratio. Comparison of the theoretical and experimental results indicated that the predictions of Ozoe et al. (1974b) can be utilized up to $Ra = 10^5$ and aspect ratios up to at least 8.4.

For other configurations the results would be different than for the parallelepipeds studied. The important implication for low-gravity buoyancy flows is that the flow patterns may not be steady or simple because of variations in the acceleration direction aboard a spacecraft.

g-Jitter Convection

It has been pointed out that, in contrast to the nonuniformities of the gravity field, some of which must be accepted as part of the natural environment of an orbital vehicle, there are transient or time-varying disturbances to the gravity field at a point. The unsteady variations, which are referred to as g-jitter, can arise from spacecraft maneuvers, crew motions, and mechanical vibrations. Evidence of g-jitter in Apollo flights is presented by Grodzka et al. (1971) and Bannister et al. (1973) and discussed in some detail by Grodzka & Bannister (1974).

In an attempt to evaluate the effects of g-jitter on fluid motions Spradley et al. (1975) performed a numerical study for various configurations. They considered three g-jitter profiles, viz. sinusoidal, absolute sinusoidal, and saw tooth, and compared the flow fields that resulted from them with that for a constant g level. They found that if the g-jitter is decomposed into a time-mean part and an oscillatory part the former is more important in determining the flow field and heat transfer.

A more comprehensive investigation was made by Kamotani et al. (1981), which also gives better physical insight into the phenomena. The body force in an orbiting vehicle is first carefully derived to define clearly the nature of g-jitter. From this it can be seen that the oscillatory part of the g-jitter is more important than the mean part which is contrary to the model used by Spradley et al. (1975). Consideration is then given to a rectangular container completely filled with a fluid. The two longer walls are maintained at constant but different temperatures and the end walls are insulated. Although the g-jitter vector actually varies randomly in direction and time, for simplicity it is first taken to act only in the direction normal to the imposed temperature gradient and to have the form

$$f = A'F(t) + F_m,$$

where A' is the amplitude of the oscillatory part and F_m is the mean g-jitter level. If the oscillatory part is caused by a periodic vibration of amplitude, a, and frequency, ω, then $A' = a\omega^2$. The relevant dimensionless parameters are found to be the Reynolds number, Re $= \omega L_x^2/v$, the Prandtl number, Pr, the aspect ratio, Ar $= L_x/L_y$, the relative amplitude of fluid oscillations, $\epsilon A' = \epsilon a/L_x$, and the ratio of the mean g-jitter level to the oscillatory level, $F_m/a\omega^2$, where L_x and L_y are the lengths of the longer and shorter sides of the container, respectively, and $\epsilon = \beta \triangle T$ with $\triangle T$ a characteristic temperature difference. The Prandtl number is taken to range from 10^{-2} for liquid metals to 10 for liquids. In spacecraft ω is typically 0.1 to 10 rad s^{-1} and values of v for liquids, gases, and liquid metals are the order of 10^{-5} to 10^{-7} m^2 s^{-1}. The Reynolds number will thus be larger than unity except for very small containers (L_x and $L_y <$ 1 cm). The aspect ratio is taken to be unity in the analysis but other aspect ratios are also studied. Since the g-jitter is considered to be predominantly oscillatory, the ratio $F_m/a\omega^2$ is very small. The parameter $\epsilon A'$ is also less than unity. It is found that the g-jitter component normal to the imposed temperature gradient is most important in generating thermal convection. It induces an unicellular oscillatory fluid motion. The oscillatory-velocity level increases with Re for a given aspect ratio and is of the order $\epsilon a\omega^2$. The temperature-oscillation level increases with Re and Pr and is of the order of $\epsilon A' \triangle T$. If $\epsilon < \epsilon A' \ll 1$, the oscillatory g-jitter can generate a mean secondary flow that is superimposed on the oscillatory flow. The secondary flow velocity is of the order of $\epsilon^2 A' a\omega$. During transient periods both the velocity and temperature oscillation levels are, in general, of the same order as those for the steady state. The g-jitter component parallel to the imposed temperature gradient does not cause appreciable changes in the oscillatory flow field but could possibly alter the thermal-stability criterion. Thus, for fully confined fluids g-jitter induces predominantly oscillatory flows in which the fluid elements oscillate in a relatively small region. The associated temperature oscillations, although very small, might influence the dynamics of the growth front in the processing of materials. The effects of g-jitter on fluids with free surfaces most likely will be much more significant, but that problem has not, as yet, received any attention.

Surface-Tension Gradient Convection

The presence of an interface between two fluid phases can influence the motion of the fluids when either the interface has a finite curvature that is different from that at equilibrium or when the interfacial or surface tension varies from point to point. Since surface tension is a function of temperature, composition, and electrical potential, gradients of any one of these or combinations of them lead to surface-tension gradients. It has been pointed out by Ostrach (1977b) that there are essentially two basic modes of flow generated by surface-tension gradients that are analogous to those induced by buoyancy. If the gradients are along the interface, the surface-tension gradients act like shear stresses applied by the interface to the adjoining fluids and thereby generate flows or affect existing ones. These flows are analogous to conventional (or stable) convection and are called thermocapillary, diffusocapillary, or thermoelectric flows according to their respective causes, viz. temperature, concentration, or electric potential gradients. These flows can affect transport processes. Of the relatively little work done on this type of problem, the emphasis has been of diffusocapillary flows until recently.

If one of the above-mentioned gradients is perpendicular to the interface a "Marangoni instability" can occur under proper conditions that can lead to cellular flows. This situation is analogous to the unstable type of buoyancy flows. Problems of this type have received more attention than those with gradients along the interface. The Marangoni instability is identified as "interfacial turbulence" by Sternling & Scriven (1959) and Kenning (1968) calls it a "form of interfacial turbulence." In accord with the discussion of unstable buoyancy flows, these flows are actually laminar although they induce greater mixing.

When surface-tension flows are considered under terrestrial conditions, the relative importance of gravity and surface-tension forces was, until recently, thought to be determined by the Bond number, $Bo = \rho g d^2/\sigma$, where ρ is the fluid density, d is a characteristic dimension, and σ the surface tension. On this basis it was concluded that gravity suppresses surface tension except in configurations with very small dimensions, such as very thin films, capillary tubes, and droplets and bubbles. Even with such a restriction there are many technologically important processes in which surface tension can be significant. Kenning (1968) and Levich & Krylov (1969) outline the work done covering such applications as boiling heat transfer, spreading of films such as oil and paint, wave phenomena, jet decay, and corrosion mechanisms. Surface tension was also studied as a mechanism of flame spreading (see for example, Sirignano & Glassman 1970).

It will now be stated, and later shown, that the Bond number given above is actually associated only with situations in which surface tension influences fluids by means of interface curvature and, therefore, it is referred to herein as the *static* Bond number. For flows induced by surface-tension *gradients*, a *dynamic* Bond number will be presented that indicates the ratio of gravity to surface-tension gradient forces.

In any event, in considering the types of flows that could occur at low-g, surface-tension gradient flows are among the most interesting, not only for their intrinsic scientific value but also for their possible technological importance in relation to containerless processing. In particular, since the flows due to gradients along the interface are the more important type for materials processing and have received relatively less attention than the other type, emphasis is given to them here.

PARAMETERS As was indicated in the discussion of buoyancy-driven flows, dimensionless parameters are useful for estimating flow velocities and obtaining qualitative features of the flows. They are also useful to indicate dominant physical factors, mathematical simplifications, data correlations, and proper theoretical and experimental models. Although a physically reasonable reference velocity for surface-tension gradient flows was indicated, or implied, by Levich (1962), Kenning (1968), and Stanek & Szekely (1964), no explicit derivation of all the parameters based on such a reference was made until recently (Ostrach 1979). The derivation will now be outlined in order that the proper parameters and their implications are clear. The procedure follows that in obtaining Equations (5) and (6). The reference velocity, U_R, is obtained from the balance of the tangential shear stresses at the free surface because the driving force is applied there. For the configuration depicted in Figure 2 with a flat interface this is

$$-\mu \frac{\partial U}{\partial Y} = \frac{\partial \sigma}{\partial X} = \frac{\partial \sigma}{\partial T} \frac{\partial T}{\partial X}. \tag{12}$$

If the Reynolds number is less than unity, the inertia effects can be seen in Equation (5) to be negligible and the flow will be a "viscous" type. Therefore, the surface traction will penetrate downward into the fluid by viscosity and the proper length scale for Y is D. In nondimensional form the above equation is

$$-\frac{\partial u}{\partial y} = \frac{(\partial \sigma / \partial T)\Delta T D}{\mu U_R L} \frac{\partial \tau}{\partial x},$$

from which, for both terms to be of the same order,

$$U_R = \frac{|\partial \sigma / \partial T|\Delta T}{\mu} \frac{D}{L}, \tag{13}$$

where ΔT represents a characteristic temperature difference. When D/L is not unity the inequality for "viscous" flows is

Re$(D/L)^2 \ll 1$

and when this is combined with Equation (12), the condition becomes

$$\frac{D}{L} \ll \left(\frac{\mu \nu}{(\partial \sigma / \partial T)\Delta T D}\right)^{1/2} \equiv (R_\sigma)^{-1/2}. \tag{14}$$

Equations (13) and (14) define the characteristic velocity for "viscous" flows driven by surface-tension gradients and the surface-tension Reynolds number.

If $\text{Re}(D/L)^2 \gg 1$ a boundary-layer flow will occur and a boundary-layer thickness, δ, is the appropriate length scale, where

$$\frac{\delta}{L} = -\frac{1}{(D/L)(\text{Re})^{1/2}}. \tag{15}$$

From Equations (12) and (15) it follows that

$$U_R = \frac{|\partial\sigma/\partial T|\Delta T}{\mu}\frac{\delta}{L} = \frac{|\partial\sigma/\partial T|\Delta T}{\mu}\left(\frac{\nu}{U_R L}\right)^{1/2}$$

so that, for boundary layer flows

$$U_R = \left[\frac{(\partial\sigma/\partial T)^2 \Delta T^2 \nu}{\mu^2 L}\right]^{1/3} \tag{16}$$

and

$$\frac{D}{L} \gg (R_\sigma)^{-1/2}.$$

Note that both reference velocities are expressed in terms of the physically important variables for establishing surface-tension gradient flows.

With the reference velocities determined the dimensionless equations are the following:

For the viscous case, $R_\sigma(\text{Ar})^2 \ll 1$,

$$0 = (\text{Ar})^2 \frac{\partial^2 u}{\partial x^2} + \frac{\partial^2 u}{\partial y^2} - \frac{\partial p}{\partial y}, \tag{17a}$$

$$0 = (\text{Ar})^2 \frac{\partial^2 v}{\partial x^2} + \frac{\partial^2 v}{\partial y^2} + \frac{\text{GrAr}}{R_\sigma}\tau - (\text{Ar})^{-2}\frac{\partial p}{\partial y}, \tag{18a}$$

$$\text{Ma}(\text{Ar})^2\left(u\frac{\partial \tau}{\partial x} + v\frac{\partial \tau}{\partial y}\right) = (\text{Ar})^2\frac{\partial^2 \tau}{\partial x^2} + \frac{\partial^2 \tau}{\partial y^2}, \tag{19a}$$

where for this case the appropriate reference pressure is $\mu U_R L/D^2$ rather than ρU_R^2 and $\text{Ar} = D/L$.

For the boundary-layer case, $R_\sigma(\text{Ar})^2 \gg 1$,

$$u\frac{\partial u}{\partial x} + v\frac{\partial u}{\partial y} = \left(\frac{\text{Ar}}{R_\sigma}\right)^{2/3}\frac{\partial^2 u}{\partial x^2} + \frac{\partial^2 u}{\partial y^2} - \frac{\partial p}{\partial x}, \tag{17b}$$

$$u\frac{\partial v}{\partial x} + v\frac{\partial v}{\partial y} = \left(\frac{\text{Ar}}{R}\right)^{2/3}\frac{\partial^2 v}{\partial x^2} + \frac{\partial^2 v}{\partial y^2}$$

$$+ \frac{\text{GrAr}}{\sigma}\tau - \left(\frac{R_\sigma}{\text{Ar}}\right)^{2/3}\frac{\partial p}{\partial y}, \tag{18b}$$

$$u\frac{\partial \tau}{\partial x} + v\frac{\partial \tau}{\partial y} = \frac{1}{\text{Pr}}\left[\left(\frac{\text{Ar}}{R_\sigma}\right)^{2/3}\frac{\partial^2 \tau}{\partial x^2} + \frac{\partial^2 \tau}{\partial y^2}\right], \tag{19b}$$

where the Marangoni number, Ma = $\text{Pr}R_\sigma$, is a modified Péclet number. Equivalent expressions, of course, follow for diffusocapillary flows. The derivation is analogous to that for natural convection, in which different reference velocities are obtained for different force balances and the resulting equations contain the parameters to various powers.

The proper influence of buoyancy for each case is determined from the coefficient of the buoyancy term in Equations (18a) and (18b). For viscous flows

$$\frac{\text{GrAr}}{R_\sigma} = \frac{\rho \beta g L^2}{(\partial \sigma / \partial T)} = \left(\frac{\rho g L^2}{(\partial \sigma / \partial T) \Delta T} \right) \beta \Delta T = \text{Bd} \beta \Delta T, \qquad (20)$$

where Bd is the dynamic Bond number. For boundary-layer flows the buoyancy term must be compared with the pressure gradient, which is the highest-order term in that equation, so that

$$\frac{\text{Gr}(\text{Ar})^{5/3}}{(R_\sigma)^{5/3}} = \text{Bd} \left(\frac{\text{Ar}}{R_\sigma} \right)^{2/3} \beta \Delta T. \qquad (21)$$

Equations (20) and (21) indicate that in a sufficiently reduced-gravity environment surface-tension gradient flows will be significant. Those equations also indicate options for reducing buoyancy under terrestrial conditions.

From the above it is clear that R_σ and Ma are the fundamental parameters for flow and transport, respectively, and that the proper reference velocity is given by Equations (13) and (16). Also, the dynamic Bond number represents the ratio of gravity to surface-tension gradient forces. In the existing analyses of thermocapillary flows, which are critically summarized by Ostrach (1977b), the reference velocity is usually taken as $U_R = \nu/L$ which results in only Gr as a factor of the buoyancy term and $(1/\text{Pr})$ as a factor of the conduction term. The situation for diffusocapillary flows is similar with $U_R = \alpha/L$ and the Schmidt number, ν/D, replacing the Prandtl number as a factor of the diffusion term. In the present formulation no parameters appear in the boundary conditions when the interface is assumed to be flat. The other approaches yield R_σ as a factor of the surface-tension gradient in the interface tangential-stress condition, which could result in an incorrect boundary condition for extreme values of that parameter.

If no terms in the equations are neglected, as in numerical solutions, any nondimensionalization can be used although there are definite advantages to working with unit-order variables that are obtained by normalization. However, to obtain a qualitative view of the phenomena or to simplify the equations by ordering procedures it is essential that the equations be normalized. Thermocapillary flows including buoyancy were shown above to depend on four parameters: R_σ, Ma, Gr, and Ar. From their definitions, it can be seen that these parameters are expressed only in terms of the fluid thermophysical properties, length scales, and the imposed temperature difference. Estimates of these parameters are presented in Table 2 for length scales of 10 cm, temperature differences of 50°C, and an aspect ratio of unity. For $R_\sigma < 1$ the flow will be of a "creeping" or "highly viscous" type and inertia effects will be negligible. Such flows appear to be possible only for fluids like oils and glass. Flow boundary layers can be expected when $R_\sigma > 1$ and thermal boundary layers when Ma > 1. If Pr $\neq 1$ the two boundary layers will be of different extent. It is thus evident that a large range of problems is possible. Furthermore, buoyancy is probably not important in a normal gravitational environment for the conditions considered for water and liquid metals.

In most work to date the interface has been assumed to be flat. However, in general, the interface may be curved and deformed as a result of the forces acting on it and, therefore, its shape must be determined as part of the solution of the problem. Proper formulation of the problem requires balances of the tangential and normal stresses at the interface.

In two dimensions (Figure 2) these are, respectively,

$$2\mu H_X(V_Y - U_X) + \mu(1 - H_X^2)(U_Y + V_X)$$
$$= \sigma_T N^{1/2}(T_X + H_X T_Y) \tag{22}$$

$$(P_\infty - P) + \frac{2\mu}{N}[H_X^2 U_X + V_Y - H_X(U_Y + V_X)]$$
$$= \sigma N^{-3/2} H_{XX} \tag{23}$$

where subscripts denote derivatives, P_∞ is a reference pressure, and $N = [1 + H_X^2]$. A kinematic condition must also be satisfied at the interface. The term on the right of Equation (22) represents the interface deformation and the corresponding term in Equation (23) represents the interface curvature. The latter is particularly important at low gravity because the static shape of the interface will not be flat unless the contact angle is 90°. Normalization of Equations (22) and (23) must be carefully made because there are two sets of length scales that must be considered, namely, one for the flow field and the other for the shape variation of the free surface. Such normalization does not seem to exist in the current

Table 2 Estimates of parametric ranges for length scales of 10 cm, temperature differences of 50°C, normal gravity, and an aspect ratio of unity.

	R_σ	Gr	Ma	Gr/R_σ	Gr/$R_\sigma^{5/3}$
Silicone oils	$10^{-1}-10^4$	10^2-10^7	10^3-10^6	10^3	10
Glass	10^{-1}	10	10^2	10^2	—
Water	10^6	10^8	10^7	—	10^{-2}
Liquid metals	10^5-10^6	10^8-10^{10}	10^3-10^5	—	10^{-2}

literature. Usually a capillary number is cited as another dimensionless parameter for free-surface problems and it is obtained from a balance of the viscous and curvature terms in Equation (23) as Ca = $\mu U_R/\sigma$. For flows driven primarily by surface-tension gradients U_R is given by Equation (13) for viscous flows so that Ca = (Bd/Bo)Ar. Properly, the capillary number alone appears for viscous-type flows. For boundary-layer flows the Weber number, $\rho L U_R^2/\sigma$, also appears from a balance of the pressure and curvature terms. When U_R is given by Equation (16) the Weber number can be expressed as the product of the ratio of the dynamic and static Bond numbers, the surface-tension Reynolds number, and the aspect ratio. Also, other geometric ratios appropriate to the interface shape appear. Thus, no new parameters are obtained from the boundary conditions except the contact angle. When the fluid is subject to more than one driving force, e.g. also rotation or buoyancy, more careful analysis is necessary to obtain the appropriate parameters.

The dimensionless parameters contain the thermophysical properties of the fluids which must be known to obtain the full benefit of their use. Unfortunately, this is not the case. Values of surface tension, σ, are the more readily available but are not well known for many fluids of current interest, such as liquid metals. Much more limited is the information on σ_T (and σ_c). These data are difficult to obtain, but their determination is a challenge that must be met as soon as possible. Dimensional analysis is somewhat helpful in this regard because it demonstrates the sensitivity of flow variables to the property values as, for example, in Equations (13) and (16), and, thereby, indicates the kind of accuracy required for the property values.

Because surface traction is the flow motivator the condition of the surface is most important. Oxide formation or other surface reactions and contamination (due to dust and contaminants in the environment or apparatus) may seriously inhibit or even eliminate surface-tension-

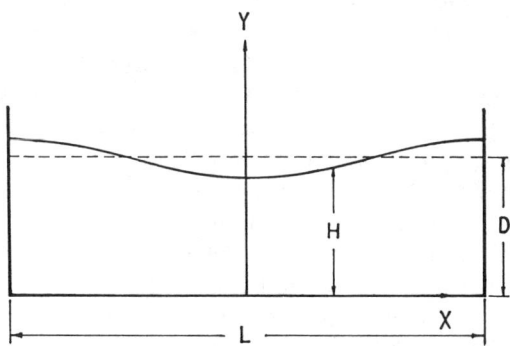

Figure 2 Surface-tension-gradient-driven flow configuration.

gradient-driven flows. This is not necessarily undesirable for practical applications but it makes the prediction and interpretation of data very difficult.

RECENT WORK Studies of flows induced by surface-tension gradients along the interface that were made up to the early 1970s were summarized by Ostrach (1977b). They consist essentially of analyses of steady flows in idealized configurations for which the interface shape is assumed to be flat (or fixed by the configurations of interest). Thus, with one possible exception, to be discussed subsequently, they have no relevance to low-gravity situations unless the contact angle is 90°. Inconsistencies exist in several of the analyses which further limit their value. An attempt to obtain experimental information in drop-tower tests was made by Ostrach & Pradhan (1978). Water was the test fluid and although considerable care was taken to avoid surface contamination subsequent experience with water makes the results somewhat questionable.

Because the inconsistencies in the analyses of surface-tension-gradient flows in even the simplest configuration, viz. a doubly infinite horizontal thin fluid layer, and because there was no treatment of the time-dependent flows, that problem was reconsidered by Pimputkar & Ostrach (1980). A formal nondimensionalization was made which indicated an explicit ordering of the equations so that ad hoc assumptions were unnecessary. The flows and surface shapes were determined for a family of different imposed surface-temperature distributions. These results, too, are limited for low-gravity conditions to 90° contact angles because the initial interface shape was taken to be flat.

Surface-tension-driven flow was numerically investigated in a vertical cylindrical melt container between two rods by Chang & Wilcox (1976). This model is intended to simulate float-zone melting which is used for purification of melts and for crystal growth. The zone is taken to be a circular cylinder with planar solid-liquid interfaces and to have an aspect ratio of unity. An important feature of the problem is neglected in this way because those shapes are closely coupled with the flow and heat transfer and should be determined from the solution instead of being imposed as boundary conditions. The effect of buoyancy on the results was investigated but inappropriate parameters led to misleading interpretations (see Ostrach 1977b). Clark & Wilcox (1980) pointed out and corrected an error in the numerical formulation of Chang & Wilcox (1976). The main feature of this numerical work is the prediction of flow cells (toroidal vortices) that encompass the entire fluid.

To simulate low-gravity thermocapillary flows terrestrially Schwabe et al. (1978) and Chun & Wuest (1978) established a bridge of transparent liquid between two disks. For the dynamic Bond number to be less than unity, the fluid column had to be less than about 5 mm. The end disks or rods could be rotated and heated or the temperature gradient could be imposed by a wire ring heater around the outside of the liquid zone. Particles were introduced to permit flow visualization and a thermocouple was inserted through the zone for temperature measurements. The test fluid used by the first group is sodium nitrate, $NaNO_3$, presumably because it is a good model for silicon; the Marangoni numbers are similar but the surface-tension Reynolds numbers are not for a given set of conditions. Chun & Wuest (1978) used water and silicone oil, methylalcohol (1979a) and octadecane, $C_{18}H_{38}$ (Chun 1980). Considerable fluid evaporation occurred except for the last of these. The particles used by both groups for flow visualization were not neutrally buoyant. Despite these shortcomings many interesting results have been obtained.

It was observed that steady flows (on the order of centimeters per second) undergo a transition to oscillatory ones above certain values of the Marangoni number, that depend on the imposed conditions. This change is identified as an "instability" and possible mechanisms are discussed by Chun & Wuest (1979a) and Schwabe et al. (1978). Since such flow and temperature oscillations are undesirable for materials processing, means of suppressing them were investigated. Chun & Wuest (1979b) examined the effect of end rotation for that purpose. Schwabe et al. (1978) show that surfactants can induce diffusocapillary flows and Schwabe et al. (1980) demonstrate that surface contamination can suppress thermocapillary convection. The effects of geometry and imposed thermal conditions on the "critical" Marangoni number are examined by Schwabe et al. (1979) and Schwabe & Scharmann (1979).

To investigate thermocapillary flows in larger fluid volumes Schwabe & Scharmann (1980, 1981) established a liquid zone (of $NaNO_3$) 20 mm long and 12.5 mm deep between two graphite block heaters contained in a rectangular quartz boat. $NaNO_3$ melts at 307°C to yield a transparent oxide melt. They postulate that in this configuration the cause of oscillations may be due to the interaction of the two basic types of surface-tension-gradient flows, i.e. the Marangoni instability is superposed on the regular flow. In an even larger volume of fluid, a layer of fluorocarbon liquid (Flourinert) 20 cm by 10 cm and 2.5 cm deep, Lowry (1981) by means of a rather unique experimental approach also observed thermocapillary flows and found them to become oscillatory. All this work provides qualitative evidence of the complex nature of flows driven by surface-tension gradients. Much more is required to obtain detailed understanding and predictability for such flows. Furthermore, the cause of the oscillations must be firmly established. Some of the mechanisms proposed in the existing literature may, in fact, be the correct ones. Others are also possible. It is clear that the inherent coupling between the imposed surface temperature or concentration gradient and the surface flows is an essential feature of the problem that must be included in analyses of the problem.

Although surface-tension flows are being investigated under normal gravitational conditions it is not possible to simulate completely on earth such flows that would occur in a low-gravity environment because of the differences in the initial shape of the interface and its damping characteristics. The small float zones may be an exception to that. Finally, it should be recognized that diffusocapillary flows can be of two types. One is caused by surface-concentration gradients caused by surfactants that do not penetrate the bulk of the fluid (Yih 1968 and Adler & Sowerby 1970). The second type of diffusocapillary flow arises as a consequence of surface-concentration gradients which are a result of concentration gradients in the bulk fluid. These can couple with thermocapillary flows and represent conditions in the containerless processing of materials.

Thermoacoustic Convection

A number of practical problems have been mentioned that require analysis of heat flow in fluids under low-gravity conditions. Among such problems are the determination of the pressure behavior in spacecraft fuel tanks and heat leaks to stored cryogens as well as the heat transfer in materials processing. Even if natural convection were virtually negligible it still may be possible that heat conduction is not the only mechanism for heat transfer. Thus, there is a need for assessing the role of driving mechanisms other than gravity. Pressure-driven convection that results when a confined compressible fluid is heated rapidly is one type of convection that is independent of gravity. Such motions have been called thermoacoustic convection because of the sonic character of the induced pressure waves.

There exist relatively few analyses of pressure-driven convection. Trilling (1955), Knudsen (1954), and Luikov & Berkovsky (1970) were among the first to investigate analytically the wave motion induced in gases by boundary-temperature gradients. They used a laminar perturbation analysis and found that a sharp rise in boundary temperature can cause pressure waves to propagate through the fluid in much the same way as does pushing a piston through a gas-filled pipe. Larkin (1967) considered one-dimensional motion of an ideal gas in a confined region at zero gravity by solving the conservation equations numerically. His computations confirm the acoustic nature of the wave motion and indicate that the heat transfer and pressure rise are much greater than would be predicted by neglecting thermally induced fluid motion. Spradley et al. (1973) used a finite-difference method to show that pressure and thermal expansion effects can be significant in determining the motion and heat transfer through confined gases that are rapidly heated. Spradley & Churchill (1975) used a modified form of the algorithm developed by Spradley et al. (1973) to compute the two-dimensional flow of an ideal gas in a rectangular enclosure including both buoyancy and pressure effects.

PROBLEM FORMULATION AND PARAMETERS The most general configuration considered is a rectangular enclosure of length L and height H that contains a compressible fluid. The fluid is initially motionless and isothermal. The upper and lower walls are thermally insulated and motion is induced by imposing a horizontal temperature difference, e.g. by suddenly raising the temperature of one wall. The fluid is assumed to be compressible, viscous, and heat-conducting and the flow to be laminar. All fluid properties except density are taken to be constant. The ideal gas law is used to relate thermodynamic variables. The gravity vector is in the direction of the $-y$ axis.

There is uncertainty about the proper characteristic velocity and time to use in obtaining the dimensionless parameters. Since the problem involves acoustic velocity waves, the isothermal velocity of sound is used for the former and the time required for a wave to travel the length of the container is taken for the latter. The dimensionless parameters are thus determined to be

$\eta = L/H$ (aspect ratio),

$\gamma = c_p/c_v$ (ratio of specific heats),

$\mathrm{Pr} = \gamma c_v \mu/k$ (Prandtl number),

$\mathrm{Re} = \rho L (RT_0/M)^{1/2} (\mu)^{-1}$ (acoustic Reynolds number),

$\mathrm{Fr} = RT_0/gHM$ (acoustic Froude number),

where R is the universal gas constant, M the molecular weight of the gas, and T_0 is the initial gas temperature.

RESULTS The first problem solved is the one-dimensional flow of helium in a large-aspect-ratio ($\eta = 1000$) container with $g = 0$. The motion is induced by suddenly raising the temperature of one wall to twice the initial value while the other wall is kept at the initial temperature. The dimensionless parameters are $\mathrm{Re} = 1.2 \times 10^6$, $\mathrm{Pr} = 0.685$, and $\gamma = 1.65$. The sudden change in wall temperature induces an expansion of the gas due to its compressibility. The expansion generates fluid motion that propagates to the other wall and then reflects. The subsequent velocity field is then influenced by the reflections of the disturbance from each of the walls. The oscillatory velocity has a maximum of about 15 meters per second and a frequency of about 1.2×10^{-4} cycles per second. The frequency of the pressure oscillations is about twice that of the velocity. The temperature distribution differs significantly from that for heat conduction alone, which indicates that thermoacoustic convection is an effective transfer mechanism and greatly enhances the heat-transfer rate.

To illustrate the combined effects of buoyancy- and pressure-driven convection Spradley & Churchill (1975) consider the flow of helium in a rectangular container. The fluid is initially motionless with $T_0 = 273°K$ and $P_0 = 1.8 \times 10^6$ dynes cm^{-2}. The motion is initiated by raising one side wall to twice the initial temperature with the opposite wall maintained at the initial temperature. The parameters are thus

$$\eta = 1.0, \quad \text{Re} = 2.07 \times 10^5, \quad \text{Pr} = 0.685, \quad \gamma = 1.67,$$

$$\text{Fr} = 2.15 \times 10^6 \text{ (based on } 1-g_0).$$

The Froude number is varied to investigate the effects of various gravity levels. At early times the motion is driven by the horizontal pressure gradient and is one-dimensional. The waves have the local period of $2(\gamma T)^{-1/2}$. The velocity amplitudes are approximately 0.02 of the sonic at the initiation of the motion but are rapidly damped with time. The pressure waves follow the velocity but with twice the effective frequency because of the effects on pressure of both left- and right-running velocity waves. This early motion is identical to the one-dimensional solution obtained by Spradley (1974).

At short times, Mach numbers are of the order of 10^2 which correspond to flow Reynolds numbers, $\text{Re}_f = VL/\nu$, of the order of 10^3 and this value is quickly reduced to $\text{Re} \sim 10^2$. As time continues the pressure waves are highly damped, density gradients develop, and buoyancy becomes the dominant driving force. Buoyancy-driven convection begins to cause circulation of the flow, which is now driven by both pressure gradients and buoyancy. To investigate the relative importance of the two driving forces, calculations were made for four cases: (a) pure conduction, (b) pure thermoacoustic convection, $g = 0$, (c) combined convection for $g = 10^{-2}g_0$, and (d) combined convection for $g = g_0$. Detailed results show that for the case considered pressure-driven convection is dominant in transferring heat at very low gravity, both pressure- and buoyancy-driven convection contribute significantly to the transient heat transfer for $g = 10^{-2}g_0$, and buoyancy-driven convection is dominant for $g = g_0$. Thus, natural convection overshadows thermoacoustic convection until the gravity level is reduced sufficiently. Then the latter can become dominant.

Phase-Change Convection

In the transition between phases of a fluid, such as in solidification or vaporization, there is usually an intrinsic change in density associated with the transition, in addition to the various physical and chemical processes that occur at the interface. For example, a shrinkage usually occurs during solidification because the density of the solid usually is higher than that of the liquid and, hence, occupies a smaller volume. This volume reduction results in a flow of liquid toward the solidifying interface. This flow (convection) modifies the advance of the solid boundary since it affects the transfer of latent heat away from the front.

The growth of a new phase in a liquid is of interest not only for solidification and freezing processes but also for bubble growth in nucleate boiling, and in mass-transfer processes. The early studies of the dynamics of phase growth in an unlimited uniform medium (see Frank 1950, for example) assumed that the densities of the two phases were equal. Chambré (1956) attempted to take the density differences into account but, unfortunately, as pointed out by Scriven (1959), the continuity equation was not satisfied by Chambré's resolutions. Horvay (1962) properly reformulated and solved the problem of the growth of a solid, with either a plane, cylindrical, or spherical boundary, starting from initially negligible dimensions into a surrounding supercooled fluid.

The dynamics of phase growth is complicated not only by the phase-change convection but also by the motion of the phase-transition front, which is unknown at first and on which boundary conditions must be specified. The nature of the problem is illustrated as follows: If the shape of the interface is assumed fixed and R denotes the position of the front, the velocity of the front is dR/dt and the liquid velocity immediately adjacent is $U(R,t)$. The net velocity causes a mass flow of liquid per unit surface area per unit time of $\rho_1[dR/dt - U(R,t)]$, where ρ_1 is the fluid density. This mass flow must equal the mass solidified so that

$$\rho_1 \left[\frac{dR}{dt} - U(R,t) \right] = \rho_2 \frac{dR}{dt}$$

or

$$U(R,t) = \frac{\rho_2 - \rho_1}{\rho_1} \frac{dR}{dt},$$

where ρ_2 is the density of the solid. This indicates the velocity due to the density difference and the movement of the front. The velocity, pressure, and transport distributions as well as the front position-time relation are then determined from the appropriate conservation equations and initial and boundary conditions.

Scriven (1959) considers the more complex problem of the growth of a single spherical vapor bubble in an unlimited body of superheated liquid in order to gain a deeper understanding of nucleate boiling. An excellent and rather general formulation of the problem of spherically symmetric phase growth precedes the treatment of the specific problem. All assumptions are clearly and explicitly given and asymptotic solutions are obtained for conditions typical of bubble growth in nucleate boiling of both pure liquids and binary mixtures.

Another version of the phenomenon is treated by Acrivos (1962) who considers mass transfer to or from a surface. In that case, the surface acts as a source or a sink of material which induces a velocity normal to the solid-fluid interface. When this velocity is appreciable it distorts the velocity profile that would exist in the absence of mass transfer and thereby affects the mass-transfer rate. Asymptotic solutions of the two-dimensional laminar boundary-layer equations for large interfacial velocities are obtained for rather general systems. The conditions under which the phase-change velocity becomes important in forced and free convection are determined. In those situations the analogy between heat and mass transfer becomes invalid.

The work described above indicates that phase-change convection can significantly influence the dynamics of phase growth even under terrestrial conditions. Since such flows are independent of gravity they can be the dominant mechanism in a low-gravity environment. In reality, the processes within and around a phase-transformation front will not allow it to assume a highly symmetric shape as has been assumed in most of the existing studies. It is, thus, evident that more research on phase-change convection is required, particularly, without the restriction of invariant interface shapes.

CONCLUDING REMARKS

A number of different types of fluid flows have been identified that can occur under low-gravity conditions. Some, such as buoyancy-driven and g-jitter convection, are acceleration-field dependent. Others, like surface-tension gradient, thermoacoustic, and phase-change convection are independent of the acceleration field. Although most of these flows could also occur under normal gravitational conditions they are usually overshadowed by other flows except under rather limited conditions. Thus, they have received relatively little attention. However, in a low-gravity environment they can become significant and, therefore, more careful study of them is required. It is also possible that two or more forces (such as buoyancy and surface-tension gradients) can act simultaneously. Their combined action can result in unusual flows. What little work that has been done on such problems pertains to very specific situations so that flows due to coupled forces are not, as yet, predictable.

Other topics like critical-state phenomena and sedimentation processes, although not discussed herein, also warrant study under reduced-gravity conditions.

It is hoped that the foregoing dispells the misconception that a spacecraft in a near-earth orbit provides a quiescent state of zero-gravity. Important and interesting fluid motions can occur even when the magnitude of the gravitational acceleration is reduced by many orders of magnitude. The determination of the nature and extent of such flows and their associated transport poses many challenging problems to the fluid dynamicist.

Acknowledgment

The author acknowledges with thanks many fruitful discussions of the problems discussed herein with his colleague, Prof. Y. Kamotani, and with Prof. Dudley A. Saville of Princeton University. The author also wishes to express appreciation to Dr. John R. Carruthers, Director, Materials Processing in Space, for research support in this area under several NASA contracts in his program.

Literature Cited

Acrivos, A. 1962. The asymptotic form of the laminar boundary-layer mass-transfer rate for large interfacial velocities. *J. Fluid Mech.* 12:337–51

Adler, J., Sowerby, L. 1970. Shallow three-dimensional flows with variable surface tension. *J. Fluid Mech.* 42:549–59

Bannister, T. C., Grodzka, P. G., Spradley, L. W., Bourgeois, S. V., Heddon, R. O., Facemire, B. R. 1973. Apollo 17 heat flow and convection experiments: Final data analyses results. *NASA TM X-64772*

Carruthers, J. R. 1973. Thermal convection instabilities relevant to crystal growth from liquids. In *Preparation and Properties of Solid State Materials*, ed. W. R. Wilcox, 3:1–121. New York: Marcel Dekker

Chambré, P. L. 1956. On the dynamics of phase growth. *J. Mech. Appl. Math.* 9:224–33

Chang, C. E., Wilcox, W. R. 1976. Analysis of surface tension driven flow in floating zone melting. *Int. J. Heat Mass Transfer* 19:335–66

Chun, Ch. -H. 1980. Experiments on steady and oscillatory temperature distribution in a floating zone due to the Marangoni convection. *Acta Astronautica* 7:479–88

Chun, Ch. -H., Wuest, W. 1978. A microgravity simulation of the Marangoni convection. *Acta Astronautica* 5:681–86

Chun, Ch. -H., Wuest, W. 1979a. Experiments on the transition from the steady to the oscillatory Marangoni-convection of a floating zone under reduced gravity effect. *Acta Astronautica* 6:1073–82

Chun, Ch. -H., Wuest, W. 1979b. Flow phenomena in rotating floating zones with and without Marangoni convection. *Proc. 3rd. European Symp. Material Sciences in Space, Grenoble, ESA SP-142*

Clark, P. A., Wilcox, W. R. 1980. Influence of gravity on thermocapillary convection in floating zone melting of silicon. *J. Crystal Growth* 50:461–69

Clever, R. M. 1973. Finite amplitude longitudinal convection rolls in an inclined layer. *J. Heat Transfer* 95:407–08

Conway, B. A. 1972. Development of Skylab experiment T-013 crew/vehicle disturbance. *NASA TN D-6584*

Dodge, F. T. 1975. Fluid physics, thermodynamics, and heat transfer experiments in space: Final report of the overstudy committee. *NASA CR-134742*. Southwest Research Institute, San Antonio, Texas

Fineblum, S. S., Haron, A. S., Saxton, J. A. 1971. Heat transfer and thermal stratification in the Apollo 14 cryogenic oxygen tanks. *MSC Cryogenics Symposium Papers, MSC-04312*. NASA-Manned Spacecraft Center, Houston, Texas

Frank, F. C. 1950. Radially symmetric phase growth controlled by diffusion. *Proc. R. Soc. London Ser. A* 201:586–99

Fu, B. -I., Ostrach, S. 1981. The effects of stabilizing thermal gradients on natural convection in a square enclosure. In *Natural Convection*, ed. I. Catton, R. N. Smith, HTD 16:91–104. ASME

Grodzka, P. G., Bannister, T. C. 1974. Natural convection in low-g environments. *AIAA Pap. No. 74-156*

Grodzka, P.G., Fan, C., Hedden, R. O. 1971. The Apollo 14 heat flow and convection demonstration experiments: Final results of data analysis. *LMSC-HREC D22533*. Lockheed Missiles Space Co., Huntsville, Ala.

Hart, J. E. 1971. Stability of flow in a differentially heated inclined box. *J. Fluid Mech.* 47:547–76

Hart, J. E. 1973. A note on the structure of thermal convection in a slightly slanted slot. *Int. J. Heat Mass Transfer* 16:747–53

Hollands, K. G. T., Konicek, L. 1973. Experimental study of the stability of differentially heated inclined air layers. *Int. J. Heat Mass Transfer* 16:1467–76

Horvay, G. 1962. Freezing into an undercooled melt accompanied by density change. *Proc. US Nat. Congr. Appl. Mech.* 2:1315–25

Kamotani, Y., Prasad, A., Ostrach, S. 1981. Thermal convection in an enclosure due to vibrations aboard spacecraft. *AIAA J.* 19:511–16

Kenning, D. B. R. 1968. Two-phase flow with nonuniform surface tension. *Appl. Mech. Rev.* 21:1101–11

Knudsen, J. R. 1954. The effects of viscosity and heat conductivity on the transmission of plane sound pulses. *J. Acoust. Soc. Am.* 26:51–57

Larkin, B. K. 1967. Heat flow to a confined fluid in zero gravity. In *Progress in Astronautics and Aeronautics Thermophysics of Spacecraft and Planetary Bodies*, ed. G. B. Heller, 20:819–32

Levich, V. G. 1962. *Physicochemical Hydrodynamics*. Englewood Cliffs, NJ: Prentice-Hall

Levich V. G., Krylov, V. S. 1969. Surface-tension-driven phenomena. *Ann. Rev. Fluid Mech.* 1:293–316

Lowry, S. 1981. An experimental study of heat induced surface tension driven flow. MS thesis. Case Western Reserve Univ. Dept. Rep. *#FTAS/TR-81-154*. Dept. Mech. and Aero. Engrg., Cleveland, Ohio

Luikov, A. V., Berkovsky, B. M. 1970. Thermoconvective waves. *Int. J. Heat Mass Transfer* 13:741–47

National Research Council. 1978. *Materials Processing in Space*. Space Applications Board, Washington, DC

Ostrach, S. 1954. Combined natural- and forced-convection laminar flow and heat transfer of fluids with and without heat sources in channels with linearly varying wall temperatures. *NACA TN 3141*

Ostrach, S. 1964. Laminar flows with body forces. In *Theory of Laminar Flows: High Speed Aerodynamic and Jet Propulsion*, ed. F. K. Moore, 4:528–718., Princeton Univ. Press

Ostrach, S. 1966. Role of analysis in the solution of complex physical problems. *Proc. Third Int. Heat Transfer Conf.* 6:31–43

Ostrach, S. 1972. Natural convection in enclosures. In *Advances in Heat Transfer*, ed. J. P. Hartnett, T. F. Irvine, Jr., 8:161–227. New York/London: Academic

Ostrach, S. 1977a. Convection phenomena of importance for materials processing in space. In *Progress in Astronautics and Aeronautics*, ed. L. Steg, 52:3–32. New York: Am. Inst. Aeronautics and Astronautics.

Ostrach, S. 1977b. Motion induced by capillarity. In *Physicochemical Hydrodynamics: V. G. Levich Festschrift*, 2:571–89. London: Advanced Publications.

Ostrach, S. 1979. Convection due to surface-tension gradients. In *(COSPAR) Space Research*, ed. M. J. Rycroft, 19:563–570. Oxford/New York: Pergamon

Ostrach, S. 1980. Natural convection with combined driving forces. *Physicochemical Hydrodynamics.* 1:233–47

Ostrach, S., Loka, R. R., Kumar, A. 1980. Natural convection in low aspect ratio rectangular enclosures. In *Natural convection in Enclosures*, ed. K. E. Torrance, I. Catton, HTD 6:1–10. *ASME*

Ostrach, S., Pnueli, D. 1963. The thermal instability of completely confined fluids inside some particular configurations. *Trans. ASME* 85:346–54

Ostrach, S., Pradhan, A. 1978. Surface-tension induced convection at reduced gravity. *AIAA J.* 16:419–24

Ostrach, S., Raghavan, C. 1979. Effect of stabilizing thermal gradients on natural convection in rectangular enclosures. *J. Heat Transfer* 101:238–43

Ozoe, H., Sayama, H., Churchill, S. W. 1974a. Natural convection in an inclined square channel. *Int. J. Heat Mass Transfer* 17:401–6

Ozoe, H., Yamamoto, K., Sayama, H., Churchill, S. W. 1974b. Natural convection in an inclined rectangular channel heated on one side and cooled on the opposing side. *Int. J. Heat Mass Transfer* 17:1209–17

Ozoe, H., Sayama, H., Churchill, S. W. 1975. Natural convection in an inclined rectangular channel at various aspect ratios and angles—experimental measurements. *Int. J. Heat Mass Transfer* 18:1425–31

Pimputkar, S. M., Ostrach, S. 1980. Transient thermocapillary flow in thin layers. *Phys. Fluids* 23:1281–85

Pimputkar, S. M., Ostrach, S. 1981. Convective effects in crystals grown from melt. *J. Crystal Growth*. In press

Rice, R. A. 1971. Apollo 14 flight support and systems performance. *MSC Cryogenics Symposium Papers, MSC-04312*. NASA-Manned Spacecraft Center, Houston, Texas

Saville, D. A., Ostrach, S. 1978. Fluid mechanics of continuous flow electrophoresis. *TR NASA-1, Final Report on Contract NAS-8-31349 Code 361*. Dept. Chem. Engrg. Princeton Univ.

Schwabe, D., Scharmann, A. 1979. Some evidence for the existence and magnitude of a critical Marangoni number for the onset of oscillatory flow in crystal growth melts. *J. Crystal Growth* 46:125–31

Schwabe, D., Scharmann, A. 1980. Thermocapillary convection in crystal growth melts. *Letters in Heat and Mass Transfer* 7:283–92

Schwabe, D., Scharmann, A. 1981. The magnitude of thermocapillary convection in large melt volumes. *Adv. Space Res.* 1:13–16

Schwabe, D., Scharmann, A., Preisser, F. 1979. Steady and oscillatory Marangoni convection in floating zones under 1-g. *Proc. 3rd European Symp. Materials Sciences in Space, Grenoble, ESA SP-142*

Schwabe, D., Scharmann, A., Preisser, F. 1980. Studies of Marangoni convection in crystal growth melts. Int. Astronautical Federation, *IAF-80-C-140*. Oxford: Pergamon

Schwabe, D., Scharmann, A., Preisser, F., Oeder, R. 1978. Experiments on surface tension driven flow in floating zone melting. *J. Crystal Growth* 43:305–12

Scriven, L. E. 1959. On the dynamics of phase growth. *Chem. Engrg. Sci.* 10:1–13

Sherman, M., Ostrach, S. 1967. Lower bounds to the critical Rayleigh number in completely confined regions. *J. Appl. Mech.* 34:308–12

Sirignano, W. A., Glassman, I. 1970. Flame spreading above liquid fuels: Surface tension flows. *Combust. Sci. and Tech.* 1:307–12

Solan, A., Ostrach, S. 1979. Convection effects in crystal growth by closed-tube chemical vapor transport. In *Preparation and Properties of Solid State Materials*, ed. W. R. Wilcox, 4:63–110. New York: Marcel Dekker

Spradley, L. W. 1974. Thermoacoustic convection of fluids in low gravity. *AIAA Paper 74-76*

Spradley, L. W., Bourgeois, S. V., Lin, F. N. 1975. Space processing convection evaluation: g-jitter convection of confined fluids in low gravity. *AIAA Paper 75-695*

Spradley, L. W., Bourgeois, S. V., Fan, C., Grodzka, P. G. 1973. A numerical solution for thermoacoustic convection of fluids in low gravity. *NASA CR-2269*

Spradley, L. W., Churchill, S. W. 1975. Pressure and buoyancy-driven thermal convection in a rectangular enclosure. *J. Fluid Mech.* 70:705–20

Stanek, V., Szekely, J. 1964. The effect of surface driven flows on the dissolution of a partially immersed solid in a liquid—Analysis. *Chem. Engrg. Sci.* 25:699–716.

Stark, J. A., Bradshaw R. D., Blatt, M. H. 1974a. Low-G fluid behavior technology summaries. *NASA CR-134746*. General Dynamics Convair Div., San Diego, Calif.

Stark, J. A., Blatt, M. H., Bennett, F. O., Campbell, B. J. 1974b. Fluid management systems technology summaries. *NASA CR-134748*. General Dynamics Convair Div., San Diego, Calif.

Sternling, C. V., Scriven, L. E. 1959. Interfacial turbulence: Hydrodynamic instability and the Marangoni effect. *AIChE J.* 5:514–23

Trilling, L. 1955. On thermally induced sound fields. *J. Acoust. Soc. Am.* 27:425–31

Unny, T. W. 1972. Thermal instability in differentially heated inclined layers. *J. Appl. Mech.* 39:41–46

Yih, C. S. 1968. Fluid motion induced by surface-tension variation. *Phys. Fluids* 11:477–80

THE STRANGE ATTRACTOR THEORY OF TURBULENCE

Oscar E. Lanford III

Department of Mathematics, University of California, Berkeley, California 94720

INTRODUCTION

It is a fact of experience almost too familiar to notice that dissipative physical systems subject to weak steady driving approach states of dynamic equilibrium that are independent of initial condition. As the strength of the driving is increased, these systems typically undergo a sequence of transitions—the details depending on the system—and arrive eventually at behavior that may be described as chaotic or turbulent. The turbulent motion is not entirely without regularity, but the regularity is statistical in character and appears only when long-term time averages are examined.

Ideally, the mechanisms producing the transition from steady to chaotic behavior, and the detailed nature of the motion in the chaotic regime, should be deducible directly from the equations of motion for the system in question, i.e. the Navier-Stokes or Boussinesq equations in the case of classical kinds of fluid systems. Direct attacks on these equations, however, meet with overwhelming difficulties. On the one hand, control over the analytic properties of the equations is not yet good enough either to prove or to disprove the existence of regular solutions for all times and arbitrary regular initial data. On the other hand, it seems quite hopeless to try to compute explicit analytic solutions with chaotic behavior, to say nothing of computing, from first principles, the statistical distribution describing the behavior of typical solutions. To circumvent the difficulties of a direct approach, a number of oblique lines of attack have been developed. One of these approaches, known as the *strange attractor* theory of turbulence, is the subject of this review.

This approach focuses on the time dependence of turbulent motion; the fundamental idea on which it is based is:

> Turbulent time dependence is not an exceptional feature of particular equations of motion but a property shared by a broad class of typical differential equations.

Adopting this point of view changes the perspective from one of studying particular—and intractable—equations to trying to answer the question:

> How does a typical solution of a typical differential equation behave over the long run?

A substantial body of deep mathematical theory is available to be applied to this question, and mathematical work in this area in recent years has been both invigorated and focused by interaction with the physical study of the chaotic behavior of dissipative systems.

The approach has at least two obvious drawbacks. One is that there is no guarantee that the Navier-Stokes equation will indeed turn out to be typical. This objection is not as serious as it might appear. The Navier-Stokes equation is after all only an approximation, albeit a very good one for most purposes. Even if it were to turn out to have nontypical properties, the very notion of "typical" means that most small perturbations on it would produce an equation with typical behavior. Furthermore, the discovery of a nontrivial exceptional qualitative property of the Navier-Stokes equation would be a great step towards understanding that equation, so the program can further our understanding even if its fundamental presupposition ultimately turns out to be wrong.

The second drawback is that, at best, the investigation of typical behavior can furnish only a list of alternatives. Which of the alternatives actually occurs for a particular equation can only be determined by a detailed study of that equation (or by performing an experiment, either a computer experiment on the equation or an actual experiment on the physical system it describes.)

Up to now, at least, this approach has not contributed very much to the solution of the traditional questions about turbulence or to the practical computation of critical parameter values, phenomenological parameters like effective turbulent viscosity, etc. Its successes have come more in suggesting new questions to be investigated experimentally than in explaining the results of prior experiments. Although the mathematical theory has developed some very powerful methods of analysis, it has generally not been possible to sum up the principal insights in a few concise theorems that can be applied without regard to the reasoning behind them. In short, it is a better source of tools than of recipes.

Terminology

We will be discussing differential equations. By a *state* for a differential equation, we mean a complete specification of initial condition; the space of all states will be called the *state space*. Thus, for a Hamiltonian system, the state space means the phase space rather than the configuration space. For an incompressible fluid system, a point of the state space is a velocity field, defined on the physical region occupied by the fluid, with vanishing divergence and satisfying appropriate boundary conditions. We will use the term *orbit* to refer to a solution to the differential equation regarded as a curve in the state space, and call *solution flow* the motion on the state space that advances each point along its respective orbit. We will say that a stationary or periodic orbit is *stable* or *attracting* if all orbits starting sufficiently near to it converge to it; this property is frequently called *asymptotic stability in the sense of Lyapunov.*

The terms *turbulent, chaotic,* and *stochastic* (applied to describe time dependence of solutions to a differential equation) will be used interchangeably. Note, however, a slight and potentially question-begging difference in connotation; *stochastic,* as normally used, implies the existence of a well-defined average behavior.

Some General References

The idea that chaotic time dependence of turbulent fluid flows might be understood as a property of fairly general differential equations was first advanced in a way that attracted widespread attention in Ruelle & Takens (1971a,b). [A very suggestive example had been pointed out earlier by Lorenz (1963), but Ruelle & Takens were not aware of Lorenz's work.] The paper of McLaughlin & Martin (1975) was very influential in popularizing these ideas. Recent general surveys include Ruelle (1978a,b, 1980a,b), Lanford (1981), and Eckmann (1981).

CHAOTIC BEHAVIOR

It is often felt that there is something paradoxical about having solutions to a deterministic equation behave in a chaotic or stochastic fashion. There is, however, no real paradox; the solutions are, in fact, uniquely determined by the initial conditions, but the effects of small changes in the initial conditions are so amplified by the equations of motion that any *finite-precision* information about the initial conditions provides no *finite-precision* information about the state of the system at much later times. In other words:

> An important element in the explanation of the chaotic behavior of solutions of deterministic equations of motion is the sensitive dependence of solutions on initial conditions.

The use of probabilistic concepts in the analysis of deterministic motion should, in any case, be familiar from classical equilibrium statistical mechanics. In fact, sensitive dependence on initial condition, in the form of the notion of ergodicity, has long played a central role in one of the standard justifications for the foundations of that subject (see, for example,

Lebowitz & Penrose 1973). It needs to be noted, however, that there are substantial differences between Hamiltonian systems—to which the usual formalism of classical statistical mechanics applies—and the sort of dissipative systems under discussion here. The key distinction lies in the volume-preserving character of the solution flow for Hamiltonian systems (Liouville's Theorem) which has as a consequence the fact that the solution flow is *recurrent* (meaning, roughly, that almost all orbits come back arbitrarily near their initial points infinitely often). For dissipative systems, on the other hand, what usually happens is that most of the points of the instantaneous state space are *transient* in the sense that the orbits that start there eventually go to and stay in another part of the state space. A simple instance is provided by the stable dynamic equilibrium that is usually set up when a dissipative system is driven gently. In this situation, all orbits, no matter where in the state space they start, converge eventually to a single stationary solution corresponding to laminar motion. In a certain sense, the system has *no* effective degrees of freedom, although the state space may have large or even infinite dimension. It seems very likely that something similar happens for more strongly driven dissipative systems, even those whose motion is chaotic, viz., that

> There are one—or possibly a few—invariant sets of relatively low dimension in the state space to which almost all orbits converge.

These sets are what are called *attractors*. One of the great surprises in this subject is that attractors, except for the very simplest ones, are typically not smooth surfaces in the state space but rather more complicated kinds of sets.

Before taking up the notion of attractor in more detail, we need to elaborate on what is meant in practice by chaotic behavior of a physical system whose equation does not depend explicitly on time. Roughly, the idea is that the system behaves, over a long period of time, in a repetitive but not strictly periodic fashion. It should be emphasized that, in this article (and in the field it surveys), the analysis is focused on understanding temporal—not spatial—chaos. There is a tendency to identify chaotic motion with spatially complicated flow patterns. This identification is not entirely mistaken, since laminar flow generally has simple geometry, but it may be misleading. It is possible to have either

> a fluid flow with extremely complicated intrinsic spatial structure and no time dependence at all (in, for example, large-aspect-ratio convection)

or

> chaotic motion with relatively simple spatial structure (small-aspect-ratio convection).

It is not even true that *apparent* temporal complexity necessarily indicates chaotic behavior. Convective systems, for example, can undergo quite complicated periodic motion before they become aperiodic. When they do become stochastic, the motion can often be decomposed, at least roughly, into a small stochastic component superimposed on a much larger periodic component. It is often not easy to distinguish between such weakly stochastic motion and complicated but purely periodic motion simply by watching the system. The standard way to detect a stochastic component in the motion is to measure the *power spectrum* of some dynamical variable. Stripped of technicalities, this means the following: Measure, at equally spaced times t_i some numerical quantity such as one component of the velocity at a particular point of the fluid. Call the measured quantities $X_1, \ldots, X_N$. Subtract the mean and take the discrete Fourier transform, i.e. form

$$\tilde{X}(\omega) = \frac{1}{\sqrt{N}} \sum_{j=1}^{N} (X_j - \overline{X}) e^{i\omega j}, \tag{1}$$

where ω is of the form $2\pi k/N$ and where

$$\overline{X} = \frac{1}{N} \sum_{j=1}^{N} X_j. \tag{2}$$

Then see whether $|\tilde{X}(\omega)|^2$ is concentrated in a series of sharp peaks. The appearance of a "broad-band" component in $|\tilde{X}(\omega)|^2$ (beyond that due to the finite precision of the measurements) is generally taken as the operational definition of stochastic behavior of the system in question.

ATTRACTORS

One of the most fruitful ways of studying the mathematical structures underlying observed stochastic behavior of physical systems has been the careful study, mostly with the aid of computers, of simplified models, and one model that has been particularly informative is the *Lorenz system*, a set of three coupled differential equations:

$$\frac{dx}{dt} = -\sigma x + \sigma y; \quad \frac{dy}{dt} = rx - y - xz;$$

$$\frac{dz}{dt} = -bz + xy, \tag{3}$$

where b, σ, and r are constants. It is hard to imagine a much simpler system that is neither linear nor two-dimensional, but the solutions to these equations nevertheless do very complicated things. E. N. Lorenz (1963) discovered numerically a striking mathematical structure which has come to be known as the *Lorenz attractor* and which occurs for these equations with

$$b = 8/3 ; \quad \sigma = 10 ; \quad r = 28 . \tag{4}$$

The exact parameter values are not crucial, but the behavior of typical solutions definitely does depend on the parameters and is quite different in other regions of parameter space. It should also be noted that, in spite of overwhelming numerical evidence, there is to my knowledge no complete proof that the structure about to be described actually does occur for these specific equations. It is not hard to see that it does occur for *some* equations.

The phenomenology is as follows: The equations admit three stationary solutions, one at the origin and the other two (which we will denote by $C_\pm$ and refer to as *centers*) at $x=y=\pm\sqrt{b(r-1)}$; $z=r-1$. All three stationary solutions are unstable. Orbits that start near the origin escape monotonically; those that start near the centers escape through growing oscillations. If a solution is computed starting from some more or less randomly chosen initial point, what is found without exception is that the orbit will, after an initial transient regime of variable length, settle down to a motion in which, most of the time, it can be thought of as performing oscillations about one of the centers. The oscillation grows in amplitude; when it reaches a critical size, the orbit abruptly makes a transition to oscillation about the other center. This oscillation again grows and the orbit eventually makes a transition back to oscillating about the first center, and so on. A representative orbit is shown in Figure 1.

The amplitude of oscillation immediately after transition varies from transition to transition, and it in turn determines the number of oscillations before the next transition. The sequence of numbers of oscillations between transitions appears random, and the power spectra of the coordinates are continuous (see Figure 2). Thus, the motion both appears chaotic and satisfies the standard operational test for chaotic behavior.

The mathematical object responsible for this behavior is sketched schematically in Figure 3. This sketch represents a family of orbits in the three-dimensional state space for the Lorenz system. It is not even approximately to scale; proportions have been distorted in the hope of making the mathematical structure more transparent. To a first approximation, the structure looks like two reasonably flat loops of ribbon, one lying above the other along a central band, and the two glued together at the bottom of that band. The motion flows around the loops, clockwise on the left and counter-clockwise on the right. Going once around the right-hand loop constitutes a single oscillation around C_+. Orbits beginning either above or below the ribbon are attracted quickly down to its immediate vicinity and then follow the flow on it. The double-loop structure is strictly invariant under the solution flow; any point on it has an orbit that can be traced both forward and backward for all time without leaving it.

The central band is divided in half by orbits that flow essentially straight down to the stationary solution at the origin; these orbits are exceptions to the pattern of growing oscillations followed by transitions displayed by typical orbits. Orbits to the left of this boundary will make their next oscillation around C_-; those to the right will go next around C_+. The fact that oscillations around the centers are growing in amplitude means that, for example, a loop around C_+ brings the orbit back to the left of where it started out. A transition from oscillation around C_+ to oscillation around C_- occurs when an orbit making a loop around C_+ comes back to the left of the dividing boundary.

The central band divides in half laterally at the bottom of this boundary, and each half, after having made a loop around the appropriate center, has become wide enough to cover almost the entire top of the band. Thus,

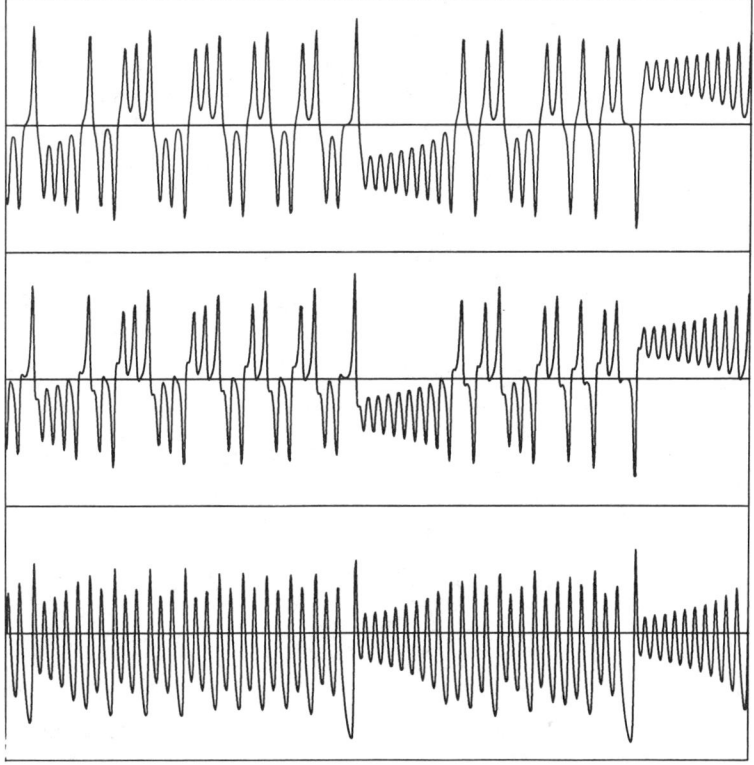

Figure 1 A representative orbit for the Lorenz system. From top to bottom: x, y, and z plotted against t.

orbits are pulled apart laterally as they flow around the loops, and this accounts for the observed sensitive dependence on initial conditions.

A typical orbit on this structure wanders over the surface, coming arbitrarily close to each point infinitely often. There are, however, a great many nontypical orbits. We have already mentioned the orbits in the middle of the central band which simply converge down toward the origin. There are also many periodic orbits, all unstable, as well as orbits with more subtle kinds of atypical behavior.

We next take a closer look at the ribbons and argue that they cannot be simple surfaces but must rather have infinitely many layers. Start at the top of the central band where there are two approximately parallel ribbons, one on top of the other. As we have drawn the picture, the upper ribbon is made up of orbits returning to the central band after a loop around C_+; the lower, around C_-. As the orbits flow down the central band, the two ribbons are drawn together. At the bottom, they form a two-sheeted surface which proceeds to split laterally in two with half going left around C_- and half right around C_+. Thus, the ribbon of orbits going around C_+ or C_- has at least two sheets, the upper one made up of orbits whose previous circuit was around C_+, the lower of orbits whose previous circuit was around C_-. These sheets are carried closer together by the flow but the separation remains nonzero, so the upper ribbon at the top of the central band actually has two layers rather than just one. The same is true for the lower ribbon, and therefore the structure at the bottom of

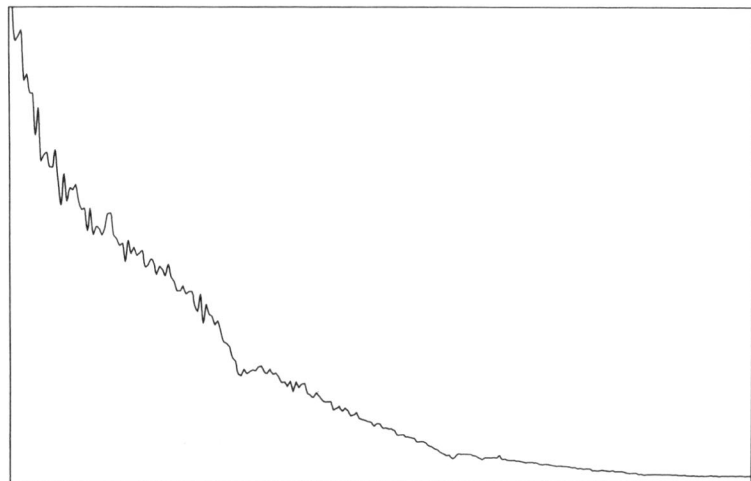

Figure 2 The power spectrum of the x coordinate of the Lorenz system (on a linear scale). The frequency ranges from 0 to 5; the oscillations apparent in Figure 1 have a frequency of about 1.5.

the central band actually has four layers rather than just two. Thus, the ribbons going around C_+ and C_- are actually four-sheeted, and so on. Continuing to argue in this way we see that all the ribbons must have infinitely many sheets.

This object is an instance of what has come to be called a *strange attractor*. The formulation of the definitive definition of the term *attractor* must await a more complete understanding of what the possible phenomena are. The general idea is clear, however: an attractor is a set that attracts nearby orbits (i.e. an orbit that starts near the attractor stays near it and converges to it as time goes to infinity). It should also be required that the set be closed and invariant under the solution flow (i.e. be made up of complete orbits defined for all time) and that the solution flow on the set be *recurrent*, i.e. that most orbits return infinitely often to the vicinity of their starting points. A situation which occurs frequently and which suffices to guarantee that the motion is adequately recurrent is that there is a single orbit in the attractor passing arbitrarily near to every point.

Stable stationary solution and limit cycles are elementary and trivial examples of attractors. It has turned out that the other kinds of attractors that occur most frequently are structures with infinitely many layers like the Lorenz attractor described above. Hence the epithet "strange." The general investigation of attractors has barely begun. Obtaining a complete classification looks, at present, extremely remote. On a less ambitious and more pragmatic level, there does not exist a convincing list of the "simplest" possibilities. Even worse, there are very simple examples—notably, the *Hénon* attractor (Hénon 1976, Hénon & Pomeau 1976, Curry 1979)— whose properties really aren't understood at all. There does exist, however,

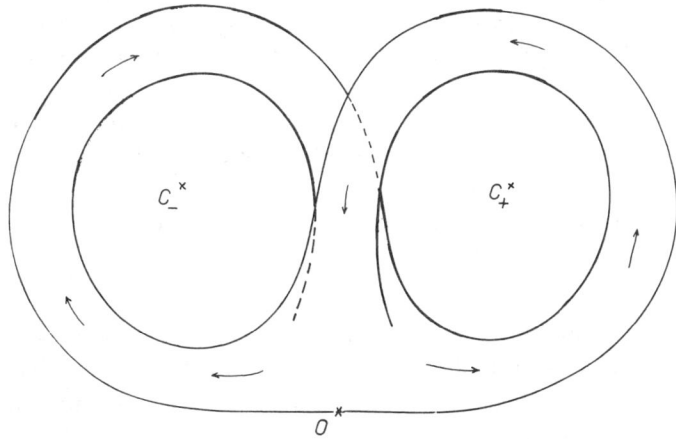

Figure 3 A schematic view of the Lorenz attractor.

one class of attractors for which there is a detailed and deep mathematical theory. These are the attractors satisfying a condition introduced by S. Smale (1967) and known as *Axiom A*. Roughly, Axiom A requires that orbits on or very near to the attractor have a strong and technically convenient form of sensitive dependence on initial condition, for both positive and negative time; it also requires that arbitrarily close to each point of the attractor there passes a periodic orbit. We will return in the final section of this review to one of the most important properties of attractors satisfying Axiom A.

For a more detailed discussion of the Lorenz attractor see Lorenz (1963), Guckenheimer (1976), Ruelle (1976a), Lanford (1978), Guckenheimer & Williams (1979), and Williams (1979). The literature on attractors satisfying Smale's Axiom A is extensive. For an introductory survey, see Bowen (1978).

SCENARIOS

In addition to asking how typical solutions of a typical differential equation behave, it is also possible to ask how this behavior varies as a parameter in the differential equation changes. Of particular interest is the study of the transition process from an equation whose solutions are asymptotically regular (i.e. stationary or periodic) to one with some kind of strange attractor. One can hope that there will turn out to be comparatively few such transition processes, each with distinctive characteristics permitting it to be identified without a detailed analysis of the underlying equations. J.-P. Eckmann has introduced the term *scenario* for such typical transition processes.

The Hopf bifurcation is an excellent classical example of a successful application of this approach. Let us review briefly how it works. The idea is to see what happens when, as a parameter is changed, a stationary solution to a differential equation loses stability. We consider, then, a differential equation depending on a parameter μ and a stationary solution for that equation which may also depend on μ. The particular situation we want to examine is that

for $\mu < \mu_c$, all the eigenvalues of the linearization of the equation at the stationary solution have strictly negative parts,

but

at $\mu = \mu_c$, a single complex-conjugate pair of simple nonreal eigenvalues crosses into the right half-plane with nonzero speed.

The Hopf Bifurcation Theorem says that the qualitative nature of the motion near the stationary solution is determined by the sign of a certain nonlinear function d of the first, second, and third derivatives of the differential equation with respect to the state variables at the stationary solution for μ_c. (The explicit formula for d is extremely complicated but need not concern us here.) For $d>0$, what happens is that, for μ slightly larger than μ_c, the equation has an attracting periodic orbit—i.e. a stable nonlinear oscillation—in the neighborhood of the now unstable stationary solution. As μ decreases to μ_c, the oscillation collapses down to the stationary solution; its amplitude is asymptotically proportional to $\sqrt{\mu-\mu_c}$. For $d<0$, something equally specific (if less interesting) happens: for μ slightly less than μ_c, there is a nonattracting periodic orbit which, as μ increases to μ_c, collapses down to the stationary solution with amplitude asymptotically proportional to $\sqrt{\mu_c-\mu}$. One of the great strengths of the theorem is that it assures us that, under our assumptions about the eigenvalues of the linearization, these are the *only* two possibilities except in the degenerate case $d=0$. (For a detailed discussion of the Hopf bifurcation, and proofs of the results cited above, see Marsden & McCracken 1976.)

The most interesting applications of the Hopf Bifurcation Theorem to physical systems do *not* proceed by verifying that the equations of motion satisfy the hypotheses. What is done, rather, is to observe experimentally the way the system in question goes from a stationary to an oscillatory regime as the parameter passes through a critical value. If the amplitude of the oscillations grows like $\sqrt{\mu-\mu_c}$ as μ passes μ_c, it is fairly safe to conclude that the transition process is a Hopf bifurcation, the square-root behavior serving as an experimentally verifiable signature. In this way, it is possible to arrive at a fairly precise picture of the transition from steady to oscillatory behavior even in situations where it is impossible to compute the stationary solution accurately, either analytically or numerically.

The Hopf bifurcation is thus an extremely successful scenario for the comparatively elementary transition from steady to periodically oscillatory motion. A few scenarios for the more interesting transition from periodic to aperiodic motion have been analyzed; a useful practical summary of this area has been given recently by Eckmann (1981). We will concentrate here on just one scenario, known as the *Feigenbaum transition*, which has been identified unequivocally in numerical studies of a number of simple models and perhaps observed in convection experiments as well.

The transition process is not a single bifurcation but an infinite sequence of them; it may be described as follows: The starting point is an attracting periodic orbit which loses stability at a first critical parameter value μ_0. In losing stability, it produces a new attracting periodic orbit in much the same way as an attracting periodic orbit is produced in the Hopf bifurcation. The new orbit tracks the old one closely but goes around it twice before closing (see Figure 4).

Thus, the period of the stable oscillation for μ just above μ_0 is twice that for μ just below μ_0. We will therefore refer to this bifurcation as the *period-*

doubling bifurcation; it is also frequently called the *pitchfork bifurcation*. As the parameter is further increased, a second critical value μ_1 is reached at which the doubled orbit itself loses stability through a period-doubling bifurcation. Thus, for μ just above μ_1, the system has an attracting periodic orbit that follows four times around the fundamental orbit before closing, i.e. a stable oscillation with period about four times the base period. This orbit in turn undergoes a period doubling bifurcation (at $\mu=\mu_2$) producing an oscillation with period about eight times the base period, and in fact the cascade continues ad infinitum in a finite parameter interval. The parameter value μ_∞ at which the sequence of doublings accumulates represents the onset of chaotic behavior.

A qualification is necessary here. The fact that such infinite sequences of period doublings do occur and cannot be eliminated by small changes in the differential equation is well established, as are a number of striking features of the accumulation process. What is not well understood is the character of the motion for μ slightly above μ_∞. Computer experiments strongly suggest that for most parameter values just above μ_∞ there is a strange attractor or perhaps several strange attractors, similar in character to the Hénon attractor. Unfortunately, nothing like this has been proved as yet; it simply isn't known for certain that a strange attractor *ever* appears at the accumulation of period doublings for a differential equation depending on a parameter. One thing that is known is that, if such an attractor exists, it must be very unstable with respect to changes in μ; there are values of μ above but arbitrarily close to μ_∞ for which there is an attracting periodic orbit.

The accumulation of period doublings has some characteristic features, discovered by M. Feigenbaum (1978, 1979b, 1980), which are consider-

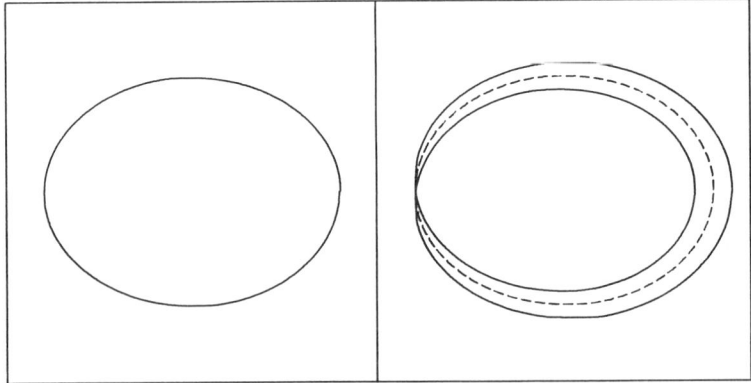

Figure 4 The period doubling bifurcation. On the left: the stable orbit for $\mu < \mu_0$. On the right: the solid curve represents the doubled stable orbit for $\mu_0 < \mu < \mu_1$; the dotted curve the now-unstable doubled orbit.

ably more distinctive than the square-root behavior of the amplitude in the Hopf bifurcation. The first, and easiest to state precisely, concerns the rate of convergence of the successive critical parameter values (μ_n) to their limit and says that, excluding degenerate cases, the convergence is geometric with a universal ratio:

$$\lim_{n \to \infty} \frac{\mu_n - \mu_{n-1}}{\mu_{n+1} - \mu_n} = 4.6692 \ldots . \tag{5}$$

This ratio has been observed in a number of model systems, and its origin and universality are well understood theoretically. Because of the relatively large value of the limiting ratio, all but the first few μ_n's can be expected to be very close together, and this makes the observation of successive period doublings and physical measurement of the ratio extremely difficult. Unless the parameter can be controlled extraordinarily well, what will be observed is one or two period doublings and then chaotic behavior. In fact, up to now, the limiting ratio has not been measured, even roughly, in any physical system.

Feigenbaum has also argued that the power spectrum for essentially any dynamical variable for μ near μ_∞ should display a universal ratio between the strengths of lines corresponding to periods of about $2^n \tau_0$ and those of nearby lines corresponding to periods of about $2^{n+1}\tau_0$. (Here, τ_0 denotes a base period). This part of Feigenbaum's analysis has not yet been put on a firm mathematical footing, but it does appear that something similar to what Feigenbaum predicts has been observed in convection experiments performed by Libchaber & Maurer (1980). This area is currently under very active investigation, and the situation should be clarified soon.

To close this section, we should point out that, although the notion of scenario is an appealing and powerful one, it does not apply to all transitions from regular to chaotic behavior. Discontinuous transitions, in which the chaotic component of the motion is large for parameter values even slightly above the critical value, can and do occur. The transition from a stable stationary solution to the Lorenz attractor is a mathematical example, and pipe flow appears to be a physical one. (These two situations are, however, not quite parallel. For the Lorenz system, there is a critical parameter value above which the stationary solution is no longer stable. The attractor and the stable stationary solution coexist for parameter values slightly below the critical one. For pipe flow, the stationary solution remains stable for arbitrarily large values of the Reynolds number, but becomes more and more sensitive to perturbations of small but finite amplitude).

For the theory of the Feigenbaum transition see, in addition to the works already cited, Feigenbaum (1979a), Collet et al. (1980, 1981), and Lanford (1980).

STATISTICAL THEORY

The investigation of strange attractors throws some light on the question of what should be meant, in a fundamental sense, by a statistical theory of a dissipative differential equation. The question needs to be turned around from the form usual in the classical statistical mechanics of Hamiltonian systems. Because of Liouville's Theorem, Hamiltonian systems—at least those with bounded and nonsingular energy surfaces—come equipped with natural time-independent probability distributions, the microcanonical ensembles of statistical mechanics. Such an ensemble is essentially just normalized area on an energy surface, and thus is a comparatively elementary and familiar construct; the investigation of whether there are other, radically different, stationary probability distributions can justifiably be dismissed as an empty mathematical exercise. Part of the reason why these ensembles are so natural is that the sets to which they assign probability zero conform to our intuitive notion of negligible sets of initial conditions. That is, if some particular behavior occurs only for a set of initial conditions of microcanonical probability zero, we can reasonably conclude that behavior will never be observed.

There is a general theorem about invariant probability distributions, the Birkhoff Pointwise Ergodic Theorem, which asserts that the long-time limit of the time average of any dynamical variable exists, except perhaps for a set of initial conditions of probability zero. (Here, "dynamical variable" simply means a function on the state space sufficiently well behaved for its integral to be defined and finite). Applied to Hamiltonian systems and the microcanonical ensemble, this theorem says that the time average of any dynamical variable will settle down if watched long enough. The existence of limiting time averages is an extremely familiar fact of experience, both for Hamiltonian and for dissipative systems, and it is commonly assumed that this is a general property of differential equations provided that their solutions remain in a bounded region of the state space. There is, however, no such general theorem for dissipative systems, and it is possible to find differential equations for which time averages do not exist. (An example is given in Ruelle 1980a).

Once the existence of time averages for Hamiltonian systems is established, attention turns to the problem of computing them. The most favorable situation is one where the time average does not depend on the initial condition, but is simply given by the ensemble average of the dynamical variable in question. It is not hard to show that this will be the case for all dynamical variables if and only if the system is *ergodic*, i.e. if and only if there is no way to decompose the energy surface into two parts, each of nonzero microcanonical probability, in a time-invariant way. Determining whether a particular Hamiltonian is ergodic is generally a delicate mathematical problem, and a great deal of important and deep work has been done on such questions over the past fifty years.

The situation for dissipative systems looks entirely different. Liouville's Theorem does *not* hold; indeed, solution flows generally contract volumes in the state space. As already noted, it is not even automatic that limiting time averages exist. There are, on the other hand, general abstract theorems asserting the existence of time-invariant probability distributions, and even of ergodic probability distributions, provided that solution curves don't run off to infinity. These probabilities are not, as in the Hamiltonian case, spread out over the state space. Consider, for example, what happens in the vicinity of an attractor. Any invariant probability distribution must assign probability zero to the set of all orbits converging to the attractor but not actually lying in it. Thus, whatever invariant probability distribution we might choose to work with, it is no longer justified to interpret a set of initial states of probability zero as physically negligible. Moreover, on a typical complicated attractor like the Lorenz attractor, there are a great many invariant probability distributions and it is not at all apparent how to go about singling out the right one—analogous to the microcanonical ensemble for Hamiltonian systems—to represent the statistics of typical orbits.

Rather than focusing on the choice of an invariant probability distribution, it is more satisfactory to start from limiting time averages of dynamical variables. These are, in any case, the quantities of most direct interest. One might hope that, in good cases, limiting time averages would exist along all orbits and be independent of orbit. This is slightly too optimistic. In the first place, there is no reason why a solution flow should have only one attractor. If there are several, the limiting time average will depend on which attractor the orbit approaches. We will therefore concentrate on a single attractor and study time averages along orbits converging to it. The second qualification is less obvious. It turns out, for the nontrivial attractors that have been studied in detail, that there are always many orbits with exceptional long-term behavior. A simple example is that these attractors generally contain many (unstable) periodic orbits. The best one can hope for, then, is that time averages will be *essentially* independent of orbit, i.e. that among the orbits converging to our attractor there may be a small set of exceptional orbits but that time averages do exist and are independent of orbit as long as the orbit does not belong to the exceptional set. A sensible meaning to give to the word "small" in this case is that the set of exceptional orbits has zero volume in the space of instantaneous states; exactly as for Hamiltonian systems, it is reasonable to suppose that such a set will never be seen in an experiment. If this favorable situation obtains, we will say that the attractor is *ergodic*. It follows from standard theorems that there is then an invariant probability distribution on the attractor—in physical terms, a stationary ensemble—such that the ensemble average reproduces the time average along nonexceptional orbits for all (continuous) dynamical variables.

Two questions now present themselves:

Are the attractors, which arise in practice, ergodic?

Can the ensemble reproducing the time average along nonexceptional orbits be described directly?

The only answer to the first question available at this time is that the few attractors whose structures are well understood—i.e. the Lorenz attractor and those satisfying Smale's Axiom A—turn out to be ergodic. For Axiom A attractors, this is the important Bowen-Ruelle Ergodic Theorem (Bowen & Ruelle 1975, Bowen 1975, Ruelle 1976b). For this same class a surprising answer to the second question is also available. Omitting numerous technicalities, this answer is roughly as follows: First take a hypersurface slicing through the attractor transversally in such a way that every orbit on the attractor crosses the hypersurface frequently. Introduce "coordinates" on the attractor by describing each point by giving the last place its orbit crossed the hypersurface, and the time since that crossing. Now cut up the intersection of the hypersurface with the attractor into a finite number of sufficiently small pieces, and describe an orbit on the attractor by saying which of these pieces it hits in which order. If the pieces are labeled with, say, $1,2,...,n$, then this procedure associates with the orbit a two-sided infinite sequence of integers in the range from 1 to n. If the cutting-up is done with sufficient care, it can be arranged that

> The set of sequences thus obtained from all orbits on the attractor can be described in a simple way; it is the set of all sequences in which certain pairs (i,j) never occur in succession;
>
> The sequences are essentially in one-to-one correspondence with points on the intersection of the attractor with the hypersurface.

The ensemble that reproduces time averages along nonexceptional orbits can be transported to an ensemble on the space of sequences. This transported ensemble turns out to be the thermodynamic equilibrium ensemble for a one-dimensional array of copies of a system with a discrete set of n states, with some nearest-neighbor exclusions and otherwise interacting through a many-body potential, that decreases exponentially as the separation goes to infinity.

The "equilibrium ensemble" for an Axiom A attractor thus looks much more complicated than the microcanonical ensemble for a Hamiltonian system. To construct it, it is necessary both to have a great deal of detailed information about the solution flow and to find the thermodynamic equilibrium ensemble for an infinite assembly of systems interacting in a nontrivial way. In simple cases at least, the description given might be used as a starting point for the construction of numerical approximations. The main interest of the Bowen-Ruelle Ergodic Theorem is, however, foundational. At least for very well behaved dissipative systems, it answers in a convincing and precise way the question of how, in principle, the equilibrium ensemble is to be defined.

ACKNOWLEDGMENTS

Preparation of this review was begun while the author was a visitor at the IHES in Bures-sur-Yvette, France. Financial support for that visit from the Volkswagen Foundation, and continuing financial support from the National Science Foundation (MCS78-06718), are gratefully acknowledged.

Literature Cited

Bowen, R. 1975. *Equilibrium States and the Ergodic Theory of Anosov Diffeomorphisms, Lecture Notes in Mathematics 470.* Berlin/Heidelberg/New York: Springer. 108 pp.

Bowen, R. 1978. *On Axiom A Diffeomorphisms, CBMS Regional Conf. Ser. 35.* Providence: Am. Math. Soc. 45 pp.

Bowen, R., Ruelle, D. 1975. The ergodic theory of Axiom A flows. *Invent. Math.* 29:181–202

Collet, P., Eckmann, J.-P., Koch, H. 1981. Period doubling bifurcations for families of maps on R^n, *J. Stat. Phys.* 25:1–14

Collet, P., Eckmann, J.-P., Lanford, O. E. 1980. Universal properties of maps on an interval. *Commun. Math. Phys.* 76:211–54

Curry, J. H. 1979. On the Hénon transformation. *Commun. Math. Phys.* 68:129–40

Eckmann, J.-P. 1981. Roads to turbulence in dissipative dynamical systems. *Rev. Mod. Phys.* In press

Feigenbaum, M. J. 1978. Quantitative universality for a class of nonlinear transformations. *J. Stat. Phys.* 19:25–52

Feigenbaum, M. J. 1979a. The universal metric properties of nonlinear transformations. *J. Stat. Phys.* 21:669–706

Feigenbaum, M. J. 1979b. The onset spectrum of turbulence. *Phys. Lett.* 74A:375–78

Feigenbaum, M. J. 1980. The transition to aperiodic behavior in turbulent systems. *Commun. Math. Phys.* 77:65–86

Guckenheimer, J. 1976. A strange, strange attractor. See Marsden & McCracken 1976, pp. 368–81

Guckenheimer, J. Williams, R. F. 1979. Structural stability of Lorenz attractors. *Publ. Math. IHES* 50:59–72

Hénon, M. 1976. A two-dimensional mapping with a strange attractor. *Commun. Math. Phys.* 50:69–77

Hénon, M., Pomeau, Y. 1976. Two strange attractors with a simple structure. In *Turbulence and Navier Stokes Equation,* ed. R. Temam, pp. 29–68. *Lecture Notes in Mathematics 565.* Berlin/Heidelberg/New York: Springer. 194 pp.

Lanford, O. E. 1978. Qualitative and statistical theory of dissipative systems. In *Statistical Mechanics: C.I.M.E. 1976,* pp. 26–98. Napoli: Liquori. 235 pp.

Lanford, O. E. 1980. Remarks on the accumulation of period doubling bifurcations. In *Mathematical Problems in Theoretical Physics,* pp. 340–42. *Lecture Notes in Physics 116.* Berlin/Heidelberg/ New York: Springer. 412 pp.

Lanford, O. E. 1981. Strange attractors and turbulence. In *Hydrodynamic Instabilities and the Transition to Turbulence,* ed. H. L. Swinney, J. P. Gollub, pp. 7-26. *Topics in Applied Physics 45.* Berlin/Heidelberg/New York: Springer. 292 pp.

Lebowitz, J. L., Penrose, O. 1973. Modern ergodic theory. *Physics Today* 26(2):23–31

Libchaber, A., Maurer, J. 1980. Une expérience de Rayleigh-Bénard de géométrie réduite. *J. Phys., Colloques C3* 41:51–56

Lorenz, E. N. 1963. Deterministic nonperiodic flow. *J.Atmos. Sci.* 20:130–41

Marsden, J. E., McCracken, M. 1976. *The Hopf Bifurcation and its Applications. Applied Mathematical Sciences* 19. New York/Heidelberg/Berlin: Springer. 408 pp.

McLaughlin, J. B., Martin, P. C. 1975. Transition to turbulence in a statically stressed fluid system. *Phys. Rev. A* 12:186–203

Ruelle, D. 1976a. The Lorenz attractor and the problem of turbulence. See Hénon & Pomeau 1976, pp. 146–58

Ruelle, D. 1976b. A measure associated with Axiom A attractors. *Am. J. Math.* 98:619–54

Ruelle, D. 1978a. Dynamical systems with turbulent behavior. In *Mathematical Problems in Theoretical Physics*, pp. 341–60. *Lecture Notes in Physics 80*. Berlin/Heidelberg/New York: Springer. 438 pp.

Ruelle, D. 1978b. Sensitive dependence on initial condition and turbulent behavior of dynamical systems. *Ann. NY Acad. Sci.* 316:408–16

Ruelle, D. 1980a. Measures describing a turbulent flow. *Ann. NY Acad. Sci.* 357:1–9

Ruelle, D. 1980b. Les attracteurs étranges. *La Recherche* 108:132–44

Ruelle, D., Takens, F. 1971a. On the nature of turbulence. *Commun. Math. Phys.* 20:167–92

Ruelle, D., Takens, F. 1971b. Note concerning our paper "On the nature of turbulence." *Commun. Math. Phys.* 23:343–44

Smale, S. 1967. Differentiable dynamical systems. *Bull. Am. Math. Soc.* 73:747–817

Williams, R. F. 1979. The structure of Lorenz attractors. *Publ. Math. IHES* 50:73–99

DYNAMICS OF OIL GANGLIA DURING IMMISCIBLE DISPLACEMENT IN WATER-WET POROUS MEDIA

A. C. Payatakes

Department of Chemical Engineering, University of Houston, Houston, Texas 77004

INTRODUCTION

A ganglion is a nodular blob of a non-wetting phase that occupies at least one and usually several adjoining chambers of the void space in a permeable medium. Strong interest in the dynamic behavior of a population of non-wetting ganglia undergoing immiscible displacement has arisen recently, mainly because this problem is central to the understanding of oil-bank formation during enhanced oil recovery by chemical flooding. More generally, the same problem arises in the analysis of the relative permeabilities to any pair of wetting and non-wetting phases, when the saturation of the wetting phase exceeds ~ 0.60. Saturation of a phase is defined as the fraction of the void space that is occupied by that phase. Many drainage or imbibition phenomena fall into this category. To fix ideas, however, we concentrate on the case where the non-wetting phase is oleic (oil-based), the wetting phase is aqueous, and the objective is enhanced oil recovery.

Origin of Oil Ganglia

After the primary and secondary oil production stages, 40 to 80% of the amount of oil originally in place remains entrapped in the pores of the oil-bearing rock (typically a sandstone or a dolomite). On the average, at the end of secondary flooding, two thirds of the oil remains entrapped in the reservoir. This residual oil exists in the form of discrete oil ganglia and occupies 25 to 50% of the porous space, whereas the rest is occupied by the aqueous phase. Typically, oil ganglia at this state have sizes ranging from one to fifteen elemental chamber volumes. These ganglia were produced through shedding of blobs by the retreating oleic phase during secondary flooding. Shedding of a blob can take place when a fingerlike protrusion of the oleic mass forms a narrow neck under the combined effect of the local pressure difference across the interface, ΔP_{neck}, and the interfacial tension γ_{ow} (Figure 1). For sufficiently large ΔP_{neck} the interface at the neck becomes unstable and the neck ruptures.

Roof (1970) studied oil-neck rupture in connection with the snap-off of oil droplets in water-wet cores. Morrow (1979) applied this idea to the snap-off of oil blobs from the retreating oleic phase during water flooding. More recently, Mohanty, Davis & Scriven (1980) extended this idea by considering rupture of oil necks in a porous-medium model similar to that developed by Payatakes, Ng & Flumerfelt (1980), namely a network of cells of the constricted-tube type. Under quasistatic deformation conditions, these authors predict that neck rupture occurs if the following approximate criterion is satisfied, $\Delta P_{\text{neck}} > 4\gamma_{\text{ow}}/d_{\text{p}}$, where d_{p} is the volumetric diameter of the pore chamber adjacent to the neck.

Oil ganglia shed by the retreating oleic phase during secondary flooding do not usually travel far. The expected distance over which a given ganglion migrates before it gets stranded is a function of the ganglion size, the

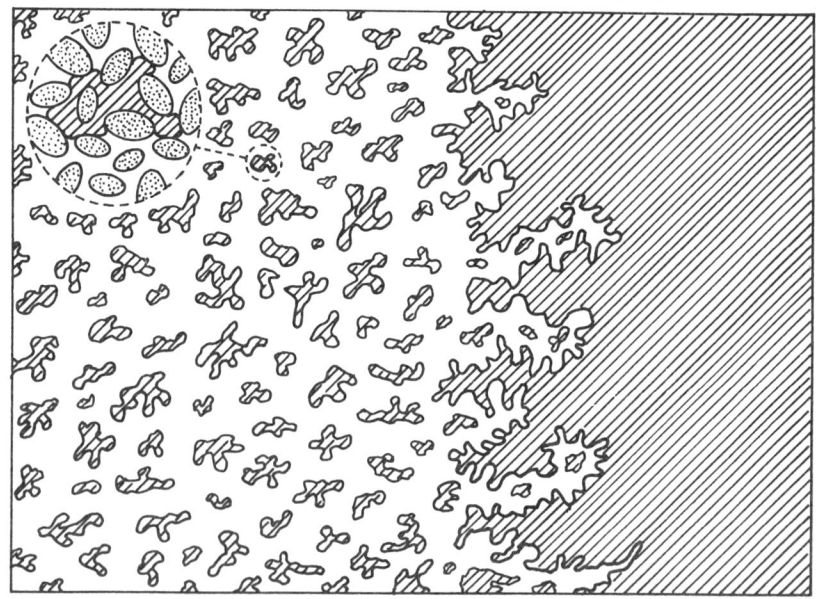

Figure 1 Schematic representation of immiscible oil displacement and formation of oil ganglia by waterflooding; hatched areas = oil, clear area = water, dotted areas = solid. The grains are only shown in the detail (inset).

porous medium geometry, the contact angle, the water saturation S_w, and the capillary number. The capillary number is defined as $N_{Ca} = \mu_w V_f / \gamma_{ow}$, where μ_w is the dynamic viscosity of the aqueous phase and V_f is the superficial velocity of the flood. During secondary flooding N_{Ca} varies in the range 10^{-7} to 10^{-5}, which means that capillary forces dominate viscous forces. In light of the Monte Carlo simulation results of Ng & Payatakes (1980), it is clear that in the more usual case when the shed ganglia are small, having volume of $\sim$ 20 chamber-volumes or less, entrapment is virtually immediate. Much larger ganglia are shed less frequently; these may travel some distance driven by the pressure gradient until they fission into smaller daughter ganglia, which in turn get trapped easily. This explains the experimental observation that the residual oil consists mainly of ganglia having size less than $\sim$ 20 chamber-volumes. It is the objective of enhanced oil recovery (EOR) processes to mobilize and produce a significant part of this residual oil at acceptable cost.

Chemical Flooding; The Problem of Oil Reconnection

One of the most promising EOR processes is chemical (or micellar) flooding. Floods of this type consist of a "chemical slug" being driven by a viscous polymer solution for mobility control, namely suppression of fingering. The chemical slug is a mixture of water, surfactant (usually a petroleum sulfonate), a cosurfactant (such as tetrabutyl alcohol), and sodium chloride. At optimal salinity, chemical flooding begins as a quasi-miscible process, since the injected chemical slug can solubilize the oil ganglia that it encounters. At this stage the flood is very efficient, missing only oil entrapped in pockets of the reservoir that are very inaccessible to flow. As the chemical slug advances, however, the displacement process deteriorates to the much less efficient immiscible mode (Healy & Reed 1974, Healy, Reed & Stenmark 1976, Nelson & Pope 1977). The deterioration is caused by compositional changes of the chemical slug, brought about by dilution, oil pick-up, loss of surfactant due to adsorption and precipitation, changes of salinity, ion exchange, etc. It is for this reason that immiscible displacement is a problem of fundamental importance in the analysis of chemical flooding.

Once the immiscible mode of displacement is established, the further success or failure of the flood depends on whether a substantial oil-bank is formed and maintained, or not. An oil-bank has oil saturation significantly higher than the residual one, and it is formed—if it is formed at all—near the moving front of the chemical slug. Even though basic studies of the oil-bank structure are lacking, researchers in this field assume that an oil-bank is composed of blobs large enough to be immune to stranding. These superganglia move nearly in tandem, and pick up additional residual ganglia in their paths through collision-coalescence. The question, then, arises naturally: how do the originally small and easy-to-trap oil ganglia get mobilized and reconnected to form the very large blobs that constitute a successful oil-bank? In the present work we are mainly concerned with this question.

To make the problem tractable, we concentrate on a sufficiently small part of the porous medium, so that compositional changes can be neglected. Further, we treat the two phases, oleic and aqueous, as single-component liquids—"oil" and "water"—keeping in mind that the physical properties of these phases may very well be functions of position on a macroscopic scale. The present analysis is designed to predict conditions of incipient oil-bank formation, and it needs to be modified once the very large blobs constituting a bank begin to appear. The method of oil-ganglion dynamics promises to evolve into a useful diagnostic and even predictive tool addressed to incipient oil-bank formation. In addition, some of its aspects will be useful in analyzing steady-state, fully developed, two-phase flow in permeable media.

The discussion of the problem is divided into the following topics: (a) theoretical modeling of the porous medium, (b) motion, fissioning, and stranding of a solitary oil ganglion, and (c) dynamics of a population of oil ganglia.

THEORETICAL MODELING OF THE POROUS MEDIUM

The fate of a solitary ganglion and, more importantly, the collective fate of a large population of interacting oil ganglia depends strongly on the geometry of the void space. The reasons for this dependence are many and will become clear as we proceed with the analysis. To illustrate the point, however, consider a single oil ganglion. For given macroscopic pressure gradient ∇P, viscosities μ_o and μ_w, interfacial tension γ_{ow}, retreating contact angle θ_r, and advancing contact angle θ_a, the motions of the oil-water interfaces depend on the geometry of the throats in which they reside. The pressure fields inside and outside the ganglion are affected not only by the position and shape of the ganglion itself but also by the detailed geometry of a large control volume of the porous medium containing the ganglion. Furthermore, once an oil ganglion is set in motion, its subsequent fate depends largely on the geometry of the track downstream of it in the direction of the macroscopic pressure gradient. The porosity, the size distribution and actual locations of the throats that connect adjoining chambers, the size distribution of the chambers, the coordination number of the chambers (which determines the degree of interconnectivity), the degree to which the medium possesses vugs (pore chambers much larger than a typical grain), and the surface roughness of the walls are all important features.

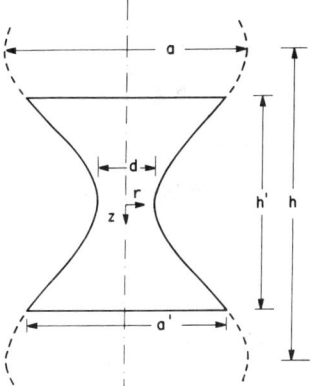

Figure 2 Typical unit cell of the porous medium model ($-h'/2 \leq z \leq h'/2$). The wall profile is a sinusoidal function in z. The segment ($-h/2 \leq z \leq h/2$) is the extended unit cell. From Payatakes et al. (1980).

No progress can be made in analyzing the fate of one or many ganglia unless a proper mathematical description of the porous matrix is first obtained. Unfortunately, oil-bearing rocks have porous structures that are much too complex for complete mathematical description (see, for example, Wardlaw & Taylor 1976, Wardlaw & Cassan 1978). The same is true, although to a lesser degree, of relatively simpler porous media used in bench-scale studies, such as sandpacks and beadpacks. Hence we must use approximate models. The objective of a porous-medium model is to provide a reasonable idealization of the geometrical structure of the prototype so that the transport process of interest can be treated mathematically. To this end, the model must incorporate the most relevant characteristics of the prototype, while its complexity should be kept at a manageable level.

Such a model, pertaining to sandpacks and beadpacks, was developed recently by Payatakes et al. (1980). It consists of a network of unit cells of the constricted-tube type. A typical unit cell is shown in Figure 2 and it can be thought of as a segment of a sinusoidal tube. The constriction diameter d, the length h', the entrance diameter a', the wavelength h, and the maximum diameter a are random variables. A method for determining the size distributions of these variables is given in Payatakes et al. (1980), and it requires knowledge of the porosity, the initial drainage curve, and the grain size distribution, all readily obtainable data. The network itself should have the same coordination number[1] as that of the prototype—

---

[1] The coordination number of a porous medium is defined as the number of neighboring chambers connected directly through throats to a given chamber.

usually five to six—and, in general, it should be random. However, to simplify the analysis, the initial studies in Payatakes et al. (1980) and Ng & Payatakes (1980) were based on a cubic network. A plane of the cubic network is shown in Figure 3; note that for convenience all unit cells are depicted with the same "bow-tie" symbol, regardless of the fact that they do not have the same size. A node of such a network represents the center of a chamber of the prototype. The six half-unit cells connected to a node represent an elemental void space, namely a chamber together with one half of each satellite throat. For this reason, the set of half-unit cells connected to one and the same node is called a conceptual elemental void space (CEVS). The lines connecting unit cells are assumed to have no volume and to present no resistance to flow; they simply indicate flow paths. Within this framework, it is simple to represent an oil ganglion, simply by filling the appropriate CEVS's with oil and the rest with water.

Before proceeding with the analysis of the motion of a solitary oil ganglion, it is pertinent to briefly discuss other porous-media models that have been considered in connection with the problem at hand. A comprehensive review of models up to 1974 was given by van Brakel (1975), who con-

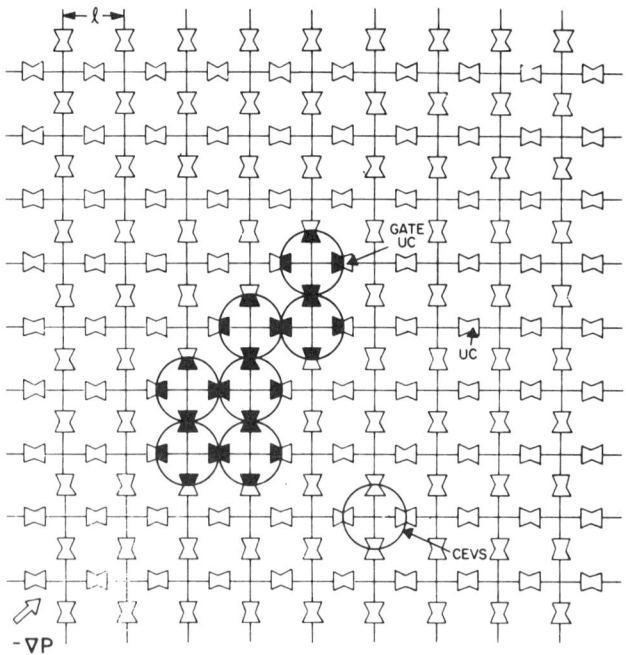

Figure 3 Two-dimensional depiction of idealized cubic network of unit cells. A conceptual elemental void space (CEVS) and a unit cell (UC) are identified. An idealized oil ganglion occupying seven adjoining pores is also shown. From Payatakes et al. (1980).

cluded that none of the models existing at that time provided a satisfactory basis for the modeling or simulation of immiscible displacement.

One of the most notable of earlier models is the random network of randomly sized capillary tubes developed by Fatt (1956) and Josseling de Jong (1958), and it has been successful as a basis for the study of some—but not all—aspects of immiscible displacement. The latest application of this model was presented by Larson, Scriven & Davis (1981) and Larson, Davis & Scriven (1981) in conjunction with percolation-theory concepts. These authors were mainly concerned with the dependence of residual non-wetting saturation on the capillary number. They obtained good agreement with data by adjusting the coordination number of the network and other parameters of the model. Models of this type suffer from the shortcoming that they omit the strongly converging-diverging character of the void space. This omission is critical and leads to some unrealistic conclusions. For instance, if a water-oil interface is not at equilibrium at a given position within a cylindrical capillary, it cannot find a position of equilibrium by advancing or retreating within the same capillary, irrespective of conditions. On the contrary, it may have to travel through several bonds of the network before it reaches a node having satellite bonds with diameters that will allow the interface to establish equilibrium. Another important defect of these models is that flow and pressure drop in cylindrical capillaries are not very representative of the situation in strongly converging-diverging conduits, even under creeping-flow conditions (Payatakes et al. 1973, Neira & Payatakes 1978, 1979). The network of constricted-tube unit cells does not suffer from these defects. Actually, the capillary network model can be thought of as a limiting case of the constricted-tube network model.

Another important class of models consists of regular packings of uniform balls, known as *ideal soil*. The ideal-soil model dates back to the pioneering studies of drainage and imbibition by Haines (1930). It was used by Snyder & Stewart (1966) to analyze one-phase flow and by Melrose (1965, 1968) to study the equilibrium of interfaces in porous media. The ideal-soil model provides a realistic representation of the geometry near throats and it inherently possesses the converging-diverging character of the flow channels. Its main shortcoming is that it is completely periodic, lacking the random nature of the prototype. Consequently, it cannot account for the stochastic aspects of the fate of oil ganglia or more generally of drainage and imbibition. The constricted-tube network model is superior in that it shares the stochastic nature of the prototype. On the other hand, it simplifies the geometry of the throats more than the ideal soil does, since the throats of the prototype are not axisymmetric. Still, this last shortcoming is not crucial.

In the remainder of this review we discuss oil-ganglion motion and dynamics using the constricted-tube network model as basis for the analysis.

MOBILIZATION, FISSIONING, AND STRANDING OF A SOLITARY OIL GANGLION

In order to understand the problems associated with the motion of oil ganglia, one must first understand the equilibrium and motion of interfaces. We divide our discussion into the following topics: (*a*) equilibrium of oil-water interfaces, (*b*) moving interfaces, (*c*) quasistatic motion of a solitary oil ganglion, and (*d*) dynamic motion of a ganglion. By quasistatic motion we mean that the ganglion creeps along driven by a flood whose capillary number equals or slightly exceeds the critical value required for mobilization of the ganglion under consideration. By dynamic motion we mean that the ganglion moves driven by a flood with capillary number that substantially exceeds the critical value. As we see below, quasistatic and dynamic motion differ in certain aspects.

Equilibrium of Oil-Water Interfaces

The equilibrium of menisci is a subject of great interest in its own right and has received much attention. A state-of-the-art review was given recently by Michael (1981). Hence, we forego a lengthy discussion, presenting only the material necessary to make this article self sufficient.

Consider an oil ganglion in a water-wet porous medium and assume that the pressure gradient is nil, so that both phases are quiescent. The system can be considered water-wet if the intrinsic (or equilibrium) contact angle, as measured from the aqueous phase, θ_e, is less than 40 deg. The ganglion has several protruding appendices residing near throats of the porous matrix (Figure 1).

Let us first concentrate on a single meniscus, separating two large masses of oil and water (Figure 4a). The conditions of static equilibrium for such an interface are

$$P_o - P_w = \gamma_{ow} J, \tag{1}$$

$$\gamma_{os} = \gamma_{ws} + \gamma_{ow} \cos\theta_e, \tag{2}$$

where J is the Gaussian curvature of the interface, P_o and P_w are the pressure values in the oil and water, respectively, γ_{os} is the interfacial tension between oil and the solid, and γ_{ws} is the interfacial tension between water and the solid. Equation (1) expresses the condition for hydrostatic equilibrium applicable to each point of the oil-water interface. Equation (2) applies in a similar fashion to the three-phase contact line. Although Equations (1) and (2) are sufficient to express hydrostatic equilibrium, a third relation is needed to ensure that the interface is also stable, that is, capable of withstanding small perturbations without collapsing. This was first realized by Gibbs who discussed a particular case in quantitative terms (Miller & Miller 1956, Melrose 1965, 1968). Melrose (1970) developed the following criterion for stability in water-wet systems, based on a thermodynamic argument.

For $\theta_e < 40$ deg the interface is stable if

$$\frac{dJ}{dV_o} \geq 0, \qquad (3)$$

where V_o is the volume of the oil in the immediate neighborhood of the interface (Figure 4a). Examples of stable and unstable configurations are given by Melrose & Brandner (1974). For a given fluid-fluid-solid system

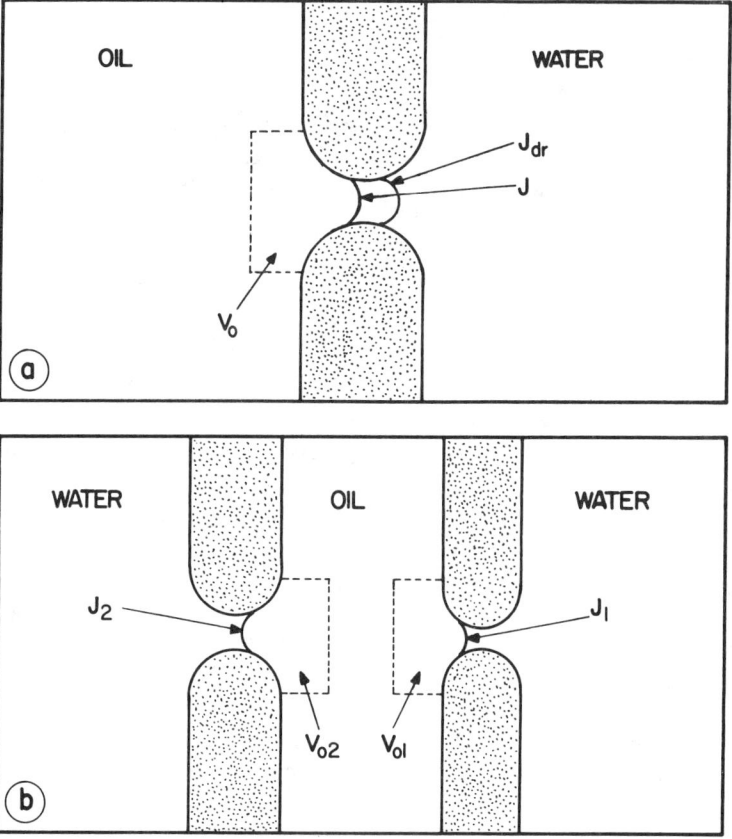

Figure 4 Equilibrium and stability of menisci. (*a*) Meniscus separating one pool of oil from one pool of water. (*b*) Two menisci separating a mass of oil from two pools of water.

and throat geometry there are two configurations of the interface such that they envelope all stable configurations. The maximum stable curvature is called drainage curvature, J_{dr}, and the minimum stable curvature is called imbibition curvature, J_{imb}. The imbibition curvature is obtained if and when the water advancing in the throat meets with water forming pendular rings at grain-grain contacts on the far side of the throat. Both J_{imb} and J_{dr} are functions of θ_e. An interface is stable if $J_{imb}(\theta_e) \leq J \leq J_{dr}(\theta_e)$.

Let us consider now a mass of oil separated from water by two interfaces having curvatures J_1 and J_2 (Figure 4b). Here, we cannot consider the stability of one interface alone; we must look at the entire system. This problem was considered by Gillette & Dyson (1974), who argued that the system is stable if positive work is required to perturb it from its present position. Neglecting gravity, this argument leads to the following stability criterion for water-wet systems,

$$J_1 = J_2 \quad \text{and} \quad \frac{dJ_1}{dV_{o1}} + \frac{dJ_2}{dV_{o2}} \geq 0 . \tag{4}$$

This criterion can be easily generalized to an oil mass with several oil-water interfaces, namely to an oil ganglion. Hence, we conclude that for $\nabla P = 0$ and neglecting gravitational effects the curvatures of all interfaces of a ganglion have the same value irrespective of the sizes and shapes of the throats in which they reside. The omission of gravity is justified by the small scale of the system; typically, a chamber has a diameter of 50 μm or less.

Moving Interfaces

This problem is very complicated and not completely understood yet. In particular, the exact nature of the boundary condition at the moving contact line is a problem not fully resolved. A recent review was given by Dussan V. (1979). For our purposes, we simply make note of the fact that when a meniscus advances or recedes the contact angle deviates from its intrinsic value θ_e. The advancing contact angle (advancing away from the wetting phase) θ_a is larger than θ_e, whereas the receding contact angle (receding towards the wetting phase) is smaller than θ_e. Morrow (1975) presented an experimental study of dynamic contact angles for wetting and non-wetting systems, with emphasis on petroleum-recovery applications. Rillaerts & Joos (1980) presented an experimental study and an approximate theoretical analysis showing that if θ is the dynamic-contact angle in a cylindrical tube and v_o is the average velocity, then (cos θ − cos θ_e) is proportional to $(\mu_w v_o/\gamma_{ow})^{1/2}$, namely proportional to the square root of the capillary number for the tube flow. In this manner they successfully correlated a large set of data from several different systems. These results may be approximately applied to menisci moving in converging-diverging tubes, by simply accounting for the fact that the average velocity v_o is a function of axial position.

Quasistatic Motion of a Solitary Oil Ganglion

RESPONSE OF AN OIL GANGLION TO AN EXTERNAL PRESSURE GRADIENT Consider a ganglion trapped in a porous medium. If the pressure gradient (neglecting hydrostatic effects) is nil, both the aqueous phase and the ganglion will be immobile, and the curvatures of all interfaces of the ganglion will have the same value (see above) (Figure 5a). Assume now that we increase the pressure gradient to a nonzero value ($-\nabla P > 0$) that is not sufficiently large to effect mobilization (Figure 5b). Since the pressure in the ganglion, P_o, is virtually uniform, and the pressure in the water decreases along the ganglion, the pressure differences ($P_o - P_w$) across downstream interfaces are now larger than those across upstream ones. The ganglion responds to the new situation as follows. Downstream interfaces advance somewhat into their respective throats, so that their curvatures increase; in addition, their contact angles decrease tending towards the limiting value of θ_r for zero velocity (cf. Dussan V. 1979), θ_r^0. At the same time, upstream interfaces retract somewhat within their respective throats, so that their curvatures decrease; in addition, their contact angles increase tending towards the limiting value of θ_a for zero velocity, θ_a^0. In this new position, Equations (1) and (2) are still satisfied at all interfaces, keeping in mind that instead of θ_e we must now use the appropriate values of the contact angle.

The resulting capillary pressure imbalance tends to move the ganglion backward, whereas the external pressure gradient (caused by viscous dissipation in the aqueous phase) tends to move it forward. The two forces cancel each other and a new equilibrium is established. Clearly, the ratio of the viscous force over the net capillary force, expressed by the capillary number N_{Ca}, is of fundamental importance here.

Assume now that we increase the pressure gradient to a critical value $-\nabla P = G_{cr}$ at which point the curvature of one of the downstream interfaces finally reaches its critical drainage curvature, the value of which depends on the geometry of the corresponding throat and the limiting receding contact angle, θ_r^0. At this point the ganglion cannot resist mobilization any longer, and it moves. This motion is relatively fast and has been called a rheon by Heller (1968) and Melrose & Brandner (1974). During the rheon, oil invades one of the downstream chambers (an event known as a xeron), while the aqueous phase invades one of the upstream chambers that used to be occupied by the oil (an event known as a hygron) (Figure 5c). In porous media with very irregular geometry, it is possible that if the xeron occurs in an unusually large chamber, two (or more) hygrons may be necessary to supply enough oil for the xeron. The reverse may also occur. Finally, if the proper conditions for a hygron develop at a site where the ganglion is one-chamber thick, then the ganglion may

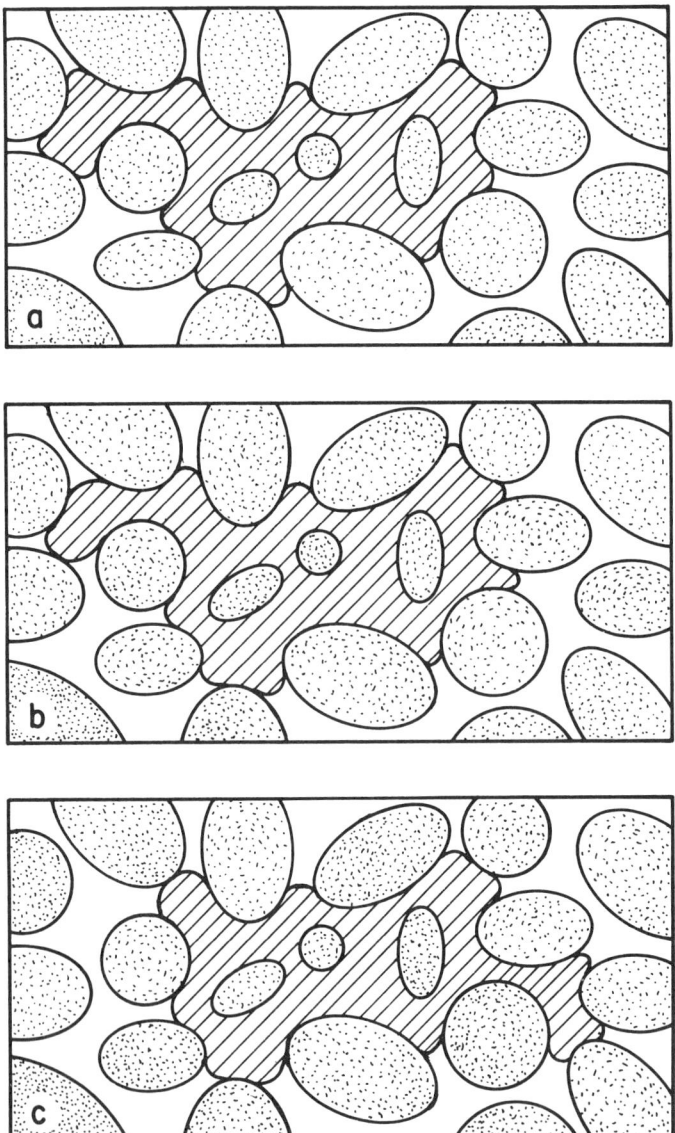

Figure 5 Mobilization of an oil ganglion. (*a*) No flow; ganglion at equilibrium. (*b*) The aqueous phase flows from left to right with pressure gradient less than the critical value required for mobilization. (*c*) If the pressure gradient exceeds the critical value for mobilization, a rheon takes place.

fission in two daughter ganglia, as follows. First, the part of the ganglion in the chamber being invaded by water will thin and take a roughly cylindrical necklike form extending across the chamber and having diameter approximately equal to the average of the diameters of the throats at its two ends. The curvature of the interface of the neck can be estimated as $(4/d_{c,\ell}) < J_{neck} < (4/d_{c,s})$ where $d_{c,\ell}$ is the diameter of the larger of the two end-throats and $d_{c,s}$ that of the smaller. If the local pressure in the water exceeds that in the oil by ΔP_{neck} and if $\Delta P_{neck} > \gamma_{ow} J_{neck}$, the neck will rupture and the ganglion will fission in two daughter ganglia. A conservative criterion for fission is $\Delta P_{neck} > 4\gamma_{ow}/d_{c,s}$. After fission, one or both of the daughter ganglia may become trapped, as the driving force acting on a ganglion is virtually proportional to its length (see below), whereas the capillary resistance is a function of the throat sizes, the interfacial tension, and the contact angle.

QUASISTATIC CRITERION FOR MOBILIZATION AND FISSIONING OF OIL GANGLIA Ng & Payatakes (1980) developed a criterion for the quasi static mobilization of oil ganglia. To this end, they assign indices 1, 2, 3, etc. to all gate unit cells, namely those that contain interfaces, starting with the one furthest downstream. Then, they make a sensitivity analysis as follows. They assume that all interfaces are locked in their unit cells, except for two, one downstream with index i and one upstream with index k. For this particular pair of gate unit cells, there is a critical pressure gradient that will cause mobilization. By repeating this calculation for all possible pairs, one can identify the particular pair ($i = I, k = K$) for which the required pressure gradient is minimum. If mobilization is to occur, it will proceed through the Ith and Kth gate unit cells. This analysis is equivalent to identifying the maximum appendix mobility factor, namely

$$\beta_{KI} = \Delta L_{KI} \cos \theta_{KI} / [J_{dr,I}(\theta_r^0) - J_{\ell b, K}(\theta_a^0)] . \tag{5}$$

Here, ΔL_{KI} is the distance between the throats with indices K and I, θ_{KI} is the angle between the line connecting throats K and I and the macroscopic flow direction, $J_{dr,I}$ is the drainage curvature in the Ith downstream gate unit cell, and $J_{\ell b, K}$ is the lower bound of the imbibition curvature, estimated as $J_{\ell b, K} = 4\cos \theta_a^0 / a_K$ (where a_K is the maximum diameter of the unit cell with index K; Figure 6; see also Figure 2). Ng & Payatakes used the further simplification that $\theta_a^0 \simeq \theta_e \simeq \theta_r^0$, but it is better to make the distinction. Note, however, that their calculated results were not affected by this approximation, as those were based on a zero contact angle (perfect wettability). If ∇P is the actual macroscopic pressure gradient, the criterion for mobilization is expressed in terms of the appendix mobilization number, N_{am}:

$N_{am} = \beta_{KI} |\nabla P| / \gamma_{ow} > 1 \rightarrow$ mobilization ,

$N_{am} = \beta_{KI} |\nabla P| / \gamma_{ow} < 1 \rightarrow$ stranding . \hfill (6)

During mobilization, the xeron takes place through the Ith gate unit cell and the hygron through the Kth gate unit cell. In terms of the more traditional capillary number, the equivalent mobilization criterion becomes

$$N_{Ca} = (\mu_w V_f/\gamma_{ow}) > (k_{rw}k/\beta_{KI}) \rightarrow \text{mobilization},$$

$$N_{Ca} = (\mu_w V_f/\gamma_{ow}) < (k_{rw}k/\beta_{KI}) \rightarrow \text{stranding}, \quad (7)$$

where k_{rw} is the relative permeability to water and k is the absolute permeability. This mobilization criterion agrees well with existing experimental data (Ng, Davis & Scriven 1978, Batycky 1979, Hinkley 1981).

If the hygron index K is located somewhere in the middle of the ganglion, and the ganglion thickness at that point consists of only one CEVS (the one about to be invaded by water), the oil ganglion is assumed to fission. This assumption is reasonable, in view of the fact that the curvature of the interface in the Kth gate unit, namely $J_{\ell b, K}$ (θ_a^0), is estimated in the most conservative manner, namely based on the chamber diameter a_K. Hence, the actual corresponding interface has an even stronger tendency to retreat and collapse. Note that this situation arises when the chamber and throats in question are small and the local tendency for imbibition large.

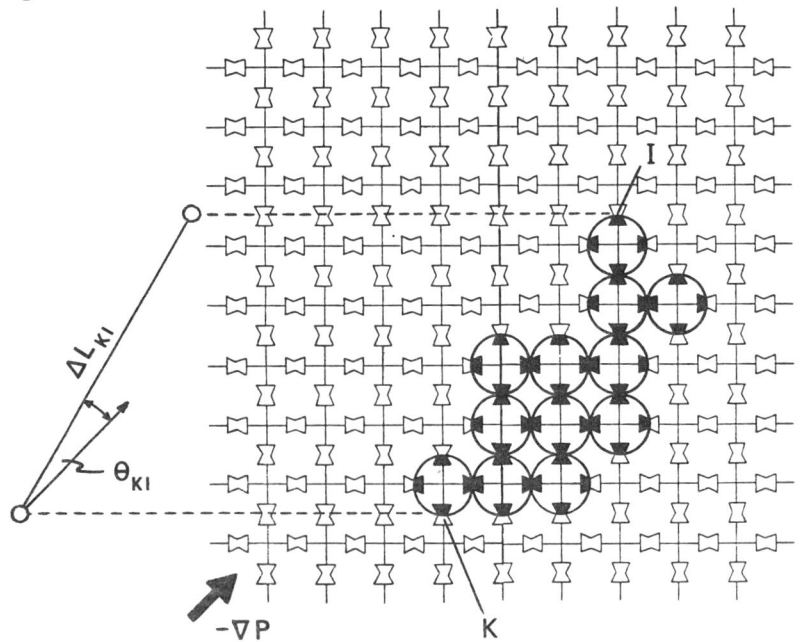

Figure 6 Concepts involved in the quasistatic criterion for mobilization.

Clearly, this analysis of mobilization and fission hinges on the determination of β_{KI}. Ng & Payatakes developed an efficient algorithm for the computer-aided determination of β_{KI}. Based on this algorithm they also developed a Monte Carlo simulation method to study the fate of solitary oil ganglia.

MONTE CARLO SIMULATION OF THE FATE OF SOLITARY GANGLIA This stochastic simulation consists in (a) assigning to the porous-medium model unit-cell dimensions corresponding to those of the prototype, (b) generating an oil ganglion of given volume with random shape at a random position, (c) specifying the capillary number, (d) specifying the contact angles θ_r^0, θ_a^0, and (e) applying the mobilization-fissioning criterion repeatedly, until stranding or fission occurs.

A sample realization of the fate of a 9-CEVS ganglion moving in a 100 × 200 sandpack of porosity $\epsilon = 0.395$ is shown in Figure 7. The calculations were based on contact angles $\theta_e = \theta_a^0 = \theta_r^0 = 0$ and a capillary number $N_{Ca} = 1.07 \times 10^{-3}$. The ganglion underwent seven rheons; as it moved it became elongated and aligned with the direction of the pressure gradient. During the seventh rheon it fissioned into two daughter ganglia, one small and one large. It so happened in this example that the smaller daughter ganglion remained stranded, whereas the larger took at least one more step. This realization is typical of thousands of similar realizations performed by Ng & Payatakes. The implications are clear. A mobilized solitary oil ganglion undergoes fission after every few (say 5 to 50) rheons, so long as it moves. All the small daughter ganglia generated through fission get stranded eventually, unless the capillary number exceeds a critical value (say, $N_{Ca} > 10^{-2}$) for which even 1-CEVS ganglia keep moving. However, such capillary values cannot be easily achieved, let alone be maintained, in practice. The same conclusions can be drawn for a large population of noninteracting ganglia. Oil ganglia may fail to interact either because their number concentration is too small and so collisions between pairs of ganglia are rare, or because collisions are not fruitful due to failure of the colliding ganglia to coalesce, or because of a combination of these factors. Coalescence may become difficult due to stabilizing films of dissolved or suspended substances at oil-water interfaces. Under such conditions negligible microdisplacement efficiency is expected (Payatakes et al. 1980). Substantial oil microdisplacement can be achieved only if oil ganglia collide and coalesce frequently. Fission tends to create many small and stranded ganglia, whereas collision-coalescence tends to create fewer and larger ganglia. These processes compete against each other, and the outcome of this competition determines whether an oil-bank will begin to form or not. In order to quantify these observations we must first put the Monte Carlo simulation results in numerical form. This is done as follows.

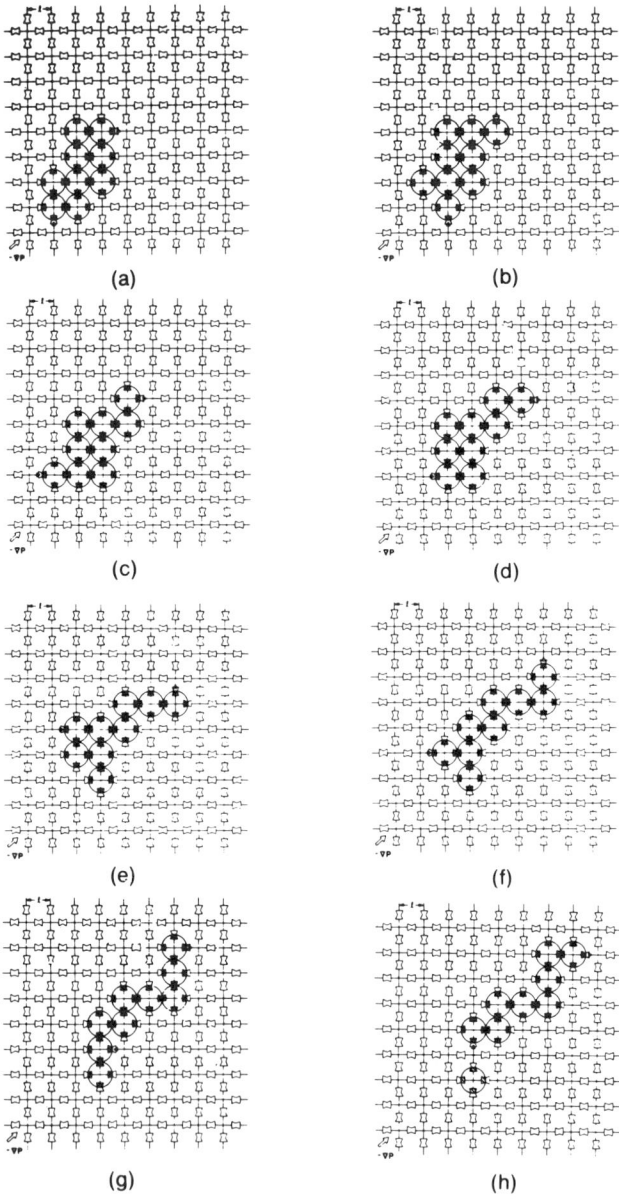

Figure 7 Typical realization of the fate of a nine CEVS ganglion introduced randomly into a 100 × 200 sandpack for capillary number $N_{Ca} = 1.07 \times 10^{-3}$. (*a*) Initial ganglion; (*b*) through (*g*) show the positions and shapes of the ganglion resulting from six successive rheons. (*h*) With the seventh rheon the ganglion fissions in two daughter ganglia. It so happens here that the small daughter ganglion remains stranded whereas the large one takes at least one more step. From Ng & Payatakes (1980).

PROBABILITIES OF MOBILIZATION, FISSION, AND STRANDING Suppose that we introduce into the porous medium a single ganglion of given volume but random shape, while a flood of capillary number N_{Ca} is in progress. There are three possible events that may follow: (a) the ganglion remains trapped at the position in which it was placed (stranding), (b) the ganglion takes at least one step without fissioning (mobilization), (c) the ganglion takes one step but fissions in the process (breakup). The probabilities of these events are denoted by S, M, and B respectively. These per-rheon probabilities can be obtained by repeating the one-step realization several hundreds of times and determining the frequency of occurrence of each event. Plots of S, M, and B versus N_{Ca} for a 100 × 200 sandpack and ganglion sizes up to 51-CEVS's, assuming $\theta_r^0 = \theta_e = \theta_a^0 = 0$, were given in Ng & Payatakes (1980). The case for 15-CEVS ganglia is shown in Figure 8. Observe that for $N_{Ca} < 10^{-4}$, the per-rheon probability of stranding S is unity. In the range $10^{-4} < N_{Ca} < 10^{-3}$, S diminishes sharply, and for $N_{Ca} > 3 \times 10^{-3}$ it becomes nil. As soon as S begins to decrease, the per-rheon probability of mobilization M begins to increase and so does the per-rheon probability of fission B. For $N_{Ca} > 3 \times 10^{-3}$, all three probabilities reach plateau values, $S = 0$, $M \simeq 0.85$, $B \simeq 0.15$. This is due to the fact that, once stranding becomes impossible due to a high N_{Ca} value, the probability of breaking of a randomly shaped ganglion depends on the local topology of the porous medium. This behavior is typical of all ganglia studied, with the only difference that for smaller ganglia the S,

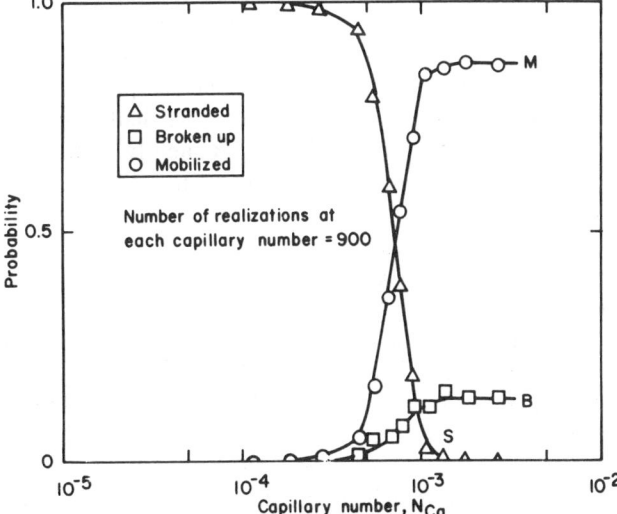

Figure 8 Plot of the probabilities of mobilization M, breakup per rheon B, and stranding per rheon S for fifteen-CEVS ganglia in a 100 × 200 sandpack versus capillary number. From Ng & Payatakes (1980).

M, B vs N_{Ca} curves are translated to the right, towards larger N_{Ca} values, whereas for larger ganglia the curves are translated to the left, towards smaller N_{Ca} values.

The effects of stranding and fissioning on oil ganglion dynamics can be conveniently expressed in terms of the stranding and breakup coefficients, which are defined by

$$\left.\frac{\partial n}{\partial z}\right)_{\text{due to stranding}} = -\lambda n,$$

$$\left.\frac{\partial n}{\partial z}\right)_{\text{due to fissioning}} = -\phi n, \tag{8}$$

where $n\Delta v$ is the number of moving v-ganglia per unit porous medium volume. Note that a v-ganglion is one having volume in the range v to $(v+\Delta v)$. Ng & Payatakes showed that λ and ϕ are given by

$$\lambda = -\frac{S}{(1-M)}\frac{\ln M}{s_z}, \qquad \phi = -\frac{B}{(1-M)}\frac{\ln M}{s_z}, \tag{9}$$

where s_z is the average length of migration of the ganglion per rheon and it is easily determined from the Monte Carlo simulation. Both λ and ϕ have dimensions of inverse length, so that $\lambda\ell$ and $\phi\ell$, where ℓ is the length of periodicity of the porous medium, are dimensionless coefficients. Plots of $\lambda\ell$ and $\phi\ell$ versus normalized ganglion size v/V_{CEVS} for several N_{Ca} values were given in Ng & Payatakes; here V_{CEVS} is the average volume of a CEVS. In general, $\lambda\ell$ decreases sharply with increasing v/V_{CEVS} and N_{Ca}; $\phi\ell$ increases with increasing v/V_{CEVS} but shows no systematic dependence on N_{Ca}. Another important observation arising from the Monte Carlo simulation is that fission is usually very asymmetric, producing one small daughter ganglion and one large one. This is of particular importance for large ganglia, which suffer fission more frequently than small ones. A large ganglion undergoing quasistatic motion usually loses through fissioning only small parts of its body.

We have already seen that as a ganglion moves, it tends to get elongated and to become aligned with the direction of the pressure gradient. Motivated by the fact that the mobilization-fission criterion depends on, among other factors, the shape and orientation of the ganglion [cf. Equations (5) and (6)], Payatakes, Ng & Woodham (1981) examined the effects of elongation on $\lambda\ell$ and $\phi\ell$. They found that $\lambda\ell$ decreases from a high original value, which corresponds to the ensemble of random shapes, to a lower, stable value (the cruising value) after just a few rheons. Ganglia with volume less than 20 V_{CEVS} reach their cruising $\lambda\ell$ value in $\sim$ ten rheons. Much larger ganglia may require more rheons. Payatakes et al. also found that $\phi\ell$ is not affected substantially or systematically by elongation. They concluded that the random-shape value of $\lambda\ell$ must be used in the boundary condition at the flood front to account for the initially random shapes, whereas the cruising value must be used in the integrodifferential ganglion population balance equations (see below) to account for the effects of elongation. Rapin (1980) made systematic calculations of random-shape and cruising values of $\lambda \ell$ and found that in general $\ln(\lambda \ell)$ is a linear function of v/V_{CEVS}. Some typical results are shown in Figure 9.

Dynamic Motion of a Solitary Oil Ganglion

DYNAMIC MOTION AND FISSION Experimental observations by Rapin (1980) and Hinkley (1981) show that, when a ganglion moves driven by a flood having capillary number that substantially exceeds the critical value for mobilization, the behavior of the ganglion is somewhat different than the one predicted under quasistatic conditions. The main difference is that the ganglion can advance at two or more interfaces simultaneously, rather

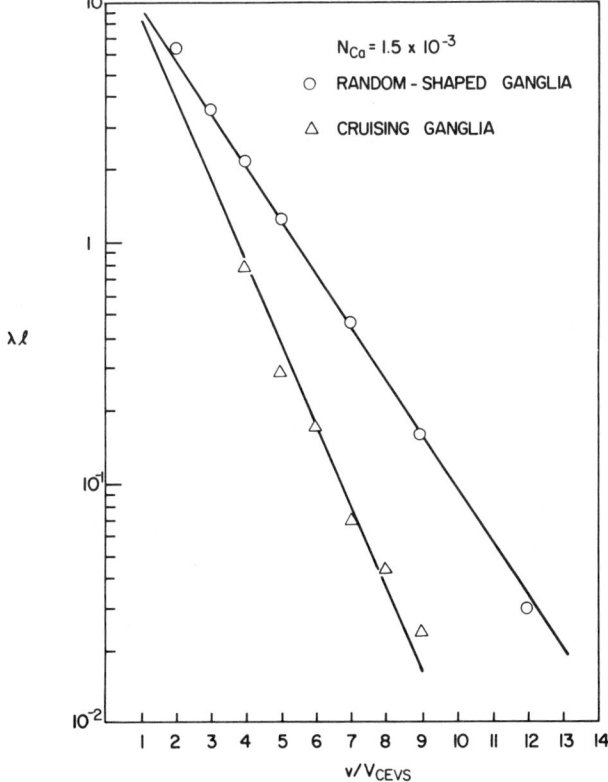

Figure 9 Typical plot of the dimensionless stranding coefficient $\lambda \ell$ versus reduced ganglion volume v/V_{CEVS} for random-shaped ganglia and for elongated (cruising) ganglia.

than at just one. This difference has some significant consequences. First, the tendencies for elongation and alignment are reduced, even though they are still present. Second, when the downstream part of a ganglion encounters a set of narrow throats through which it cannot proceed rapidly, the ganglion may grow a new branch at some appropriate site along its body (Figure 10a), thus bypassing the obstruction. This maneuver, however, often results in folding the ganglion into a U-shape (Figure 10b), which in turn leads to fissioning at the base of the U (Figure 10c and d). Such a fission creates daughter ganglia of comparable size, as opposed to the asymmetric fission usually associated with quasistatic deformation. A similar type of fission, made possible by the simultaneous advancement of the ganglion at two fronts, is shown in Figure 11. Here, the porous medium is composed of a monolayer of identical beads arranged in parallel rectilinear rows and held between two flat plates. Initially, the ganglion occupies symmetric positions in two adjacent tracks. There is no obstruction to the motion of the ganglion on either side; both frontal interfaces advance in tandem, and the ganglion divides into two prongs. The two prongs fail to meet and coalesce again (due to lack of a lateral driving force) and the ganglion eventually fissions into similar daughters.

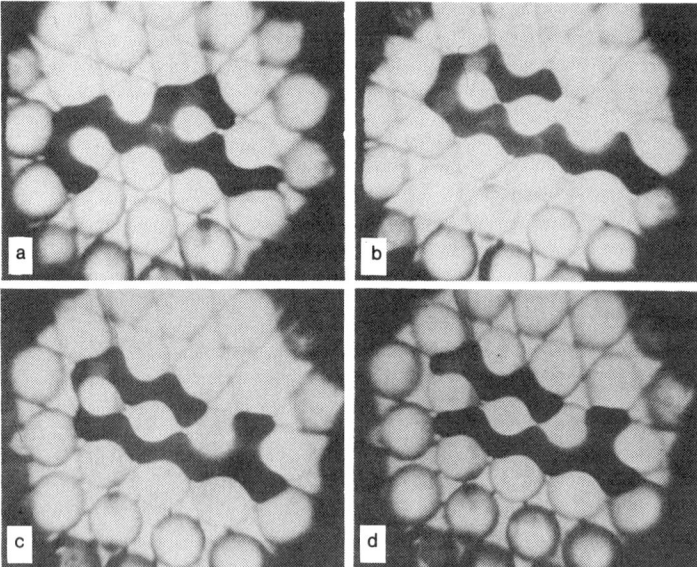

Figure 10 Fission of an oil ganglion in dynamic motion. The porous medium consists of a monolayer of randomly arrayed glass beads, with diameters in the range 1.0 ± 0.2 mm, held between two glass plates. (*a*) The downstream end of the ganglion has encountered a set of narrow throats and slows down; the ganglion grows a new branch. (*b*) The ganglion takes a U shape. (*c*) The base of the U collapses. (*d*) The ganglion has fissioned into two daughters. From Rapin's (1980) movie films.

In addition to these aspects of fissioning, another important feature of dynamic ganglion motion is the explicit appearance of time as an independent variable. The average velocity with which the centroid of a ganglion migrates downstream, $\bar{u}_z$, is a variable of fundamental importance on the relative permeability to oil and on the rate of collisions among ganglia.

AVERAGE GANGLION VELOCITY A dimensional analysis of the problem shows that under creeping-flow conditions the reduced velocity $\bar{u}_z/V_f$ depends on N_{Ca}, v/V_{CEVS}, μ_o/μ_w, θ_r, θ_a, and on the dimensionless geometric parameters defining the porous-medium geometry. The Bond number must be added to this list, if gravitational effects are included in the analysis. Clearly, the problem is quite complex. At present, there is no satisfactory theoretical analysis capable of predicting the velocity of a ganglion. Slattery (1974) developed a linear momentum balance for a non-wetting ganglion in a sinusoidal tube. Oh & Slattery (1976) extended that work, concentrating on the possible effects of interfacial viscosities on the motion of oil ganglia in straight and sinusoidal tubes. However, this work needs to be extended to oil ganglia moving in networks of constricted-tube unit cells, if it is to be applied to ganglion dynamics studies. Such an effort is still missing.

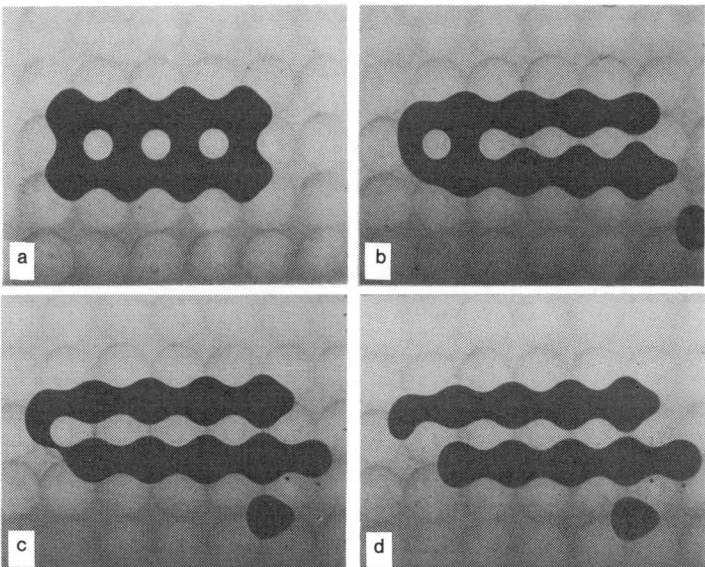

Figure 11 Fission of an oil ganglion in dynamic motion. The porous medium consists of a monolayer of regularly arrayed glass beads with diameter 3 mm, held between two glass plates. (*a*) Initial position. (*b*) After three rheons by each prong; the small ganglion at the lower right corner is stranded and can be used as a marker. (*c*) The ganglion is fissioning. (*d*) The two daughter ganglia move on. From Hinkley's (1981) movie films.

A measure of understanding of the dependence of $\bar{u}_z/V_f$ on N_{Ca} and v/V_{CEVS} with all other physical and geometrical parameters fixed can be obtained from the experimental results of Rapin (1980) and Hinkley (1981). Both researchers used a monolayer of uniform glass beads held between two glass plates as a porous-medium model. Rapin's packings were random, whereas Hinkley's were regular. Both researchers used a system of dead crude oil equilibrated with various mixtures of water/surfactant (petroleum sulfonate; WITCO TRS 10-80)/cosurfactant (tetrabutyl alcohol)/sodium chloride. Their results can be summarized as follows:

1. For $\mu_o/\mu_w = 7$, the reduced velocity $\bar{u}_z/V_f$ varies in the range from 0.0 to 0.25, depending on the reduced volume v/V_{CEVS} and the capillary number N_{Ca} (Figure 12.)
2. For fixed volume v/V_{CEVS}, the reduced velocity $\bar{u}_z/V_f$ is a monotonically increasing function of N_{Ca}.
3. For values of the capillary number near the critical mobilization value, the reduced velocity $\bar{u}_z/V_f$ increases monotonically with v/V_{CEVS}, reaching an asymptotic value for v/V_{CEVS} larger than $\sim$ 15 to 20. This value depends on N_{Ca} (among other factors).

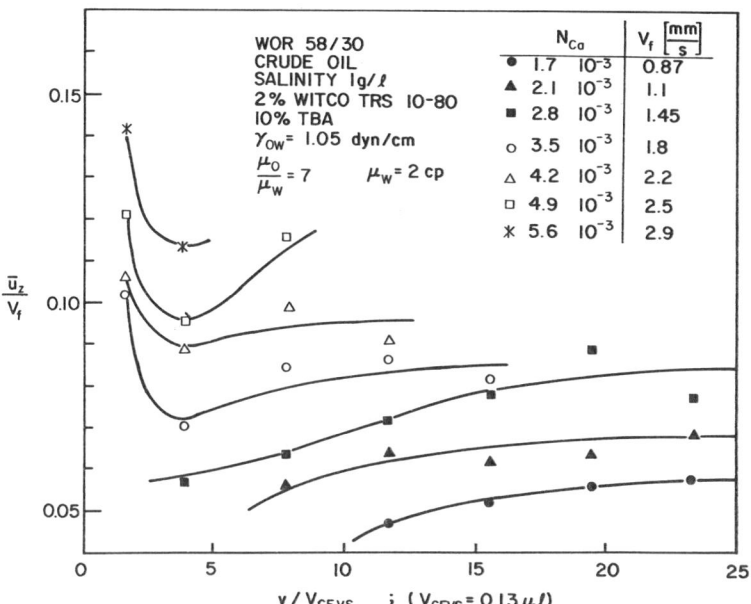

Figure 12 Plot of reduced average ganglion velocity versus reduced ganglion volume for several typical capillary number values. Here, the viscosity ratio is $\mu_o/\mu_w = 7$. Data from Rapin (1980).

4. For values of the capillary number that substantially exceed the critical mobilization value, the reduced velocity $\bar{u}_z/V_f$ often has a weak or moderate minimum in the range $v/V_{CEVS} = 3$ to 5; 2-CEVS ganglia move substantially faster than larger ones, whereas the velocity of ganglia larger than 5-CEVS increases slowly with size, tending to an asymptotic value that depends on N_{Ca} (among other factors).
5. For $\mu_o/\mu_w = 0.6$, the reduced velocity $\bar{u}_z/V_f$ varies in the range from 0.0 to 2.5, depending on v/V_{CEVS} and N_{Ca}. This means that ganglia having viscosity smaller than that of the aqueous phase can move much faster than the aqueous phase itself. This is due to the fact that the pressure difference between the ends of a ganglion depends strongly on the porous structure and the viscosity of the connected (here the aqueous) phase. Another important factor seems to be that a fast-moving ganglion does not fill the channels that it occupies completely but is surrounded by an aqueous layer of substantial thickness. This, in turn, gives the ganglion a less sinusoidal form and makes its motion easier.
6. In the case of very fast ganglion motion induced by a high capillary number, the upstream interface deforms drastically due possibly to the onset of inertial effects (Figure 13).

An analysis of the dynamic motion of a ganglion in a porous medium requires the simultaneous solution of the flow problem both inside and outside the ganglion, also taking into account the capillary effects. Such an analysis must be based on a network approach that accounts for the cooperative behavior of all pertinent physical and geometrical factors.

DYNAMICS OF OIL-GANGLION POPULATIONS

Consider an immiscible flood of given capillary number advancing in the z direction in a given porous medium. Consider now the subpopulation of v-ganglia, namely ganglia with volume between v and $(v+\Delta v)$. Only a fraction of this subpopulation gets mobilized by the flood. Mobilized v-ganglia may become restranded, they may fission, or they may collide and coalesce with other stranded or moving ganglia. Payatakes et al. (1980) developed a balance equation of the birth-death type for the number concentration of moving v-ganglia, $n(z,t;v)$, and one for the number concentration of stranded v-ganglia, $\sigma(z,t;v)$. Neglecting axial dispersion, these equations become

$$\frac{\partial n(z,t;v)}{\partial t} + \bar{u}_z(v;\mathbf{a}_1)\frac{\partial n(z,t;v)}{\partial z}$$
$$= -n(z,t;v)\{\bar{u}_z(v;\mathbf{a}_1)[\lambda(v;\mathbf{a}_2) + \phi(v;\mathbf{a}_2)]$$

$$+ \int_0^\infty K_{11}(u, v; \mathbf{a}_4) n(z, t; u) du$$

$$+ \int_0^\infty K_{01}(u, v; \mathbf{a}_4) \sigma(z, t; u) du \Big\}$$

$$+ 1/2 \int_0^v K_{11}(u, v - u; \mathbf{a}_4) n(z, t; u) n(z, t; v - u) du$$

$$+ [1 - S(v; \mathbf{a}_2)] [\int_0^v K_{10}(u, v - u; \mathbf{a}_4)$$

$$n(z, t; u) \sigma(z, t; v - u) du$$

$$+ \int_v^\infty \bar{u}_z(u; \mathbf{a}_1) \phi(u; \mathbf{a}_2) W(u, v) n(z, t; u) du] , \quad (10)$$

$$\frac{\partial \sigma(z, t; v)}{\partial t} = [\bar{u}_z(v; \mathbf{a}_1) \lambda(v; \mathbf{a}_2)] n(z, t; v)$$

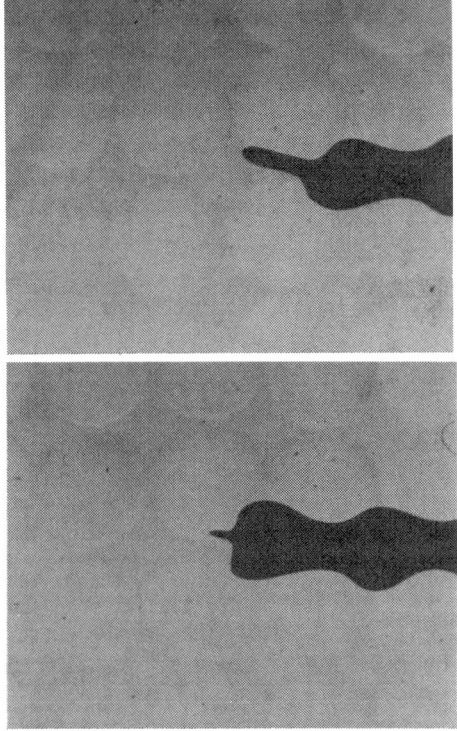

Figure 13 Upstream ends of oil ganglia showing large deformations at very high capillary number ($N_{Ca} > 1$). From Hinkley's (1981) movie films.

$$+ S(v \, ; \, \mathbf{a}_2)|[\int_0^v K_{10}(u \, , \, v-u \, ; \, \mathbf{a}_4)$$

$$n(z \, , \, t \, ; \, u)\sigma(z \, , \, t \, ; \, v-u)du$$

$$+ \int_v^\infty \bar{u}_z(u \, ; \, \mathbf{a}_1)\phi(u \, ; \, \mathbf{a}_2)W(u \, , \, v)n(z \, , \, t \, ; \, u)du] \quad (11)$$

$$- \sigma(z \, , \, t \, ; \, v)\int_0^\infty K_{10}(u \, , \, v \, ; \, \mathbf{a}_4)n(z \, , \, t \, ; \, u)du \, ,$$

where

$$K_{ij}(u \, , \, v \, ; \, \mathbf{a}_4) = R_{ij}(u \, , \, v \, ; \, \mathbf{a}_1)C_{ij}(u \, , \, v \, ; \, \mathbf{a}_3) \quad (12)$$

are the collision-coalescence kernels. $R_{ij}(u,v;\mathbf{a}_1)$ is the rate of collisions between u-ganglia and v-ganglia and $C_{ij}(u,v;\mathbf{a}_3)$ is the probability of coalescence given a collision. The index i is set equal to 1 or 0 depending on whether the u-ganglion is moving or not (the index j pertains to the v-ganglion). The variables $\mathbf{a}_1$, $\mathbf{a}_2$, $\mathbf{a}_3$, and $\mathbf{a}_4$ are parameter vectors.

The physical meanings of the terms appearing in Equation (10) are as follows (from left to right): (a) rate of accumulation of moving v-ganglia, (b) net efflux of v-ganglia due to convection, (c) rate of loss of moving v-ganglia due to stranding, fission, collision-coalescence with other moving ganglia, and collision-coalescence with stranded ganglia, (d) rate of generation of moving v-ganglia through collision-coalescence of appropriately sized smaller moving ganglia, (e) rate of generation of moving v-ganglia through collision-coalescence of smaller moving ganglia with stranded ones, and through fission of larger ganglia.

The physical meanings of the terms of Equation (11) are (from left to right): (a) rate of accumulation of stranded v-ganglia, (b) rate of stranding of mobilized v-ganglia, (c) rate of generation of stranded v-ganglia through collision-coalescence of smaller ganglia, one of which is stranded, and through stranding of v-ganglia formed by fission of larger ganglia, (d) rate of loss of stranded v-ganglia by collision-coalescence with oncoming ganglia.

The stranding coefficient λ, the fission coefficient ϕ, the per-rheon probability of stranding S, and the mode-of-fission probability W are obtained from Monte Carlo simulation of the fate of solitary oil ganglia. Note that in order to account for the local oil saturation on the values of these parameters, the value of N_{Ca}/k_{rw} should be read on the abscissa of plots such as Figure 8, rather than the value of N_{Ca} (Ng & Payatakes 1980). Theoretical expressions for K_{ij} were given in Payatakes et al. (1980). The probability of coalescence given a collision, C_{ij}, is a very important parameter, whose value ranges, in principle, from 0 to 1. At present, there are neither experimental nor theoretical methods for determining C_{ij} directly. In view of the importance of this parameter in reconnecting oil ganglia to form an oil-bank, the study of C_{ij} is a task of high priority.

Equations (10) and (11) are to be integrated with the appropriate boundary conditions at the flood front. In the simple case when $V_i > \bar{u}_z$ for all ganglion sizes we have:

For $z = V_i t$, $t > 0$:

$$n = F(z, v)[1 - S(v; \mathbf{a}_2)] / \left[1 - \frac{\bar{u}_z(v; \mathbf{a}_1)}{V_i}\right], \tag{13}$$

$$\sigma = F(z, v)S(v; \mathbf{a}_2).$$

Here $V_i = V_f/\epsilon S_w$ is the interstitial velocity of the flood. Note that in Payatakes & Ng (1981) V_f appeared mistakenly, instead of V_i, in the counterpart of Equation (13).

A collocation method for this integration was developed by Ng (1980). The utility of the method can be illustrated by sample calculations from Ng's work for the following conditions: porous-medium model corresponding to a 100 × 200 sandpack; $N_{Ca} = 3 \times 10^{-3}$; $\mu_o/\mu_w = 7$; initial size distribution given by $F(z,v) = 0.25 [1 - (v/11V_{CEVS})]$ for $v \leq 11\ V_{CEVS}$ and $z > 0$ and by $F(z,v) = 0$ for $v > 11\ V_{CEVS}$ and $z > 0$. The average ganglion velocity for these conditions was estimated from Rapin's (1980) data. Assuming an initial residual oil saturation value of $S_{or} = 0.504$, and the values $C_{10} = C_{01} = 0.50$, $C_{11} = 0.10$ for the probabilities of coalescence given a collision, Ng obtained the results plotted in Figure 14. These plots give the spatial distribution of $n\Delta v$ and $\sigma \Delta v$ for various ganglion sizes at the instant when the flood has reached the position $z/\ell = 50$. The concentrations of mobilized small ganglia ($v < 8\ V_{CEVS}$) are maximum at the flood front and drop rapidly with distance away from it, due to stranding, coalescence, and fission (Figure 14). On the other hand, the concentrations of larger ganglia actually increase away from the flood front. Observe that some large ganglia which did not exist in the original population ($v = 13$ to $15\ V_{CEVS}$) also begin to form. These observations indicate that here the process of collision-coalescence is vigorous enough to overcome fission and stranding and that favorable conditions for oil-bank formation exist. The $\sigma \Delta v$ vs z/ℓ curves for $v = 3$ to $5\ V_{CEVS}$ (Figure 14) have valleys. This is due to the fact that mobilized ganglia following the flood front at some distance collide and coalesce with stranded ganglia they encounter, thus reducing the amount of trapped oil.

Ng (1980) repeated the above calculations setting $C_{10} = C_{01} = C_{11} = 0$ (no coalescence) and found, as expected, that only a negligible amount of oil was produced. Oil ganglia that got mobilized fissioned quickly into daughter ganglia that got stranded anew. This helps to underline the crucial importance of C_{ij}. Ng also repeated the above calculations with $C_{10} = C_{01} = 0.50$ and $C_{11} = 0.10$, but with residual oil saturation S_{or}

= 0.252, namely only one half that of the first example. He found that the valleys in the $\sigma\Delta v$ vs z/ℓ curves disappeared, and that a larger fraction of the oil was left behind. This inferior performance was due to the reduced rate of collisions among ganglia, K_{ij}. This indicates that for given porous medium, oil, and flood conditions, the efficiency of the flood is higher, if the value of S_{or} is higher. Note, however, that in practice a larger S_{or} value usually indicates a reservoir with very unfavorable pore structure, which in turn means that flooding may not be very effective. Larger S_{or} leads to better microdisplacement efficiency provided that all other factors are equal.

PROBLEMS FOR FUTURE WORK

The major unsolved problems emerging from this review are: (*a*) theoretical analysis of the dynamic motion of ganglia with emphasis on the fission coefficient ϕ and the velocity $\bar{u}_z/V_f$, (*b*) theoretical analysis of the probabilities of coalescence given a collision, C_{ij}, (*c*) development of more efficient algorithms for the integration of the population balance equations, (*d*) application of the ganglion-dynamics method to the study of relative

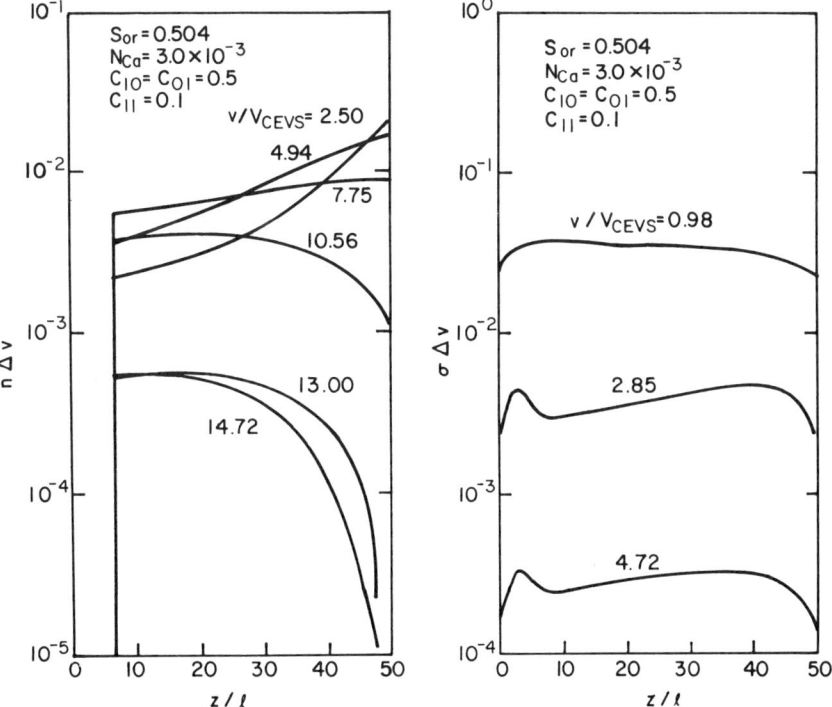

Figure 14 Spatial profiles of the number concentrations of moving and stranded *v*-ganglia for various values of the reduced ganglion volume v/V_{CEVS}. From Ng (1980).

permeabilities, and (*e*) generalization of the porous-medium model so that it applies not only to sandpacks but also to reservoir rocks.

ACKNOWLEDGMENTS

Work performed at the University of Houston was supported by U.S. Department of Energy, Grant No. E(40-1)-5075, and with support from the Shell Oil Company and Marathon Oil Company. Thanks are due to Mr. Richard Hinkley for his help in producing the photographs.

Literature Cited

Batycky, J. 1979. Dependence of residual oil mobilization on wetting and roughness. *Res. Rep. RR-41, Petrol. Recov. Inst.* Calgary, Canada. 58 pp.

Dussan V., E. B. 1979. On the spreading of liquids on solid surfaces: Static and dynamic contact lines. *Ann. Rev. Fluid Mech.* 11:371–400

Fatt, I. 1956. The network model of porous media. *Petrol. Trans. Am. Inst. Mining Engrs.* 207:144–81

Gillette, R. D., Dyson, D. C. 1974. Stability of static configurations with applications to the theory of capillarity. *Arch. Rational Mech. Analysis* 53:150–77

Haines, W. B. 1930. Studies in the physical properties of soil. Part V. *J. Agricult. Sci.* 20:97–116

Healy, R. N., Reed, R. L. 1974. Physicochemical aspects of microemulsion flooding. *Soc. Petrol. Eng. J.* 14:491–501

Healy, R. N., Reed, R. L., Stenmark, D. G. 1976. Multiphase microemulsion systems. *Soc. Petrol. Eng. J.* 16:147–60

Heller, J. P. 1968. The drying through the top surface of a vertical porous column. *Soil Sci. Soc. Am. Proc.* 32:778–86

Hinkley, R. 1981. *Oil Ganglion Motion.* MS thesis. Univ. Houston, Houston, Tex.

Josselin de Jong, G. 1958. Longitudinal and transverse diffusion in granular deposits. *Trans. Am. Geophys. Union* 39:67–74

Larson, R. G., Scriven, L. E., Davis, H. T. 1981. Percolation theory of two phase flow in porous media. *Chem. Eng. Sci.* 36:57–73

Larson, R. G., Davis, H. T., Scriven, L. E. 1981. Displacement of residual nonwetting fluid from porous media. *Chem. Eng. Sci.* 36:75–85

Melrose, J. C. 1965. Wettability as related to capillary action in porous media. *Soc. Petrol. Eng. J.* 5:259–71

Melrose, J. C. 1968. Thermodynamic aspects of capillarity. *Ind. Eng. Chem.* 60(3):53–70

Melrose, J. C. 1970. Interfacial phenomena as related to oil recovery mechanisms. *Can. J. Chem. Engrg.* 48:638–44

Melrose, J. C., Brandner, C. F. 1974. Role of capillary forces in determining microscopic displacement efficiency for oil recovery by water flooding. *Can. J. Petrol. Technol.* 13(4):54–62

Michael, D. H. 1981. Meniscus stability. *Ann. Rev. Fluid Mech.* 13:189–215

Miller, E. E., Miller, R. D. 1956. Physical theory for capillary flow phenomena. *J. Appl. Phys.* 27:324–32

Mohanty, K., Davis, H. T., Scriven, L. E. 1980. Physics of oil entrapment in water-wet rock. *SPE 9406, Soc. Petrol. Engrg.,* Dallas, Tex. 16 pp.

Morrow, N. 1975. The effects of surface roughness on contact angle with special reference to petroleum recovery. *J. Can. Petrol. Technol.* 14(4):42–53

Morrow, N. 1979. Interplay of capillary, viscous and buoyancy forces in the mobilization of residual oil. *J. Can. Petrol. Technol.* 18(3):35–46

Neira, M. A., Payatakes, A. C. 1978. Collocation solution of creeping Newtonian flow through periodically constricted tubes with piecewise continuous wall profile. *AIChE J.* 24:43–54

Neira, M. A., Payatakes, A. C. 1979. Collocation solution of creeping Newtonian flow through sinusoidal tubes. *AIChE J.* 25:725–30

Nelson, R. C., Pope, G. A. 1977. Phase relationships in chemical flooding. *SPE 6773, Soc. Petrol. Engrg.,* Dallas, Tex. 12 pp

Ng, K. M. 1980. *Oil ganglion dynamics in flow through porous media.* PhD thesis. Univ. Houston, Houston, Tex. 180 pp

Ng, K. M., Davis, H. T., Scriven, L. E. 1978. Visualization of blob mechanics in flow through porous media. *Chem. Eng. Sci.* 33:1009–17

Ng, K. M., Payatakes, A. C. 1980. Stochastic simulation of the motion, breakup and stranding of oil ganglia in water-wet granular porous media during immiscible displacement. *AIChE J.* 26:419–29

Oh, S. G., Slattery, J. C. 1976. Interfacial tension required for significant displacement of oil. *Proc. 2nd ERDA Symp. on Enhanced Oil and Gas Recovery, Tulsa Oklahoma*, pp. D-2/1–D-2/19

Payatakes, A. C., Ng, K. M., Flumerfelt, R. W. 1980. Oil ganglion dynamics during immiscible displacement: Model formulation. *AIChE J.* 26:430–43

Payatakes, A. C., Ng, K. M., Woodham, G. 1981. On the fate of oil ganglia during immiscible displacement in water-wet granular porous media. *Proc. 3rd Intern. Conf. Surface and Colloid Sci., Stockholm, Sweden*. In press

Payatakes, A. C., Tien, C., Turian, R. M. 1973. A new model for granular porous media. Part II. numerical solution of steady state incompressible Newtonian flow through periodically constricted tubes. *AIChE J.* 19:67–76

Rapin, S. 1980. *Behavior of non-wetting oil ganglia displaced by an aqueous phase*. MS thesis. Univ Houston, Houston, Tex. 160 pp

Rillaerts, E., Joos, P. 1980. The dynamic contact angle. *Chem. Eng. Sci.* 35:883–87

Roof, J. G. 1970. Snap-off of droplets in water-wet cores. *Soc. Petrol. Engrg. J.* 10:85–90

Slattery, J. C. 1974. Interfacial effects in the entrapment and displacement of residual oil. *AIChE J.* 20:1145–54

Snyder, L. J., Stewart, W. E. 1966. Velocity and pressure profiles for Newtonian creeping flow in regular packed beds of spheres. *AIChE J.* 12:167–73

van Brakel, J. 1975. Pore space models for transport phenomena in porous media. Review and evaluation with special emphasis on capillary liquid transport. *Powder Technol.* 11:205–36

Wardlaw, N. C., Cassan, J. P. 1978. Estimation of recovery efficiency by visual observation of pore systems in reservoir rocks. *Bull. Can. Petrol. Geol.* 26:572–85

Wardlaw, N. C., Taylor, R. P. 1976. Mercury capillary pressure curves and the interpretation of pore structure and capillary behavior in reservoir rocks. *Bull. Can. Petrol. Geol.* 24:225–62

NUMERICAL METHODS IN FREE-SURFACE FLOWS

Ronald W. Yeung

Department of Applied Mathematics[1], University of Adelaide, Adelaide, South Australia 5001 and Department of Ocean Engineering, Massachusetts Institute of Technology, Cambridge, Massachusetts 02139

1. INTRODUCTION

Free boundaries occur in a wide variety of physical phenomena—jets, cavities, seepage of groundwater, ice melting in water, gravity waves, just to name a few. All share the common feature that the domain of interest has an unknown boundary, on which a double condition is to be imposed. In this category of problems, gravity waves stand out with the characteristic that both restoring and inertial forces are important. This results in a complex, yet often fascinating, interplay between potential and kinetic energies. It is in the context of surface gravity waves, or water waves, that we use the term free surface in this article. A more general title is chosen in hope of stimulating some mutually beneficial exchange with other free-surface workers.

Even with the neglect of surface tension and viscosity, the theory of water waves is not a simple subject. Traditionally, the theory is classified into two areas: one based on systematic expansions, with the amplitude-to-wave-length ratio taken as a small parameter (Stokes 1847), the other based on the assumption of small depth-to-wavelength ratio (Boussinesq 1871, Rayleigh 1876). A comprehensive account of the more recent developments in each may be found in the reviews of Yuen & Lake (1980) and Miles (1980). The complexity of the subject of water waves compounds with the introduction of a body into the wave field, and direct numerical solution is often necessary.

We focus our attention on both linear and nonlinear problems, the former in the context of infinitesimal-wave theory. Existence and uniqueness theorems do not generally exist for these problems, particularly the

---

[1] Australian-American Educational Foundation exchange visitor, Spring 1981.

nonlinear ones. Unfortunately, these problems also define an area of a great deal of engineering interest. Nevertheless, driven by an increasing demand for the utilization and the resources of the seas, hydrodynamicists have moved ahead to develop the necessary computational tools and have provided some useful engineering solutions. Many problems, physical and numerical, remain unsolved. The days for full-scale numerical simulation are still ahead.

The nontrivial nature of some of the free-surface flow problems to be discussed is perhaps well illustrated by a concluding remark of Southwell & Vaisey (1946). Noted for his pioneering works in the development and application of relaxation methods to a large variety of physical problems, Southwell remarked after attempting several "free-streamline" problems in the cited paper: "Problems concerned with 'free' stream-lines (with a double condition) are among the hardest yet attempted in this series; an essential instability (on this boundary) makes any tentative solution liable to diverge; 'graded nets' have proved indispensable." Southwell & Vaisey were considering the nonlinear steady flow about a gliding plate in an era of "human computers." They were perplexed by the absence of waves downstream. Southwell's views must still be shared in part by many modern workers. Even today, the precise role of a numerical radiation condition in nonlinear flow is not clearly understood.

This survey covers and contrasts the numerical methods that are actively being used to solve free-surface flow problems. The literature cited is slanted towards body-wave problems because of the author's own interests. However, relevant techniques related to "free-wave" calculations are also mentioned to illustrate different facets of a methodology. The companion article of Schwartz & Fenton (1982) in this volume, addressing various aspects of highly nonlinear waves, is also relevant to the general subject area of free-surface flow. In order to make the scope of the work more manageable, and to provide a more distinct contrast of the methodologies, we restrict our attention to problems based on a potential or stream-function formulation. The use of the velocity potential is justifiable only when the inertial forces dominate the viscous forces. Cases of bluff bodies in steady motion and small bodies in a wave field of large amplitude are well-known exceptions. Methods based on the use of the primitive-variables equation (velocity-pressure formulations) are not included here, but merit a separate review.

Three major categories of methods are reviewed here: finite differences, finite elements, and boundary-integral equations. In an era of continuous evolution of numerical methods, it is sometimes difficult (and risky) to classify them. The more outstanding methods, in fact, have a hybrid character that incorporates better features of the others. Besides, finite difference with respect to time is used inevitably in most methods for unsteady problems; discretization techniques in integral-equation methods could be regarded as a finite-element treatment of the boundary. But the confusion appears to disappear when the categorization is taken to mean the manner in which the field equation is tackled in the computational (interior) domain. Spectral methods, because of their limitations in dealing with arbitrary body geometries, are not extensively covered. For completeness, the relevant equations are first recalled in Section 2, followed by sections that describe the individual features of each type of method. The emphasis is on formulation and techniques. Details regarding implementation are generally available from the cited references or text materials.

2. GOVERNING EQUATIONS

We assume at the outset that the fluid is incompressible, inviscid, and lacking surface tension. The flow is assumed to be irrotational except in regions behind the body where a thin vortex sheet may exist because of the generation of lift. In this latter case, an appropriate "cut" in the fluid domain Ω must be chosen to render the velocity potential single-valued. Let $Oxyz$ be a moving coordinate system with absolute velocity $\mathbf{U} = (U,0,0)$. The Oxz plane is taken to coincide with the "still" water surface, $\mathcal{F}_0$, and the y axis points upwards. The fluid disturbance at time t is thus described by a velocity potential $\phi(x,y,z,t)$ with the fluid velocity $\mathbf{u}$ given by $\nabla\phi$. It is well known that ϕ is a solution of Laplace's equation and that the momentum equations yield Bernoulli's integral. Hence

$$\nabla^2\phi(\mathbf{x}, t) = 0 \quad \text{for } \mathbf{x} = (x, y, z) \text{ in } \Omega, \tag{2.1}$$

$$\frac{p(\mathbf{x}, t)}{\rho} + \phi_t - U\phi_x + \tfrac{1}{2}|\nabla\phi|^2 + gy = 0, \tag{2.2}$$

where p is the fluid pressure, ρ the density, and g the acceleration of gravity. In (2.2) and henceforth, subscripts will be used to denote partial derivatives, and the arguments shown for the first field variable are understood to apply to all others.

The fluid is bounded on top by a free surface $\mathcal{F}$ described by, say, $y = \eta(x,z,t)$, where η is the free-surface elevation, internally by the surface of a rigid body $\mathcal{H}$, which may or may not intersect $\mathcal{F}$, and below by a bottom surface $\mathcal{B}$, defined by $y = -h(x,z,t)$, with the depth h normally taken as constant. The boundary contours that define Ω will be denoted by $\partial\Omega$. The kinematic boundary conditions on $\mathcal{H}$ or $\mathcal{B}$ are

$$\left.\frac{\partial\phi}{\partial n}\right|_{\mathcal{B},\mathcal{H}} = V_n, \tag{2.3}$$

where $\mathbf{n}$ is an exterior unit normal to Ω. V_n in (2.3) is the normal velocity of the boundary surface, which can be considered prescribed on $\mathcal{H}$, but vanishes on $\mathcal{B}$. We note in passing that it is convenient to take **U** as the translational velocity of the body, in which case $\mathcal{H}$ is time independent if the body motion is steady. On the free surface $\mathcal{F}$, the kinematic and dynamic boundary conditions can be obtained respectively by noting that $y = \eta$ is a material surface, and by setting $p(x,\eta,t)$ to the applied pressure (in excess of atmosphere) p_0. Hence,

$$\frac{D\eta}{Dt} = \phi_y \qquad \text{on } \mathcal{F}, \tag{2.4a}$$

$$\frac{D\phi}{Dt} = \tfrac{1}{2}|\nabla\phi|^2 - g\eta - p_0/\rho \qquad \text{on } \mathcal{F}, \tag{2.4b}$$

where $D/Dt = \partial/\partial t + (\nabla\phi - \mathbf{U}) \cdot \nabla$ is the material derivative in a moving frame. Note that (2.4a) is essentially a statement of (2.3) with V_n being the normal velocity of the free surface. A more traditional formulation in a stationary frame can be recovered simply by setting $\mathbf{U} = 0$ in the above. For the purpose of later usage, it is worthwhile to express (2.4) in their full form:

$$\eta_t(x, z, t) = \phi_y(x, \eta, z, t) + (U - \phi_x)\eta_x - \phi_z\,\eta_z, \tag{2.5a}$$

$$\phi_t(x, \eta, z, t) = -g\eta(x, z, t) + U\phi_x - \tfrac{1}{2}|\nabla\phi|^2, \tag{2.5b}$$

where we have taken $p_0 = 0$ for simplicity. Equation (2.1) with boundary conditions (2.3) and (2.4) defines the nonlinear initial-boundary-value problem of the motion of water waves in the presence of a rigid body. Existence proofs for the nonlinear problem are available only for a few special cases. The order of the differential equations, however, suggests that initial conditions are appropriate for the following: $\phi(\mathbf{x},0^-)$, $\eta(x,z,0)$ and η_t. Once ϕ is solved, (2.2) can be used to evaluate $p(\mathbf{x},t)$, in particular on the body surface, whereby physical quantities of interest such as force or moment can be calculated. Various specializations of the foregoing general formulation are next reviewed.

STEADY-MOTION PROBLEMS If the motion has achieved a steady state in the moving frame, the free-surface conditions are simply (2.5), with the left-hand side set equal to zero:

$$\phi_n(x, \eta, z) = -U\eta_x, \qquad U\phi_x = g\eta + \tfrac{1}{2}|\nabla\phi|^2. \tag{2.6a,b}$$

However, since the problem in this situation is kinematically equivalent to that due to a uniform flow of velocity $-U$ about a fixed $\mathcal{H}$, it is common to introduce a total potential Φ defined by

$$\Phi = -Ux + \phi.$$

The condition (2.3) and (2.6) are now replaced by

$$\frac{\partial \Phi}{\partial n}\bigg|_{\mathcal{B},\mathcal{H}} = 0, \qquad (2.7)$$

$$\frac{\partial \Phi}{\partial n}\bigg|_{\mathcal{F}} = 0, \qquad \tfrac{1}{2}|\nabla \Phi|^2 + g\eta = \tfrac{1}{2}U^2. \qquad (2.8\text{a,b})$$

In two dimensions, similar formulation can be written in terms of a stream function Ψ, where $\mathbf{u} = (\Psi_y - U, -\Psi_x, 0)$. The equivalent conditions are then:

$$\Psi|_{\mathcal{F}} = 0, \qquad \tfrac{1}{2}\left(\frac{\partial \Psi}{\partial n}\right)^2\bigg|_{\mathcal{F}} + g\eta = \tfrac{1}{2}U^2, \qquad (2.9\text{a,b})$$

$$\Psi|_{\mathcal{B}} = Uh(x = +\infty). \qquad (2.10)$$

However, the body $\mathcal{H}$ takes on a stream-function value that is unknown.

In steady-state problems where the flow is subcritical, physical observations indicate that waves occur only downstream. Thus "asymptotic conditions" of the following form may be stated:

$$\phi \to o(1) \qquad \text{as } x \to \infty \qquad (2.11)$$

and

$$\phi \to W(x, y, z) + C \qquad \text{as } x \to -\infty, \qquad (2.12)$$

where $W(x,y,z)$ is a wave-like solution, C a constant, both of which are not a priori known. Such conditions are not necessary in the initial-value formulation.

LINEARIZED STEADY-MOTION PROBLEMS If U is considered to be of $O(1)$, ϕ and η of $O(\epsilon)$, where the small number ϵ is either the Froude number based on the submergence of the disturber (cases of deep submergences) or the longitudinal surface slope of the disturber (cases of thin or slender bodies), the nonlinear terms involving ϕ and η in (2.6) could be discarded. The linearized free-surface conditions to be satisfied on $y = 0$ are therefore

$$-U\phi_x + g\eta = 0, \qquad U\eta_x + \phi_y = 0 \qquad \text{on } \mathcal{F}_o, \qquad (2.13\text{a,b})$$

which can be combined to yield

$$\phi_{xx} + \kappa\phi_y = 0 \qquad \text{on } \mathcal{F}_o, \qquad (2.14)$$

where $\kappa \equiv g/U^2$. If the body is not deeply submerged, a consistent linearization procedure requires that the condition (2.3) on $\mathcal{H}$ be satisfied in some linearized manner also. A detailed account of the formalism of such a perturbation expansion can be found in Wehausen (1973) and more recently in Dern (1977). Many workers, however, prefer to use (2.3) in its exact form together with (2.14). If $\mathcal{H}$ intersects $\mathcal{F}_o$, the boundary-value problem (2.1), (2.3), and (2.12) is known as the Neumann-Kelvin problem, for which neither existence nor uniqueness proofs have been established.

In the last decade, another type of linearized theory has been developed (Ogilvie 1968, Baba & Takekuma 1975, Newman 1976) as an attempt to model low-speed steady flow over full bodies, such as a tanker hull. The fully nonlinear equation (2.8) may be rewritten as follows:

$$\Phi_s \left(\tfrac{1}{2}\Phi_s^2 \right)_s = -g\Phi_y \qquad \text{on } y = \eta, \tag{2.15}$$

where s is the arc-length along the streamline under consideration. If Φ is decomposed as a sum of $\Phi^{(r)}$ and $\phi^{(w)}$ where $\Phi^{(r)}$ satisfies (2.1), (2.7) and $\Phi_y^{(r)} = 0$ on $y = 0$, the so-called double-body potential, and $\phi^{(w)}$ is a wave-like potential, the resulting linear free-surface condition for $\phi^{(w)}$ reduces to

$$\begin{aligned}&\frac{\partial}{\partial s}(\Phi_s^{(r)2}\phi_s^{(w)}) + g\phi_y^{(w)} \\ &= -(\Phi_s^{(r)2}\Phi_{ss}^{(r)} + g\eta^{(r)}\Phi_{yy}^{(r)}) \qquad \text{on } y = 0,\end{aligned} \tag{2.16}$$

where $\eta^{(r)} = (U^2 - \Phi_s^{(r)2})/2g$ and s is now the double-body streamline coordinate. The variable coefficients in (2.16) are indicative of a physical process in which the wave-like motion is being convected downstream by the double-body potential; the latter can be determined by a number of well-established means.

LINEARIZED OSCILLATORY-MOTION PROBLEMS If the disturber has no forward velocity and conducts oscillatory motion about some equilibrium position, the small parameter for linearization is the amplitude of the body motion. Equation (2.5) then reduces to

$$\eta_t(x, z, t) = \phi_y, \qquad \phi_t(x, 0, z, t) = -g\eta \tag{2.17a,b}$$

or

$$\phi_{tt}(x, 0, z, t) + g\phi_y = 0, \tag{2.18}$$

which are the linearized conditions of classical water-wave theory. The body condition (2.3) may be satisfied at its equilibrium position.

If the motion is time-harmonic of the $e^{-i\omega t}$ type, where $i = \sqrt{-1}$, and ω is the angular frequency, the standard decomposition of the form $\phi(\mathbf{x},t) = \varphi(\mathbf{x})e^{-i\omega t}$, $V_n(\mathbf{x},t) = v_n(\mathbf{x})e^{-i\omega t}$, etc. can be employed. Whence it follows from (2.3) and (2.18) that the boundary conditions for the (time-) complex spatial potential $\varphi(\mathbf{x})$ are

$$\varphi_n = v_n, \qquad \text{on } \mathcal{B}, \mathcal{H}, \tag{2.19}$$

$$-\frac{\omega^2}{g}\varphi(x, 0, z) + \varphi_y = 0, \tag{2.20}$$

the latter being a condition of mixed type on $\mathcal{F}_0$. As before, φ must satisfy (2.1). Furthermore, in order to obtain outgoing waves, a radiation condition must be imposed. This can be stated as

$$\varphi_x \mp ik\varphi \to 0 \quad \text{as } x \to \pm\infty, \tag{2.21}$$

$$(kr)^{1/2}(\varphi_r - ik\varphi) \to 0 \quad \text{as } r(=\sqrt{x^2+z^2}) \to \infty, \tag{2.22}$$

for the two- and three-dimensional cases, respectively. The wave number k in (2.21, 2.22) is the solution of the equation $\omega^2 = gk \tanh kh$. We remark that (2.19)–(2.22) represents the class of radiation and diffraction problems that one customarily encounters in the study of motion of floating bodies, and wave forces on structures. An Existence and Uniqueness theorem for this class of problem has been given by John (1950) under rather restrictive hypotheses of the body geometry. A more general proof has recently been given by Lenoir & Martin (1981).

Not considered specifically in this article is the case of a steadily translating body that undergoes periodic oscillations about some mean position, the so-called ship-motion problem. A recent review of Newman (1978) covered this in some detail.

3. FINITE-DIFFERENCE METHODS

The use of finite-difference techniques in solving partial-differential equations is a well-established one. There is a substantial and rapidly growing amount of literature addressing aspects of stability, convergence, and efficiency of the various types of differencing schemes available. The literature also abounds in fluid-mechanics applications. Indeed, finite-difference methods may very well be regarded as the backbone of many of today's flow-simulation codes (Patterson 1978, D. Chapman 1978). General features of such numerical simulation techniques are well discussed by Emmons (1972), Orszag & Israeli (1974), and MacCormack & Lomax (1979). Details of the methods and relevant analysis are available from texts such as Richtmyer & Morton (1967), Roache (1976), and Ames (1977). The present section covers only those aspects that are particularly relevant to the implementation of finite-difference methods in free-surface flow problems. By inspection, it is evident that the treatment of the field equation in a potential or stream-function formulation is relatively straightforward. The complication is associated with the determination of the free-surface location in conjunction with the satisfaction of the boundary conditions and the field equation, all preferably simultaneously.

Finite-difference methods are most suitable, or at least simplest to implement, for boundary geometry that is rectilinear. For nonlinear problems, the free-surface boundary will not usually intersect the mesh system at grid points that are regularly spaced. This is similar to the difficulty caused by the presence of an arbitrarily shaped body in the fluid, but with the further complication that such intersections are time dependent. The use of "irregular stars" near the fluid boundaries is generally called for, but they are inferior in accuracy compared with the regular ones in the rest of the field. To achieve an accuracy near the boundary that is the same order as the field, one normally would have either to refine the mesh locally or use a difference formula of a higher-order accuracy. Either approach will complicate the numerical procedure substantially. Boundary conditions are known to have a strong influence on the accuracy of the flow solutions; their proper treatments are thus imperative. With the presence of a free surface that is unknown a priori, we thus have the unfortunate situation that the regions that demand the greatest accuracy are precisely those where it is hardest to achieve. The use of boundary-fitted coordinates, which allow the problem to be solved in a mapped domain composed solely of rectangles, will overcome this type of difficulty but at the expense of introducing some other complexities.

In many practical applications, one is interested only in the steady-state solution. Because of the difficulty of implementing a *nonlinear* radiation condition, if one indeed exists, at the outflow boundary, most workers prefer to solve an initial-value problem and obtain the large-time asymptotic behavior. The assumption is that such an asymptotic solution is equivalent to the steady state. For nonlinear problems, there is no reason a priori why the asymptotic solutions are independent of initial condition. Besides, it is conceivable that phenomena such as wave breaking may interrupt the solution process, before steady state is reached.

The initial-value approach has the advantage that the free-surface conditions can be used to advance the solution in time. The choice of the finite-difference form is critical; seemingly minor modifications can sometimes have drastic effects on the stability and convergence characteristics of the system. Stability is not equivalent to accuracy. Overly stable finite-difference forms can introduce such a large amount of artificial viscosity that the final results are completely distorted.

To monitor the accuracy of the computations, it is useful to apply an energy check on the numerical solution as time progresses. Let E be the total energy of the fluid *motion* in the domain Ω. In a fixed reference frame ($U=0$) the following energy theorem has been given by John (1949):

$$\frac{1}{\rho}\frac{dE}{dt} = \frac{d}{dt}\int_\Omega [\tfrac{1}{2}|\nabla\phi|^2 + gy]d\Omega \qquad (3.1a)$$

$$= \int_\Omega \nabla\phi_t \cdot \nabla\phi \, d\Omega - \int_{\partial\Omega} \left(\frac{p}{\rho} + \phi_t\right) V_n \, d\partial\Omega \qquad (3.1b)$$

$$= \int_{\partial\Omega} \phi_t(\phi_n - V_n) \, d\partial\Omega - \int_{\partial\Omega} \frac{p}{\rho} V_n \, d\partial\Omega \,, \qquad (3.1c)$$

where the Transport Theorem and (2.2) have been used in arriving at the second equality, and the divergence theorem at the third. V_n here is the normal velocity of the boundary surface. Note that because of the free-surface conditions, the boundary $\mathcal{F}$ does not contribute to the right-hand side of (3.1c), as is to be expected from physical reasoning. Since the second integral of (3.1c) over $\mathcal{H}$ is the rate of work done by the body $\dot{W}_\mathcal{H}$, we obtain

$$\dot{W}_\mathcal{H} = \dot{E} + \int_\Sigma [pV_n - \rho\phi_t(\phi_n - V_n)]d\partial\Omega$$

$$= \rho\frac{d}{dt}\left[\frac{1}{2}\int_{\partial\Omega}\phi\phi_n\,d\partial\Omega + \frac{g}{2}\int_{\mathcal{F}_0}\eta^2\,dxdz\right] \quad (3.2)$$

$$- \int_\Sigma [pV_n - \rho\phi_t(\phi_n - V_n)]d\partial\Omega \,.$$

The last integral of (3.2) vanishes when Σ is a stationary rigid surface, otherwise it accounts for the rate of work by Σ and the energy flux out of the control volume Ω. Regardless of the type of numerical method, (3.2) is particularly convenient for computation, since it involves boundary quantities that are normally either known or being sought after anyway. Linearized forms of (3.2) have been used by Haussling & Van Eseltine (1974) in two dimensions, and Ohring & Telste (1977) in three dimensions. One useful specialization of (3.2) is the case of steady motion of a body, for which $\dot{E} = 0$ in the moving frame of reference. In this case, if we choose a vertical plane Σ^- perpendicular to the direction of motion downstream and take advantage of the vanishing of disturbances far upstream, (3.2) can be simplified to

$$R = \tfrac{1}{2}\rho\int_{\Sigma^-}(\phi_y^2 + \phi_z^2 - \phi_x^2)d\partial\Omega + \tfrac{1}{2}\rho g\int_{\Sigma^-\cap\mathcal{F}_0}\eta^2\,dz\,, \quad (3.3)$$

where R is the steady-state resistance, and the location of Σ^- is arbitrary. This equation can provide a consistent check of the local calculations around the body with the downstream-outflow behavior. An alternate derivation of (3.3) based on momentum considerations was given by Wehausen (1973). Equation (3.3), like (3.2), is exact within the assumptions of potential flow. Neither one has been implemented in checking nonlinear calculations.

Finite-Difference Forms

In the interest of notational simplicity, we discuss here the treatment of two-dimensional problems only; three-dimensional analogues of what follow are often quite self-evident. The notation $\phi(i\Delta x, j\Delta y, n\Delta t) \equiv \phi_{ij}^n$ will be used; in particular $j = j^*$ will denote the grid points on the free surface (or $y = 0$, in the case of linearized problems). The five-point central-differencing formula of Laplace's equation (at any time index) is

$$\phi_{ij} = \tfrac{1}{4}(\phi_{i+1,j} + \phi_{i-1,j} + \phi_{i,j-1} + \phi_{i,j+1}) + O(\Delta x^2), \tag{3.4}$$

where Δx and Δy are assumed to be the same. The coefficients in front of the "neighborhood points" will be more complex (see, for example, Forsythe & Wasow 1960), if the points are not equally spaced around (i,j) and the subsequent five-point formula of the same points is only of $O(\Delta x)$ accuracy. Equation (3.4) is generally solved by successive over-relaxation based on the following rearranged formula:

$$\phi_{ij}^{(k+1)} = \phi_{ij}^{(k)} + \frac{\omega}{4}[\phi_{i+1,j}^{(k)} + \phi_{i-1,j}^{(k+1)} + \phi_{i,j+1}^{(k)} + \phi_{i,j-1}^{(k+1)}$$
$$- 4\phi_{ij}^{(k)}], \qquad i\uparrow, j\uparrow \tag{3.4a}$$

where the superscript in parenthesis indicates the stage of iteration, with i,j being taken in an ascending sequence, and ω is the *relaxation* parameter. Dirichlet conditions on the boundary are thus easily handled by (3.4a); but Neumann or mixed-type conditions require a redefinition of the coefficients on the right-hand side. Because of its ease of implementation, (3.4a) is preferred by most workers (von Kerczek & Salvesen 1974, Chan & Hirt 1974, R. Chapman 1976). Higher-order methods have been used by Ohring (1975), who also utilized a direct method due to Buzbee et al. (1971) for handling the irregular stars on the solid boundary.

Being first-order in time, the free-surface conditions (2.5a,b) are particularly useful for advancing η and ϕ on the free boundary. Various difference forms are conceivable. For illustration, we consider the case of linearized conditions (2.17a,b). A few obvious schemes are

$$\eta^{n+1} = \eta^n + \Delta t\, \phi_y^n, \qquad \phi^{n+1} = \phi^n - \frac{g\Delta t}{2}(\eta^{n+1} + \eta^n); \tag{3.5a}$$

$$\eta^{n+\frac{1}{2}} = \eta^{n-\frac{1}{2}} + \Delta t\, \phi_y^n, \qquad \phi^{n+1} = \phi^n - g\Delta t\, \eta^{n+\frac{1}{2}}; \tag{3.5b}$$

$$\overline{\eta^{n+1}} = \eta^n + \Delta t\, \phi_y^n, \qquad \overline{\phi^{n+1}} = \phi^n - g\Delta t\, \eta^n, \tag{3.5c}$$

$$\eta^{n+1} = \eta^n + \frac{\Delta t}{2}(\phi_y^n + \overline{\phi_y^{n+1}}), \qquad \phi^{n+1} = \phi^n - \frac{g\Delta t}{2}(\overline{\eta^{n+1}} + \eta^n);$$

$$\eta^{n+1} = \eta^n + \frac{\Delta t}{2}(\phi_y^n + \phi_y^{n+1}), \qquad \phi^{n+1} = \phi^n - \frac{g\Delta t}{2}(\eta^{n+1} + \eta^n); \tag{3.5d}$$

where all values of ϕ are understood to be at j^*. The first two are one-step explicit schemes that use the current information at n or $n-\tfrac{1}{2}$ to advance the solution. (3.5b) is a staggered system, which is actually centered at n and is thus $O(\Delta t^2)$ in accuracy. (3.5c) is a two-step predictor-corrector method without iteration. $\overline{\phi_y^{n+1}}$ however has to be determined based on

$\overline{\phi^{n+1}}$ of the predictor step. (3.5d) is the so-called Euler modified method, which is implicit, thus requiring a simultaneous solution of ϕ and η via ϕ_y at $n+1$.

An assessment of the stability characteristics of these schemes may be made by a *simplified* von Neumann analysis. Let $\phi(x, y, n\Delta t)$ be of the form $\phi^n e^{ikx+ky}$ where k is a wave number; thus $\phi_y^n = k\phi^n$. If (ϕ^{n+1}, η^{n+1}) is now written in terms of an amplification matrix G multiplying (ϕ^n, η^n), the necessary condition for stability is that the spectral radius of G be less than 1 (Lax & Richtmyer 1956). The eigenvalues for the various cases of (3.5) can be calculated and are shown below, with the dimensionless number $f \equiv kg\Delta t^2/2$ as parameter:

$$|\lambda| \geq (1 + f)^{1/2}, \qquad (3.6a)$$

$$|\lambda| = 1, \text{ for } f \leq 2; \quad |\lambda| \leq f - 1 + \sqrt{f(f-2)}, \text{ for } f > 2, \quad (3.6b)$$

$$|\lambda| = (1 + f^2)^{1/2}, \qquad (3.6c)$$

$$|\lambda| = 1, \text{ for all } f. \qquad (3.6d)$$

This simple, yet original, analysis is rather informative. It can be seen that the implicit scheme is the only unconditionally stable scheme for all k. In particular, we observe that (3.5a) is always unstable. The staggered system (3.5b), which has been used by Chapman (1976), is conditionally stable if short waves are not excited. Chapman actually made some quasi-three-dimensional calculations for a vertical surface-piercing plate moving steadily at an angle of attack. Within the assumptions of high-Froude-number theory, the solution at each station was treated two-dimensionally in the cross plane, with "time" being taken as the longitudinal position of the station divided by the forward speed. The two-step scheme of (3.5c) has been used by Yen et al. (1977) to advance a two-dimensional finite-element solution for a moving pressure distribution. Though only marginally unstable, $|\lambda| = 1 + O(\Delta t^4)$, this scheme is clearly unsuitable for long-time (steady-state) calculations. Such instability is apparently the cause of the sudden drag increase in Yen's calculation at large value of time.

Implicit schemes similar to (3.5d) are preferred by most workers (Chan & Hirt 1974, Ohring & Telste 1977, Haussling & Coleman 1979) because of the stability property. Further, since $|\lambda| = 1$ there is no artificial damping introduced, at least according to linear analysis. Zero damping, however, does not imply the absence of phase errors, the latter being critical in phenomena involving interference effects such as wave resistance. The fully nonlinear equations (2.5a, 2.5b) are best treated in a manner similar to (3.5d). In a moving coordinate system, strongly convective terms are present. Upwind differencing is usually used to "preserve" the transportative property, but its effect has been seldom analyzed. The judgement is often an indirect one based on a comparison of physical quantities of

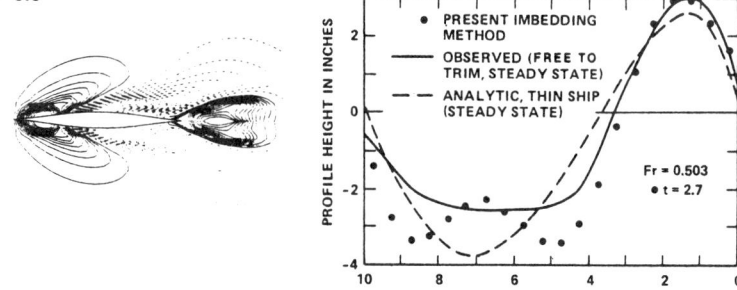

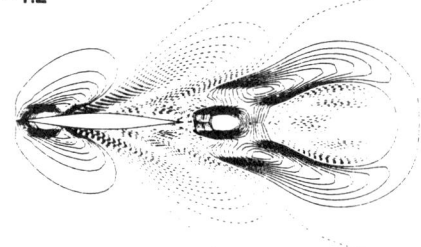
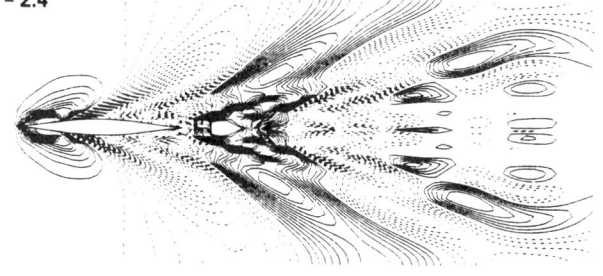
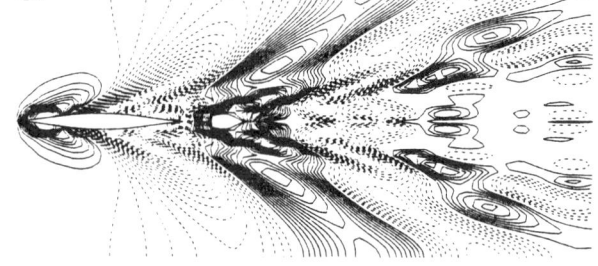

interest, such as force and wave-elevation, with experimental results, which naturally include other factors not modeled in the original equations.

The general nonlinear conditions (2.5a,b) can be solved iteratively using the following algorithm or minor variants of it. The difference equations corresponding to (3.5d) are now written as

$$(\eta^{n+1,\,k+1} - \eta^n)/\Delta t = \tfrac{1}{2}(\phi_y^n + \phi_y^{n+1,\,k}) + F(\eta^n, \eta^{n+1,\,k}, \phi^{n+1,\,k}),$$
(3.7a)

$$(\phi^{n+1,\,k+1} - \phi^n)/\Delta t = -\tfrac{1}{2}g(\eta^{n+1,\,k+1} + \eta^n) + G(\eta^n, \eta^{n+1,\,k+1}, \phi^{n+1,\,k}),$$
(3.7b)

where F and G are the remaining terms in (2.5a,b). The second superscript in (3.7) is the stage index of the iterations within a time step. Thus, starting with known values of ϕ in Ω and values of η at step n, η and ϕ (on $\mathscr{F}$) are first updated successively by (3.7). The new values of ϕ are next used as a Dirichlet condition on $\mathscr{F}$, and (3.4a) is solved with the body boundary condition (2.3) and some appropriate conditions upstream and downstream. A new ϕ_y^{n+1} is now calculable and (3.7) can thus be repeated to complete another stage of iteration, and so on, until converged values of η and ϕ are achieved for the time step $n+1$.

Figure 1, taken from Ohring & Telste (1977), gives free-surface contour plots generated by a moving ship that starts from rest. The free-surface conditions used by the authors are linearized; the body condition is, however, exact. A rigid computation box of dimensions 6 × 1.6 × 1 body lengths was used. The development of the Kelvin wave pattern is clearly visible in these plots, but the authors had difficulty obtaining steady-state results. The oscillatory behavior of the wave-resistance coefficient reported in this work could be caused by the sloshing modes of the "tank." Ohring & Telste (also Haussling & Coleman 1979) reported the presence of "wiggles" in their solution and numerical filtering was necessary.

If the ultimate solution desired is that corresponding to a steadily translating body in or near a free surface, it would appear that a direct attack on Equations (2.8a,b) or (2.9a,b) is perhaps more appropriate. This approach has been apparently less successful. In two-dimensions, as far as the free-surface conditions are concerned, it is more convenient to use either a stream-function or potential formulation that includes the free stream. The kinematic condition is then merely a Neumann or Dirichlet type. A unique procedure was given by von Kerczek & Salvesen (1974),

---

Figure 1 Time-sequence plots of the free-surface contours generated by an impulsively starting ship (from Ohring & Telste 1977). Numerical calculations satisfy an exact body boundary condition whereas "analytic" results are based on Equation (5.8a). Fr is the Froude number based on forward speed, and **t** the number of ship lengths travelled divided by Fr.

who considered the nonlinear problem of steady flow about a submerged point vortex. Proceeding in a downstream direction, these authors successively adjust each "free-surface node" so that the dynamic boundary condition can be satisfied. Since interaction occurred among nodes, they utilized an interpolation polynomial, extending about one half of a wavelength downstream, to represent the free surface and determine the amount of correction based on the pressure residual at the node under consideration. Each node adjustment required a repetitive solution of (3.4a) until the pressure residual was acceptable. The process is conceivably not the most efficient, but was apparently convergent. The authors reported that they were able to generate waves to within 15 percent of the steepest periodic waves of Stokes (1880a). In a subsequent paper on subcritical flow in shallow water (Salvesen & von Kerczek 1978) these authors noted the necessity of accounting for an upstream surge when comparing numerical results with experiments based on the depth Froude number. Also noted was the interesting result that for a given Froude number, downstream waves tend to shorten rather than lengthen, as in the case of deep water, when a disturber increases in strength.

One obvious approach to solving the nonlinear conditions (2.9) is to use Newton's method iteratively. This is best carried out using a numerical formulation that has implicitly satisfied the field equation, e.g. an integral-equation method or a spectral method. Finite-difference or finite-element methods are not particularly suitable in this respect. Newton's method used in conjunction with a spectral representation based on Fourier series has recently been used by Reinecke & Fenton (1981) for calculating profiles of free waves (see also Schwartz 1981 in Section 5). They report that convergence to 12 decimal places within a few iterations was possible! The initial "guess" was taken simply as linear Stokes waves, but other extrapolative means were found necessary for the case of very steep waves.

Since Dean's (1965) work, the stream-function formulation with a Fourier representation has been used extensively for evaluating properties of nonlinear periodic waves (see also Dean 1974, Chaplin 1980). This particular approach becomes ineffective when a physical body is present in the wave field, in which case a simple spectral representation is no longer available. Further, neither the value of the stream function itself nor its normal derivative is known on the body boundary.

Boundary-Fitted Coordinates

The basic idea of boundary-fitted coordinates is to transform the physical boundaries of a problem to coordinate lines in a mapped space, wherein finite-difference computations can be conveniently carried out in a regular mesh system without extensive interpolation along the boundary. In estuary hydrodynamics, the simple transformation in the vertical direction,

$\zeta = (\eta-y)/(\eta+h)$, has been used quite successfully for some time (Freeman et al. 1972, Noye et al. 1981). More elaborate techniques were considered by Winslow (1966) for interior flows. Thompson et al. (1974) have further developed the idea for exterior and multi-body flows. If (ξ,ζ) denote the mapped coordinates in two dimensions, the generating system is typically taken as the solution of the Poisson equation (Thompson et al. 1976):

$$\xi_{xx} + \xi_{yy} = P(\xi,\zeta), \qquad \zeta_{xx} + \zeta_{yy} = Q(\xi,\zeta), \qquad (3.8)$$

with Dirichlet conditions (viz. ξ,ζ being constant) on $(x,y) \in \mathcal{B}, \Sigma$, and $\mathcal{F}$. Here P,Q are preselected functions that control the density of coordinate lines. Since the objective is to work in the mapped domain, (3.8) is actually solved with the dependent and independent variables interchanged:

$$\alpha x_{\xi\xi} - 2\beta x_{\xi\zeta} + \gamma x_{\zeta\zeta} = -J^2[x_\xi P(\xi,\zeta) + x_\zeta Q(\xi,\zeta)],$$

$$\alpha y_{\xi\xi} - 2\beta y_{\xi\zeta} + \gamma y_{\zeta\zeta} = -J^2[y_\xi P(\xi,\zeta) + y_\zeta Q(\xi,\zeta)], \qquad (3.9)$$

where α, β, and γ are relatively simple functions of the first derivatives of x and y with respect to ξ and ζ, and J is the Jacobian $\partial(x,y)/\partial(\xi,\zeta)$. The coupled quasi-linear elliptic system (3.9) must be solved at every instant of time if the boundary contours, such as $\mathcal{F}$, are time dependent. Figure 2, taken from Haussling & Coleman (1979), shows an example of the mapping of a doubly connected region of the flow onto a H-shaped (ξ,ζ) plane.

The formulation of the problem is completed by transforming the field equation and boundary conditions into the (ξ,ζ) plane using the usual rules for implicit functions. Equation (2.1), for example, would now read as

$$\alpha \phi_{\xi\xi} - 2\beta \phi_{\xi\zeta} + \gamma \phi_{\zeta\zeta} + J^2(P\phi_\xi + Q\phi_\zeta) = 0, \qquad (3.10)$$

which is fundamentally similar to (3.9). At each time step, the *transformed* expressions of the free-surface condition (2.5a,b) can be used to advance η and ϕ as described earlier. However, this must now be followed by a successive over-relaxation solution of (3.9), and then of (3.10), the latter being subject to condition (2.3), and it is necessary to apply this procedure iteratively because of the implicit nature of (3.7). It is evident that the original problem with difficult boundary conditions has been exchanged for one having complicated field equations but straight-line boundaries. It is not clear which method is more efficient for the same degree of accuracy. For free-surface related flows no comparisons have been made on a one-to-one basis at the writing of this article.

The boundary-fitted coordinate techniques have been successfully applied by Shanks & Thompson (1977), Haussling & Coleman (1979), and Chan & Chan (1980). Shanks & Thompson sought a direct solution of the Navier-Stokes equation (primitive-variable form) for a translating hydrofoil at very low Reynolds numbers (< 100). Although the practical implications of conducting flow calculations at such low Reynolds numbers is unclear, the results were impressive and physically plausible.

Using the velocity-potential formulation, Haussling & Coleman calculated the fluid motion due to a submerged circular cylinder undergoing either translating or swaying motion. These authors reported difficulties in obtaining steady-state results for the case of translatory motion because incipient breaking waves were encountered in the transient stage. We note in passing that the convective terms in the free-surface conditions of Haussling & Coleman (1979) were of the wrong signs. This could be merely typographical since the numerical results appear reasonable from the physical standpoint.

As an alternative to boundary-fitted coordinates method, it is worthwhile to mention that Multi-Level Adaptive Techniques, which provide very efficient means of solving flexible multi-level grid-structure systems, can be quite effective in dealing with boundary contours that require fine resolution. A more detailed account may be found in Brandt et al. (1980).

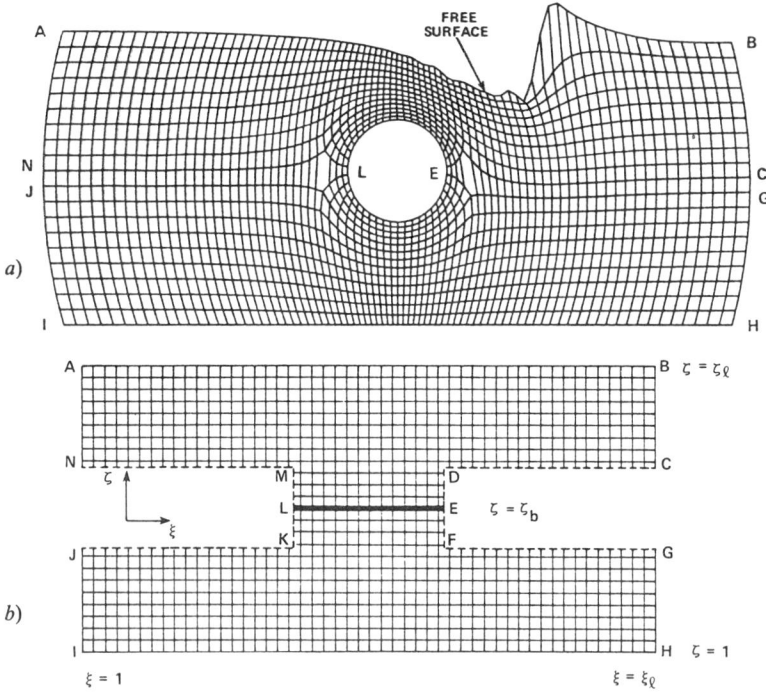

Figure 2 (*a*) Free-surface profile and boundary-fitted coordinates in the physical plane for a cylinder translating in water of finite depth. (*b*) The corresponding transformed computational region. (From Haussling & Coleman 1979)

Radiation Conditions

In the interest of keeping the computation region to a minimum, it is always desirable to truncate Ω by some control boundary Σ. Surely, a simple approach is to take a sufficiently large Σ beyond which disturbances can be considered vanishingly small. But such a boundary must necessarily expand in time as the disturbances propagate outwards. An alternative approach is to map the point at infinity to a finite value. Grosch & Orszag (1977) indicated that such mappings were not particularly useful for solutions that oscillate out to infinity. The ideal Σ would be one that is close to the region of interest but at the same time "transparent" to the flow. In practice, it is necessary to prescribe some sort of boundary conditions. What then is the appropriate boundary condition that will have a minimal effect on the interior solution? This has always been a "sore point" in computational fluid mechanics. There is no absolutely satisfactory answer to this nontrivial difficulty and the search still continues (see, for example, Hedstrom 1979, Rudy & Strikwerda 1980, Carmerlengo & O'Brien 1980).

For steadily propagating free waves that are periodic in space, planes of symmetry serve well as truncation boundaries on which symmetry conditions can be applied. In unsteady flow a helpful artifice is to assume the solution to be of the form $Q(x,y,t) = Q(x-ct,y)$, with Q representing ϕ or η. This leads to the well-known one-dimensional Sommerfeld condition (1949):

$$\frac{\partial Q}{\partial t} + c \frac{\partial Q}{\partial x} = 0, \tag{3.11}$$

where c is the phase velocity of the right-propagating waves. If the process is time harmonic $Q = \overline{Q}e^{-i\omega t}$, $\overline{Q}$ being complex in time, (3.11) reduces to (2.21) for $\overline{Q}$, which has been applied directly as an open-boundary condition by Chenault (1970), Bai & Yeung (1974), Zienkiewicz & Bettess (1975). It is noteworthy that (3.11) is not indicative of the presence of other solutions that "propagate" at infinite velocity because of the elliptic nature of the field equation. A successful way of accounting for them is by a matching procedure discussed in Sections 4 and 5.

Equation (3.11) is a standard transportive equation, for which forward-time upwind-differencing is known to be stable if the Courant number is less than unity (Thom & Apelt 1961). Its implementation is straightforward if c is known. However, in nonlinear problems the dispersion characteristics is generally unknown, and hence also c. Nevertheless, one may assume that (3.11) is locally correct with c being a slowly varying function of space and time. Orlanski (1976) proposed determining c numerically based on (3.11) at neighborhood points just interior to the boundary Σ and at an earlier time step. The boundary values of Q on Σ can then be advanced by applying (3.11) at the present time step but this time on Σ, using the value of c just determined. Orlanski derived a three-time-level formula based on centered-time and upwind-space differencing as an approximation to (3.11). Following Orlanski, Chan (1977) used a similar approximation but with upwind-centered time differencing, and noted that his scheme was non-dissipative for constant c. By taking advantage of this condition and body-fitted coordinates in three-dimensions Chan & Chan (1980) attempted the unsteady problem of nonlinear flow about a general ship hull. Although it is premature to judge the success of the reported computations at this time, the work is perhaps the most ambitious undertaking of all those reported here.

Two additional points concerning (3.11) are noteworthy. Because of stability considerations, the *numerically* feasible value of c must be less than $\Delta x/\Delta t$ (the Courant number $c\Delta t/\Delta x$ being unity). This imposes an upper limit on the fastest wave that a given spatial and temporal gridsize can represent. Equation (3.11) is physically reflective for two-dimensional plane waves at oblique incidence to Σ. Engquist & Majda (1977) have developed a hierarchy of *absorbing* boundary conditions for hyperbolic equations in two dimensions. Their leading-order result was (3.11), which is exact for one-dimensional waves. The next order, however, was given by the condition

$$c\frac{\partial^2 Q}{\partial x \partial t} + c^2 \frac{\partial^2 Q}{\partial t^2} - \frac{1}{2}\frac{\partial^2 Q}{\partial y^2} = 0. \tag{3.12}$$

Equation (3.12) was shown to reduce the reflection coefficient from 17% to 3% at 45° incidence. Procedures for implementing this are apparently more complex but seems worth the effort.

It is of interest to note that in a moving coordinate system of velocity U, (3.11) becomes $Q_t + (c-U)Q_x = 0$, which yields no useful information in the steady-state limit. The difficulty of imposing a "radiation condition" in a steady-state formulation is hence apparent. However, in their calculations of two-dimensional waves, von Kerczek & Salvesen (1974) simply imposed a uniform-flow condition upstream and an extrapolative condition at the downstream (outflow) boundary. Although such an outflow condition did not actually satisfy the field equations, the results appeared to suggest that the irregularities created at the outflow boundary had little upstream effects. It is unclear to what extent this was associated with the "upwind" iteration algorithm used by these authors in the successive overrelaxation procedure.

An alternative technique to absorbing conditions is the use of artificial damping in part of the fluid domain or boundary. This has received some attention in the literature (e.g. Arakawa & Mintz 1974, Israeli & Orszag

1974, Chan 1975). The technique usually gives rise to a nontrivial problem of "designing" a damping pad, whose absorption spectrum must be wide, but whose effects on the phase of the fluid response is to be minimal.

4. THE METHODS OF FINITE ELEMENTS

Of the three major categories of methods discussed in this paper, the finite-element method is generally considered the newest comer. Introduced and developed in the 1960s, it has established itself as the mainstream computational tool for stress analysis of structures of complicated geometry. It was soon adapted to numerous other areas of applied mechanics. Extensive reviews of its general successes in fluid-flow problems have been given by Shen (1977) and Norrie & de Vries (1978). Applications pertaining to water-wave radiation and diffraction were also covered recently by Mei (1978).

The finite-element and finite-difference methods share the common feature that both tackle the field equation directly. In fact, if linear interpolative elements based on a uniform grid system were used, Zienkiewicz (1975) showed that the variational form of the Laplace operator yields an expression identical with the five-point central-differencing formula. However, with the introduction of curvilinear or isoparametric elements, the finite-element approach allows one to cope with arbitrary boundary geometry with little loss of accuracy. Although this particular advantage over finite-differencing appears to be modestly reduced when boundary-fitted coordinates are exploited as described earlier, finite-element methods still retain superiority in flexibility, particularly in cases where "superelements" can be introduced to overcome any local irregularities or difficult behavior at infinity.

In its most general form, the underlying principle of the finite-element method is one based on the method of weighted residuals. The usual procedure consists of first subdividing the domain of interest into a mesh of finite-sized subregions, within each of which the solution is represented by some convenient choice of trial functions, usually polynomials, with unknown parameters to be determined. The trial functions are determined by substituting them into the field equation and requiring the integrated error or residuals based on certain weighting functions to vanish. An integration by parts is normally performed to reduce the inter-element continuity requirement of the trial functions and to incorporate any inhomogeneous boundary conditions that are as yet not satisfied. For the Laplace operator, the integrability requirement after such steps amounts merely to an inter-element continuity of the function being sought. The weighting functions, also known as test functions, can be chosen in a variety of ways; each leads to a different type of method of weighted residuals. One particular choice is to make the space of test functions identical with that of the trial functions. This is known as Galerkin's method. Since each member of the test-function space yields one condition from the residual criterion, there are just enough equations to determine the unknown parameters in the problem. The formulation described in the foregoing is commonly referred to as the "weak formulation." In contrast, a "strong formulation" is one based on the existence of a variational principle where a functional is made stationary. Corresponding to each variational principle, it is possible to find the equivalent Galerkin procedure, but not always vice versa. For linear problems, the matrix resulting from a variational principle is always symmetric, thus reducing storage requirement and solution time. Such an advantage disappears when the problem is nonlinear, although the variational principle itself can be used to provide valuable insight into the dynamics of the physical system.

Variational Methods

The fully nonlinear problem defined by Equations (2.1), (2.3), and (2.5) with $U = 0$ can be considered equivalent to taking the variation of the following Lagrangian function:

$$J(\phi, \eta) = -\int dt \left\{ \int_\Omega p \, d\Omega + \int_{\mathcal{B} \cup \Sigma \cup \mathcal{H}} V_n \phi \, d\partial\Omega \right\}, \qquad (4.1)$$

where p is the pressure in excess of atmospheric, as given by (2.2). Bateman (1932) has used the pressure as the Lagrangian to obtain the equations of motion of an inviscid incompressible fluid. However, Luke (1967) was the first to point out that the free-surface boundary conditions follow from (4.1). If one considers ϕ and η as independent variations and keeps in mind that η is a function of x, z, and t, then

$$\delta J = \int dt \left\{ -\int_\Omega \nabla^2 \phi \delta\phi \, d\Omega + \int_{\mathcal{B} \cup \Sigma \cup \mathcal{H}} (\phi_n - V_n) \, \delta\phi \, d\partial\Omega \right.$$

$$\left. + \int_{\mathcal{F}} \left(\phi_n - \frac{\eta_t}{[1 + \eta_x^2 + \eta_z^2]^{1/2}} \right) \delta\phi \, d\partial\Omega - \int_{\mathcal{F}} p \, \delta\eta \, d\partial\Omega \right\}, \qquad (4.2)$$

where $\delta\phi$ is subject to no variations on the time boundaries. The last term in (4.2) arises from a variation in the physical domain Ω. It can be seen that all boundary conditions are satisfied *naturally* in (4.2). Luke's original form does not contain the second term in (4.1) since he assumed $\delta\phi = 0$ on the boundary Σ. In spite of its apparent simplicity, (4.1) has not been used for numerical computations. However, extensive use of this variational principle has been made by Whitham (1967, 1970) to examine the dispersion characteristics of nonlinear water waves. An alternative form that is dynamically equivalent to (4.1) has been given by Miles (1977).

If the flow is *steady* in a moving frame of reference and ϕ is the disturbance potential, the functional J in (4.1) can be written as

$$J(\phi, \eta) = \int_\Omega (\tfrac{1}{2}|\nabla\phi|^2 - U\phi_x)d\Omega \qquad (4.3)$$
$$- \int_\Sigma (Un_x - V_n)\phi\, d\partial\Omega + \int_{\mathcal{F}_0} \tfrac{1}{2}g\eta^2\, dxdz,$$

where V_n is the efflux associated with Σ, which is generally unknown. By taking the variation of (4.3), it is easy to show that conditions (2.1), (2.3), and (2.6) are satisfied naturally. If a boundary Σ could be found such that $\delta\phi = 0$, i.e. if ϕ were satisfied as an essential boundary condition on Σ, then (4.3) implies

$$J = \tfrac{1}{2}\phi^T \mathbf{A}(\eta)\phi + \mathbf{B}(\eta)\phi + \mathbf{C}(\eta), \qquad (4.4)$$

where ϕ and η represent the unknown vectors of nodal values in Ω and $\mathcal{F}$ and the superscript T represents the transpose. In (4.4) $\mathbf{A}$ is the global matrix after assembly, $\mathbf{B}$ and $\mathbf{C}$ are vectors. Taking advantage of the quadratic nature of (4.4) in ϕ, we can extremize J with respect to ϕ, holding η constant, and obtain

$$\mathbf{A}(\eta)\phi = -\mathbf{B}(\eta), \qquad (4.5)$$

which is linear in ϕ and can be solved to obtain $\phi = -\mathbf{A}^{-1}\mathbf{B}$. If this is substituted back in (4.4) and the resulting expression is extremized with respect to η, the following nonlinear equation in η results:

$$\tfrac{1}{2}\mathbf{B}'(\eta)\phi = \mathbf{C}'(\eta), \qquad (4.6)$$

which can be solved iteratively or by linearization. It is useful to note that the quantities $\mathbf{B}'(\eta)$ and $\mathbf{C}'(\eta)$ will involve only the layer of elements on the free surface, but their dependence on η is algebraically complicated. If the values of η obtained from (4.6) differ from those assumed in (4.5), it is necessary to adjust the finite-element mesh in the neighborhood of the free surface and repeat the procedure again. This is basically the so-called moving-net of computations (see, for example, Sarpkaya & Hiriart 1975).

The difficulty of finding a proper truncation boundary is obvious. To circumvent this, Yim (1975) extended (4.3) to a hybrid formulation that accounts for the flow exterior to Σ. The exterior flow is based on linear theory, but the compatibility of the solutions at the interfacing surface Σ was not addressed. No calculations were presented by Yim.

It is sometimes more convenient to consider a total potential Φ that includes the incident stream. If $\tfrac{1}{2}U^2$ is the Bernoulli constant, the functional corresponding to (4.3) for Φ is given by

$$J(\Phi, \eta) = \int_\Omega \tfrac{1}{2}|\nabla\Phi|^2\, d\Omega - \int_\Sigma V_n\Phi\, d\partial\Omega + \tfrac{1}{2}\int_{\mathcal{F}_0} (g\eta^2 - U^2\eta)dxdz, \qquad (4.7)$$

where we note that homogeneous boundary conditions on $\mathcal{F}$, $\mathcal{H}$, and $\mathcal{B}$ are all satisfied naturally. Apparently unaware of (4.7), Chan & Larock (1973) and Larock & Taylor (1976) calculated the exit flow from an orifice under gravity by using (4.7) without the last term. Instead, an essential condition for the potential on the free surface was imposed. This potential was estimated by integrating the velocity, whose variation was determined by using the dynamic boundary condition. The position of the free surface was successively adjusted to satisfy the no-flux condition. In their examples, Chan & Larock reported that convergence to four decimal places of accuracy occurs within ten iterations.

Complementary to (4.7), a variational principle exists in terms of the stream function in two-dimensional flow. This can be obtained by considering the Lagrangian $p + \frac{1}{2}|\nabla\Psi|^2$, where Ψ is the stream function. The functional now is given by

$$I(\Psi, \eta) = \int_\Omega \tfrac{1}{2}|\nabla\Psi|^2 \, d\Omega + \tfrac{1}{2}\int_{\mathcal{F}_0} (U^2\eta - g\eta^2)dx, \tag{4.8}$$

where Ψ is either prescribed or its normal derivatives assumed to vanish on $\partial\Omega$. If Ψ and η are treated as independent variations then

$$\begin{aligned}\delta I = &-\int_\Omega \nabla^2\Psi \,\delta\Psi \, d\Omega + \int_{\partial\Omega} \frac{\partial\Psi}{\partial n}\delta\Psi \, d\partial\Omega \\ &+ \int_{\mathcal{F}_0} [\tfrac{1}{2}(U^2 + |\nabla\Psi|^2) - g\eta]\delta\eta \, dx,\end{aligned} \tag{4.9}$$

which does not appear to satisfy the dynamic boundary condition on $\mathcal{F}$. However, to preserve mass conservation on the free surface, any variation in Ψ on $\mathcal{F}$ must be made in relation to $\delta\eta$. Hence, by noting that $\delta\Psi ds = -\Psi_n \,\delta\eta \, dx$ on $\mathcal{F}$, we obtain

$$\int_{\mathcal{F}} \frac{\partial\Psi}{\partial n}\delta\Psi \, d\partial\Omega = -\int_{\mathcal{F}_0} |\nabla\Psi|^2 \delta\eta \, dx, \tag{4.10}$$

which yields the correct dynamic condition when combined with the third term of (4.9).

Related forms of (4.8) were first given by O'Carroll (1976) and Betts (1979). Computations based on this formulation have been made by Betts (1979) and Aitchison (1980) for nonlinear flow over weirs. The normal procedure is first to determine the critical flux value for a given U and weir geometry. The flux is, in fact, the value of the stream function on $\mathcal{B}$, if Ψ is taken to be zero on the free surface $\mathcal{F}$; both can be specified as essential conditions. Aitchison argued that the critical flux could be determined by observing that the upstream waves reversed in phase as the critical value was approached. For any flux less than critical, she next repeatedly relocated the downstream truncation boundary until the upstream waves disappeared. In the present formulation, (4.8) requires the tangential velocity Ψ_n on the truncation boundary to vanish completely. These are essentially surfaces of constant potential alluded to earlier. In the presence of disturbances generated by an obstruction, such surfaces need not be planar, as Aitchison had assumed. Very reasonable results were obtained by Betts & Assaat (1980) when they applied the procedure to the calculation of nonlinear *free* waves that are periodic in space, and vertical planes of symmetry do exist.

Hybrid Formulations

The term hybrid method is used here to designate methods that employ different solution techniques in two or more subregions of the problem. A common approach is to take advantage of the availability of analytical solutions in flow regions where the geometry is relatively simple. Regions where the geometry is arbitrary or complicated will still be handled by finite elements in the usual context. Such an approach generally allows the user to reduce the number of mesh points and unknowns, with a resulting decrease in storage requirement and computational time. A further advantage is that the analytical solutions can be chosen to permit a rational treatment of the effects of radiation. However, the solutions of the different subdomains must be properly matched together. The idea is quite similar to the analytical theory of matched asymptotics, except that the matching process is often a direct one across the common boundaries of the "inner" and "outer" solution. The hybrid concept as described is not only restricted to finite-element methods. Similar applications can also be found in integral-equation (Section 5) and finite-difference methods (Shaw 1975).

The more successful applications of hybrid methods have so far been restricted to linearized problems where analytical solution in the exterior regions could be obtained without too much difficulty. In particular, treatments for steady flow in a uniform stream or time-harmonic flows with the linearized conditions (2.14) or (2.20) have been quite well established. A convenient way of classifying them is based on their formulation: a "weak" one or a "strong" one in the context of finite-element terminology. We shall discuss them in just that order.

The majority of the methods are based on Galerkin's formulation. Consider the fluid domain to be defined by an interior region Ω^i and an exterior region Ω^e with Σ as the separating boundary. One particular approach to the steady-flow problem, as given by Bai (1977), is to require the *interior* potential $\phi^{(i)}$ to satisfy the following Galerkin statement:

$$\int_{\Omega^i} \nabla^2 \phi^{(i)} \, \psi^{(i)} \, d\Omega + \int_{\partial\Omega} \phi_n^{(i)} \, \psi^{(i)} \, d\partial\Omega - \int_{\Omega^i} \nabla\phi^{(i)} \cdot \nabla\psi^{(i)} \, d\Omega$$

$$= \int_{\mathcal{H}} V_n \phi^{(i)} \, \psi^{(i)} \, d\partial\Omega + \int_{\Sigma} \phi_n^{(e)} \, \psi^{(i)} \, d\partial\Omega + \kappa^{-1} \int_{\mathcal{F}_0} \phi_x^{(i)} \, \psi_x^{(i)} \, dxdz$$

$$- \kappa^{-1} \left[\oint_{\mathcal{F}_0 \cap \mathcal{H}} \phi_x^{(i)} \, \psi^{(i)} \, n_x ds + \oint_{\mathcal{F}_0 \cap \Sigma} \phi_x^{(e)} \, \psi^{(i)} \, n_x ds \right]$$

$$= 0, \qquad \text{for all } \psi^{(i)} \text{ in } \Omega^i, \tag{4.11}$$

where $\psi^{(i)}$ is a member of the test-function space. The terms with the factor κ^{-1} are consequences of integration by parts of the free-surface condition (2.14). We observe that the derivatives of $\phi^{(i)}$ on Σ are matched with those of $\phi^{(e)}$ in the exterior region as if the latter were given. We note that in two-dimensional problems, the notation $\oint [\ ]n_x ds$ in (4.11) should be interpreted as an evaluation of the quantity in [ ] at the designated intersection points. Bai went through a development similar to (4.11) in Ω^e for the *exterior* potential $\phi^{(e)}$ and test function $\psi^{(e)}$, both assumed to take the form of eigenfunctions, but his results can be simplified and stated as

$$\int_{\Sigma} (\phi_n^{(e)} \, \psi^{(e)} - \psi_n^{(e)} \, \phi^{(i)}) d\partial\Omega - \kappa^{-1} \int_{\mathcal{F}_0 \cap \Sigma} [\phi_x^{(e)} \, \psi^{(e)} - \psi_x^{(e)} \, \phi^{(i)}] n_x ds$$

$$= 0, \qquad \text{for all } \psi^{(e)}. \tag{4.12}$$

To obtain a better understanding of this condition, we restrict our attention to the case of two-dimensional flow, but what follows can be generalized to three dimensions without much difficulty. First, we observe that in the exterior region(s), a complete and orthonormal set of eigenfunctions $\psi_p^{(e)}$, $p=1,2,\ldots$ exists with the following property:

$$<\psi_p^{(e)}, \psi_q^{(e)}> \equiv \int_{-h}^{0} \psi_p^{(e)} \psi_q^{(e)} \, dy - \kappa^{-1} \psi_p^{(e)} \psi_q^{(e)} \Big|_{y=0}$$

$$= \delta_{pq}, \qquad p, q \geq 1, \tag{4.13}$$

where δ_{pq} is the Kronecker delta. If $\phi^{(e)}$ is expressed as an eigenfunction expansion of the form:

$$\phi^{(e)} = \sum_{j=1}^{\infty} \alpha_j \psi_j^{(e)}, \tag{4.14}$$

where the α_j's are unknown coefficients, it is rather straightforward to show, using (4.13), that (4.12) is completely equivalent to the condition

$$\alpha_j = <\psi_j^{(e)}, \phi^{(i)}\big|_{\Sigma}>, \qquad j = 1, 2, \ldots. \tag{4.15}$$

In simple terms, (4.15) or (4.12) states that the exterior potential has been expanded in a series of orthogonal eigenfunctions, using the interior potential at Σ as a Dirichlet condition. The determination of $\phi^{(i)}$ on Σ can be completed by noting that the space of $\psi^{(i)}$ in (4.11) is identical with that of $\phi^{(i)}$, and $\phi_n^{(e)}$ (as well as $\phi_x^{(e)}$) is expressible in terms of the α_j's. A more explicit illustration of this methodology of matching can be found in Yeung (1981). The radiation condition in this problem can be imposed as follows: the absence of waves in the upstream exterior region is enforced by requiring the particular coefficient associated with the wave mode to vanish; this yields an extra condition in the form of (4.15) that compensates for the *apparent* indeterminancy of the phase angle of the downstream waves at Σ. An analogous situation arose in the hybrid integral-equation method of Yeung & Bouger (1979). The Galerkin formulation of (4.11) and (4.12) has been used by Bai (1977, 1978) to calculate flow about two-dimensional bodies under a free surface and a steadily moving ship in a canal. With this method, Bai (1979) also derived an approximate formula that corrects for blockage effects of ship models in a towing-tank experiment.

A rather different sort of approach for this very same problem was used by Yamamoto & Kagemoto (1976). Instead of starting with the linearized boundary-value problem, they made use of the functional (4.3) for the interior region Ω^i with the assumptions that V_n and η were known on Σ, the former being satisfied as a natural condition, the latter as an essential condition. If the contributions from the free-surface integral in (4.3) were expanded about $\mathscr{F}_0$ (i.e. $y = 0$), and only the leading-order terms (ϕ, η assumed small) were kept, the following functional was obtained:

$$J^{(i)}(\phi^{(i)}, \eta^{(i)}) = \int_{\Omega^i} (\tfrac{1}{2}|\nabla\phi^{(i)}|^2 - U\phi_x^{(i)})d\Omega + \int_{\Sigma} (Un_x - \phi_n^{(e)})\phi^{(i)}\,d\partial\Omega$$
$$+ \int_{\mathscr{F}_0} \left(\frac{g}{2}\eta^{(i)2} - U\phi_x^{(i)}\eta^{(i)}\right)d\partial\Omega. \qquad (4.16)$$

The first variation of (4.16) with respect to $\phi^{(i)}$ and $\eta^{(i)}$ can now be taken and manipulated to yield:

$$\delta J^{(i)} = -\int_{\Omega^i} \nabla^2\phi^{(i)}\delta\phi^{(i)}\,d\Omega + \int_{\mathscr{B}\cup\mathscr{H}} (\phi_n^{(i)} - Un_x)\,\delta\phi^{(i)}\,d\partial\Omega$$
$$+ \int_{\Sigma} (\phi_n^{(i)} - \phi_n^{(e)})\delta\phi^{(i)}\,d\partial\Omega$$
$$+ \int_{\mathscr{F}_0} [(\phi_y^{(i)} + U\eta_x^{(i)})\delta\phi^{(i)} + (g\eta^{(i)} - U\phi_x^{(i)})\delta\eta^{(i)}]d\partial\Omega$$
$$+ \oint_{\mathscr{F}_0\cap(\Sigma\cup\mathscr{H})} (Un_x\eta^{(i)})\delta\phi^{(i)}\,ds. \qquad (4.17)$$

Thus, it is clear that all boundary conditions of the problem are satisfied naturally, except for the last term. Yamamoto & Kagemoto pointed out that this "inconsistent" term was caused by the linearization process, which makes mass conservation impossible along the intersection contours $\mathcal{F}_0 \cap (\Sigma \cup \mathcal{H})$. By examining the nonlinear terms discarded, they argued the necessity of introducing an auxilliary condition along these contours to supplement (4.17). The final form used by Yamamoto & Kagemoto is a Galerkin-type equation given by

$$\delta J^{(i)} - \oint_{\mathcal{F}_0 \cap \mathcal{H}} \phi_n^{(i)} \, \eta^{(i)} \, \delta \phi^{(i)} \, ds - \oint_{\mathcal{F}_0 \cap \Sigma} U n_x \, \eta^{(e)} \, \delta \phi^{(i)} \, ds = 0 , \quad (4.18)$$

where the additional integrals of (4.18) can be seen to combine with the last term of (4.17) to yield mass conservation. No details on the treatment of the second term of (4.18), which is nonlinear, were given. In the exterior region, a solution representation based on (4.14) was used by these authors. The matching of the interior and exterior potentials across Σ was accomplished by a collocation method on Σ. The authors reported some results for the two-dimensional flow about a vertical plate piercing the free surface. Their formulation appears to generate a wave profile of equal heights on both surfaces of the plate. No results were available to allow a direct comparison with those based on Bai's procedure.

The linearized steady flow about a submerged two-dimensional body was solved also by Mei & Chen (1976) using hybrid elements. The formulation was one that utilized the variational principles for the time-harmonic problems described below. Fictitious radiation and scattering problems based on the free-surface condition (2.14) were introduced. The unknown "incident-wave" amplitude of the scattering problem was determined so that the upstream waves vanished completely when combined with the radiation problem. Mei & Chen have also outlined how the procedure can be generalized to three-dimensional problems.

For the class of *time-harmonic* problems defined by (2.3), (2.19), and (2.20), a variety of hybrid techniques based on the Galerkin procedure have been successfully implemented. The usual approach is to construct a functional of the weak form for the interior potential. For this purpose, it is customary to consider $\varphi_n^{(i)}$ as prescribed (and given by $\varphi_n^{(e)}$) on Σ. This will match the normal derivatives and couple the interior problem with the unknown parameters of the exterior representations, which could be in the form of a source distribution (Berkhoff 1972), or an eigenfunction expansion (Chenot 1975). The additional conditions for the determination of the exterior parameters can then be obtained by equating the exterior and interior potential at Σ, with a criterion based on either collocation, least square, or Galerkin. For the last case, if an orthonormal expansion exists in Ω^e, (4.15) results.

An alternative approach based on Galerkin's procedure has been adopted by Lenoir & Jami (1978), who considered φ on Σ as an essential condition, and related this potential to that on $\mathcal{H}$ by using the Green function of the problem (see Section 5), which, by definition, satisfies the radiation condition as required. The idea is originally due to McDonald & Wexler (1972) and has the advantage that if Σ is sufficiently far from $\mathcal{H}$, the singular terms of the Green functions are more amenable to numerical treatment. This coupling condition between Σ and $\mathcal{H}$, however, brought back the unpleasant feature of irregular frequencies associated with the use of Green functions (see section 5 for details). Further, the choice of such a coupling condition is not unique. One can presumably relate the normal derivative on Σ to that on $\mathcal{H}$, instead of the potential, or in fact any linear combination of them.

Variational principles specifically applicable to a *hybrid formulation* have been obtained for time-harmonic problems by Bai & Yeung (1974) and Chen & Mei (1974). These have been applied successfully to a variety of two- and three-dimensional problems of water-wave radiation and diffraction (see Mei 1978 for more references). The aforementioned principles can be shown to be special cases of the following:

$$J(\varphi^{(i)}, \varphi^{(e)}) = \tfrac{1}{2}\int_\Omega (\nabla\varphi^{(i)})^2 d\Omega - \frac{\omega^2}{2g}\int_{\mathcal{F}_0} \varphi^{(i)2}\, d\partial\Omega - \int_{\mathcal{H}} v_n\varphi^{(i)}\, d\partial\Omega$$
$$- \int_\Sigma [\varphi_n^{(e)}(\varphi^{(i)} - \tfrac{1}{2}\varphi^{(e)}) + k(\varphi^{(i)} - \varphi^{(e)})(\varphi_n^{(i)} - \varphi_n^{(e)})] d\partial\Omega ,$$

(4.19)

where k could be any arbitrary value in $[0,1]$ and $\varphi^{(e)}$ is assumed to satisfy all conditions in Ω^e. In spite of the apparently non-unique form of J, the extremum is always given by the third integral. The variation of (4.19) can be easily shown to yield all boundary conditions, including the matching of φ and φ_n, naturally. In particular, the first and last term can be combined to yield the following Galerkin form:

$$\int_\Sigma |(\varphi_n^{(i)} - \varphi_n^{(e)})[\delta\varphi^{(i)}(1-k) + \delta\varphi^{(e)}k]$$
$$- (\varphi^{(i)} - \varphi^{(e)})[\delta\varphi_n^{(i)}k + \delta\varphi_n^{(e)}(1-k)]| d\partial\Omega ,$$

(4.20)

the interpretation of which in the context of weighted residuals is rather obvious. The computational capabilities of the type of hybrid methods discussed here is illustrated by Figure 3. Taken from Yue et al. (1978), the figure shows the (time-) complex free-surface amplitude around an elliptic island on a circular base due to unit-amplitude waves at two different angles of incidence. The formulation used is one related to (4.19) with $k = 0$.

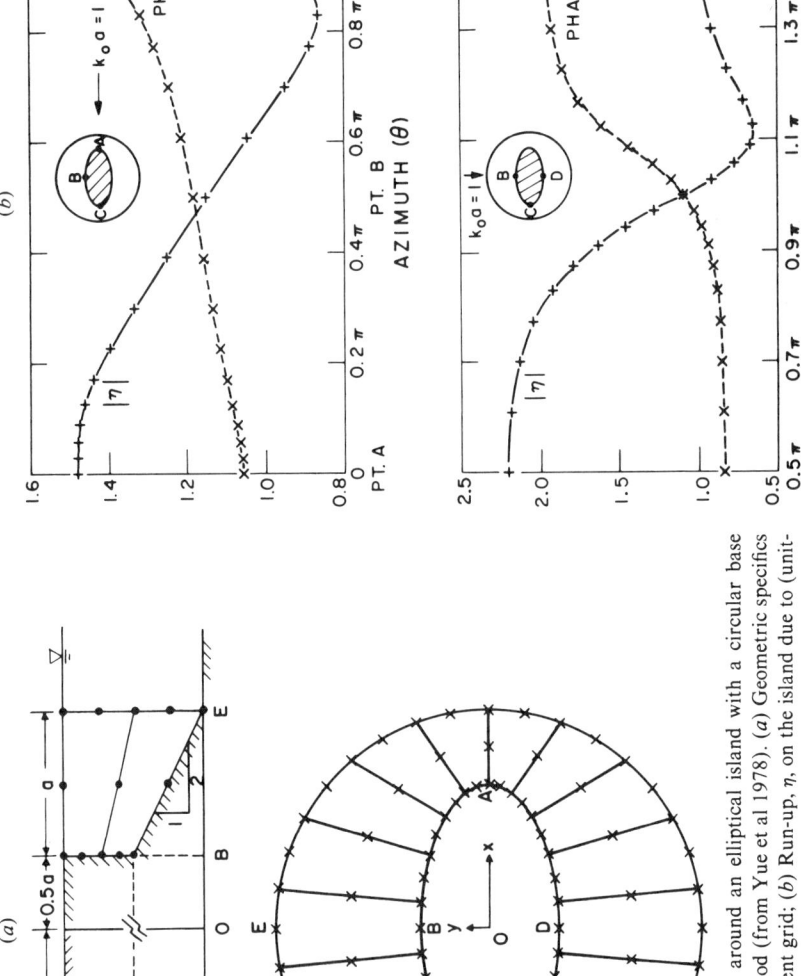

Figure 3 Wave diffraction around an elliptical island with a circular base using a hybrid-element method (from Yue et al 1978). (*a*) Geometric specifics of the island and finite-element grid; (*b*) Run-up, η, on the island due to (unit-amplitude) waves at two different angles of incidence, k_0 being the wave number.

The use of special boundary elements for tackling time-dependent domains has received considerable attention lately in finite-element methods (Banerjee & Butterfield 1979). For unsteady water-wave problems, Wellford & Ganaba (1980) have implemented a pseudo-variational principle due to Pian & Tong (1968).

$$I(\phi, V_n) = \frac{1}{2}\int_\Omega |\nabla\phi|^2 d\Omega - \int_\mathcal{F} V_n (\phi - \tilde{\phi}) d\partial\Omega + \int_{\mathcal{H} \cup \mathcal{B}} \phi V_n \, d\partial\Omega,$$
(4.23)

where $\tilde{\phi}$ and V_n on $\mathcal{F}$ is the potential and its normal derivative respectively. Equation (4.23) can be used to determine V_n on $\mathcal{F}$ at successive time steps if $\mathcal{F}$ and $\tilde{\phi}$ are known. Starting with known initial values, Wellford & Ganaba advanced $\mathcal{F}$ and $\tilde{\phi}$ in time by an explicit scheme similar to (3.5a), which gave rise to numerical instabilities. Artificial damping was introduced as an attempt to eliminate unwarranted oscillations. The approach used here appears promising but the method needs further refinements. Licht (1980) considered the simpler unsteady problem with the linearized free-surface condition. A fully implicit scheme is used to advance ϕ and η, with (3.11) being used as the open-boundary condition. He experimented with a number of constant values of c to compare his numerical results with existing ones, but offered no insight on the choice of c for a general problem.

5. BOUNDARY-INTEGRAL EQUATION METHODS

The treatment of problems in potential theory by integral-equation methods is classical. The use of "single-layer" and "double-layer" distributions to solve problems of Neumann and Dirichlet type, respectively, are well known from texts such as Kellogg (1929). These formulations lead to Fredholm integral equations of the second kind, for which a rather complete mathematical theory of existence and uniqueness has been developed (see Mikhlin 1957). Less seems to be known about integral equations of the first kind, but it is generally accepted that a well-posed problem will lead to physically plausible numerical solutions. A distinct advantage of an integral-equation formulation over the space-discretization formulations of Sections 3 and 4 is that the space dimension of the problem is reduced by one. Of course, some "fundamental" solution must first be obtained. Free-surface flows seem particularly suitable for integral-equation treatments since physical quantities of primary interest, such as wave height and fluid pressure, are required only on the boundaries. Space-discretization techniques would appear inefficient and wasteful, since they yield massive amounts of interior data that are normally of minor use.

Integral equations for linearized free-surface flows in the presence of bodies were available in the 1950s from the works of John (1950) and Stoker (1957). Very little progress was made towards solving them, except for the few cases of geometry that were described by special coordinates. The works of Hess & Smith (1964, 1967) and their collaborators marked the beginning of a new era of computation. With the advent of the high-speed computer and of the discretization techniques developed by these authors, calculations of flows about arbitrarily shaped bodies in an infinite fluid became possible. The successes in aerodynamics were not matched at an equal pace by those in free-surface hydrodynamics. The delay was associated with the need to develop accurate and efficient means of evaluating and integrating (in space) the complicated Green functions associated with the free-surface conditions. Earlier works were often marred by uncertainties in numerical accuracy, particularly those in three dimensions.

A small departure from tradition was made by Yeung in 1973 (see Bai & Yeung 1974), who pointed out that one could obtain the "usual" wavelike solution by a direct application of Green's theorem, using just the infinite-fluid source. This has, in fact, opened up a new avenue of formulation in body-wave problems. To distinguish the use of such "simple-source" functions from traditional Green functions that satisfy the linearized free-surface conditions and the radiation condition, we shall use the term *Green function* to mean specifically the latter. Literature in both areas is growing. The power of integral-equation methods in treating nonlinear problems has been well demonstrated by Longuet-Higgins & Cokelet (1976). These authors developed a time-stepping algorithm, based on a Lagrangian description of the free-surface fluid particles, for studying the evolution of the profiles of breaking waves. Inverse methods, based on reversing the roles of dependent and independent variables, have recently been found to be quite effective in treating a special class of nonlinear two-dimensional problems, where the body is part of the fluid boundary.

Are boundary-integral equation methods more efficient than space-discretization methods? We shall provide a guideline for answering this nontrivial, but recurrent, question. Let us assume that the major phases of the computational efforts consist of 1. the generagion of a matrix associated with the discretization and 2. the direct "inversion" of this matrix. Suppose that K and N are the total number of unknowns corresponding to integral-equation and space discretization respectively, and that m is the bandwidth associated with the latter method (m being K for the former, which gives rise to a full matrix). Suppose further that β (and γ) is the ratio of the time required to generate a typical matrix coefficient to that required in an inversion step for the integral-equation (and space-discretization) method. Then, clearly, we could argue that the former is superior if

$$K^2(K+3\beta) \leq 3Nm(m+\gamma),$$

where the first term on each side represents the effort involved in the inversion process, and the second in the generation process. Noting that m is $O(N^{1/2})$ and $O(N^{2/3})$ for two- and three-dimensional problems respectively, we obtain

$$(K+3\beta) \leq 3e^2 \begin{cases} 1+\gamma N^{-1/2}, & \text{for two dimensions}, \quad (5.2a) \\ N^{1/3}+\gamma N^{-1/3}, & \text{for three dimensions}, \quad (5.2b) \end{cases}$$

where $e \equiv N/K$. Equation (5.2) says that for small K, corresponding to the situation of using Green functions, integral-equation methods are superior if the coefficient generation index β is less than $O(e^2)$ in two dimensions and $O(e^2 N^{1/3})$ in three dimensions. This is certainly true in the two-dimensional case. In three dimensions, an inevitable increase in β, together with a moderate decrease in e, actually makes the various methods more competitive. In the other limit where K is $O(N^{1/2})$ and $O(N^{2/3})$, in two and three dimensions, integral-equation formulation is always a viable alternative when β is $O(K^2)$ and $O(K^{3/2})$. This simplified analysis does not, of course, account for a host of other factors, such as the use of relaxation iterative techniques, the non-constant value of β within the matrix, the effective use of a hybrid formulation to minimize the computational domain, and the manner in which the double condition on the free surface can be satisfied. Nevertheless, (5.2) should provide a rational basis for making comparisons among various methods. For example, one can then speak of the typical β value in an integral-equation formulation, or the different γ values for various space-discretization methods. We emphasize that boundary-integral methods are always superior in data-storage requirement. Space-discretization techniques saturate the rapid-access memory very quickly; peripheral data storage and transfer are costly in time and effort.

Methods Based on Green's Functions

The method of Green's functions is quite powerful in linearized problems whenever a Green function satisfying the appropriate linearized free-surface conditions and other auxilliary conditions can be derived. A rather complete collection of them is available from Wehausen & Laitone (1960). Integral equations for the velocity potential can generally be derived from Green's third identity:

$$\frac{1}{2}\phi(P) = \int_{\partial\Omega} \phi(Q)G(P,Q)\, d\partial\Omega_Q - \int_{\partial\Omega} \phi G_\nu\, d\partial\Omega_Q, \quad P \in \partial\Omega, \quad (5.3)$$

where Q is the "source" point, and P the field point, assumed to approach $\partial\Omega$ from the interior of Ω. Here, the subscript ν denotes the operator $\mathbf{n}(Q)\cdot\nabla$. The Green function is assumed to be of the form $[r^{-1}+H(P,Q)]/4\pi$ in three dimensions, $[-\log r +H]/2\pi$ in two dimensions, with r being $|PQ|$ and H a harmonic function chosen to satisfy the appropriate boundary conditions of the problem. As an example, for time-harmonic problems, if H is constructed so that G satisfies (2.20), (2.19) on $\mathcal{B}$, and the radiation condition (2.21), the boundary $\partial\Omega$ in (5.3) can be shown to reduce to $\mathcal{H}$ only.

A variety of integral equations exist in the literature of body-wave hydrodynamics. These variants are all related to the assumptions made about the behavior of the potential $\bar{\phi}$ inside the body. We shall denote this interior domain by $\bar{\Omega}$; it "intersects" with $\mathcal{F}$ if $\mathcal{H}$ is surface-piercing. For illustration, consider the *time-harmonic* problems just mentioned. Applications of (5.3) in Ω and $\bar{\Omega}$, with P in Ω, can be combined to yield

$$\frac{1}{2}\begin{pmatrix}\varphi(P)\\ \bar{\varphi}(P)\end{pmatrix} = \int_{\mathcal{H}}(\varphi_\nu-\bar{\varphi})G\,d\partial\Omega - \int_{\mathcal{H}}(\varphi-\bar{\varphi})G_\nu d\partial\Omega\,,\quad\begin{cases}P\in\partial\Omega\\ P\in\partial\bar{\Omega}\end{cases}. \quad (5.4)$$

Since the interior potential has no physical meaning, one could choose $\bar{\varphi}$ identically zero in $\bar{\Omega}$, hence also $\bar{\varphi}_\nu$. This choice yields the so-called Green's mixed distribution, or, equivalently,

$$\frac{1}{2}\varphi(P) + \int_{\mathcal{H}}\varphi G_\nu d\partial\Omega = \int_{\mathcal{H}} v_n G\,d\partial\Omega\,,\quad P\in\partial\Omega\,, \quad (5.5)$$

which is a Fredholm integral equation of the second kind. Alternatively, with $\bar{\varphi} = \varphi$ a source distribution of strength $\sigma = \varphi_\nu - \bar{\varphi}_\nu$ results, whereas with $\bar{\varphi}_\nu = \varphi_\nu$, a normal-dipole distribution of strength $\mu = \bar{\phi} - \phi$ follows. If the Neumann condition on $\mathcal{H}$ is imposed, the integral equation for σ is

$$\frac{1}{2}\sigma(P) + \int_{\mathcal{H}}\sigma G_n\,d\partial\Omega = v_n(P)\,,\quad P\in\partial\Omega\,, \quad (5.6)$$

whose kernel is related to that of the mixed distribution by a similarity transformation. Equation (5.6) is the version "preferred" by most workers (Frank 1967, Lebreton & Margnac 1968, Garrison & Rao 1971, Faltinsen & Michelsen 1974, and others), probably because of the influence from earlier aerodynamic literature. Fundamentally, (5.5) and (5.6) involve the same amount of computational effort (G and G_n or G_ν), since φ, being proportional to the pressure, is the desired quantity. Actually, (5.5) has a slight advantage over (5.6), and we shall address that momentarily. The mixed distribution has been used by Potash (1971) and Macaskill (1977); the latter treated the two-dimensional problem of unequal depths by matching at a common boundary. A representative work using dipole distribution is that of Chang & Pien (1975), who noted that (5.5) yields the following integral equation of the first kind for μ, in terms of the interior potential $\bar{\varphi}$:

$$\int_{\mathcal{H}} \mu G_\nu \, d\partial\Omega = \frac{1}{2} \bar{\varphi}(P), \tag{5.7}$$

where $\bar{\varphi}(P) = Uy$ for $\varphi_\nu = \bar{\varphi}_\nu = n_y$, and so on for other rigid-body motions. For an arbitrary φ_ν this procedure breaks down since $\bar{\varphi}$ is unknown. Note, however, in the case of rigid-body motion, $\varphi(P)$ ($= \bar{\varphi} - \mu$) is known immediately once μ is obtained by solving (5.7). Thus, in this sense, the dipole-distribution method can be 50% more efficient than the other two methods. Being dominant *off* the diagonal, the matrix equations of (5.7) are much less amenable to iterative solutions than those of (5.5) and (5.6). An integral equation for μ can also be obtained by taking the normal derivative of (5.4), but this results in second derivatives of the Green function, which compounds the already difficult task of evaluating a highly oscillatory Green function.

The integral equations (5.5–5.7) are all plagued by the presence of irregular frequencies when $\mathcal{H}$ is surface-piercing. This was first pointed out by John (1950) in the context of source distributions for floating-body problems. More recent references related to similar difficulties in accoustic radiation may be found in Jones (1974) and Mei (1978). Here, it suffices to mention that the difficulty is associated with the vanishing of the Fredholm determinant of (5.5–5.7). We recall that a representation based on either a mixed or source distribution imposes a Dirichlet condition on $\partial\bar{\Omega}$, and that the dipole distribution imposes a Neumann condition on $\partial\bar{\Omega}$. If the homogeneous equations associated with these interior problems have either unbounded or nontrivial solution, which happens at a discrete set of "resonant frequencies" in a closed "basin," these representations break down. Note that (5.5) and (5.6) have the same irregular frequencies since the kernel of one is the "transpose" of the other. Further, a subtle distinction between the mixed and source representation is that the inhomogeneous term of the former is orthogonal to the interior eigensolutions at these frequencies, whereas that of the latter is generally not. By the Fredholm alternatives (see, for example, Delves & Walsh 1974), Equation (5.5) actually has a solution, though not unique, whereas (5.6) has none. Neither situation is entirely desirable, but it has been observed that in actual numerical computations (Adachi & Ohmatsu 1979), the mixed distribution yields less "perturbed" results than the source one. After all, it is difficult to "hit" the irregular frequencies precisely, since they are unknown a priori.

Various means of circumventing this difficulty exist. Overspecifying the interior problem by putting a "lid" on $\mathcal{F}$, is one effective alternative (Kobus

1976). Modifying the kernel function G by adding other concentrated singularities in $\overline{\Omega}$ (Ogilvie & Shin 1978, Sayer & Ursell 1977) is another. Recent work of Ursell (1980a) shows that the choice of the coefficient multiplying the concentrated singularities in the modified G is not entirely arbitrary.

For linearized *steady-motion problems*, the integral equation analogous to (5.4) is

$$\frac{1}{2}\begin{pmatrix}\phi(P)\\ \overline{\phi}(P)\end{pmatrix} = \int_{\mathcal{H}} [(\phi_\nu - \overline{\phi}_\nu)G - (\phi - \overline{\phi})G_\nu]d\partial\Omega$$

$$+ \kappa^{-1}\oint_{\mathcal{H} \cap \mathcal{F}_0} [(\phi - \overline{\phi})G_x - (\phi_x - \overline{\phi}_x)G]n_x \, ds \,, \qquad (5.8)$$

where G is assumed to satisfy (2.3) on $\mathcal{B}$, (2.14), and the radiation condition (2.12). The additional term, known commonly as the *line integral* in the Neumann-Kelvin problem of ship hydrodynamics, is associated with an integration by parts of (2.14). Note now that the requirement $\overline{\phi}_\nu = \phi_\nu$ does not result in purely dipole terms, since $\overline{\phi}_x = \phi_x$ is not subsequently implied. On the other hand, a representation consisting purely of sources (surface and "line") is possible. Numerous attempts have been made to solve (5.8), but none made to investigate the possible existence of irregular frequencies. A common misconception has been that the line integral is highly singular. Actually, when the "source" point of G is on $\mathcal{F}_0$, the Green function is merely a weakly singular pressure point. Furthermore, such a singularity may not even exist because of possible cancellation with the edge behavior of the surface distribution. Practical implications or logical inconsistency put aside, it seems that the Neumann-Kelvin problem deserves more thorough attention on the analytical side, especially in view of the amount of effort invested in solving it. Ursell (1980b) examined the two-dimensional problem in some detail and concluded that a "least singular solution" can exist if the source density at the edge of the "surface" is the same as that which arises from the intersection point $\mathcal{H} \cap \mathcal{F}_0$. The implication of this in three dimensions is not immediately obvious, because the normal derivatives and the longitudinal derivatives are not identical, as in the case of two dimensions. However, it is precisely this kind of analysis that would elucidate the exact mathematical difficulty of such a formulation. The works of Guével et al. (1977) and Tsutsumi (1979) are, perhaps, representative of the state of the art of such numerical endeavors. Paradoxically, no two independent works exist, regardless of the numerical methods, that are in agreement with each other.

The "difficulty" associated with the line integral disappears if $\mathcal{H}$ does not intersect with $\mathcal{F}_0$ or if $\mathcal{H}$ is taken as a vertical plate on the centerplane where a linearized body condition is applied. In the latter case, (5.8) can be easily shown to yield

$$\phi(P) = 2 \int_{\mathcal{H}} V_n(Q) G(P, Q) \, dx dz(Q) , \qquad (5.8a)$$

which formed the basis of the well-known work of Michell (1898) on the wave-resistance of thin bodies.

An integral-equation formulation for *time-dependent* problems involving the condition (2.18) was developed only relatively recently (Finkelstein 1957). While (5.3) is still the basis of the formulation, the extra parameter (time) leads to a memory integral, the computational effort of which is quite extensive. Aside from (2.18), the unsteady Green function satisfies two homogeneous initial conditions on the free surface. With the assumptions that the fluid disturbances were initially zero, it can be shown (Yeung 1982) that the integral equation for $\phi(P, t)$ is given by

$$\frac{1}{2}\phi(P, t) - \int_{\mathcal{H}} \phi(Q, t) G_\nu^\circ (P, Q) \, d\partial\Omega_Q =$$

$$- \int_{\mathcal{H}} V_n(Q, t) G^\circ (P, Q) \, d\partial\Omega_Q \qquad (5.9)$$

$$+ \int_0^t d\tau \int_{\mathcal{H}} [\phi(Q, \tau) H_{\tau\nu}(P, Q, t-\tau) - V_n(Q, \tau) H_\tau] d\partial\Omega ,$$

for $P \in \partial\Omega$,

where $G^\circ = 4\pi(1/r - 1/r')$, for three dimensions, with r' being the distances from P to the image point of Q about $y=0$, and a similar modification of G applies also to two dimensions. The unsteady-wave effects are all imbedded in the harmonic function H_τ, which has the property $H_\tau(P,Q,0) = 0$. Thus, the "memory effects" occur only on the right-hand side of the integral equation, but (5.9) needs to be solved at each t. Yeung discussed also the physical meaning of some of the initial conditions that were not quite clearly stated in Finkelstein's original work. Figure 4 shows the transient heave motion of a freely floating cylinder with unit initial displacement. Since $V_n(t)$ is unknown, it must be determined simultaneously from the rigid-body dynamics equation. The fine agreement with experiment seen here is consistent with the fact that hydrodynamic forces in the frequency domain have been well predicted, in the past, by time-harmonic linear theory. An existence and uniqueness theorem for such transient-motion problems has been given by Beale (1977).

Earlier work of Daoud (1975) considered a related formulation for an expanding wedge, as well as the difficulties involved in evaluating the wave kernel. R. B. Chapman (1979) considered the same problem as Yeung, but represented the fluid motion by a discrete set of wave harmonics. The method used to advance the wave motion was quite efficient, but the wave number and frequency distribution must be carefully chosen to represent a non-reflecting exterior medium.

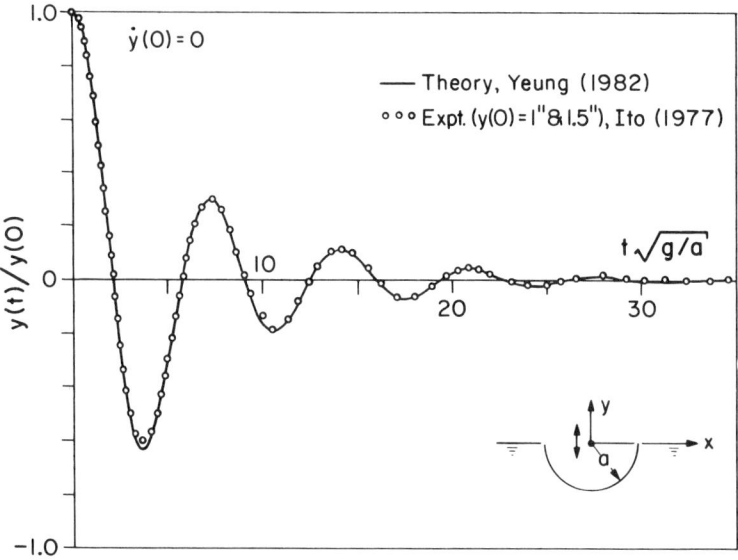

Figure 4 Time history of the transient response of a freely-floating semicircular cylinder with given initial displacement (from Yeung 1982); comparison between linearized theory and experiments.

Simple-Source Formulations

If the wave function H in (5.3) is taken simply as zero, the following relation results in two dimensions:

$$\pi\phi(P) = \int_{\partial\Omega} [\phi_\nu(Q)\log\frac{1}{r}(P,Q) - \phi\frac{\partial}{\partial\nu}\log\frac{1}{r}]d\partial\Omega_Q, \quad P \in \partial\Omega,$$

(5.10)

which represents a distribution of simple (free-space) sources and dipoles on the entire boundary. Alternatively, if we introduce the complex potential $W(Z) = \phi+i\psi$, with $Z = x+iy$, Cauchy's integral formula yields

$$\pi i W(Z) = \oint_{\partial\Omega} \frac{W(Z')}{Z'-Z} dZ', \quad Z', Z \in \partial\Omega,$$

(5.11)

where Z is the field point and the integral $\oint$ is to be interpreted in the sense of Cauchy principal values. Equation (5.11) is completely equivalent to stating that ϕ and ψ each satisfies (5.10). This can be shown easily by letting $Z'-Z = re^{i\theta}$, θ being a polar angle, and by making use of the Cauchy-Riemann relations. Equation (5.11) is, of course, not applicable in three dimensions whereas (5.10) holds with $\log r^{-1}$ replaced by $\frac{1}{2} r^{-1}$.

The simple relation (5.10) is the basis of the hybrid methods developed by Yeung (Bai & Yeung 1974, Yeung 1975) for linearized free-surface flows. Actually, Yeung treated the exterior domain as if it were interior, but with an appropriate matching of the interior and exterior solutions across the open boundary Σ. Consider the time-harmonic problem; then $\varphi_\nu^{(i)}$ is either known because of (2.19) or expressible in terms of $\varphi^{(i)}$ because of (2.20) on $\mathcal{B} \cap \mathcal{H} \cap \mathcal{F}_0$. Thus (5.10) is a Fredholm integral equation of the second kind on those boundaries. If the exterior potential $\varphi^{(e)}$ is given in terms of an eigenfunction series of the form (4.14), a direct matching of $\varphi^{(e)}$ with $\varphi^{(i)}$ as well as their normal derivatives on a vertical boundary Σ yields

$$\int_\Sigma [\varphi^{(i)} \frac{\partial}{\partial \nu} \log r - \varphi_\nu^{(i)} \log r] d\partial\Omega$$

$$= \sum_{j=1}^\infty \alpha_j \int_\Sigma [\psi_j^{(e)} \frac{\partial}{\partial \nu} \log r - \psi_{jx}^{(e)} \log r] d\partial\Omega ,$$

(5.12)

which occurs on the right-hand side of (5.10). The resulting integral over the vertical boundary is relatively easy to evaluate. A similar substitution of $\varphi^{(i)}$ on the left-hand side of (5.10) leads to an identity that can be used to determine the coefficients α_j by collocation on Σ. The first term in the eigenfunction series corresponds to a propagating-wave solution. Note that if only one term is taken in (5.12), it is effectively the same as applying the radiation condition (2.21). In its most general form, (5.12) accounts for all nonpropagative disturbances.

This *hybrid* integral-equation formulation was applied to both two- and three-dimensional problems by Yeung. It is free from irregular frequencies. It can handle bottoms of unequal asymptotic depths in two dimensions. It is also obvious for the case of constant depth that the integral over $\mathcal{B}$ is completely eliminated (see Bai & Yeung 1974) if $\log r$ in (5.10) is simply replaced by $\log rr''$, where r'' is the distance from P to the image point of Q about $y = -h$. The disadvantage of this method, when compared with the traditional Green-function formulation, is that K in (5.2) is now considerably larger, but this is offset by a much smaller β. Yeung & Bouger (1979) also demonstrated that this method is equally applicable to the linearized steady-motion problems defined by (2.3), (2.14), and (2.12). The boundary condition (2.14), however, requires a higher-order interpolation function than before. In this work, the steady-state radiation condition was rationally treated by simply omitting the wave-like eigenfunction upstream and keeping it downstream.

Berhault (1980) considered obtaining the solution of the integral equation of Yeung by a variational formulation. His procedure yields a symmetric matrix but requires a double integration of the source function. Harten & Efrony (1978) proposed a further subdivision of $\Omega^{(i)}$, with similar matchings as described, to generate a block structure in the coefficient matrix. Efficiency could be improved, they reported, by an order of magnitude. Domain subdividing formulated with (5.10) in mind appears to combine the best features of boundary-integral methods (simplicity) and space-discretization techniques (banded matrix). Earlier, Harten (1975) extended this simple-source formulation to tackle time-dependent linearized flows in an *enclosed* domain. Such interior problems, of course, do not have the difficulty associated with open boundaries. Soh (1976, 1980) considered a vortex-sheet representation of the linearized unsteady problem by using (5.11) with $W(Z)$ replaced by the complex velocity $(u-iv)$. In terms of these velocities, (2.17) implies

$$\eta_t(x, t) = v(x, 0), \qquad u_t(x, t) = -g\eta_x, \qquad (5.13a,b)$$

which permits the advancement of η and u at each value of t, with v related to u by (5.11). Soh used an explicit scheme for (5.13), which appeared to cause a steady increase in phase error. Quantities outside of the truncation boundary were treated by extrapolation. A related approach had been proposed earlier by Zaroodny & Greenberg (1973), but the computations presented were too crude for drawing any conclusions. The simple-source formulation was used in a similar manner by Mattioli (1978) for solving the two-dimensional shallow-water equations.

The most successful calculations of unsteady two-dimensional waves were carried out by Longuet-Higgins & Cokelet (1976), who used the integral-equation formulation (5.10) for ϕ, the dynamic free-surface condition (2.4b) (with $U = 0$), and a Lagrangian description of the free-surface fluid particles:

$$\frac{D\mathbf{x}}{Dt} = \nabla \phi. \qquad (5.14)$$

We note in passing that the vertical component of (5.14) is basically (2.4a). The periodic behavior of the wave system in x was accounted for by using the following transformation (Nekrasov 1921): $\zeta = e^{ikZ}$, where ζ is the complex variable in the mapped plane, and k the wave number associated with the periodicity. While the use of (5.10) is not new, their time-stepping procedure is quite interesting. With the wave profile and the potential on $\mathcal{F}$ given initially by a procedure such as Schwartz's (1974) for steady progressive waves, ϕ_n for a number of "marker particles" on $\mathcal{F}$ can be obtained by solving (5.10). This is an integral equation of the first kind. The Lagrangian values of ϕ and $\mathbf{x}$ of these marker points are next advanced by using (2.4b) and (5.14), since their right-hand sides, which involve velocities, are now known. Actually, these authors used a fourth-order predictor-corrector method, which required another solution of (5.11) at the corrector stage, using the predictor value. With the new values of ϕ and new location of $\mathcal{F}$, the cycle can be repeated. The procedure is remarkably simple, for a complicated nonlinear problem as such. The authors exercised care in attaining a consistent order of accuracy in all of their intermediate calculations. Figure 5 shows the development of a plunging breaker when the authors apply a surface pressure [p_0 in (2.4b)] for half a period. Since a Lagrangian description of the surface is used, the calculations can proceed without any difficulty even after the wave front becomes vertical. Note that the straining motion of the primary waves near the wave crests causes the marker particles to "bunch together," thus providing the very desirable (but unexpected) effect of improving resolution near a sharp-curvature region. The authors pointed out that grid-scale oscillations occurred in their calculations, but conjectured that they could be physical in origin. Numerical smoothing was used to eliminate these unwarranted oscillations.

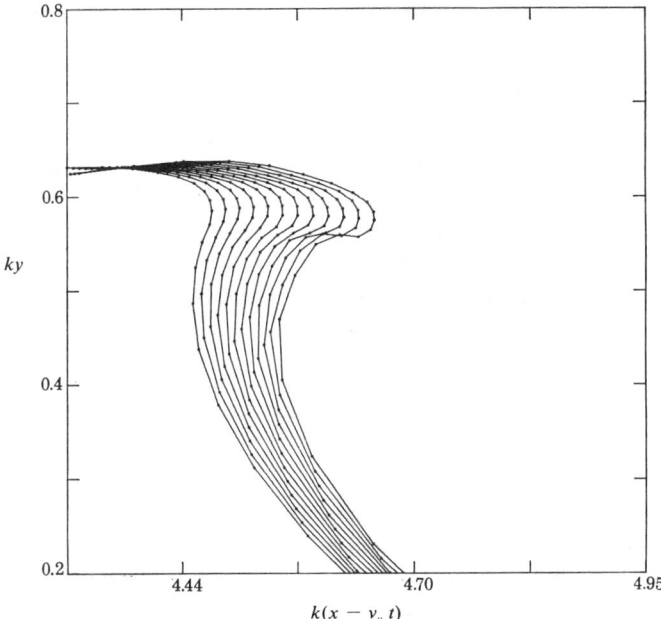

Figure 5 Theoretical wave profiles of a plunging wave breaker generated by applying a pressure distribution on a progressive wave train for one half of a wave period (from Longuet-Higgins & Cokelet 1976). The non-dimensional surface pressure used was $kp/\rho g = 0.146$. Profiles shown correspond to $4.712 < (kg)^{1/2}t < 4.928$, where k is the wave number. v_p is the phase velocity given by $(g/k)^{1/2}$. Marker particles on the free surface were initially distributed evenly along the wave length.

This Lagrangian-Eulerian integral-equation formulation was applied quite successfully by Fenton & Mills (1976) in shallow-water calculations. Faltinsen (1977) considered an extension of this method to the problem of an oscillating body in water of an infinite extent, but encountered much difficulty in eliminating the effects of the truncation boundary. Vinje & Brevig (1980) conducted a similar calculation by assuming that the body was periodic in space, but using (5.11) instead of (5.10). Their preliminary results appeared to be quite different from Faltinsen's, particularly for large-amplitude body motions.

Methods based on simple-source distributions have been used by Gadd (1976) and Dawson (1977) for three-dimensional *steady-motion* of ships. Of particular interest is Dawson's work which utilized the double-body approximation. Dawson assumed that $\phi^{(w)}$ in (2.16) can be represented by a simple source-sheet on $\mathcal{H} \cap \mathcal{F}_o$. This source strength on $\mathcal{H}$ was determined, as usual, by (2.3), but on $\mathcal{F}_o$ by (2.16), using a four-point upstream differencing scheme for the streamwise velocity. Actually, the last term of (2.16) was "neglected" by Dawson in his formulation. However, the computed wave resistance and wave profiles show definite improvement over traditional linear theory. Dawson claimed that upstream differencing was all that needed to satisfy the "radiation condition."

Inverse Formulations

The inverse formulation takes advantage of the fact that the free surface is a stream surface [see (2.9a)]. If the usual roles of the dependent (Φ, Ψ) and independent (x,y) variables are now interchanged, the dynamic free-surface condition (2.9b), though still nonlinear, can be satisfied on a *known* boundary. Stokes (1880b) noted this simplification in his work on steady periodic waves. In terms of the complex potential $W(=\Phi+i\Psi)$, this inverse representation for waves of phase velocity c and wavelength k is simply given by

$$Z(W) = -\frac{W}{c} + i \sum_{j=1}^{\infty} \alpha_j \left(e^{ikW/c} - e^{-2kh} e^{-ikW/c}\right), \quad \text{Im}(\alpha_j) = 0,$$

(5.15)

which yields the free-surface elevation when evaluated at $\Psi = 0$. In fact, considerable effort by many workers in the ensuing years has been devoted to the determination of the coefficient α_j's so that the dynamic condition can be satisfied. More notable recent work is by Thomas (1968), Schwartz (1974), and Cokelet (1977).

The inverse formulation has also been much used in jets and sluice flows (Larock 1970, Moayeri 1973). Conformal mapping is normally used in problems with physical boundaries to transform both the physical and free boundaries to constant Ψ lines. The resulting problem can be solved by

Hilbert-Riemann techniques or by direct numerical methods. The need for such a mapping is evident from the fact that stagnation points are now singular points in the inverse formulations. Thus, if ζ is the mapped complex variable, the dynamic condition (2.9b) now becomes

$$\left|\frac{d\zeta}{dW}\right|^{-2}\left|\frac{dZ}{d\zeta}\right|^{-2} - U^2 + 2g\,\text{Im}(Z) = 0\,, \tag{5.16}$$

which is more amenable to numerical treatment if $Z'(\zeta)$ is chosen to absorb the singular behavior. The type of geometry that can be handled is therefore rather restrictive. Inverse formulations for three-dimensional flows in a general context were considered by Jeppson (1972), but the resulting field equation was so complex that it was no longer amenable to boundary-integral equation treatment.

Using an inverse formulation, Vanden Broeck & Tuck (1977) considered the problem of steady flow about a semi-infinite body of finite draft. The authors employed the following transformation for a rectangular stern or bow profile

$$\frac{d\zeta}{dW} = -1 + \left(\frac{W}{W+1}\right)^{1/2}\frac{dZ}{dW}\,, \tag{5.17}$$

where all variables were so normalized that $Z = 0$ ($W = 0$) was the contact point between $\mathcal{F}$ and $\mathcal{H}$ and $Z = -id$ ($W = -1$) was the "keel" corner, d to be determined as part of the solution. The radical factor in (5.17) was chosen based on the behavior of the zero Froude-number solution at these points. If Cauchy's integral formula is applied to $\zeta'(W)$ in the complex W plane, it follows that

$$\zeta'(W) = \frac{-1}{2\pi i}\int_{-\infty}^{\infty}\frac{\zeta'(\tilde{\Phi})}{(\tilde{\Phi}-W)}d\tilde{\Phi}\,, \tag{5.18}$$

where $\zeta'(\tilde{\Phi})$ is the inverse velocity along the streamline $\Psi = 0$. Or equivalently, after accounting for the body condition and making use of (5.17),

$$x_\Phi(\Phi, 0) = \left(\frac{\Phi+1}{\Phi}\right)^{1/2}\left[1 - \frac{1}{\pi}\int_0^\infty\left(\frac{\tilde{\Phi}}{\tilde{\Phi}+1}\right)^{1/2}\frac{y_\Phi(\tilde{\Phi}, 0)}{\Phi-\tilde{\Phi}}d\tilde{\Phi}\right]. \tag{5.19}$$

Equation (5.19) provides a relation between the horizontal and vertical slopes of the inverse velocity on the free surface because $(x_\Phi, y_\Phi) = (u,v)/|\mathbf{u}|^2$. The determination of these two unknown functions requires the use of the dynamic condition (5.16), which reduces to

$$y(x_\Phi^2 + y_\Phi^2) - (d-H) = 0\,, \tag{5.20}$$

where H is the nondimensional draft of the body. The nonlinear integro-differential equation (5.19), (5.20) was solved by Vanden Broeck & Tuck for stern flows. A low-Froude-number series was developed and summed near the body, this was used to "seed" and match with a numerical solution of the integral equation based on Newtonian iteration. The stern-flow results presented marked the first published nonlinear body-wave computations, in which both free-surface and body conditions were satisfied exactly. It is also significant to note that these authors were unable to obtain a convergent solution for the bow-flow problem. It was conjectured (also by Dagan & Tulin 1972) that a jet structure developed near the stagnation point. Some ideas on how to handle such a jet situation were also discussed in this paper.

Inverse formulations following this line of approach were recently used by Vanden Broeck & Schwartz (1979) for finite-depth water waves, by Schwartz (1981) for a moving pressure distribution, and by Forbes (1981) for bottom-mounted obstacles. These last two works noted the possibility that wave-free subcritical flows exist at certain geometries, as in the case of linear theories. Both applied zero wave-elevation and uniform-velocity conditions at the upstream end of the integral equation, which was also truncated downstream abruptly. These authors noted that the truncation, as in the original work of Vanden Broeck & Tuck (1977), had only indiscernible effects on the interior solution, except for the occasional presence of grid-scale upstream waves. Is this rather successful procedure restricted to the Cauchy-type integral-equation that is being solved? Or is it a methodology that can be applied to steady problems in general? The question remains unanswered.

6. SUMMARY REMARKS

A variety of numerical methods and techniques for treating linear and nonlinear free-surface flow problems have been reviewed in this article. With the exception of the method of Green functions, all are faced with the difficulty of imposing an effective open boundary condition of some kind. Hybrid methods based on matching an interior numerical solution with an exterior analytical representation appear to be the most rational. Unfortunately, these have been successfully developed only for linearized problems. A one-dimensional Sommerfeld-type condition has been employed quite successfully in a number of instances, but questions related to effects of oblique incidence and of the presence of non-propagative type solutions still remain. In steady-translation problems "asymmetrical" techniques, such as upstream differencing, appear to give the flow enough of a "preferred" direction that waves normally occur only downstream. Work that uses a large computation domain, with zero-disturbance condition at the outer boundary, generally pushes the computation hardware to such a limit that attention to accuracy and convergence seems always lacking.

We have made no attempts to compare the computer time of the various methods, since such numbers will depend on a variety of "behind-the-scenes" factors. But we have proposed in Section 5 a simple formula based on which such comparisons can be made on a more rational basis, at least for linear problems. Although the nonlinear free-surface conditions can be handled by a variety of iterative techniques that are strongly method-dependent, it seems evident that an iterative solution on just the free-boundary surface or contour should be considerably more convenient and efficient than one that couples such iterations with the solution of the field equation. Boundary-integral-equation methods have precisely this advantage, apart from the usual superiority in storage requirement and ease of implementation. Space discretization techniques are generally regarded more suitable for problems where the field is inhomogeneous. Nevertheless, all methods have contributed to the understanding of free-surface flows about bodies.

The complexities associated with free-surface flows are amazing. Recent discoveries related to the modulational instability of weakly nonlinear wavetrains, the absence of a definite end state during the evolution of such unstable wavetrains, and the bifurcation of large-amplitude waves serve well as examples. A thorough understanding of physical phenomena involving nonlinear body-wave problems requires more than direct numerical solutions. Parallel analytical investigation and experimental confirmation of numerical results are highly desirable. The "best" numerical methods to come may well be those that exploit analytical simplifications that are appropriate for the physical phenomenon being examined.

ACKNOWLEDGMENT

Preparation of this article has been supported primarily by the Office of Naval Research, under task NR062-611. Partial support by the AAEF and the NSF is also gratefully acknowledged. It is a pleasure to thank Professor E. O. Tuck for his comments on the manuscript. I would like to dedicate this work to the late Mr. K. C. Chow and his family. This review was completed April, 1981.

Literature Cited

Adachi, H., Ohmatsu, S. 1979. On the influence of irregular frequencies in the integral equation solutions of the time-dependent free-surface problems. *J. Soc. Naval Architects Jpn.* 145:127–36

Aitchison, J. M. 1980. A finite element solution for critical flow over a weir. *Proc. Int. Conf. Finite Elements in Flow Problems, Banff, Alberta, Canada* 2:52–59

Ames, W. F. 1977. *Numerical Methods for Partial Differential Equations.* 2nd ed. New York: XIV+365 pp.

Arakawa, A., Mintz, Y. 1974. The UCLA atmospheric general circulation model. *Univ. Calif. Los Angeles Workshop Notes*

Baba, E., Takekuma, K. 1975. A study on free-surface flow around the bow of slowly moving full forms. *J. Soc. Naval Architects Jpn.* 137:1–10

Bai, K. J. 1977. A localized finite-element method for steady, three-dimensional free-surface flow problems. *Proc. 2nd Int. Conf. Numer. Ship Hydrodyn., Berkeley, Calif.* pp. 78–87

Bai, K. J. 1978. A localized finite-element method for two-dimensional steady potential flows with a free surface. *J. Ship Res.* 22:216–30

Bai, K. J. 1979. Blockage correction with a free surface. *J. Fluid Mech.* 94:433–52

Bai, K. J., Yeung, R. W. 1974. Numerical solutions to free-surface flow problems. *10th Symp. Nav. Hydrodyn., Cambridge, Mass,* pp. 609–33; disc. pp. 634–47

Banerjee, P. K. B., Butterfield, R., eds. 1979. *Developments in Boundary-Element Methods,* Vol. 1. London: Applied Science Publisher. xii+292 pp.

Bateman, H. 1932. *Partial Differential Equations.* Cambridge: Univ. Press. New York: Dover, 1944. xxii+522 pp.

Beale, J. T., 1977. Eigen function expansion for objects floating in an open sea. *Comm. Pure Appl. Math.* 30:283–313

Berhault, C. 1980. An integro-variational method for interior and exterior free surface flow problems. *Appl. Ocean Res.* 2:33–38

Berkhoff, J. C. W. 1972. Computation of combined refraction-diffraction. *Proc. 13th Coastal Engrg. Conf., Vancouver.* pp 471–90

Betts, P. L. 1979. A variational principle in terms of stream function for free-surface flows and its application to the finite element method. *Comput. Fluids.* 7:145–53

Betts, P. L., Assaat, M. I. 1980. Finite element solutions of large amplitude water waves. *Proc. Int. Conf. Finite Elements in Flow Problems, Banff, Alberta, Canada* 2:24–32

Boussinesq, J. 1871. Théorie de l'intumescence liquide appelée onde solitaire ou de translation se propageant dans un canal rectangulaire. *C. R. Acad. Sci.* Paris 72:755–59

Brandt, A., Dendy, J. E., Ruppel, H. 1980. The multigrid method for semi-implicit hydrodynamics codes. *J. Comput. Phys.* 34:348–70

Buzbee, B. L., Dorr, F. W., George, J. A., Golub, G. H. 1971. The direct solution of the discrete Poisson equation on irregular regions. *SIAM J. Numer. Anal.* 8:722–36

Carmerlengo, A. L., O'Brien, J. J. 1980. Open boundary conditions in rotating fluids. *J. Comput. Phys.* 35:12–35

Chan, R. K. C. 1975. Two-dimensional time-dependent calculations of large amplitude surface gravity waves due to a surface disturbance. *Proc. 1st Int. Conf. Numer. Ship Hydrodyn. Gaithersburg, Md.,* pp. 315–32

Chan, R. K. C. 1977. Finite difference simulation of the planar motion of a ship. *Proc. 2nd Int. Conf. Numer. Ship Hydrodyn., Berkeley, Calif.,* pp. 39–52

Chan, R. K. C., Chan, F. W-K. 1980. Numerical solutions of transient and steady free-surface flows about a ship of general hull shape. *13th Symp. Naval Hydrodyn., Tokyo, Jpn.,* pp. 257–80

Chan, R. K. C., Hirt, C. W. 1974. Two-dimensional calculations of the motion of floating bodies. *10th Symp. Naval Hydrodyn., Cambridge, Mass.,* pp. 667–82

Chan, S. T. K., Larock, B. E. 1973. Free-surface ideal fluid flows by finite elements. *ASCE J. Hydraul. Div.* 99:(HY6):959–74

Chang, M. S., Pien, P. C. 1975. Hydrodynamic forces on a body moving beneath a free surface. *Proc. 1st Int. Conf. Numer. Ship Hydrodyn.* pp. 530–60

Chaplin, J. R. 1980. Developments of stream-function wave theory. *Coastal Engrg.* 3:179–205

Chapman, D. R. 1978. Computational aerodynamics development and outlook. *AIAA J.* 17:1293–1313

Chapman, R. B. 1976. Free-surface effects for yawed surface-piercing plates. *J. Ship Res.* 20:125–36

Chapman, R. B. 1979. Large amplitude transient motion of two-dimensional floating bodies. *J. Ship Res.* 23:20–31

Chen, H. S., Mei, C. C. 1974. Oscillation and wave forces in a man-made harbor in the open sea. *10th Symp. Naval Hydrodyn., Cambridge, Mass,* pp.573–96

Chenault, W. 1970. *Motion of a ship at the free surface.* MS thesis. Naval Postgraduate School, Monterey, Calif.

Chenot, J. L. 1975. Méthode numérique de calcul du mouvement d'un corps flottant soumis à l'influence d'une houle périodique en théorie linéaire.*Rev. Inst. Fr. Pet.* 30:779–802

Cokelet, E. D. 1977. Steep gravity waves in water of arbitrary uniform depth. *Philos. Trans. R. Soc. London, Ser. A* 286:183–230

Dagan, G., Tulin, M. P. 1972. Two-dimensional free-surface gravity flow past blunt bodies. *J. Fluid Mech.* 51:529–43

Daoud, N. 1975. Potential flow near to a fine ship's bow. *Univ. Mich., Dept. Naval Arch. Mar. Engrg. Rep. No. 177* ii+58pp.

Dawson, C. W. 1977. A practical computer method for solving ship-wave problems. *Proc. 2nd Int. Conf. Numer. Ship Hydrodyn., Berkeley, Calif.,* pp. 32–38

Dean, R. G. 1965. Stream function representation of nonlinear ocean waves. *J. Geophys. Res.* 70:4561–72

Dean, R. G. 1974. Evaluation and development of water wave theories for engineering applications. *U.S. Army Corps Engrs. Coastal Engrg. Res. Rep. Sp. Rep. No. 1.* 534 pp.

Delves, L. M., Walsh, J. 1974. *Numerical Solution of Integral Equations*. Oxford: Clarendon 339 pp.

Dern, J. C. 1977. Existence, uniqueness and regularity of the solution of Neumann-Kelvin problem for two or three dimensional submerged bodies. *Proc. 2nd Int. Conf. Numer. Ship. Hydrodyn., Berkeley, Calif.*, pp. 57–77.

Emmons, H. W. 1972. Critique of numerical modeling of fluid-mechanics phenomena. *Ann. Rev. Fluid Mech.* 2:15–36

Engquist, B., Majda, A. 1977. Absorbing boundary conditions for the numerical simulation of waves. *Math. Comput.* 31(139):629–51

Faltinsen, O. M. 1977. Numerical solution of transient nonlinear free-surface motion outside or inside moving bodies. *Proc. 2nd Int. Conf. Numer. Ship Hydrodyn., Berkeley, Calif.* pp. 347–57

Faltinsen, O. M., Michelsen, F. C. 1974. Motions of large structures in waves at zero Froude number. *Int. Symp. Dynamics of Marine Vehicles and Structures in Waves, Univ. College, London,* pp. 99–114

Fenton, J. D., Mills, D. A. 1976. Shoaling waves: numerical solution of exact equations. *Proc. IUTAM. Symp. Waves on Water of Variable Depth, Canberra. Lecture Notes in Physics* 64:94–101

Finkelstein, A. B. 1957. The initial-value problem for transient water-waves. *Comm. Pure Appl. Math.* 10:511–22

Forbes, L. K. 1981. On the wave resistance of a submerged elliptical body. *J. Engrg. Math.* In press

Forsythe, G. E., Wasow, W. R. 1960. *Finite-Difference Methods for Partial Differential Equations*. New York: Wiley. 444 pp.

Frank, W. 1967. Oscillations of cylinders in or below the free surface of deep fluids. *Naval Ship Res. Dev. Cent. Rep. 2375*. vit 40 pp.

Freeman, N. G., Hale, A. M., Danard, M. B. 1972. A modified sigma equation approach to the numerical modeling of Great Lakes' hydrodynamics. *J. Geophys. Res.* 77:1050–60

Gadd, G. E. 1976. A method of computing the flow and surface wave pattern around full forms. *R. Inst. Naval Arch.* 118:207–19

Garrison, C. J., Seetharama Rao, V. 1971. Interaction of waves with submerged objects. *Proc. ASCE. J. Waterways, Harbors, Coastal Engrg. Div.* 97:259–77

Grosch, C. E., Orszag, S. A. 1977. Numerical solutions of problems in unbounded regions: coordinate transforms. *J. Comput. Phys.* 25:273–96

Guével, M., Delhommeau, G., Cordonnier, J. P. 1977. Numerical solution of the Neumann-Kelvin problem by. the method of singularities. *Proc. 2nd Int. Conf. Numer. Ship Hydrodyn., Berkeley, Calif.,* pp. 107–23

Harten, A. 1975. An efficient differentio-integral equation technique for time-dependent free-surface flows. *Proc 1st Int. Conf. Numer. Ship Hydrodyn. Gaithersburg, Md.,* pp. 717–28

Harten, A. Efrony, S. 1978. A partition technique for the solution of potential flow problems by integral equations methods. *J. Comput. Phys.* 27:71–87

Haussling, H. J., Coleman, R. M. 1979. Nonlinear water waves generated by an accelerated circular cylinder. *J. Fluid Mech.* 92:767–81

Haussling, H. J., Van Eseltine, R. T. 1974. A combined spectral finite-difference method for linear and nonlinear wave problems. *Naval Ship Res. Dev. Cent., Carderock, Md., Rep. 4580.* vii+35 pp.

Hedstrom, G. W. 1979. Nonreflecting boundary conditions for nonlinear hyperbolic systems. *J. Comput. Phys.* 30:222–37

Hess, J. L., Smith A. M. O. 1964. Calculation of non-lifting potential flow about arbitrary three-dimensional bodies. *J. Ship Res.* 8:22–44

Hess, J. L., Smith, A. M. O. 1967. Calculation of potential flow about arbitrary bodies. *Prog. Aeronaut. Sci.* 8:1–137

Israeli, M., Orszag, S. A. 1974. Numerical simulation of radiation boundary conditions by damping. *Flow Res. Kent. Wash. Rep. #30.*

Jeppson, R. W. 1972. Inverse solution to three-dimensional potential flow. *Proc. ASCE, J. Engrg. Mech. Div.* 98:789–812

John, F. 1949. On the motion of floating bodies -I. *Comm. Pure Appl. Math.* 2:13–57

John, F. 1950. On the motion of floating bodies -II. *Comm. Pure Appl. Math.* 3:45–10

Jones, D. A. 1974. Integral equations for the exterior acoustic problem. *Q. J. Mech. Appl. Math.* 27:129–42

Kellogg, O. D. 1929. *Foundations of Potential Theory*. Berlin: Springer, ix+384 pp.

Kobus, J. M. 1976. *Application de la méthode des singlularités au problème des flotteurs cylindriques soumis à des oscillations harmoniques forcées de faible amplitude*. These (Docteur-Ingénieru). Univ. Nantes ix+156 pp.

Larock, B. E. 1970. A theory for free outflow beneath radial gates. *J. Fluid Mech.* 41:851–64

Larock, B. E., Taylor, C. 1976. Computing three-dimensional free-surface flows. *Int. J. Numer. Meth. Engrg.* 10:1143–1152

Lax, P. D., Richtmyer, R. D. 1956. Survey of the stability of linear finite difference equations. *Comm. Pure Appl. Math.* 9:267–93

Lebreton J.-C. Margnac, A. 1968. Calcul des mouvements d'un navire ou d'une platforme amarrés dans la houle. *La Houille Blanche* 23:379–89

Lenoir, M., Jami, A. 1978. A variational formulation for exterior problems in linear hydrodynamics. *Comput. Method Appl. Mech. Engrg.* 16:341–59

Lenoir, M., Martin, D. 1981. An application of the principle of limiting absorption to the motions of floating bodies. *J. Math. Anal. Appl.* 79:370–83

Licht, C. 1980. *Etude théorique et numérique de l'évolution d'un système fluideflotteur.*, Thèse (Docteur-ingénier). Univ. Nantes

Longuet-Higgins, M. S., Cokelet, E. D. 1976. The deformation of steep surface waves on water. I. A numerical method of computation. *Proc. R. Soc. London Ser. A* 350:1–26

Luke, J. C. 1967. A variational principle for a fluid with a free surface *J. Fluid Mech.* 27:395–97

Macaskill, C. C. 1977. *Numerical solutions of some fluid flow problems by boundary integral-equation techniques.* PhD thesis. Dept. Appl. Math., University of Adelaide, S. A.

MacCormack, R. W., Lomax, H. 1979. Numerical solution of compressible viscous flows. *Ann. Rev. Fluid Mech.* 11:289–316

Mattioli, F. 1978. Wave induced oscillations in harbours of variable depth. *Comput. Fluids,* 6:161–72

McDonald, B. H., Wexler, A. 1972. Finite element solution of unbounded field problems. IEEE Trans. Microwave Theory and Technique, MTT-20:841–47

Mei, C. C. 1978. Numerical methods in water-waves diffraction and radiation. *Ann. Rev. Fluid Mech.* 10:393–416

Mei, C. C., Chen, H. S. 1976. A hybrid element method for steady linearized free-surface flows. *Int. J. Numer. Meth. Engrg.* 10:1153–75

Michell, J. H. 1898. The wave resistance of a ship. *Philos. Mag.* 45(5):106–23

Mikhlin, S. G. 1957. *Integral Equations and their Applications to Certain Problems in Mechanics, Mathematical Physics and Technology.* New York: Macmillan xiv+341 pp.

Miles, J. W. 1977. On Hamilton's principle for surface waves. *J. Fluid Mech.* 83:153–58

Miles, J. W. 1980. Solitary waves. *Ann. Rev. Fluid Mech.* 12:11–43

Moayeri, M. S. 1973. Flow in open channels with smooth curved boundaries. *J. Hydraul. Div., Proc. ASCE* pp. 2217–32

Nekrasov, A. I. 1921. On waves of permanent type. *Izv. Ivanovo-Voznesensk. Politekhn. Inst. I.* 3:52–65

Newman, J. N. 1976. Linearized wave resistance theory. *Proc. Int. Seminar on Wave Resistance, Tokyo, Jpn.* pp 31–44

Newman, J. N. 1978. The theory of ship motions. *Adv. Appl Mech.* 18:221–83

Norrie, D. H., de Vries, G. 1978. A survey of the finite element application in fluid mechanics. *Finite Elements in Fluids vol. 3,* pp. 363–96

Noye, J., May, R. L., Teubner, M. D. 1981. Three-dimensional numerical model of tides in Spencer Gulf. *Ocean Management* 6:137–47

O'Carroll, M. J. 1976. Variational technique for free-streamline problems. *Proc. 2nd Symp. Finite Element Methods in Flow Problems, Ligure, Italy,* pp. 489–95

Ogilvie, T. F. 1968. Wave resistance—The low-speed limit. *Univ. Michigan, Dept. Naval Arch. & Marine Engrg. Rep. No. 2*

Ogilvie, T. F., Shin, Y. S. 1978. Integral-equation solutions for time-dependent free-surface problems. *J. Soc. Naval Architects Jpn.* 143:41–51

Ohring, S. 1975. A fast fourth-order Laplace solver for application to numerical three-dimensional water wave problems. *Proc. 1st Int. Conf. Numer. Ship Hydrodyn., Gaithersburg, Md.,* pp. 641–63

Ohring, S., Telste, J. 1977. Numerical solution of transient three-dimensional ship-wave problems. *Proc. 2nd Int. Conf. Numer. Ship Hydrodyn, Berkeley, Calif.* pp. 88–103

Orlanski, I. 1976. A simple boundary condition for unbounded hyperbolic flows. *J. Comput. Phys.* 21:251–69

Orszag, S. A., Israeli, M. 1974. Numerical simulation of viscous incompressible flows. *Ann. Rev. Fluid Mech.* 6:281–318

Patterson, G. S. 1978. Prospects for computational fluid mechanics. *Ann. Rev. Fluid Mech.* 10:289–300

Pian, T. H., Tong, P. 1968. Basis of finite element analysis for solid continua. *Int. J. Numer. Math. Engrg.* 1:3–28

Potash, R. L. 1971. Second-order theory of oscillating cylinders. *J. Ship. Res.* 14:295–324

Rayleigh, Lord. 1876. On waves. *Philos. Mag.* 1(5):257–79

Reinecke, M., Fenton, J. D. 1981. A Fourier approximation method for steady water waves. *J. Fluid Mech.* 104:119–37

Richtmyer, R. D., Morton, K. W. 1967. *Difference Methods for Initial-Value Problems.* 2nd ed. New York: Wiley Interscience. 405 pp.

Roache, P. J. 1976. *Computational Fluid Dynamics.* Albuquerque, NM: Hermosa. vii+446 pp.

Rudy, D. H., Strikwerda, J. C. 1980. A nonreflecting outflow boundary for subsonic Navier-Stokes Calculations. *J. Comput. Phys.* 36:55–70

Salvesen, N., von Kerczek, C. 1978. Nonlinear aspects of subcritical shallow-water flow past two-dimensional obstructions. *J. Ship Res.* 22:203–11

Sarpkaya, T., Hiriart, G. 1975. Finite element analysis of jet impingement on axisymmetric curved deflectors. In *Finite Elements in Fluids, Vol. 1*, New York: Wiley pp. 265–79

Sayer, P., Ursell, F. 1977. Integral-equation methods for calculating virtual mass in water of finite depth. *Proc. 2nd Int. Conf. Numer. Ship Hydrodyn., Berkeley, Calif.*, pp. 176–84

Schwartz, L. W. 1974. Computer extension and analytic continuation of Stokes' Expansion for gravity waves. *J. Fluid Mech.* 62:553–78

Schwartz, L. W. 1981. Nonlinear solution for an applied overpressure on a moving stream. *J. Engrg. Math.* 15:147–156

Schwartz, L. W., Fenton, J. D. 1982. Strongly nonlinear waves. *Ann. Rev. Fluid Mech.* 14:39–60

Shanks, S. P., Thompson, J. F. 1977. Numerical solution of the Navier-Stokes equations for 2-D hydrofoils in or below a free surface. *Proc. 2nd Int. Conf. Ship Hydrodyn., Berkeley, Calif.*, pp 202–20

Shaw, R. P. 1975. An outer boundary integral equation applied to transient wave scattering in an inhomogeneous medium. *J. Appl. Mech.* 42:147–52

Shen, S.-F. 1977. Finite-element methods in fluid mechanics. *Ann. Rev. Fluid Mech.* 9:421–45

Soh, W. K. 1976. Vortex sheet method for calculation of linearized free surface waves. *Rep. Nav/Arch 76/5*. Sch. Mech. & Indust. Engrg., Univ. New South Wales, 17 pp.

Soh, W. K. 1980. Computer simulation of water waves and the frequencies of a submerged cylinder. *Proc. 7th Australas. Hydraul. Fluid Mech. Conf., Brisbane*, pp. 60–63

Sommerfeld, A. J. W. 1949. *Vorlesungen über theoretische Physik, II. Mechanik der deformierbaren Medien.* Leipzig: Akademie. xi+335 pp.

Southwell, R. V., Vaisey, G. 1946. Relaxation methods applied to engineering problems. XII. Fluid Motion Characterized by free stream-lines. *Philos. Trans. R. Soc. London Ser. A* 240:117–61

Stoker, J. J. 1957. *Water Waves.* New York: Wiley Interscience. xxviii+567 pp.

Stokes, G. G. 1847. On the theory of oscillatory waves. *Trans. Cambridge Philos. Soc.* 8:441–55

Stokes, G. G. 1880a. Consideration relative to the greatest height of oscillatory waves which can be propagated without change of form. *Math. Phys. Pap.* 1:225–28. Cambridge: Univ. Press

Stokes, G. G. 1880b. Supplement to a paper on the theory of oscillatory waves. *Math. Phys. Pap.* 1:314–26. Cambridge: Univ. Press

Thom, A., Apelt, C. J. 1961. *Field Computations in Engineering and Physics.* New York: Van Nostrand 65 pp.

Thomas, J. W. 1968. Irrotational gravity waves of finite height: A numeric study. *Mathematika* 15:139–48

Thompson, J. F., Thames, F. C., Mastin, C. W. 1974. Automatic numerical generation of body-fitted curvilinear coordinate system for field containing any number of arbitrary two-dimensional bodies. *J. Comput. Phys.* 15:299–319

Thompson, J. F., Thames, F. C., Hodge, J. K., Shanks, S. P., Reddy, R. N. Mastin, C. W. 1976. Solutions of Navier-Stokes equations in various flow regimes on fields containing any number of arbitrary bodies using boundary-fitted coordinate systems. *Lecture Notes in Physics* 59:421–27

Tsutsumi, T. 1979. Calculation of the wave resistance of ships by the numerical solution of Neumann-Kelvin problem. *Proc. Workshop on Ship Wave-Resistance Computations, Naval Ship Res. Dev. Cent., Carderock, Md.*, pp. 162–201

Ursell, F. 1980a. Irregular frequencies and the motion of floating bodies. Manuscript, Dept. Mathematics, Manchester Univ.

Ursell, F. 1980b. Mathematical notes on the two-dimensional Kelvin-Neumann problem of ship hydrodynamics. *13th Symp. Naval Hydrodyn., Tokyo*, pp. 245–51

Vanden Broeck, J.-M. Schwartz, L. W. 1979. Numerical computation of steep gravity waves in shallow water. *Phys. Fluids* 22:1868–71

Vanden Broeck, J.-M., Tuck, E. O. 1977. Computation of near-bow or stern flows, using series expansion in the Froude number. *Proc. 2nd Int. Conf. Numer. Ship Hydrodyn., Berkeley, Calif.*, pp. 371–81

Vinje, T. Brevig, P. 1980. Nonlinear, two-dimensional ship motions. *Norwegian Inst. Tech., Rep. on Ships in Rough Seas.* 92 pp.

von Kerczek, C. H., Salvesen, N. 1974. Numerical solutions of two-dimensional nonlinear wave problems. *10th Symp. Naval Hydrodyn., Cambridge, Mass.*, pp. 649–65

Wehausen, J. V. 1971. The motion of floating bodies. *Ann. Rev. Fluid Mech.* 3:237–68

Wehausen, J. V. 1973. Wave resistance of ships. *Adv. Appl. Mech.* 13:93–244

Wehausen, J. V., Laitone, E. V. 1960. Surface waves. *Handbuch der Physik* 9:446–778 Berlin: Springer.

Wellford, C. L., Ganaba, T. 1980. Finite element procedures for fluid mechanics problems involving large free surface motion. *Proc. 3rd Int. Conf. Finite Elements in Flow Problems, Banff, Alberta, Canada*, 2:13–23

Whitham, G. B. 1967. Nonlinear dispersion of water waves. *J. Fluid Mech.* 27:399–412

Whitham, G. B. 1970. Two-timing, variational principles and waves. *J. Fluid Mech.* 44:373–95

Winslow, A. M. 1966. Numerical solution of the quasilinear Poisson equation in nonuniform triangular mesh. *J. Comput. Phys.* 1:149–72

Yamamoto, Y., Kagemoto, H. 1976. Finite-element treatments for free surface waves caused by a body in uniform flow. *Proc. 26th Jpn. Natl. Cong. Appl Mech.*, pp. 549–561

Yen, S. M., Lee, K. D., Akai, T. J. 1977. Finite element and finite difference solutions of nonlinear free surface wave problems. *Proc. 2nd Int. Conf. Numer. Ship Hydrodyn., Berkeley, Calif.*, 305–18

Yeung, R. W. 1975. A hybrid integral-equation method for time-harmonic free-surface flows. *Proc. 1st Int. Conf. Numer. Ship Hydrodyn., Gaithersburg, Md.*, pp. 581–608

Yeung, R. W. 1981. Added mass and damping of a vertical cylinder in finite-depth waters. *Appl. Ocean Res.* 3:119–33

Yeung, R. W. 1982. The transient heaving motions of floating cylinders. *J. Engrg. Math.* In press

Yeung, R. W., Bouger, Y. C. 1979. A hybrid integral-equation method for steady two-dimensional ship waves. *Int. J. Numer. Meth. Engrg.* 14:317–36

Yim, B. 1975. A variational principle associated with a localized finite element technique for steady ship-wave and cavity problems. *Proc. 1st Int. Conf. Numer. Ship Hydrodyn., Gaithersburg, Md.*, pp. 137–53

Yue, D. K. P., Chen, H. S., Mei, C. C. 1978. A hybrid element method for diffraction of water waves by three-dimensional bodies. *Int. J. Numer. Methods in Engrg.* 12:245–66

Yuen, H. C., Lake, B. M. 1980. Instabilities of waves on deep water. *Ann. Rev. Fluid Mech.* 12:303–34

Zaroodny, S. J., Greenberg, M. D. 1973. On a vortex sheet approach to the numerical calculation of water waves. *J. Comput. Phys.* 11:440–46

Zienkiewicz, O. C. 1975. Why finite elements? In *Finite Elements in Fluids Vol. 1*, pp. 1–23. New York: Wiley

Zienkiewicz, O. C., Bettess, P. 1975. Infinite elements in the study of fluid-structure instruction problems. *Proc. 2nd Int. Symp. Computing Meth. Appl. Sci. Engrg, Versailles, France*, pp.

AUTHOR INDEX

A

Abdelwahed, M. S. T., 205
Abell, C. J., 62, 72-74
Abraham, G., 203
Ackers, P., 22
Acrivos, A., 229, 341
Adachi, H., 428
Adams, B. A., 231
Adams, J. C. Jr., 79
Adamson, T. C. Jr., 262
Adler, J., 338
Aitchison, J. M., 417
Akai, T. J., 405
Akatnov, N. J., 208
Algert, J. H., 30
Allen, C. M., 226
Allen, J., 162
Allen, J. R. L., 35, 215, 227
Allender, J. H., 181
Alpert, R. L., 198, 199
Amen, R., 230
Ames, W. F., 401
Amiet, R. K., 290
Anderson, J. L., 203
Andrews, D. G., 149
Andronov, A. A., 62, 75
Angelini, J. J., 300
Antonia, R. A., 192, 193, 196, 200
Apelt, C. J., 412
Arakawa, A., 144, 413
Arlinger, B. G., 265
Armi, L., 166-68
Armstrong, R. R., 192
Ashley, H., 288, 297-99
Assaat, M. I., 417
Assur, A., 92
Atassi, H., 292

B

Baba, E., 400
Bagnold, R. A., 21
Bai, K. J., 411, 418, 419, 423, 425, 431, 432
Bailey, F. R., 264, 275
Bailey, H. E., 301
Baines, P. G., 149
Baldwin, B. S., 263
Ball, F. K., 201
Ballhaus, W. F., 264, 270, 275
Banerjee, P. K. B., 424
Bannister, T. C., 318, 328
Barbarossa, N., 18
Barcilon, A. I., 35
Barenblatt, G. I., 230

Barnes, P., 92
Barr, D. I. H., 220, 222
Bashir, J., 196, 197
Basu, B. C., 294
Batchelor, G. B., 191
Bateman, H., 414
Batycky, J., 378
Bauer, F., 265, 274, 275
Bauer, S. N., 182
Baum, W., 230
Beale, J. T., 430
Beavers, G. S., 191, 196
Becker, H. A., 190, 193, 194, 196
Becker, R. M., 239, 240, 242
Bedard, A. J., 225
Beghin, P., 219
Bellhouse, B. J., 240, 243, 247, 249
Bellhouse, F. H., 243, 247
Benedict, B. A., 203
Benjamin, T. B., 62, 75, 215, 216, 229
Bennett, F. O., 314
Bennett, J. R., 179, 180
Bennetts, D. A., 149
Beran, D. W., 225
Bergeron, T., 8, 10, 11, 133
Berghuis, J., 250
Berhault, C., 432
Berkhoff, J. C. W., 421
Berkovsky, B. M., 338
Bettess, P., 411
Bettess, R., 20
Betts, A. K., 225
Betts, P. L., 417
Beuther, P. D., 198
Bill, R. G., 200
Bindschadler, R. A., 125
Birch, A. D., 193
Birchfield, G. E., 180
Bjerknes, J., 8, 10
Bjerknes, V., 2, 8, 10
Björk, V. O., 240
Blanchard, D. C., 10
Bland, S. R., 299
Blanton, J. O., 162
Blatt, M. H., 314
Blumen, W., 142, 144, 149
Blümke, A., 120
Bocquet, G., 91
Bodvardsson, G., 124
Boguslawski, L., 196
Bonchek, L. I., 236
Borg, M., 240
Borland, C. J., 296
Bosenberg, U., 179

Bouger, Y. C., 419, 432
Bourgeois, M. J., 237
Bourgeois, S. V., 328, 338
Boussinesq, J., 395
Bowen, R., 356, 362
Bowman, M. J., 154
Boyce, F. M., 180
Boyd, W. N., 298, 299
Bradbury, L. J. S., 191, 196
Bradshaw, P., 33, 191, 196
Bradshaw, R. D., 314
Brandner, C. F., 373, 375
Brandt, A., 266, 271, 411
Bratkovich, A., 215
Brennen, C., 55
Bretherton, F. P., 133, 137, 140-43, 146, 149
Brevig, P., 434
Brewer, L. A. III, 236
Briggs, G. A., 206, 207, 230
Briscoe, M. G., 154, 164, 165, 168, 169
Britter, R. E., 215-18, 222, 225, 228
Britz, D., 192
Brooks, N. H., 154, 156, 166-68, 171, 174, 195, 201, 203-5
Brooks, T. F., 295
Brown, A. L. Jr., 250
Brown, D. H., 203
Brown, D. R., 193
Brown, G. B., 190, 191, 196
Brubaker, J. M., 168, 173
Brun, P., 238
Bruun, H. H., 192
Brzustowski, T. A., 207
Buchan, S. J., 173
Buckney, R. T., 166, 170
Budd, W. F., 111, 120
Burdges, K. P., 264, 275
Butterfield, R., 424
Butterworth, I. J., 230
Buzbee, B. L., 404
Byatt-Smith, J. G. B., 49, 55, 57
Byers, H. R., 225

C

Caldwell, D. R., 163, 166, 168
Callegari, A., 98
Camarero, R., 266
Campbell, B. J., 314
Caracena, F., 225
Carl, J. R., 236, 239, 242, 256
Carlson, L. A., 264

443

AUTHOR INDEX

Carmack, E. C., 171, 173, 174, 178
Carmerlengo, A. L., 411
Carmody, T., 191
Carr, J. F., 231
Carr, L. W., 305, 306
Carruthers, J. R., 314
Carta, F. O., 288, 294
Cassan, J. P., 369
CAUGHEY, D. A., 261-83; 266-68, 270, 273-75, 277, 279
Cederwall, K., 207
Chabert-d'Hières, G., 53
Chambers, A. J., 193
Chambré, P. L., 341
Champagne, F. H., 190
Chan, F. W-K., 409, 412
Chan, R. K. C., 55, 404, 405, 409, 412, 413
Chan, S. T. K., 416
Chan, Y. Y., 191
Chang, C. E., 336
Chang, M. S., 427
Chaplin, J. R., 408
Chapman, D. R., 301, 401
Chapman, R. B., 404, 405, 430
Chappelear, J. E., 42, 45, 47
Charney, J. G., 10, 131
Chassaing, P., 204
Chen, A. W., 273
Chen, B., 47, 49, 51
Chen, C. F., 178
Chen, C. J., 198, 199, 206
Chen, C. W., 154
Chen, H. S., 421-23
Chen, H. Y., 28
Chen, J.-C., 203, 222
Chen, L. T., 266, 267
Chenault, W., 411
Cheng, H. K., 273
Chenot, J. L., 421
Cherno, J., 220
Chernyshova, R. T., 224
Chevray, R., 192, 193, 196, 200
Chorin, A. J., 247, 253, 256
Chow, R. R., 274, 275, 277
Chriss, T. M., 166
Christie, D. R., 230
Chu, V. H., 205, 206
Chun, Ch.-H., 336, 337
Churchill, S. W., 327, 328, 338, 340
Chyu, W. J., 299
Claria, A., 204
Clark, P. A., 336
Clarke, G. K. C., 109, 113
Clarke, R. H., 225, 230
Clever, R. M., 327
Cokelet, E. D., 41, 44, 48, 54, 55, 425, 433-35
Colbeck, S. C., 90, 92
Cole, J. D., 264

Coleman, R. M., 405, 407, 409-11
Collet, P., 359
Collins, I. F., 116
Commerford, G. L., 294
Concus, P., 53
Conway, B. A., 319
Cooper, R. I. B., 53
Corcoran, W. H., 236, 239, 242, 256
Corcos, G. M., 154, 163, 170
Cordonnier, J. P., 429
Corrsin, S., 191
Couston, M., 300
Covert, E. E., 294
Cox, C. S., 173
Crabb, D., 204, 205
Crapper, G. D., 50
Crawford, T. V., 203
Crighton, D. G., 191
Crow, S. C., 190
Csanady, G. T., 154-56, 179, 181, 208
Curry, J. H., 355

D

Dafermos, C., 121
Dagan, G., 436
Daley, R. J., 173, 174
Daly, B. J., 224
Danard, M. B., 409
Daniels, P. G., 294
Danielsen, E. F., 131
Dantan, P., 238
Daoud, N., 430
D'Asaro, E., 165, 167
Davey, A., 65, 74
Davies, P. O. A. L., 191, 192
Davis, H. T., 366, 371, 378
Davis, M. R., 192
Davis, R. E., 161, 163, 229
Davis, S. S., 295, 298, 300
Dawson, C. W., 435
De, S. C., 42
Dean, R. G., 47, 408
de Gortari, J. C., 192, 196
Delhommeau, G., 429
Delichatsios, M. A., 197, 198
Delisi, D. P., 172
Dellsperger, K. C., 239
Delves, L. M., 428
Dendy, J. E., 411
Denton, R. A., 162
de Quervain, M. R., 228
Dern, J. C., 399
Desopper, A., 292, 294, 300
de Szoeke, R., 161, 163
Devik, O., 11
de Vries, G., 45, 413
Dillon, T. M., 161-63, 168
Dodge, F. T., 315

Dodson, M. G., 193
Dorr, F. W., 404
Dorsey, N. E., 95
Dowell, E. H., 299
Draghici, I., 144, 145
Drake, L. D., 98
Duffy, D. G., 149
Dulikravich, D. J., 267
Durao, D. F. G., 204, 205
Durbin, P. A., 163
Dussan V, E. B., 374, 375
Dyson, D. C., 374
Dzulynski, S., 227

E

Eady, E. T., 131
Eckmann, J.-P., 349, 357, 359
Edelmann, W., 144
Edelsten, D. J., 227
Edge, R. D., 52
Edwards, A., 227
Efrony, S., 432
Einstein, H. A., 17, 18, 21
Eiseman, P. R., 265
Ekman, V. W., 3
Eliasen, E., 149
ELIASSEN, A., 1-11; 10, 135, 137-39, 144
Elle, B. J., 83
Ellis, F. H. Jr., 250
Ellison, T. H., 204
Emerson, S., 165
Emery, K. O., 179
Emmons, H. W., 401
ENGELUND, F., 13-37; 18, 21, 26-29, 32, 33, 35
England, W. G., 224
Engquist, B., 412
Esaias, W. E., 154
Eskinazi, S., 204
Evans, R. J., 92
Everitt, K. W., 196
Eysink, W. D., 203

F

Facemire, B. R., 328
Fairlie, B. D., 62, 72
Faller, A. J., 131
Faltinsen, O. M., 427, 434
Fan, C., 328, 338
Fan, F. N., 164
Fan, L.-N., 203, 206
Fandry, C., 158
Farmer, D. M., 162, 164
Fatt, I., 371
Fay, J. A., 205, 206, 219, 223
Fearn, R. L., 205
Fedorenko, R. P., 271
Fedorov, K. N., 154, 172, 230
Feigenbaum, M. J., 358, 359

AUTHOR INDEX 445

FENTON, J. D., 39-60; 43, 45-47, 56, 57, 396, 408, 434
Ferriss, D. H., 191
Fiedler, H., 191, 193-96
Fife, P., 51
Fineblum, S. S., 326
Finkelstein, A. B., 430
Finlayson, B. L., 250
Finsterwalder, S., 120
Fischer, H. B., 154, 156, 166-68, 171, 174, 195, 201, 203, 205
Fischer, K. H., 165
Fisher, M. J., 191, 192, 196, 199
Fjortoft, R., 144
Fleeter, S., 295
Flumerfelt, R. W., 366, 369, 370, 379, 387, 389
Forbes, L. K., 437
Ford, D. E., 163, 173
Forney, L. J., 205, 206
Forster, G. R., 226
Forsythe, G. E., 404
Foster, T. D., 173
Fowler, A. C., 91, 93, 95-98, 100, 101, 108, 113, 116-22
Fowler, N. O., 236
Fox, D. G., 203
Fox, M. J. H., 46, 49
Frank, F. C., 341
Frank, W., 427
Frater, R. W. M., 236, 237, 239, 240, 242, 250
Frazier, T. V., 225
FREDSØE, J., 13-37; 28, 29, 33-35
Freeman, B. E., 224
Freeman, N. G., 409
Freeman, R. W., 190
Freymuth, P., 190
Frick, J., 275
Friedrichs, K. O., 121
Fromme, J. A., 292
Fu, B.-I., 321, 325
Fuchs, H. V., 192
Fujita, T. T., 225
Fultz, D., 53, 131
Fung, K. Y., 278, 300
Fung, Y. C., 287
Fung, Y.-T., 192

G

Gabbay, S., 237, 239, 240, 242
Gadd, G. E., 435
Gallagher, B., 173
Ganaba, T., 424
Garabedian, P. R., 265, 267, 274, 275
Garner, H. C., 292

Garrett, C., 154, 164, 166-68
Garrick, I. E., 289
Garrison, C. J., 427
Gartrell, G., 204
Gartrell, G. Jr., 168
Garvine, R. W., 226
Gaster, M., 191
Gebhart, B., 200
Gelhar, L. W., 29
Gent, P. R., 145
George, J., 204
George, J. A., 404
George, W. K., 198, 199
Georgeson, E. M. H., 218
Gibson, C. H., 164, 168, 170
Gidel, L. T., 148
Giesing, J. P., 292
Gill, A. E., 140, 191
Gillette, R. D., 374
Girodoux-Lavigne, Ph., 300
Glassman, I., 330
Glen, J. W., 91, 93
Godske, C. L., 11
Goff, R. C., 225
Golberg, M. A., 292
Goldberg, M. B., 205, 206
Goldschmidt, V. W., 192, 196
Golub, G. H., 404
Gordon, D. A., 237, 238
Gordon, I. I., 62, 75
Gottlieb, A., 256
Graf, W. H., 182
Grant, A. J., 191
Grant, M. A., 48, 49
Grant, W. D., 167
Gray, C. B. J., 173, 174
Green, T., 231
Greenberg, M. D., 433
Gregg, M. C., 154, 164, 165, 168, 169
Gregory, N., 79
Grigoryan, S. S., 105
Grodzka, P. G., 318, 328, 338
Grosch, C. E., 411
Gross, M. G., 231
Grossman, B., 264, 265
Grotjhan, R., 149
Guckenheimer, J., 356
Guével, M., 429
Gutmark, E., 191, 196
Guy, H. P., 14

H

Hafez, M., 270-72
Hafez, M. M., 273
Haines, W. B., 371
Hale, A. M., 409
Hales, A. L., 230
Hall, A. J., 162
Hall, F. F., 225
Hall, M. G., 82
Halpern, D., 161, 163

Hamad, G., 292
HAMBLIN, P. F., 153-87; 174, 179-81, 195, 200, 201, 203, 205
Han, T., 67
Hancock, G. J., 294
Haney, W. P., 278
Hansen, C. G., 161
Hansen, E., 21
Harleman, D. R., 177
Haron, A. S., 326
Harrison, E. C., 236, 239, 242, 256
Hart, J. E., 321, 327
Harten, A., 270, 432, 433
Hauser, L. E., 224
Haussling, H. J., 403, 405, 407, 409-11
Havelock, T. H., 48
Hawkes, I., 91
Hayashi, T., 28, 205
Healy, R. N., 367
Hebbert, B., 166, 167, 169, 173
Heckley, W. A., 147, 148
Hedden, R. O., 328
Hedstrom, G. W., 411
Heller, J. P., 375
Henderson, Y., 249
Hendrickson, J., 42
Henne, P. A., 278, 279
Hénon, M., 355
Henze, A., 240
Hersey, J. B., 131
Heskestad, G., 196
Hess, J. L., 425
Hester, D. D., 205
Hickie, B. P., 180
Hicks, R. M., 278
Hinkley, R., 378, 383, 385, 386, 388
Hino, M., 31
Hinson, B. L., 264, 275
Hirata, M., 198-200, 206
Hiriart, G., 415
Hirt, C. W., 404, 405
Ho, R. T., 29
Hobbs, P. V., 91, 92, 228
Hodge, J. K., 409
Hodgson, T. H., 295
Hofer, K., 203, 206
Hogan, S. J., 51
Hollan, E., 171, 179, 182
Hollands, K. G. T., 327
Holst, T. L., 266, 270
Holyer, J. Y., 174, 228
Homicz, G., 290
Hooke, W. H., 225
Hopfinger, E. J., 218, 219
Horvay, G., 341
HOSKINS, B. J., 131-51; 133, 137, 140-47, 149
Hotovy, S. G., 273
Hottel, H. C., 193, 194, 196

AUTHOR INDEX

Hoult, D. P., 205, 206, 213, 219
Huang, J. C. K., 179
Hui, W. H., 291
Humphreys, H. W., 198, 200
Hunt, J. C. R., 62, 72-74
Huppert, H. E., 174, 220, 228
Hurdis, D. A., 229
Hussain, A. K. M. F., 193
HUTTER, K., 87-130; 90, 92, 93, 97, 98, 100-3, 105, 106, 108, 109, 111, 115-20, 122-24, 203, 206

I

Idso, S. B., 225
Iken, A., 125
IMBERGER, J., 153-87; 154-56, 158, 160, 162-71, 173-75, 177, 178, 180
Imboden, D. M., 165
Indergand, R. F., 305, 306
Ingram, R. S., 225
Ionescu, M. I., 236
Israeli, M., 401, 413
Ives, D. C., 265
Ivey, G., 166
Ivey, G. N., 168, 178

J

Jackson, R. G., 30
Jain, S. C., 31
James, W. P., 203
Jameson, A., 265-71, 273-75, 279
Jami, A., 421
Jassby, A., 165
Jeppson, R. W., 436
Jirka, G. H., 174, 177, 223
John, F., 401, 402, 424, 428
Johnson, C. L., 173
Johnson, F. E., 249
Johnson, I. R., 100, 111, 114, 116, 118, 122-24
Johnson, M. C., 173
Johnson, M. W. Jr., 126
Johnson, N. M., 173
Johnson, R. R., 278
Johnston, R. F., 191
Jones, D. A., 428
Joos, P., 374
Joseph, D. D., 62
Josselin de Jong, G., 371
Joyce, T. M., 171

K

Kacprzynski, J. J., 275
Kagemoto, H., 419
Kalmanson, D., 236

Kamb, W. B., 95, 113
Kamotani, Y., 328
Kanari, S., 180
Kao, T. W., 178, 224, 229, 230
Katz, E. J., 131
Kawall, J. G., 196
Keady, G., 41
Keen, C. S., 225
Keffer, J. F., 196
Keller, J. B., 50, 53
Keller, J. D., 273
Kellogg, O. D., 424
Kemp, N. H., 290
Kennedy, J. F., 23, 28, 30, 31
Kenning, D. B. R., 330, 331
Keulegan, G. H., 215, 220
Keyfitz, B. L., 264
Killworth, P. D., 171
Kimura, G., 45
Kinnersley, W., 50
Klaasen, K., 230
Kline, S. J., 30
Knudsen, J. R., 338
Kobus, J. M., 428
Koenigsberg, M., 236, 237, 250
Koh, R. C. Y., 154, 156, 166-68, 171, 174, 195, 201, 203-5, 222
Köhler, J., 253
Komar, P. D., 227, 228
Konicek, L., 327
Korn, D. G., 265, 267, 274, 275
Korteweg, D. J., 45
Kotler, M., 237
Kotsovinos, N. E., 192, 196-202
Kranenberg, C., 218
Krasovskii, Yu. P., 41
Krass, M. S., 103, 105, 109
Kraus, E. B., 154, 160, 163
Krausche, D., 205
Krylov, V. S., 179
Kullenberg, G., 165
Kumar, A., 326
Küssner, H. G., 292
Kvon, V. I., 224
Kwak, D., 300

L

Laitone, E. V., 41, 45, 426
Lake, B. M., 54, 395
Lambert, O., 305, 306
Lambourne, N. C., 291, 301, 302
LANFORD, O. E. III, 347-64; 349, 356, 359
Laniado, S., 236, 237, 250
Laporte, J. P., 238
Larkin, B. K., 338
Larock, B. E., 416, 435

Larson, D. A., 91, 93, 98, 100, 101, 113, 116, 118-22
Larson, R. G., 371
Lau, J. C., 196, 199
Lau, J. P., 35
Laurent, F., 238
Lax, P. D., 405
Lazier, J., 180
Lean, G. H., 222
Le Balleur, J. C., 300
LeBlond, P. H., 154
Lebowitz, J. L., 350
Lebreton, J.-C., 427
Lee, A. C., 162, 181
Lee, C. S. F., 240, 243
Lee, K. D., 405
Legendre, R., 61, 62, 65, 69, 70
Legerer, F., 111
Lehn, H., 165
Lenau, C. W., 45, 48
Lenoir, M., 401, 421
Leonard, A. S., 203
Leontovich, E. A., 62, 75
Leppert, E. L. Jr., 287, 289
Levich, V. G., 330, 331
Levy, L., 237
Levy, L. L. Jr., 300
Lewellen, W. S., 206
Libchaber, A., 359
Licht, C., 424
Lick, W., 120, 154
Lighthill, M. J., 62, 64-67, 74, 117, 279
Lile, R. C., 93
Lin, F. N., 328
Lin, T. Y., 240
Linden, P. F., 219
Lipton, I., 237, 238
LIST, E. J., 189-212; 154, 156, 166-68, 171, 174, 195, 200-3, 205, 208, 222
Liu, J. T. C., 191
Liu, P. C., 179
Lliboutry, L. A., 90, 92, 95, 97, 109, 116, 120
Loh, I., 166, 167, 169, 173
Loka, R. R., 326
Lomax, H., 296, 298, 300, 301, 401
Long, C. E., 166
Longinov, V. V., 227
Longuet-Higgins, M. S., 41, 43-46, 49, 50, 53-56, 425, 433, 434
Lorber, P. F., 294
Lorenz, E. N., 349, 351, 356
Lowry, S., 337
Lowson, M. V., 82, 83
Luikov, A. V., 338
Luitermoza, J. F., 265
Luke, J. C., 414
Luti, F. M., 207

AUTHOR INDEX 447

Lynch, F. T., 278
Lyne, V. D., 174, 178
Lyons, W. A., 225

M

Mabey, D. G., 301
Macaskill, C. C., 427
MacCormack, R. W., 401
MacVean, M. K., 143
Madsen, O. S., 167
Maestrello, L., 192
Maier, A. G., 62, 75
Majda, A., 412
Malcolm, G. N., 83, 298, 300
Maltby, R. L., 61
Manins, P. C., 207, 230
Mann, M. J., 277
Mansfield, D. A., 225, 230
Margnac, A., 427
Marmorino, G. O., 180
Marsden, J. E., 357
Martin, D., 401
Martin, P. C., 349
Marvin, J. G., 300
Massaro, T. A., 190
Mastin, C. W., 265, 409
Masuda, A., 230
Mathieu, Y., 237, 238
Matsumoto, M., 236, 237, 250
Mattioli, F., 433
Maurer, J., 350
Maxworthy, T., 56, 57, 162, 178, 229, 230
May, R. L., 409
McAlister, K. W., 305, 306
McCarthy, D. R., 271
McClimans, T. A., 231
McCowan, J., 48
McCracken, M., 357
McCracken, M. F., 256
MCCROSKEY, W. J., 285-311; 287, 288, 292, 294, 298, 303, 305-7
McDevitt, J. B., 79
McDonald, B. H., 421
McIlhenny, W. F., 203
McInnes, B. J., 120
McIver, P., 55
McLaughlin, J. B., 349
McMahon, H. M., 205
McQuaid, J., 223
McQueen, D. M., 236, 237, 239, 240, 242, 250-56
McWilliams, J. C., 145
Mead, H. R., 274, 275, 277
Mehta, U., 296, 298, 300, 301
Mei, C. C., 413, 421-23, 428
Meier, M. F., 92, 120
Mellenthin, J. A., 79
Mellor, G. L., 163
Mellor, M., 91

Melnik, R. E., 264, 265, 274, 275, 277, 294, 300
Melrose, J. C., 371, 373, 375
Melson, W. G., 228
Mendez, R. H., 256
Meng, S. Y., 273
Merritt, D. H., 173
Messiter, A. F., 262
Michael, D. H., 372
Michalke, A., 192
Miche, R., 53
Michel, B., 91
Michell, J. H., 48, 429
Michelsen, F. C., 427
Middleton, G. V., 218, 227
Middleton, J. H., 173
Mikhlin, S. G., 424
Miles, J. W., 45, 56, 57, 395, 415
Milford, J. R., 225, 230
Miller, E. E., 373
Miller, H., 237
Miller, J. E., 132
Miller, M. J., 224, 225
Miller, R. D., 373
Mills, D. A., 56, 434
Mintz, Y., 10, 413
Mirie, R. M., 57
Moayeri, M. S., 435
Mohanty, K., 366
Mollendorf, J. C., 200
Møllo-Christensen, E., 191
Moncrieff, M. W., 224
Monk, J. D., 226
Mooers, C. N. K., 163
Moore, C. J., 192
Moore, D. J., 205
Moore, J. G., 228
Morland, L. W., 93, 95, 100, 111, 113, 114, 116, 118, 122, 123
Morris, E. M., 97
Morris, N. C. G., 226
Morris, P. J., 192, 196
Morrison, G. K., 226
Morrow, N., 366, 374
Mortimer, C. H., 154, 156, 178
Morton, B. R., 201, 203, 206
Morton, J. B., 193
Morton, K. W., 401
Moum, J. N., 196
Moussa, Z. M., 204
Mudrick, S. E., 132, 144
Muhleisen, R., 179
Muirhead, K. J., 230
Müller, F., 91
Multer, R. H., 55
Munk, W., 154
Munnich, K. O., 165
Murakami, T., 49
Murman, E., 270-72
Murman, E. M., 264

Murphy, K. M., 230
Murthy, C. R., 165, 171
Mysak, L. A., 154, 162, 181

N

Nachman, A., 98
Nagata, Y., 230
Nakagome, H., 198-200, 206
Namias, J., 10
Nash, A. A., 250
Nayfeh, A. H., 51
Neal, A. B., 230
Neff, W. D., 225
Neira, M. A., 371
Nekrasov, A. I., 433
Nelson, R. C., 367
Netter, F. H., 235
Newberger, P. A., 168
Newman, F. C., 170
Newman, J. N., 400, 401
Ng, K. M., 366, 367, 369, 370, 377-82, 389-91
Niccolls, W. O., 191, 196
Nicolis, G., 77
Nieuwland, G. Y., 262
Niiler, O., 161, 163
Niiler, P. P., 163
Nikitopoulos, C. P., 206
Nitsan, U., 109, 113
Nixon, D., 296-98
Norbury, J., 41
Nordin, C. F., 28, 30-32
Norrie, D. H., 413
Norton, F. H., 92
Noye, J., 409
Nye, J. F., 93, 95-98, 108, 113, 116, 117, 120, 124, 125

O

O'Brien, J. J., 411
O'Brien, M. P., 220
O'Carroll, M. J., 417
Oddou, C., 238
Oden, J. T., 126
Oeder, R., 336, 337
Officer, C. B., 226
Ogilvie, T. F., 400, 428
Oh, S. G., 385
Ohmatsu, S., 428
Ohring, S., 403, 404, 407
Oikawa, M., 57
Oka, Y., 240
Okabe, J. I., 45
Olfe, D. B., 47, 49
Olsen, J. J., 296
Olunloyo, V. O. S., 98
O'Malley, C. D., 249
Ono, H., 229
Orlanski, I., 149, 172, 412
Orlob, G. T., 154

448 AUTHOR INDEX

Orszag, S. A., 401, 411, 413
Osborn, T. R., 169
Osmidov, R. V., 165
OSTRACH, S., 313-45; 314, 318, 320-22, 324-26, 328, 330, 331, 333, 336
Oswatitsch, K., 62
Ou, H. W., 179
Ozoe, H., 327, 328

P

Padiyar, R., 237
Palfery, J. G., 205
Pao, H-P., 178, 224, 229
Papanicolaou, P. N., 203
Paris, E., 20
Park, C., 224
Parker, F. L., 203
Parker, G., 28
Patel, V. C., 67
Paterson, W. S. B., 89, 90, 105, 109, 113, 124
Patterson, G. S., 401
Patterson, J., 166, 167, 169, 173, 178
Patterson, J. C., 163
Paulson, C. A., 168
Pavlov, A. M., 230
PAYATAKES, A. C., 365-93; 366, 367, 369-71, 377, 379-82, 387, 389
PEAKE, D. J., 61-85; 62, 71
Pearson, M. D., 161
Pedder, M. A., 146
Pedlosky, J., 133
Penney, W. G., 52
Pera, L., 200
Peregrine, D. H., 55
Perrot, P., 238
Perry, A. E., 62, 72
PESKIN, C. S., 235-59; 236, 237, 239, 242, 247, 250-56
Peterka, J. A., 62, 72-74
Petot, D., 308
Pharo, C. H., 173, 174
Phillips, O. M., 178
Pian, T. H., 424
Pien, P. C., 427
Pierson, W. J., 51
Pimputkar, S. M., 314, 336
Pingree, R. D., 226
Plaschko, P., 190, 192
Plate, E., 33
Plotkin, J., 135
Pnueli, D., 321
Poincaré, H., 62, 63
Pomeau, Y., 355
Pope, G. A., 367
Popiel, C. O., 196
Post, A., 120
Postel, E. E., 287, 289
Poston, T., 97

Potash, R. L., 427
Powell, T., 165
Powell, T. M., 162, 163
Prabhu, A., 193, 196, 200
Pracht, W. E., 224
Pradhan, A., 336
Prandtl, L., 216
Prasad, A., 328
Preisser, F., 336, 337
Price, A. T., 52
Price, J. F., 162, 163
Priestley, C. H. B., 201, 205
Prigogine, I., 77
Pritchard, J. M., 225
Prost, J. P., 182
Pucci, S. L., 292, 303, 305, 306
Pykhov, N. V., 227

R

Radke, L. F., 228
Raghavan, C., 321, 325
Rainey, R. C., 225
Rapin, S., 383, 384, 386, 390
Rastelli, G. C., 237
Raudkivi, A. J., 15, 18, 20
Raustein, E., 144
Rayleigh, Lord, 52, 395
Raymond, C. F., 125, 126
Rayner, K. N., 162, 164
Reamer, H. H., 239
Reddy, J. N., 126
Reddy, R. N., 409
Redekopp, L. G., 229
Redhed, D. D., 273
Reed, R. J., 131
Reed, R. L., 367
Reed, R. O., 203
Reed, W. H. III, 289
Reid, R. O., 179
Reimers, C. E., 228
Reinecke, M., 408
Revelle, R., 7
Reyhner, T. A., 264, 271
Reynaud, L., 125
Reynolds, A. J., 23
Reynolds, W. C., 30
Rice, R. A., 326
Richards, K. J., 29, 30
Richardson, E. V., 14, 15
Richman, J. G., 161
Richtmyer, R. D., 401, 405
Riddell, J. C., 217
Rienecker, M. M., 45, 47, 56, 57
Rillaerts, E., 374
Rizzetta, D. P., 300
Roache, P. J., 401
Robarts, R. D., 165
Roberts, W. C., 236
Robin, G. de Q., 108, 109, 116, 120

Robins, A. G., 196
Rockwell, D. O., 191, 196
Rodi, W., 198, 199
Roof, J. G., 366
Roper, A. T., 228
Ross, B. B., 149
Rottman, J. W., 47, 49
Rouse, H., 21, 191, 198, 200
Ruddick, B. R., 171, 174
Rudy, D. H., 411
Ruelle, D., 349, 356, 360, 362
Runstadler, P. W., 30
Ruppel, H., 411
Rushmer, R. F., 250

S

Saffman, P. G., 47, 49, 51
Salvesen, N., 404, 408, 412
Sami, D., 191
Sananes, F., 204
Sandstrom, H., 180
Sankar, N. L., 307
Sarpkaya, T., 415
Sasaki, T. K., 49
Sattinger, D. H., 62, 77
Satyanarayana, B., 295
Saunders, J. B. de C. M., 249
Saville, D. A., 326
Sawyer, J. S., 135, 137, 139
Saxton, J. A., 326
Sayama, H., 327, 328
Sayer, P., 428
Saylor, J. H., 179
Scarpace, F. L., 231
Schaefer, G. W., 226, 230
Scharmann, A., 336, 337
Schatzmann, M., 206, 207
Schiff, L. B., 83, 299
Schlichting, H., 102
Schmidt, W., 214, 264
Schneider, P. E. M., 190
Schooley, A. H., 51
Schraub, F. A., 30
Schubert, G., 108, 113
Schwab, D. J., 181
Schwabe, D., 336, 337
Schwartz, J. T., 256
SCHWARTZ, L. W., 39-60; 42, 44, 46, 48-53, 396, 408, 433, 435, 437
Scorer, R. S., 205
Scriven, L. E., 330, 341, 366, 371, 378
Sears, W. R., 286
Seebass, A. R., 278
Seebass, R., 261, 289, 296-98
Seegmiller, H. L., 300
Seetharama Rao, V., 427
Segur, H. S., 206
Sells, C. C. L., 265
Serruya, S., 173
Shanks, S. P., 409

AUTHOR INDEX 449

Shapiro, M. A., 139, 140, 144, 148, 149
Shaughnessy, E. J., 193
Shaw, R. P., 417
Shen, H. W., 228
Shen, S. F., 261, 413
Sherman, F. S., 154, 163, 170
Sherman, M., 321
Shin, Y. S., 428
Shiotani, T., 48
Shore, D., 236, 237, 250
Shreve, R. L., 98
Shumskiy, P. A., 90, 103, 105, 109
Simons, D. B., 14, 15
Simons, T. J., 181, 182
SIMPSON, J. E., 213-34; 174, 215-18, 220, 223, 225, 228-31
Simpson, J. H., 226
Sirignano, W. A., 330
Skjelbreia, L., 42
Slattery, J. C., 385
Slawson, P. R., 208
Smale, S., 356
Smith, A. M. O., 425
Smith, D. M., 192, 199
Smith, J. D., 27, 33, 166
Smith, J. H. B., 62, 72, 74
Sneck, H. J., 203
Snyder, L. J., 371
Sobieczky, H., 278
Soh, W. K., 433
Solan, A., 314
Solberg, H., 8, 10
Sommerfeld, A. J. W., 411
Sonnenblick, E. H., 237
South, J., 270-72
South, J. C. Jr., 271, 273
Southwell, R. V., 396
Sowerby, L., 338
Spee, B. M., 262
Spigel, R. H., 155, 162, 163, 170, 180
Spradley, L. W., 328, 338, 340
Spring, U., 93, 111
Sreenivasan, K. R., 192
Stadler, J., 237
Stanek, V., 331
Stark, J. A., 314
Steele, T. D., 154
Stefan, H., 154, 163
Steger, J. L., 263, 301
Steinemann, S., 91
Stenmark, D. G., 367
Stephenson, S. E., 193, 196, 200
Stern, M. E., 174
Sternling, C. V., 330
Steward, F. R., 206
Stewart, I., 97
Stewart, K. M., 171
Stewart, W. E., 371

Stith, J. L., 228
Stoker, J. J., 425
Stokes, G. G., 41, 42, 395, 408, 435
Stone, P. H., 135
Street, R. L., 55
Strikwerda, J. C., 411
Strom, J., 240
Stuart, J. T., 79
Su, C. H., 57
Svensson, U., 161, 163, 174
Szekely, J., 331

T

Tabor, D., 92
Tadjbakhsh, I., 53
Takekuma, K., 400
Takens, F., 349
Talbot, L., 240, 243, 247, 249
Tamanini, F., 198, 199
Tassa, Y., 307
Tavlarides, L. L., 190
Taylor, C., 162, 416, 435
Taylor, D. E. M., 238
Taylor, G. I., 52, 201, 206
Taylor, R. P., 369
Telste, J., 403, 405, 407
Temple, L. J., 239
Tennankore, K. N., 206
Terdiman, R., 237
Teske, M. E., 206
Teubner, M. D., 409
Teuscher, L. H., 224
Thames, F. C., 265, 409
Theakstone, W. H., 91
Theodorsen, T., 286, 287
Thom, A., 412
Thomas, J. R., 193
Thomas, J. W., 435
Thompson, D. E., 92, 112
Thompson, E. F., 179
Thompson, J. F., 265, 409
Thompson, R., 158
Thompson, R. O. R. Y., 155, 162
Thorpe, A. J., 224
Thorpe, S. A., 162, 168, 169, 217
Thorpe, T., 230
Thwaites, B., 279
Tien, C., 371
Tijdeman, H., 261, 289, 296-98
Titus, J. L., 237
TOBAK, M., 61-85; 62, 71, 79, 83, 291
Tochon-Danguy, J. C., 218
Toland, J. F., 41
Tong, P., 424
Toole, J. M., 173
Tran, C. T., 308
Tranen, T. L., 279
Trilling, L., 338

Trischka, J. W., 204
Tsakiris, A. G., 237, 238
Tsutsumi, T., 429
Tuck, E. O., 436, 437
Tulin, M. P., 436
Turian, R. M., 371
Turner, J. S., 44, 164, 170, 171, 174, 178, 197, 198, 201, 203-6, 224
Tutu, N. K., 192, 193, 196, 200
Tyler, P. A., 166, 170

U

Uberoi, M. S., 191, 196, 197
Uda, M., 226
Unny, T. W., 327
Ursell, F., 44, 428, 429

V

Vaisey, G., 396
van Brakel, J., 370
Vanden-Broeck, J. M., 46, 48-51, 53, 436, 437
van der Bel-Kahn, J. M., 236
van de Vel, H., 292
van de Vooren, A. I., 292
van Dongen, M. E. H., 240, 243, 249
Van Eseltine, R. T., 403
Van Leer, J. C., 163
Vanoni, V. A., 18, 20
van Steenhoven, A. A., 240, 243, 249
van Ulden, A. P., 224
Vasiliev, O. F., 224
Vastano, A. C., 179
Velikanov, M. A., 30
Vergara, I., 203
Vialov, S. S., 124
Vinje, T., 434
Vivian, B. A., 91
Voellmy, A., 227
von Bernuth, G., 237
von Kármán, T., 216, 264, 286
von Kerczek, C., 408
von Kerczek, C. H., 404, 408, 412
Voorhis, A. D., 131
Voropayev, S. I., 230
Vreugdenhil, C. B., 224

W

Wade, J. D., 238
Walker, J. C. F., 92
Walker, W. S., 79
Wallace, R. B., 203
Walsh, J., 428
Walters, G., 52
Wang, K. C., 66, 67, 81
Ward, P. R. B., 165, 167, 168

Wardlaw, N. C., 369
Wasow, W. R., 404
Watanabe, M., 174, 177
Weertman, J., 95, 96, 103, 109, 120
Wehausen, J. V., 41, 399, 403, 426
Wehrmann, O., 190
Weidman, P., 229
Weidman, P. D., 56
Weinstock, J., 168, 169
Weiss, W., 165
Wellford, C. L., 424
Wellman, J., 230
Werlé, H., 69, 70, 82
West, N. V., 145, 146
Westerberg, H., 165
Weston, R. P., 205
Wexler, A., 421
Wexler, H. R., 237
Whiffen, M. C., 192, 199
Whillock, A. Z., 222
Whitaker, R. E., 179
White, W. R., 20, 22
Whitelaw, J. H., 204, 205
Whitham, G. B., 117, 415
Whitney, A. K., 52, 55
Wiegand, R. C., 171
Wieting, D. W., 239
Wilcox, W. R., 336
Wilkinson, D. L., 222
Wille, R., 190
Williams, F. M., 103, 109
Williams, G. C., 193, 194, 196
Williams, G. P., 22
Williams, M. H., 299
Williams, R. F., 356
Williams, R. T., 133, 135, 142, 149
Wilson, C. J. N., 228
Wilson, T. A., 191, 196
Wilton, J. R., 42, 50
Winant, C. D., 215
Winslow, A. M., 409
Witting, J., 46
Wolfe, A. W., 247, 256
Wong, F., 33
Woo, H., 62, 72-74
Wood, E. H., 237
Wood, I. R., 218, 222
Woodham, G., 382
Woods, J. D., 143
Wright, J. T. M., 239
Wright, S. J., 203, 205, 206
Wu, J., 230
Wuest, W., 336, 337
Wunderlich, W. O., 164
Wunsch, C. I., 179
Wygnanski, I., 191, 193-96

Y

Yajima, N., 57
Yalin, M. S., 30
Yamada, H., 45, 48
Yamamoto, K., 327, 328
Yamamoto, Y., 419
Yang, C. T., 22
Yates, E. C. Jr., 296
Yates, J. E., 294
Yellin, E. L., 236, 237, 239, 240, 242, 250, 251, 253-56
Yen, S. M., 405
YEUNG, R. W., 395-443; 411, 419, 423, 425, 430-32
Yih, C.-S., 198, 200, 220, 338
Yim, B., 415
Yoganathan, A. P., 236, 239, 242, 256
Yoran, C., 236, 237, 250
Yoshihara, H., 300
Younis, M., 266
Yu, N. J., 265, 278, 279
Yue, D. K. P., 422, 423
Yuen, D. A., 108, 113
Yuen, H. C., 54, 395
Yule, A. J., 191

Z

Zaroodny, S. J., 433
Zatsepin, A. G., 230
Zeitoun, M. A., 203
Zienkiewicz, O. C., 411, 413

CUMULATIVE INDEXES

CONTRIBUTING AUTHORS, VOLUMES 10–14

A

Adamson, T. C. Jr., 10:3–38
Allen, J. S., 12:389–433
Antonia, R. A., 13:131–56
Arndt, R. E. A., 13:273–328
Ashton, G. D., 10:369–92

B

Baker, G. R., 11:95–122
Berman, N. S., 10:47–64
Binnie, A. M., 10:1–10
Bird, G. A., 10:11–31
Bogy, D. B., 11:207–28
Buchhave, P., 11:443–503
Busse, F. H., 10:435–62

C

Callander, R. A., 10:129–58
Cantwell, B. J., 13:457–515
Caughey, D. A., 14:261–83
Chiang, A. S., 13:351–78
Christensen, J., 12:139–58
Corcos, G. M., 10:267–88
Crighton, D. G., 11:11–33

D

Denn, M. M., 12:365–87
Dickinson, R. E., 10:159–95
Dowson, D., 11:35–66
Dussan V., E. B., 11:371–400
Dwyer, A. A., 13:217–29

E

Eliassen, A., 14:1–12
Emmons, H. W., 12:223–36
Engelund, F., 14:13–37
Evans, D. V., 13:157–87

F

Fenton, J. D., 14:39–60
Fletcher, N. H., 11:123–46
Fredsøe, J., 14:13–37

G

Garrett, C., 11:339–69
George, W. K. Jr., 11:443–503
Griffith, W. C., 10:93–105

H

Hamblin, P. F., 14:153–87
Hanratty, T. J., 13:231–52
Hart, J., 11:147–72
Hasimoto, H., 12:335–63
Herczyński, R., 12:237–69
Horikawa, K., 13:9–32
Hoskins, B. J., 14:131–51
Hutter, K., 14:87–130

I

Imberger, J., 10:267–88,
    14:153–87

J

Jenkins, J. T., 10:197–219

K

Kazhikhov, A. V., 13:79–95
Keller, H. B., 10:417–33

L

Lai, W. M., 11:247–88
Lake, B. M., 12:303–34
Landweber, L., 11:173–205
Lanford, O. E., 14:347–64
Laws, E. M., 10:247–66
Leal, L. G., 12:435–76
Lebovitz, N. R., 11:229–46
Leibovich, S., 10:221–46
Leith, C. E., 10:107–28
Lightfoot, E. N., 13:351–78
Lin, C. C., 13:33–55
Lin, J.-T., 11:317–38
List, E. J., 14:189–212
Livesey, J. L., 10:247–66
Lumley, J. L., 11:443–503

M

Macagno, E. O., 12:139–58
MacCormack, R. W.,
    11:289–316
Martin, S., 13:379–97
Maxworthy, T., 13:329–50
McCroskey, W. J., 14:285–311
McIntire, L. V., 12:159–79
Mei, C. C., 10:393–416
Messiter, A. F., 12:103–38
Michael, D. H., 13:189–215
Miles, J. W., 12:11–43
Mow, V. C., 11:248–88
Munk, M. M., 13:1–7
Munk, W., 11:339–69
Mysak, L. A., 12:45–76

N

Naudascher, E., 11:67–94
Noble, P. T., 13:351–78

O

Ostrach, S., 14:313–45

P

Pao, Y.-H., 11:317–38
Parlange, J.-Y., 12:77–102
Patel, V. C., 11:173–205
Patterson, G. S. Jr., 10:289–300
Payatakes, A. C., 14:365–93
Peake, D. J., 14:61–85
Peskin, C. S., 14:235–59
Peterlin, A., 8:35–55
Pieńkowska, I., 12:237–69
Plesset, M. S., 9:145–85
Prosperetti, A., 9:145–85

R

Raupach, M. R., 13:97–129
Reethof, G., 10:333–67
Rhines, P., 11:401–41
Roberts, W. W. Jr., 13:33–55

451

Rockwell, D., 11:69–94
Russel, W. B., 13:425–55
Ryzhov, O. S., 10:65–92

S

Saffman, P. G., 11:95–122
Sano, O., 12:335–63
Schwartz, L. W., 14:39–60
Sears, M. R., 11:1–10
Sears, W. R., 11:1–10
Seebass, R., 12:181–222
Sherman, F. S., 10:267–88
Simpson, J. E., 14:213–34
Solonnikov, V. A., 13:79–95

T

Takahashi, T., 13:57–77
Taub, A. H., 10:301–32
Taylor, C. M., 11:35–66
Thom, A. S., 13:97–129
Tijdeman, H., 12:181–222
Tobak, M., 14:61–85
Tuck, E. O., 10:33–44

U

Uhlenbeck, G. E., 12:1–9

W

Winant, C. D., 12:271–301
Wyngaard, J. C., 13:399–423

Y

Yaglom, A. M., 11:505–40
Yeung, R. W., 14:395–442
Yuen, H. C., 12:303–34

Z

Zeman, O., 13:253–72

CHAPTER TITLES, VOLUMES 10-14

HISTORY

Some Notes on the Study of Fluid Mechanics in Cambridge, England	A. M. Binnie	10:1–10
The Kármán Years at GALCIT	W. R. Sears, M. R. Sears	11:1–10
Some Notes on the Relation Between Fluid Mechanics and Statistical Physics	G. E. Uhlenbeck	12:1–9
My Early Aerodynamic Research—Thoughts and Memories	M. M. Munk	13:1–17
Vilhelm Bjerknes and his Students	A. Eliassen	14:1–12

FOUNDATIONS

Relativistic Fluid Mechanics	A. H. Taub	10:301–32
Existence Theorems for the Equations of Motion of Compressible Viscous Fluid	V. A. Solonnikov, A. V. Kazhikhov	13:79–95
Topology of Three-Dimensional Separated Flows	M. Tobak, D. J. Peake	14:61–85

INCOMPRESSIBLE, INVISCID FLUIDS

Vortex Interactions	P. G. Saffman, G. R. Baker	11:95–112
Low-Gravity Fluid Flows	S. Ostrach	14:313–45

COMPRESSIBLE FLUIDS

Viscous Transonic Flows	O. S. Ryzhov	10:65–92
Transonic Flow Past Oscillating Airfoils	H. Tijdeman, R. Seebass	12:181–222
Existence Theorems for the Equations of Motion of Compressible Viscous Fluid	V. A. Solonnikov, A. V. Kazhikhov	13:79–95
The Computation of Transonic Potential Flows	D. A. Caughey	14:261–83

MAGNETOHYDRODYNAMICS, PLASMA FLOW, ELECTROHYDRODYNAMICS

Magnetohydrodynamics of the Earth's Dynamo	F. H. Buse	10:435–62

VISCOUS FLUIDS

Viscous Transonic Flows	O. S. Ryzhov	10:65–92
Stokeslets and Eddies in Creeping Flow	H. Hasimoto, O. Sano	12:335–63
Particle Motions in a Viscous Fluid	L. G. Leal	12:435–76
Existence Theorems for the Equations of Motion of Compressible Viscous Fluid	V. A. Solonnikov, A. V. Kazhikhov	13:79–95

BOUNDARY-LAYER THEORY

Numerical Methods in Boundary-Layer Theory	H. B. Keller	10:417–33
Ship Boundary Layers	L. Landweber, V. C. Patel	11:173–205
Analysis of Two-Dimensional Interactions Between Shock Waves and Boundary Layers	T. C. Adamson Jr., A. F. Messiter	12:103–38
Some Aspects of Three-Dimensional Laminar Boundary Layers	H. A. Dwyer	13:217–29
Progress in the Modeling of Planetary Boundary Layers	O. Zeman	13:253–72

STABILITY OF FLOW

The Structure of Vortex Breakdown	S. Leibovich	10:221–46

Self-Sustained Oscillations of Impinging Free Shear Layers	D. Rockwell, E. Naudascher	11:67–94
Finite Amplitude Baroclinic Instability	J. E. Hart	11:147–72
Meniscus Stability	D. H. Michael	13:189–215
Stability of Surfaces that are Dissolving or Being Formed by Convective Diffusion	T. J. Hanratty	13:231–52

TURBULENCE

Turbulence and Mixing in Stably Stratified Waters	F. S. Sherman, J. Imberger, G. M. Corcos	10:267–88
Geostrophic Turbulence	P. B. Rhines	11:401–41
The Measurement of Turbulence with the Laser-Doppler Anemometer	P. Buchhave, W. K. George Jr., J. L. Lumley	11:443–503
Similarity Laws for Constant-Pressure and Pressure-Gradient Turbulent Wall Flows	A. M. Yaglom	11:505–40
Turbulence in and Above Plant Canopies	M. R. Raupach, A. S. Thom	13:97–129
Conditional Sampling in Turbulence Measurement	R. A. Antonia	13:131–56
Cup, Propeller, Vane, and Sonic Anemometers in Turbulence Research	J. C. Wyngaard	13:399–423
Organized Motion in Turbulent Flow	B. J. Cantwell	13:457–515
Turbulent Jets and Plumes	E. J. List	14:189–212
The Strange Attractor Theory of Turbulence	O. E. Lanford III	14:347–64

COMBUSTION, FLOWS WITH CHEMICAL REACTION

Dust Explosions	W. C. Griffith	10:93–105
Scientific Progress on Fire	H. W. Emmons	12:223–36

SHOCK WAVES, EXPLOSIONS

Analysis of Two-Dimensional Interactions Between Shock Waves and Boundary Layers	T. C. Adamson Jr., A. F. Messiter	12:103–38

AERO- AND HYDRODYNAMIC SOUND, ACOUSTICS

Turbulence-Generated Noise in Pipe Flow	G. Reethof	10:333–67
Model Equations of Nonlinear Acoustics	D. G. Crighton	11:11–33
Air Flow and Sound Generation in Musical Wind Instruments	N. H. Fletcher	11:123–46

FLOWS IN HETEROGENEOUS AND STRATIFIED FLUIDS, ROTATING FLOWS

Turbulence and Mixing in Stably Stratified Waters	F. S. Sherman, J. Imberger, G. M. Corcos	10:267–88
Wakes in Stratified Fluids	J.-T. Lin, Y.-H. Pao	11:317–38
Geostrophic Turbulence	P. B. Rhines	11:401–41
Water Transport in Soils	J.-Y. Parlange	12:77–102

FREE-SURFACE FLOWS (WATER WAVES, CAVITY FLOWS)

Numerical Methods in Water-Wave Diffraction and Radiation	C. C. Mei	10:393–416
Solitary Waves	J. W. Miles	12:11–43
Topographically Trapped Waves	L. A. Mysak	12:45–76
Instability of Waves on Deep Water	H. C. Yuen, B. M. Lake	12:303–34
Power from Water Waves	D. V. Evans	13:157–87
Strongly Nonlinear Waves	L. W. Schwartz, J. D. Fenton	14:39–60
Numerical Methods in Free-Surface Flows	R. W. Yeung	14:395–442

BUBBLES, FILMS, SURFACE, BUBBLY FLOWS, CAVITATION

Cavitation in Bearings	D. Dowson, C. M. Taylor	11:35–66
Drop Formation in a Circular Liquid Jet	D. B. Bogy	11:207–28
On the Spreading of Liquids on Solid Surfaces: Static and Dynamic Contact Lines	E. B. Dussan V.	11:371–400

Meniscus Stability	D. H. Michael	13:189–215
Cavitation in Fluid Machinery and Hydraulic Structures	R. E. A. Arndt	13:273–328
Low-Gravity Fluid Flows	S. Ostrach	14:313–45

DIFFUSION, FILTRATION, SUSPENSIONS

Drag Reduction by Polymers	N. S. Berman	10:47–64
Toward a Statistical Theory of Suspension	R. Herczyński, I. Pieńkowska	12:237–69
Coastal Sediment Processes	K. Horikawa	13:9–32
Stability of Surfaces that are Dissolving or Being Formed by Convective Diffusion	T. J. Hanratty	13:231–52
Brownian Motion of Small Particles Suspended in Liquids	W. B. Russel	13:425–55
Dynamics of Oil Ganglia During Immiscible Displacement in Water-Wet Porous Media	A. C. Payatakes	14:365–93

MATHEMATICAL METHODS

Existence Theorems for the Equations of Motion of Compressible Viscous Fluid	V. A. Solonnikov, A. V. Kazhikhov	13:79–95
The Strange Attractor Theory of Turbulence	O. E. Lanford III	14:347–64

NUMERICAL METHODS

Monte Carlo Simulation of Gas Flows	G. A. Bird	10:11–31
Prospects for Computational Fluid Mechanics	G. S. Patterson Jr.	10:289–300
Numerical Methods in Boundary-Layer Theory	H. B. Keller	10:417–33
Numerical Solution of Compressible Viscous Flows	R. W. MacCormack, H. Lomax	11:289–316
The Computation of Transonic Potential Flows	D. A. Caughey	14:261–83
Numerical Methods in Free-Surface Flows	R. W. Yeung	14:395–442

EXPERIMENTAL METHODS

The Measurement of Turbulence with the Laser-Doppler Anemometer	P. Buchhave, W. K. George Jr., J. L. Lumley	11:505–540
Conditional Sampling in Turbulence Measurement	R. A. Antonia	13:131–56
Field-Flow Fractionation (Polarization Chromatography)	E. N. Lightfoot, A. S. Chiang, P. T. Noble	13:351–78
Cup, Propeller, Vane, and Sonic Anemometers in Turbulence Research	J. C. Wyngaard	13:399–423

BIOLOGICAL FLUID DYNAMICS

Mechanics of Animal Joints	V. C. Mow, W. M. Lai	11:247–88
Fluid Mechanics of the Duodenum	E. O. Macagno, J. Christensen	12:139–58
Dynamic Materials Testing: Biological and Clinical Applications in Network-Forming Systems	L. V. McIntire	12:159–79
The Fluid Dynamics of Insect Flight	T. Maxworthy	13:329–50
The Fluid Dynamics of Heart Valves: Experimental, Theoretical, and Computational Methods	C. S. Peskin	14:235–59

FLUID DYNAMICS OF MACHINERY

Cavitation in Fluid Machinery and Hydraulic Structures	R. E. A. Arndt	13:273–328

FLUID DYNAMICS OF AIRBORNE VEHICLES

Unsteady Airfoils	W. J. McCroskey	14:285–311

FLUID DYNAMICS OF WATERBORNE VEHICLES

Hydrodynamic Problems of Ships in Restricted Waters	E. O. Tuck	10:33–44

Ship Boundary Layers	L. Landweber, V. C. Patel	11:173–205
Power from Water Waves	D. V. Evans	13:157–87

FLUID DYNAMICS OF HYDRAULIC STRUCTURES AND OF THE ENVIRONMENT

Coastal Sediment Processes	K. Horikawa	13:9–32
Debris Flow	T. Takahashi	13:57–77
Turbulence In and Above Plant Canopies	M. R. Raupach, A. S. Thom	13:97–129
Cavitation in Fluid Machinery and Hydraulic Structures	R. E. A. Arndt	13:273–328
Sediment Ripples and Dunes	F. Engelund, J. Fredsøe	14:13–37
Dynamics of Lakes, Reservoirs, and Cooling Ponds	J. Imberger, P. F. Hamblin	14:153–87
Dynamics of Oil Ganglia During Immiscible Displacement in Water-Wet Porous Media	A. C. Payatakes	14:365–93

GEOPHYSICAL FLUID DYNAMICS

Objective Methods for Weather Prediction	C. E. Leith	10:107–28
River Meandering	R. A. Callander	10:129–58
Rossby Waves—Long-Period Oscillations of Oceans and Atmospheres	R. E. Dickinson	10:159–95
River Ice	G. D. Ashton	10:369–92
Magnetohydrodynamics of the Earth's Dynamo	F. H. Busse	10:435–62
Internal Waves in the Ocean	C. Garrett, W. Munk	11:339–69
Coastal Circulation and Wind-Induced Currents	C. D. Winant	12:271–301
Models of Wind-Driven Currents on the Continental Shelf	J. S. Allen	12:389–433
Progress in the Modeling of Planetary Boundary Layers	O. Zeman	13:253–72
Frazil Ice in Rivers and Oceans	S. Martin	13:379–97
Sediment Ripples and Dunes	F. Engelund, J. Fredsøe	14:13–37
Dynamics of Glaciers and Large Ice Masses	K. Hutter	14:87–130
The Mathematical Theory of Frontogenesis	B. J. Hoskins	14:131–51
Gravity Currents in the Laboratory, Atmosphere, and Ocean	J. E. Simpson	14:213–34

ASTRONOMICAL FLUID DYNAMICS

Relativistic Fluid Dynamics	A. H. Taub	10:301–32
Rotating, Self-Gravitating Masses	N. R. Lebovitz	11:229–46
Some Fluid-Dynamical Problems in Galaxies	C. C. Lin, W. W. Roberts Jr.	13:33–55

OTHER APPLICATIONS

Flow through Screens	E. M. Laws, J. L. Livesey	10:247–66
Continuous Drawing of Liquids to Form Fibers	M. M. Denn	12:365–87

MISCELLANEOUS

Flows of Nematic Liquid Crystals	J. T. Jenkins	10:197–219
Relativistic Fluid Mechanics	A. H. Taub	10:301–32
Debris Flow	T. Takahashi	13:57–77
Frazil Ice in Rivers and Oceans	S. Martin	13:379–97
Brownian Motion of Small Particles Suspended in Liquids	W. B. Russel	13:425–55
Sediment Ripples and Dunes	F. Engelund, J. Fredsøe	14:13–37

ORDER FORM — ANNUAL REVIEWS INC.

Please list the volumes you wish to order. If you wish a standing order (the latest volume sent to you automatically each year), indicate volume number to begin order. Volumes not yet published will be shipped in month and year indicated. Prices subject to change without notice.

ANNUAL REVIEW SERIES — Prices Postpaid, per volume	Regular Order Please send:	Standing Order Begin with:
Annual Review of ANTHROPOLOGY Vols. 1–8 (1972–79): $17.00 USA; $17.50 elsewhere Vols. 9–10 (1980–81): $20.00 USA; $21.00 elsewhere Vol. 11 (avail. Oct. 1982): $22.00 USA; $25.00 elsewhere	Vol(s). _____	Vol. _____
Annual Review of ASTRONOMY AND ASTROPHYSICS Vols. 1–17 (1963–79): $17.00 USA; $17.50 elsewhere Vols. 18–19 (1980–81): $20.00 USA; $21.00 elsewhere Vol. 20 (avail. Sept. 1982): $22.00 USA; $25.00 elsewhere	Vol(s). _____	Vol. _____
Annual Review of BIOCHEMISTRY Vols. 28–48 (1959–79): $18.00 USA; $18.50 elsewhere Vols. 49–50 (1980–81): $21.00 USA; $22.00 elsewhere Vol. 51 (avail. July 1982): $23.00 USA; $26.00 elsewhere	Vol(s). _____	Vol. _____
Annual Review of BIOPHYSICS AND BIOENGINEERING Vols. 1–9 (1972–80): $17.00 USA; $17.50 elsewhere Vol. 10 (1981): $20.00 USA; $21.00 elsewhere Vol. 11 (avail. June 1982): $22.00 USA; $25.00 elsewhere	Vol(s). _____	Vol. _____
Annual Review of EARTH AND PLANETARY SCIENCES Vols. 1–8 (1973–80): $17.00 USA; $17.50 elsewhere Vol. 9 (1981): $20.00 USA; $21.00 elsewhere Vol. 10 (avail. May 1982): $22.00 USA; $25.00 elsewhere	Vol(s). _____	Vol. _____
Annual Review of ECOLOGY AND SYSTEMATICS Vols. 1–10 (1970–79): $17.00 USA; $17.50 elsewhere Vols. 11–12 (1980–81): $20.00 USA; $21.00 elsewhere Vol. 13 (avail. Nov. 1982): $22.00 USA; $25.00 elsewhere	Vol(s). _____	Vol. _____
Annual Review of ENERGY Vols. 1–4 (1976–79): $17.00 USA; $17.50 elsewhere Vols. 5–6 (1980–81): $20.00 USA; $21.00 elsewhere Vol. 7 (avail. Oct. 1982): $22.00 USA; $25.00 elsewhere	Vol(s). _____	Vol. _____
Annual Review of ENTOMOLOGY Vols. 7–25 (1962–80): $17.00 USA; $17.50 elsewhere Vol. 26 (1981): $20.00 USA; $21.00 elsewhere Vol. 27 (avail. Jan. 1982): $22.00 USA; $25.00 elsewhere	Vol(s). _____	Vol. _____
Annual Review of FLUID MECHANICS Vols. 1–12 (1969–80): $17.00 USA; $17.50 elsewhere Vol. 13 (1981): $20.00 USA; $21.00 elsewhere Vol. 14 (avail. Jan 1982): $22.00 USA; $25.00 elsewhere	Vol(s). _____	Vol. _____
Annual Review of GENETICS Vols. 1–13 (1967–79): $17.00 USA; $17.50 elsewhere Vols. 14–15 (1980–81): $20.00 USA; $21.00 elsewhere Vol. 16 (avail. Dec. 1982): $22.00 USA; $25.00 elsewhere	Vol(s). _____	Vol. _____
Annual Review of MATERIALS SCIENCE Vols. 1–9 (1971–79): $17.00 USA; $17.50 elsewhere Vols. 10–11 (1980–81): $20.00 USA; $21.00 elsewhere Vol. 12 (avail. Aug. 1982): $22.00 USA; $25.00 elsewhere	Vol(s). _____	Vol. _____
Annual Review of MEDICINE: Selected Topics in the Clinical Sciences Vols. 1–3, 5–15, 17–31 (1950–52, 1954–64, 1966–80): $17.00 USA; $17.50 elsewhere Vol. 32 (1981): $20.00 USA; $21.00 elsewhere Vol. 33 (avail. Apr. 1982): $22.00 USA; $25.00 elsewhere	Vol(s). _____	Vol. _____
Annual Review of MICROBIOLOGY Vols. 15–33 (1961–79): $17.00 USA; $17.50 elsewhere Vols. 34–35 (1980–81): $20.00 USA; $21.00 elsewhere Vol. 36 (avail. Oct. 1982): $22.00 USA; $25.00 elsewhere	Vol(s). _____	Vol. _____
Annual Review of NEUROSCIENCE Vols. 1–3 (1978–80): $17.00 USA; $17.50 elsewhere Vol. 4 (1981): $20.00 USA; $21.00 elsewhere Vol. 5 (avail. Mar. 1982): $22.00 USA; $25.00 elsewhere	Vol(s). _____	Vol. _____
Annual Review of NUCLEAR AND PARTICLE SCIENCE Vols. 9–29 (1959–79): $19.50 USA; $20.00 elsewhere Vols. 30–31 (1980–81): $22.50 USA; $23.50 elsewhere Vol. 32 (avail. Dec. 1982): $25.00 USA; $28.00 elsewhere	Vol(s). _____	Vol. _____
Annual Review of NUTRITION Vol. 1 (1981): $20.00 USA; $21.00 elsewhere Vol. 2 (avail. July 1982): $22.00 USA; $25.00 elsewhere	Vol(s). _____	Vol. _____

(continued on reverse)

Annual Review of PHARMACOLOGY AND TOXICOLOGY
Vols. 1–3, 5–20 (1961–63, 1965–80): $17.00 USA; $17.50 elsewhere
Vol. 21 (1981): $20.00 USA; $21.00 elsewhere
Vol. 22 (avail. Apr. 1982): $22.00 USA; $25.00 elsewhere   Vol(s). _____ Vol. _____

Annual Review of PHYSICAL CHEMISTRY
Vols. 10–21, 23–30 (1959–70, 1972–79): $17.00 USA; $17.50 elsewhere
Vols. 31–32 (1980–81): $20.00 USA; $21.00 elsewhere
Vol. 33 (avail. Nov. 1982): $22.00 USA; $25.00 elsewhere   Vol(s). _____ Vol. _____

Annual Review of PHYSIOLOGY
Vols. 18–42 (1956–80): $17.00 USA; $17.50 elsewhere
Vol. 43 (1981): $20.00 USA; $21.00 elsewhere
Vol. 44 (avail. Mar. 1982): $22.00 USA; $25.00 elsewhere   Vol(s). _____ Vol. _____

Annual Review of PHYTOPATHOLOGY
Vols. 1–17 (1963–79): $17.00 USA; $17.50 elsewhere
Vols. 18–19 (1980–81): $20.00 USA; $21.00 elsewhere
Vol. 20 (avail. Sept. 1982): $22.00 USA; $25.00 elsewhere   Vol(s). _____ Vol. _____

Annual Review of PLANT PHYSIOLOGY
Vols. 10–31 (1959–80): $17.00 USA; $17.50 elsewhere
Vol. 32 (1981): $20.00 USA; $21.00 elsewhere
Vol. 33 (avail. June 1982): $22.00 USA; $25.00 elsewhere   Vol(s). _____ Vol. _____

Annual Review of PSYCHOLOGY
Vols. 4, 5, 8, 10–31 (1953, 1957, 1959–80): $17.00 USA; $17.50 elsewhere
Vol. 32 (1981): $20.00 USA; $21.00 elsewhere
Vol. 33 (avail. Feb. 1982): $22.00 USA; $25.00 elsewhere   Vol(s). _____ Vol. _____

Annual Review of PUBLIC HEALTH
Vol. 1 (1980): $17.00 USA; $17.50 elsewhere
Vol. 2 (1981): $20.00 USA; $21.00 elsewhere
Vol. 3 (avail. May 1982): $22.00 USA; $25.00 elsewhere   Vol(s). _____ Vol. _____

Annual Review of SOCIOLOGY
Vols. 1–5 (1975–79): $17.00 USA; $17.50 elsewhere
Vols. 6–7 (1980–81): $20.00 USA; $21.00 elsewhere
Vol. 8 (avail. Aug. 1982): $22.00 USA; $25.00 elsewhere   Vol(s). _____ Vol. _____

SPECIAL PUBLICATIONS

Prices Postpaid, per volume

Regular Order
Please send:

Annual Reviews Reprints: Cell Membranes, 1975–1977
(published 1978)   Soft cover: $12.00 USA; $12.50 elsewhere   _____ copy(ies)

Annual Reviews Reprints: Cell Membranes, 1978–1980
(published 1981)   Hardcover $28.00 USA; $29.00 elsewhere   _____ copy(ies)

Annual Reviews Reprints: Immunology, 1977–1979
(published 1980)   Softcover $12.00 USA; $12.50 elsewhere   _____ copy(ies)

History of Entomology
(published 1973)   Clothbound $10.00 USA; $10.50 elsewhere   _____ copy(ies)

Intelligence & Affectivity: Their Relationship During Child Development, by Jean Piaget
(published 1981)   Hardcover $8.00 USA; $9.00 elsewhere   _____ copy(ies)

Telescopes for the 1980s
(avail. Aug. 1981) Hardcover $27.00 USA; $28.00 elsewhere   _____ copy(ies)

The Excitement & Fascination of Science, Volume 1
(published 1965)   Clothbound $6.50 USA; $7.00 elsewhere   _____ copy(ies)

The Excitement & Fascination of Science, Volume 2
(published 1978)   Hardcover $12.00 USA; $12.50 elsewhere   _____ copy(ies)
                   Soft cover $10.00 USA; $10.50 elsewhere   _____ copy(ies)

To: ANNUAL REVIEWS INC, 4139 El Camino Way, Palo Alto, CA 94306 USA   (Tel. 415-493-4400)

Please enter my order for the publications checked above.

Amount of remittance enclosed $ _____   California residents, please add applicable sales tax.
Please bill me ☐   Prices subject to change without notice.
Institutional purchase order # _____

Name _____

Address _____

_____ Zip Code _____

Signed _____ Date _____

☐ Please send free copy of the current *Prospectus* each year.
☐ Send free brochure listing contents of recent back volumes for Annual Review(s) of _____